90 06164

D1758194

ESTUARINE PERSPECTIVES

Produced by

The Estuarine Research Federation

Sponsored by

U.S. Environmental Protection Agency
U.S. Fish and Wildlife Service
U.S. Geological Survey
National Oceanic and Atmospheric Administration
Marine Ecosystems Analysis Program
National Marine Fisheries Service

Proceedings of the Fifth Biennial International
Estuarine Research Conference,
Jekyll Island, Georgia, October 7-12, 1979

ESTUARINE PERSPECTIVES

Edited by

VICTOR S. KENNEDY

University of Maryland
Center for Environmental and Estuarine Studies
Horn Point Environmental Laboratories
Cambridge, Maryland

ACADEMIC PRESS   1980
A Subsidiary of Harcourt Brace Jovanovich, Publishers
New York   *London*   *Sydney*   *Toronto*   *San Francisco*

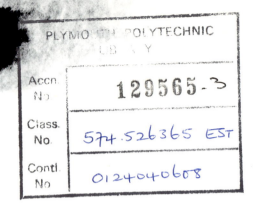
ACADEMIC PRESS, INC.
111 Fifth Avenue, New York, New York 10003

United Kingdom Edition published by
ACADEMIC PRESS, INC. (LONDON) LTD.
24/28 Oval Road, London NW1 7DX

Library of Congress Cataloging in Publication Data

International Estuarine Research Conference, 5th,
    Jekyll Island, Ga., 1979.
    Estuarine perspectives.

    Papers presented at a conference held by the Estua-
rine Research Federation and sponsored by U.S. Environ-
mental Protection Agency and others.
    Includes bibliographies and index.
    1. Estuarine ecology—Congresses. 2. Estuaries—
Congresses. I. Kennedy, Victor S. II. Estuarine
Research Federation. III. United States. Environmen-
tal Protection Agency. IV. Title.
QH541.5.E8I56 1979   574.5'26365   80-23291
ISBN 0-12-404060-8

PRINTED IN THE UNITED STATES OF AMERICA

80 81 82 83   9 8 7 6 5 4 3 2 1

CONTENTS

Chemical Cycles and Fluxes

Arctic Estuaries

Estuarine Sediment: Physical and Biological Factors

Ecosystem Dynamics

Hypotheses of Estuarine Ecology

LIST OF CONTRIBUTORS

Ackleson, S. G., College of Marine Studies, University of Delaware, Lewes, DE 19958

Amezcua Linares, F., Centro de Ciencias del Mar y Limnologia, Universidad Nacional Autónoma de México, Apartado Postal 70-305, México 20, D.F.

Bartlett, D. S., College of Marine Studies, University of Delaware, Newark, DE 19711

Batie, S. S., Department of Agricultural Economics, Virginia Polytechnic Institute and State University, Blacksburg, VA 24061

Beal, K. L., Fisheries Management Operations Branch, National Marine Fisheries Service, State Fish Pier, Gloucester, MA 01930

Black, L. F., Zoology Department, Jackson Estuarine Laboratory, University of New Hampshire, Durham, NH 03824

Boardman, D., Marine Biology, Lamont-Doherty Geological Observatory, Palisades, NY 10964

Bobbie, R. J., Department of Biological Science, Florida State University, Tallahassee, FL 32306

Botkin, D. B., Environmental Studies, University of California, Santa Barbara, CA 93106

Boynton, W. R., Chesapeake Biological Laboratory, Center for Environmental and Estuarine Studies, University of Maryland, Box 38, Solomons, MD 20688

Bozzo, W. E., Department of Biology, The American University, Washington, DC 20016

Budgell, W. P., Ocean and Aquatic Sciences, Department of Fisheries and Oceans, Canada Center for Inland Waters, 867 Lakeshore Road, Burlington, Ontario, Canada L7R 4A6

Capone, D. G., Marine Science Research Center, State University of New York, Stony Brook, NY 11794

Champ, M. A., Department of Biology, The American University, Washington, D.C. 20016

Dame, R., Coastal Carolina College and Belle W. Baruch Institute for Marine Biology and Coastal Research, University of South Carolina, Conway, SC 29526

Davis, W. M., Department of Biological Science, Florida State University, Tallahassee, FL 32306

Day, J. W., Jr., Coastal Ecology Laboratory, Center for Wetland Resources, Louisiana State University, Baton Rouge, LA 70803

de la Cruz, A. A., Department of Zoology, Mississippi State University, Mississippi State, MS 39762

DeLaune, R. D., Laboratory for Wetland Soils and Sediments, Center for Wetland Resources, Louisiana State University, Baton Rouge, LA 70803

Dunn, M. L., Department of Marine Science and Engineering, North Carolina State University, Raleigh, NC 27650

Evans, M. S., Great Lakes Research Division, University of Michigan, Ann Arbor, MI 48109

Fazio, S. D., Department of Biological Science, Florida State University, Tallahassee, FL 32306

Findlay, R. H., Department of Biological Science, Florida State University, Tallahassee, FL 32306

Galloway, G. E., Jr., Department of Earth, Space and Graphic Sciences, U.S. Military Academy, West Point, NY 10996

Glibert, P. M., The Biological Laboratories, Harvard University, Cambridge, MA 02138

Gould, G. A., III, Department of Biology, The American University, Washington, D.C. 20016

Grainger, E. H., Arctic Biological Station, Ste. Anne de Bellevue, Quebec, Canada, H9X 3L6

Greve, W., Biologische Anstalt Helgoland, Meerestation, Haus A, 2192 Helgoland, Federal Republic of Germany

Hackney, C. T., Department of Biology, University of Southwestern Louisiana, Lafayette, LA 70524

Hansen, W. J., Environmental Laboratory, U.S. Army Engineer Waterways Experiment Station, P.O. Box 631, Vicksburg, MS 39180

Hardisky, M. A., College of Marine Studies, 103A Robinson Hall, University of Delaware, Newark, DE 19711

Healy, W. B., Minister (Scientific), New Zealand High Commission, New Zealand House, Haymarket, London, England SW14 4TQ

Heck, K. J., Jr., The Academy of Natural Sciences of Philadelphia, Benedict Estuarine Research Laboratory, Benedict, MD 20612

Jeffries, H. P., Graduate School of Oceanography, University of Rhode Island, Kingston, RI 02881

Katsinis, C., Department of Electrical Engineering, University of Rhode Island, Kingston, RI 02881

Kelley, J. T., Department of Earth Sciences, University of New Orleans, New Orleans, LA 70122

Kemp, W. M., Horn Point Environmental Laboratories, Center for Environmental and Estuarine Studies, University of Maryland, Box 775, Cambridge, MD 21613

Kjerfve, B., Belle W. Baruch Institute for Marine Biology and Coastal Research, Department of Geology, and Marine Science Program, University of South Carolina, Columbia, SC 29208

Klemas, V., College of Marine Studies, University of Delaware, Newark, DE 19711

Klose, P. N., Environmental Resources Management, Inc., 999 West Chester Pike, West Chester, PA 19380

Kumpf, H. E., National Marine Fisheries Service, Southeast Fisheries Center, Office of Fishery Management, 75 Virginia Beach Drive, Miami, FL 33149

Levinton, J. S., Department of Ecology and Evolution, State University of New York, Stony Brook, NY 11794

Linthurst, R. A., Department of Botany, North Carolina State University, P.O. Box 5186, Raleigh, NC 27650

Livingston, R. J., Department of Biological Science, Florida State University, Tallahassee, FL 32306

Loder, T. C., Department of Earth Sciences and Ocean Process Analysis Laboratory, University of New Hampshire, Durham, NH 03824

Maguire, B., Jr., Department of Biology, University of Texas, Austin, TX 78712

Malone, T. C., Marine Biology, Lamont-Doherty Geological Observatory, Palisades, NY 10964

Martz, R. F., Department of Biological Science, Florida State University, Tallahassee, FL 32306

Maurer, R., National Marine Fisheries Service, Northeast Fisheries Center, Narragansett, RI 02882

McKellar, H. N., Jr., Belle W. Baruch Institute for Marine Biology and Coastal Research, and Department of Environmental Health Sciences, University of South Carolina, Columbia, SC 29208

Miller, J. M., Department of Zoology, North Carolina State University, Raleigh, NC 27650

Moore, B., III Complex Systems Group, O'Kane House, University of New Hampshire, Durham, NH 03824

Morowitz, H., Department of Molecular Biophysics and Biochemistry, Yale University, New Haven, CT 06520

Neale, P. J., Marine Biology, Lamont-Doherty Geological Observatory, Palisades, NY 10964

Nelson, D., Coastal Carolina College and Belle W. Baruch Institute for Marine Biology and Coastal Research, University of South Carolina, Conway, SC 29526

Nichols, M. M., Virginia Institute of Marine Science, College of William and Mary, Gloucester Point, VA 23062

Nickels, J. S., Department of Biological Science, Florida State University, Tallahassee, FL 32306

Odum, E. P., Institute of Ecology, University of Georgia, Athens, GA 30601

Officer, C. B., Earth Sciences Department, Dartmouth College, Hanover, NH 03755

Orth, R. J., Virginia Institute of Marine Science, and School of Marine Science, College of William and Mary, Gloucester Point, VA 20612

Osborne, C. G., Chesapeake Biological Laboratory, Center for Environmental and Estuarine Studies, University of Maryland, Box 38, Solomons, MD 20688

Patrick, W. H., Jr., Laboratory for Wetland Soils and Sediments, Center for Wetland Resources, Louisiana State University, Baton Rouge, LA 70803

Peterson, C. H., Institute of Marine Sciences, University of North Carolina at Chapel Hill, P.O. Drawer 809, Morehead City, NC 28557

Pett, R. J., Department of Zoology, College of Biological Science, University of Guelph, Guelph, Ontario, Canada, N1G 2W1

Phillips, J. H., Georgia Department of Natural Resources, Coastal Resources Division, 1200 Glynn Avenue, Brunswick, GA 31520

Philpot, W. D., College of Marine Studies, University of Delaware, Newark, DE 19711

Reed, J. P., The Ecosystems Center, Marine Biological Laboratory, Woods Hole, MA 02543

Reimold, R. J., Georgia Department of Natural Resources, Coastal Resources Division, 1200 Glynn Avenue, Brunswick, GA 31520

Reiners, F., Biologische Anstalt Helgoland, Meerestation, Haus A, 2192 Helgoland, Federal Republic of Germany

Reppert, R. T., U.S. Army Engineer Institute for Water Resources, Kingman Building, Fort Belvoir, VA 22060

Richardson, S. E., Environmental Laboratory, U.S. Army Engineer Waterways Experiment Station, P.O. Box 631, Vicksburg, MS 39180

Roff, J. C., Department of Zoology, College of Biological Science, University of Guelph, Guelph, Ontario, Canada, N1G 2W1

Rogers, G. F., Department of Zoology, College of Biological Science, University of Guelph, Guelph, Ontario, Canada, N1G 2W1

Schamberger, M. L., U.S. Fish and Wildlife Service, Office of Biological Services, Western Energy and Land Use Team, 2625 Redwing Road, Fort Collins, CO 80525

Shabman, L. A., Department of Agricultural Economics, Virginia Polytechnic Institute and State University, Blacksburg, VA 24061

Sherman, K., National Marine Fisheries Service, Northeast Fisheries Center, Narragansett, RI 02882

Slobodkin, L. B., Department of Ecology and Evolution, State University of New York, Stony Brook, NY 11794

Smith, D. W., School of Life and Health Sciences, University of Delaware, Newark, DE 19711

Smith, G. A., Department of Biological Science, Florida State University, Tallahassee, FL 32306

Smith, N. P., Harbor Branch Foundation, Inc., RR 1, Box 196, Fort Pierce, FL 33450

Stevenson, H., Biology Department, and Belle W. Baruch Institute for Marine Biology and Coastal Research, University of South Carolina, Columbia, SC 29208

Stout, J. P., Dauphin Island Sea Lab, University of South Alabama, P.O. Box 386, Dauphin Island, AL 36528

Taylor, B. F., Rosenstiel School of Marine and Atmospheric Science, University of Miami, Miami, FL 33149

Tippie, V. K., Center for Ocean Management Studies, University of Rhode Island, Kingston, RI 02881

Vierra, K. C., College of Marine Studies, University of Delaware, Lewes, DE 19958

Virnstein, R. W., Harbor Branch Foundation, Inc., RR 1, Box 196, Fort Pierce, FL 33450

Webb, K. L., Virginia Institute of Marine Science, College of William and Mary, Gloucester Point, VA 23062

Wetzel, R. L., Virginia Institute of Marine Science, College of William and Mary, Gloucester Point, VA 23062

White, D. C., Department of Biological Science, Florida State University, Tallahassee, FL 32306

Wolaver, T. G., Department of Environmental Sciences, University of Virginia, Charlottesville, VA 22903

Yáñez-Arancibia, A., Centro de Ciencias del Mar y Limnologia, Universidad Nacional Autónoma de México, Apartado Postal 70-305, México 20, D.F.

Zieman, J. C., Department of Environmental Sciences, University of Virginia, Charlottesville, VA 22903

Zingmark, R., Biology Department, and Belle W. Baruch Institute for Marine Biology and Coastal Research, University of South Carolina, Columbia, SC 29208

FOREWORD

Estuaries have a special and historic importance to man. The dense population centers that occur on virtually every major estuary attest to this fact. They are used as fishing grounds, transportation arteries, a source of industrial cooling waters, for waste disposal, and as recreational areas. The continued and increasing development of these areas has presented many problems. Dr. F. John Vernberg, past president of the Estuarine Research Federation, has stated "Estuaries and adjacent environments are ecological systems that are subjected to continual stress by natural and man-induced perturbations." The truth of this statement is obvious.

In order to ameliorate or solve these problems, it is necessary to understand the complex systems which interact within an estuary. To that end a group of scientists, students, engineers, resource managers and other specialists met at Jekyll Island, Georgia, in 1964 to hold a symposium on estuaries. The objectives were to provide an opportunity for the exchange of ideas between the various disciplines and individuals interested in estuarine research, to summarize the present knowledge of the characteristics of estuaries, and to delineate the direction of current research. Estuaries were defined, described, categorized, systematized, and discussed throughout this historic four-day meeting. About 1000 individuals participated in this first major estuarine symposium in North America.

Since that meeting, the Estuarine Research Federation has arranged four more symposia on estuaries. In 1973 a symposium considering "Recent Advances in Estuarine Research" was held in Myrtle Beach, South Carolina. The next symposium examined "Estuarine Processes" and was held in Galveston, Texas in 1975, followed by "Estuarine Interactions" held at Mt. Pocono, Pennsylvania in 1977. In 1979 the Fifth Biennial International Estuarine Research Conference was held October 7-12 at Jekyll Island, Georgia. Perhaps our returning to Jekyll Island has completed some sort of cycle and indicates a greater perspective on all our parts about the importance of estuaries. Appropriately, the title of this conference was "Estuarine Perspectives."

Investigations of estuaries are continuously being carried out by a great many scientists from many disciplines and from many research and education institutions. It is to facilitate in the transfer of information from this ongoing research that these symposia are dedicated. The papers presented in this volume are most of the invited papers presented at this Fifth Biennial International Research Conference on Estuarine Research.

The editing by Victor Kennedy and rigorous reviews by the referees have resulted in this volume that will not only be a useful book for the estuarine scientist, but also for those decision makers whose task it is to decide on the utilization and regulations necessary to protect these environments for the survival and enjoyment of those organisms living within or near these estuaries. It is with a great deal of pride that the Estuarine Research Federation presents this book.

Michael Castagna
President
Estuarine Research Federation
1977-1979

PREFACE

"... who will reveal to our waking ken
The forms that swim and the shapes that creep
under the waters of sleep?
And I would I could know what swimmeth below
when the tide comes in
On the length and the breadth of the
marvelous marshes of Glynn."

--Sidney Lanier (1878)

One hundred and one years after Sidney Lanier composed his poem celebrating the Marshes of Glynn, the Estuarine Research Federation (ERF) held a meeting in the midst of these same marshes in order that we might learn "what is abroad in the marsh and the terminal sea." This meeting came 15 years after an earlier, seminal meeting in the same location focussed the attention of scientists on estuarine systems. The proceedings of that 1964 symposium appeared as the volume ESTUARIES, edited by George Lauff. The continued utility of that compilation of 71 papers is attested to by the fact that many authors of papers in the present volume refer to papers in ESTUARIES in their text.

Since 1973, ERF has sponsored biennial meetings centering on estuarine research, and has published the proceedings of these meetings. The present volume, ESTUARINE PERSPECTIVES, represents a continuation of that practice and has been produced with a determination that standards of quality be high in order that its usefulness should emulate that of ESTUARIES. To that end I owe much to 120 referees who provided peer review of the 58 manuscripts that were submitted for consideration by participants at the invited sessions. I appreciate their prompt assistance, and the careful and rapid response of the successful authors to the various criticisms and suggestions offered.

It is difficult to compare ESTUARINE PERSPECTIVES and ESTUARIES in an effort to determine areas in which progress has been made in our understanding since 1964. Differences might be due to the fortuitous choosing of topics for both meetings, and may not say anything about changes in direction in estuarine science or about the maturing of a branch of research. However, in one field not treated in ESTUARIES but represented in ESTUARINE PERSPECTIVES, that of economic valuation and resource management, progress has certainly been made, although there is much yet to learn. I hope that the 8 papers in this book on that topic will help in the development of this important area of study.

That is not to slight the other 33 papers, each of which makes a contribution in its field. Studies on nutrient chemistry, primary productivity, sedimentation, and ecosystem dynamics continue to be important. We include information on one tropical and two Arctic estuaries in this volume. A number of contemporary techniques as applied to estuarine research are described. Finally, some hypotheses of estuarine ecology are discussed by scientists active in the fields under consideration.

The Marshes of Glynn as a physical backdrop for the meeting stimulated an appropriate sense of wonder and pleasure. Were any symposiasts to tire of the deliberations, they had only to step away from the meeting areas and be confronted with the sight of redstarts and lazuli buntings, woodstorks, and the remnants of the fall warbler migration, or the famed shrimp fleet of Brunswick trawling within a few hundred meters of shore. It was just a short walk to the "glooms of the live oaks" and the "world of marsh that borders on a world of sea." As the contents of this volume demonstrate, although our knowledge is now more extensive and our questions include some of greater sophistication, the marshes and the estuary provide continued inspiration as we echo Lanier's desire to understand them while we share his delight at the intricate beauty of these environments.

Victor S. Kennedy, Editor

ACKNOWLEDGMENTS

The papers in this volume were presented in invited sessions. The conveners of these sessions included: Elizabeth Bauereis, William Burbanck, Conrad Kirby, John Kraeuter, Mallory May, Judy Stout, Robert Virnstein and Richard Wetzel. These individuals were most helpful in the initial aspects of communication with authors and in collection of manuscripts.

I am indebted to 120 referees for their prompt assistance in the evaluation of the 58 manuscripts that were submitted for consideration.

Martin Wiley provided helpful guidance in the early stages of my job as editor. The work was made much simpler because of the trail-blazing he had accomplished in the process of editing earlier volumes in this series.

Ida Marbury quickly and competently performed most of the secretarial work involved. The staff of Economy Printing in Easton, Maryland was always cooperative and helpful during the lengthy process of providing camera-ready copy.

In his book, *Hymns of the Marshes*, Sidney Lanier provides the appropriate aesthetic background for the intellectual process of understanding estuarine structure and function. I am thankful for the pleasure his verbal pictures have provided.

VALUE AND MANAGEMENT
OF WETLANDS

ESTIMATING THE ECONOMIC VALUE OF COASTAL WETLANDS: CONCEPTUAL ISSUES AND RESEARCH NEEDS

L. A. Shabman and S. S. Batie

Department of Agricultural Economics
Virginia Polytechnic Institute and State University
Blacksburg, Virginia

Abstract: Estimates of the economic value of coastal wetlands can better establish their social worth and provide an improved focus for debates over wetlands' preservation. This has been recognized by many authors, and several approaches to economic valuation now appear in the literature on wetlands. The most prominent techniques are the transferences of energy flows to monetary equivalents and the estimation of the market value of harvestable species and other direct user services derived from wetlands. The results reported are often based upon approaches which have conceptual weaknesses. However, valid estimates of wetlands' economic values are difficult to obtain at this time due to a dearth of physical and biological data relevant for economic analysis. Cooperative research between physical and economic scientists can begin to provide the necessary information for sound economic analysis. Nonetheless, in the near future we can only expect limited success in establishing the economic value of natural coastal wetlands. Until that time, decisions on use of wetlands must be made under conditions of scientific uncertainty.

Introduction

Marine wetlands, as a natural environment, may yield numerous valuable ecological services such as provision of fish and wildlife habitat and assimilation of waste. On the other hand, these same tracts of land may be developed as residential, commercial or industrial sites. In the past, the decision to develop wetland areas has been made by private individuals acting in response to the price incentives present in the land market. Development of wetlands at the appropriate time and place meant that the owner was able to sell such sites at a positive return. However, the values to man of the many ecological services of wetlands are not considered by either the buyer or seller in such a market transaction. Because property rights for these ecological services are ill-defined, and because ecological services are collective consumption goods, there are no markets where owners of wetlands can sell ecological services to willing buyers (Mills 1978). As a result, the market price for wetlands will not reflect the value of these ecological services; so, when wetlands are developed, it will be with little or no recognition by the private buyers and sellers of the value of the ecological services foregone.

In recognition of this market failure problem, public policies and programs have been instituted during the last decade to reduce the rate at

3

which natural wetland areas are developed (Environmental Law Institute 1979). Each program prohibits alteration of natural areas unless all the benefits from alteration are judged to exceed all the costs. For example, the Virginia Wetlands' Act states that a wetland's alteration should not be allowed unless ". . . the anticipated public and private benefits of the proposed activity exceeds the anticipated public and private detriment . . ." (Virginia Code Annotated 1972).

These program guidelines have stimulated interest in obtaining monetary measures of the value of natural wetland services in order to compare the benefits of preservation with the benefits of development. Unfortunately, many of the monetary value estimates currently available in the wetlands' literature have been developed with conceptually invalid procedures. The main purpose of this paper is to illustrate some of the basic economic principles that must be understood and followed to obtain conceptually valid estimates of nonmarket values. The discussion will highlight the errors in many current studies as well as emphasize where future cooperative research between economic and biological sciences is needed. Illustrative examples will be drawn from a three-year study on the management of Virginia's coastal wetlands (Park and Batie 1978). While it is not possible to describe the full scope of economic science in a single manuscript, this paper will provide an introduction for noneconomists to the economic perspective on wetlands' valuation and management. However, we do not argue that monetary measurement of wetlands' values should be the only information used in wetlands' management programs. To the extent that multiple social objectives such as equity and ecosystem diversity exist, other measurements of benefits and costs in terms of those objectives are essential to an informed decision process.

Prices and Economic Value

Within well functioning markets, sellers and buyers exchange money for resources, goods and services (hereafter referred to as products). The amount of money that must be exchanged for each unit of a product is its market price. In a market, buyers are willing to pay money for a product if they value that product more highly than other products which may be purchased with the same money. Conversely, sellers will sell the product for monetary considerations if they value the other products the money could buy more highly than the product being considered for sale. The price at which the product is exchanged is, therefore, a measure of the value of the product to the buyer. However, the market price cannot fall below the value to the seller. Thus, market prices are not arbitrary but rather they reflect the value of the product in question to buyers and/or sellers.

However, the observed market price of a product cannot be a basis for establishing its value if the exchange process is flawed. As stated above, markets for the ecological services of natural environments, such as wetlands, either fail to exist or do not operate according to theoretically ideal criteria. As a result, observed market prices for wetland areas will not reflect the total value of the services they provide. One author describes the problem as follows:

> ". . . there are many cases where exchanges occur without money passing hands; where exchanges occur but they are not freely entered into; where exchanges are so constrained by institutional rules that it would be dubious to infer that the terms were satisfactory; and where imperfections in the conditions of exchange would lead us to conclude that the price ratios do not reflect appropriate social judgements about values. Each of these cases gives rise to deficiencies in the use of existing price data as the basis of evaluation of inputs or outputs." (Margolis 1969; p. 534).

As a result, an important economic research area is to develop "shadow values" for the services of natural wetlands where no market or only limited market information exists. A shadow value should be the equivalent of the price which would have been generated by people buying and selling the resource and its services in a market, if such a market were able to function under theoretically ideal conditions.

Economic Values from Energy Analysis

Perhaps the most widely known study of wetlands' values is one by Gosselink, Odum and Pope. They attempted to convert primary productivity of marshland into a dollar measure of value by multiplying calories of energy resulting from primary production of an acre of representative marsh by a dollar value per calorie (Gosselink et al. 1974). This dollar value is obtained by dividing the gross national product (GNP) by the national energy consumption index to calculate average GNP produced per unit of energy use in the United States.

The approach used by Gosselink et al. (1974) was based upon a procedure suggested in 1972 by E. P. Odum and H. T. Odum (Odum and Odum 1972). The basic premise of this procedure is that society's use of resources is at an optimum when the energy flow in the total environment is maximized. However, it is argued that a stronger basis for justifying the preservation of natural environments is obtained if these energy flows are converted from calories to dollars. This division of GNP by calories is justified by an argument that market prices are determined by energy use patterns in the economy (Odum and Odum 1976).

In another paper we questioned the validity of this procedure and the arguments which are presented to support it (Shabman and Batie 1978). Both Howard Odum and Eugene Odum have prepared rebuttals to our critique (H. T. Odum 1979; E. P. Odum 1979). In his rebuttal, Eugene

Odum defends the energy based valuation approach by expressing the concern that "the natural environment's goods and services are grossly undervalued in conventional economic accounting" (E. P. Odum 1979; p. 231). We support his concern. Indeed, economists have spent a good deal of their professional time in recent years attempting to develop procedures for estimating shadow values for natural environments where existing market prices are not sufficient measures of value (Freeman 1979). However, the need to improve economic valuation of natural environments does not justify the particular energy based procedure utilized by Gosselink et al. (1974). Rather than repeating the specific technical arguments from our previous critique of that procedure, the following discussion describes the fundamental premise which underlies economic valuation procedures. This discussion can clarify our basic reason for not accepting the energy based valuation procedure. Of more importance, this discussion can improve future dialogue between the disciplines.

As stated above, prices for goods and services in a well functioning market are a money measure of the subjective value buyers and sellers place on those goods and services. In short, prices are determined by people's preferences and values and not by the energy content or energy cost of goods, as is implied by the energy based valuation procedure. Thus, the search for shadow prices for nonmarket services (such as wetland services) is a search for a measure of people's values for those goods and services. These values are, of course, based upon the preferences and knowledge of the current population. People's values may change over time as they gain knowledge about the importance of certain goods and services (such as natural environments) and, as a result, they may be willing to pay more of their money income for the services of the natural environment relative to other goods. As this occurs shadow prices can be expected to rise. However, as Kenneth Boulding has observed:

> "Economics, as such, does not contribute very much to the formal study of human learning . . . our main contribution as economists is the description of what is learned; the preference functions which embody what is learned in regard to values . ." (Boulding 1969; p. 4).

The economic analyst accepts the existing structure of individual human values as the basis for calculating shadow prices. However, the concern is often expressed that these existing human values are based, in part, upon an ignorance of the contribution natural ecosystems make to human welfare (Westman 1977; E. P. Odum 1979). Eugene Odum suggests that resource scientists should educate people about ecosystem values. He then defends the energy to dollar conversion procedure as one such education effort for "extending economic cost-accounting to include these externalities on the grounds that the public will 'reveal a preference' for environmental quality only if high values are recognized for natural ecosystems in their natural state" (E. P. Odum 1979; p. 235). Thus, the defense offered for the energy based valuation procedure is that the high monetary values derived from it can change people's values. However, to the economist this is at odds with the basic thrust of the discipline. Economics is not a tool for changing values, but rather can be one tool for measuring existing values. We feel that Eugene Odum's quarrel is not with economists, but rather with the existing structure of values reflected in the price system and through correctly applied shadow pricing methods. We subscribe to the concern that people may not fully appreciate the importance of natural environments to their welfare. This is not, however, a flaw in economic analyses which only measure existing preferences. To attack economics for reporting on people's preferences and values is blaming the messenger for the message.

Pitfalls in the Use of Market Information

Although no market prices for natural wetland services exist, price and cost data for related goods and services may be used for shadow pricing. However, such data should be analyzed according to conceptually sound economic principles. The wetlands' literature reports on several attempts to utilize market information, but, in most instances, these results are derived from unsound procedures.

Derived Values: One approach to valuation is based upon the contribution of wetlands to provision of a product which is traded in the market. For example, since natural wetlands are productive of marine life, a value estimate may be derived from the market price for seafood. A typical approach has been to divide the dockside value of the fish harvest by the total wetland acres available to calculate a value per acre (Wass and Wright 1969; Gosselink et al. 1974). This ascribes the gross value of the catch to the wetlands. The reported logic of this calculation is that without the wetlands there would be no fishery. However, there are two basic errors in this approach.

First, the calculation implicitly assumes that any lost wetland acreage will directly appear as reduced marketable fish harvest. However, if other factors, such as temperature and salinity, are limiting fish population, then destruction of some wetlands may not affect fish catch. Also, by dividing total market value of the catch by total acres, the methodology implies that there is no difference between wetland acres in their productive capability. Proper valuation should identify the values associated with incremental changes in the area of wetlands of differing biological productivity. For example, it would seem likely that loss of one acre of low productivity wetlands from an area where thousands of acres are available would result in a smaller loss in value than would destruction of a high productivity wetland in an area where few wetland acres exist. Of course, establishing these value differences will require quite detailed and sophisticated technical information which links wetlands' quantity and quality to fish populations. This point is discussed further in the next section.

Second, allocating the total value of the catch to wetlands fails to recognize that the labor and capital resources employed in fish harvesting have a value in an alternative use. Therefore, the price paid to fishermen for their catch must (in the long run) be sufficient to pay a return to all the factors used in the harvest which is at least equal to their value in an alternative use. Thus, the value of the fish (and ultimately wetlands) is correctly calculated by taking the dockside value of the catch and allocating to labor and capital an amount equal to its value in an alternative use; then, the residual value is imputed to the fish (Batie and Wilson 1979). This point should be an obvious one—fish don't harvest themselves. While it is a tautology that without fish there would be no fishery, it does not follow that without fish the resources employed in their capture could not be employed in an alternative enterprise.

Alternative Cost: The alternative cost procedure argues that the estimated value of wetlands in providing a service is equal to the cost of the next best alternative way of providing the same service. The proper use of alternative cost techniques should be governed by three considerations: (1) the alternative considered should provide the same services as the wetlands; (2) the alternative selected for the cost comparison should be the least-cost alternative; and (3) there should be substantial evidence that the service would be demanded by society if its price were equal to the cost of that least-cost alternative (Howe 1971). Most of the alternative cost estimates reported in the wetlands' literature have not been subjected to any of these important tests (Gosselink et al. 1974; Westman 1977).

The study of Gosselink et al. (1974) illustrates these points quite well. They valued wetlands for waste assimilation by arguing that the waste degradation services of marsh areas can be replicated by tertiary treatment plants. They calculated the cost of such a plant and ascribed that cost as the waste assimilation value of wetlands. The first problem with this approach is that the type and level of waste treatment services they used as the alternative would not be provided by all wetlands. To apply the approach properly, the type and level of waste assimilation provided by specific areas of wetlands must be determined. Such services will differ according to the characteristics of the wetlands and the amount of waste received by the wetlands' area. Second, the alternative chosen may not be the least-cost waste treatment technology available. Perhaps a combination of land treatment, changes in production technologies and different waste treatment technology would be less expensive in particular areas.

A third serious flaw is the implicit assumption that the demand for tertiary waste treatment exists. The costs associated with the removal of each unit of additional waste can be characterized as sharply increasing, particularly for tertiary treatment (Kneese and Bower 1973). The implicit assumption is that these sharply increased costs provide additional natural values for which society would be willing to pay. This assertion must be carefully documented in order to utilize properly the alternative cost approach, but it seldom is. The possible fallacy of this assumption can be stressed by reference to the following example. Assume that an acre of wetlands can produce a ton of marine worms per year. Further, assume that a ton of marine worms could be propagated artificially in a laboratory at a cost of $100,000. Could we then conclude that wetland services which produce a ton of marine worms are worth $100,000 to society? The answer is no, unless we can convincingly demonstrate that society would be willing to pay $100,000 per ton for marine worms.

Toward Improved Economic Valuation

Proper estimation of the economic values of wetlands requires basic information from the biological and physical sciences (Midwest Research Institute 1979; Batie and Shabman 1980). This information should document the linkages between the existence of wetlands and particular services such as increased wildlife population density. While research of this general nature has been done, it is often not of sufficient detail or of proper design for use in economic valuation. We feel confident in making these statements after three years of research on coastal wetlands' management in Virginia. In the limited space available here, we can only summarize our general conclusions for some of the many services often attributed to coastal wetlands in the Chesapeake Bay.

Flood Control: Conceptually, wetlands could provide flood control in at least three ways. One is by the wetlands acting as a peat sponge. Alternatively, wetland vegetation could serve to reduce the velocity of flood water. Finally, coastal wetlands could provide flood control by acting as a reservoir, that is, by being a low lying area. The evidence that we have been able to collect for marine wetlands in the Chesapeake Bay suggests that these flood control services do not exist. First, coastal wetlands are subject to periodic inundation by tides; therefore, even where the composition of wetlands is a peat substrate, this peat would already be saturated with water and thus unable to act as a sponge. There is also little evidence to suggest that vegetation actually does reduce the velocity of flood water after the flooding has achieved a height that submerges the vegetation. Marine wetlands provide protection from coastal flooding to adjoining land parcels in the sense that any open area between housing and the ocean provides flood protection. However, it is not the wetlands in their natural state that provides that protection, since a filled wetlands would also protect neighboring parcels from flood damage as would a parking lot, an open field, or a forested area (Owens, Park and Batie in preparation).

Erosion Control: In our study of wetlands in the Chesapeake Bay, we were unable to find technical data that supported the assertion that wetlands provide erosion protection. Therefore, with the assistance of the Virginia Institute of Marine Science, we selected experimental case study areas and examined historical rates of erosion in areas identical to one another with the exception that some water frontage consisted of wetlands and some of fastland. Our findings are that the Chesapeake Bay wetlands erode at the same rate as fastlands when subjected to similar winds, tides, currents and storms. Thus, the rate of erosion is the same whether the wetlands are filled or left in their natural state. In addition, our evidence suggests that wetlands exist more frequently in those areas with low erosion potential. This can explain the common observation that where wetlands exist there is often little erosion. However, this is not because wetlands provide superior erosion protection, but rather they exist where erosion forces are minimal (Owens, Park and Batie in preparation).

Oyster Population: Although oysters are well studied, the linkages between their availability for commercial harvest and wetlands are not well established. Although oysters do ingest primary production products, it is not known to what extent wetlands provide the main source, some of the source, or little of the source of this primary production. It is not even known if primary production is the limiting factor for oyster propagation and growth. Conceivably, oyster populations are not limited by food supply, but rather by predation, oxygen, temperature or light. Thus, we do not know if there is a critical acreage of wetlands necessary for oyster populations, below which oyster populations will decline. Also, differing wetlands' qualities may or may not be significant (Batie and Wilson 1979).

Wildfowl: Wildfowl are valuable to humans for hunting, viewing, and their contributions to the food web. Yet little is known concerning the relationship between wetlands and wildfowl populations. Walker (1973) reports that a great percentage of wetlands in the Chesapeake area and the Mid-Atlantic region generally is not heavily utilized by migrating birds. Furthermore, waterfowl appear to be "flexible in seeking out staging and feeding areas, and they adapt to change more easily than other organisms" (Walker 1973; p. 81). Indeed, on Virginia's Eastern Shore a popular area with many waterfowl is dredge spoil banks.

Waste Assimilation: Tidal wetlands are reputed to provide a valuable ecological service in the form of water quality maintenance (Walker 1973). Wetlands' waste assimilative capacity may function in basically three ways. The first is that wetlands may serve as tertiary treatment system when they are artificially loaded with sewage sludge. Although some research studies suggest wetlands have a significant capacity for assimilating nutrients (Valiela, Teal and Sass 1973), the ability of wetlands to continue this service with high rates of loading for an extended period of time is questionable (Bender and Correll 1974).

The second manner in which wetlands may improve water quality is in removing pollutants, particularly nutrients, from the estuary during natural tidal flushing of marsh. There is inconclusive evidence as to the extent of this service, although research suggests nutrients are changed from particulate to dissolved forms in some cases (Aurand and Daiber 1973; Axelrad, Moore and Bender 1976; Stevenson et al. 1976).

The third way in which wetlands may affect water quality is in relation to pollutant loading from nonpoint runoff. Wetlands apparently act as a trap for sediment, nutrients, and other materials adsorbed to sediment particles (Boto and Patrick 1979). Unfortunately, no completed studies have attempted to quantify these processes, although research currently underway on the Rhode River watershed in Maryland is addressing this issue (Correll, D. L., Aug. 1979, Chesapeake Bay Center for Environmental Studies, Edgewater, MD. pers. comm.).

Wetlands' Services: In viewing the broad range of services attributed to wetlands, Walker concluded:

> "Thus far I have shown that the scientific justifications for coastal wetlands preservation are not quite as clear cut as they appear at first blush. The primary productivity of marshes is evident, but little can be said about the dependence of important species on marshes, nor the response of the estuarine ecosystem to marsh destruction. Similarly, water quality seems to be improved by wetlands, but the dynamics of nutrient cycling is too poorly understood to predict the impact of wetlands on overall estuarine water quality. The erosion, sediment and flood control capacities of wetlands may only be modest, and are rather unpredictable" (Walker 1973; p. 90).

The point of this discussion is not to suggest that the ecological services of wetlands are nonexistent; however, there is a high degree of uncertainty about those services. As a result, improved estimation of the technical linkages between wetlands and natural services must exist before sound economic values can be estimated. Figure 1 illustrates this with respect to oyster populations. Resource management inputs such as fishermen, oyster dredges, and even the property rights associated with harvesting oyster beds combine with the biological and physical inputs, such as wetlands and salinity conditions. These inputs enter a production process which ultimately determines total oyster population and oyster harvest. The value of the harvest depends upon the level of consumer demand and oyster supply. The objective of economic analyses is to isolate the contribution of each of these inputs, including wetlands, to the final value of the harvest. To accomplish this objective, it is necessary to understand how changes in the level of one or more of these factors, taken together or separately, will translate into changes in oyster supply and hence value. However, this valuation process can proceed only if the economic, biological and physical linkages are understood and are quantifiable.

Policy and Research: The Next Steps

The state of the art in wetlands' valuation will lag behind the need to make wetlands' management decisions for the foreseeable future. Given the current lack of full knowledge, we support adoption of management

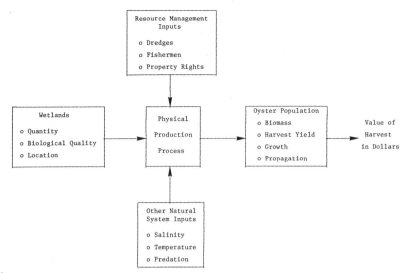

Figure 1. Factors contributing to the value of oyster harvest.

guidelines which stress wetlands' preservation, unless the expected value of foregone development is "unacceptably" large. Of course, what is deemed unacceptable must be a broad social decision, but is one in which economic analysis can have the important role of identifying the values of wetlands for development. If the value of wetlands' development was clarified, public decisions to preserve wetlands, given the current uncertainty about natural wetland values, could be more easily defended. Specifically, lower development values make the argument for denial of a development permit more compelling (Shabman et al. 1980). An important reason for concentrating on the estimation of development values is that land market information is available for estimating values. In fact, we have conducted such analyses of development values during the past three years. The procedures used are reported elsewhere (Shabman and Bertelsen 1979; Shabman et al. 1980), and only the results are reported here for purposes of this discussion.

In Virginia Beach, Virginia, development values foregone by wetland preservation (if development proposal is for residential lots) are not remarkably high. Development values vary according to location and type of development, but for illustrative purposes a 3/4-acre site with 150 feet of frontage on an open bay would have a development value of $14,000 (Shabman and Bertelsen 1979). Development values for water access through a private marina and for vacation home sites on wetlands in rural Accomack County, Virginia, were calculated for the situation when no fastland alternative site was available. Development values for a marina were $5.8 million per acre for five acres of wetlands. If a fastland alternative site was available, there were no positive returns to marina development in wetland areas (Shabman et al. 1980). Here, some allowance for wetland development might be acceptable, especially since virtually the whole county shoreline is wetland and few, if any, comparable fastland alternatives exist. However, the value of developing additional acres for marinas will fall sharply as additional marinas are built. This decline of value with additional development is also true for the returns for second home development; these were estimated to be $40,000 per acre for one of the few Accomack recreational developments that utilized wetlands.

It will not be possible to conduct a detailed development value analysis for each wetland allocation decision; however, these research results suggest some general guidelines that might be followed. First, due to the uncertainty of natural wetland values, the development should move forward only upon the demonstration of "large" development values. In short, the burden of proof in the public decision process should be shifted from those who wish to preserve wetlands to those who wish to develop them. Second, the provision of water access to a large group of lot owners (or the general public) by development of small areas of wetlands may have a high social value, especially in areas where water access is limited. However, more intensive management of existing water access facilities should also be considered as a means of reducing the need for marsh development for water access. Third, the value of marsh filling for creation of waterfront lots (especially in areas with extensive waterfront) appears to have a relatively low value when compared with provision of water access.

In the meantime, the research community should continue to improve our understanding of the role of wetlands in the natural and economic systems. We believe that conceptually valid economic estimates of wetland values are possible where physical wetland linkages to wetland services are established. However, since in most cases the technical linkages between wetlands and natural services are not well established, there should be additional research focused on alternative development values of wetlands. Such research has the attractive attributes of a high probability of successful completion and considerable utility in improving public decisions. Alternative development uses of wetlands worthy of investigation, in addition to the uses discussed in this paper, include: commercial uses, such as restaurants; industrial uses, such as manufacturing enterprises and ports; and recreational uses, such as public parks and beach access.

At the same time, research programs should be developed for ascertaining economic values of natural wetlands. To be fruitful this research should be conducted through cooperative efforts among the various disciplines. As the previous discussion of wetlands' contributions to oyster populations has noted, there must be an appreciation of the nature of the production process which ultimately yields the wetland services of interest. There is every reason to be optimistic that research projects that reflect such an appreciation will generate information of considerable utility for wetlands' management.

Acknowledgments

This research reported in this paper was partially sponsored by NOAA Office of Sea Grant, Department of Commerce, under Grant Nos. 04-6-158-44086 and 04-7-158-44086. The U.S. Government is authorized to produce and distribute reprints for governmental purposes not withstanding any copyright notation that may appear hereon.

References Cited

Aurand, D. and F. C. Daiber. 1973. Nitrate and nitrite in the surface waters of two Delaware salt marshes. *Chesapeake Sci.* 14:105-111.

Axelrad, D. M., K. A. Moore and M. E. Bender. 1976. *Nitrogen, Phosphorus and Carbon Flux in Chesapeake Bay Marshes,* Bulletin 79, Virginia Water Resources Research Center, Blacksburg, VA. 182 pp.

Batie, S. S. and L. A. Shabman. 1980. Valuing non-market goods: conceptual and empirical issues—a discussion. *Amer. J. Agric. Econ.* (in press).

Batie, S. S. and J. R. Wilson. 1979. *Economic Values Attributable to Virginia's Coastal Wetlands as Inputs in Oyster Production.* Research Division Bulletin No. 150, Virginia Tech, Blacksburg, VA. 27 pp.

Bender, M. E. and D. L. Correll. 1974. The use of wetlands as nutrient removal systems. Chesapeake Research Consortium Report No. 29, Virginia Institute of Marine Sciences Contribution 624, Gloucester Point, VA. 12 pp.

Boto, K. G. and W. H. Patrick, Jr. 1979. The role of wetlands in the removal of suspended sediments, pp. 479-489. *In:* P. E. Greeson, J. R. Clark, and J. E. Clark (eds.), *Wetlands Functions and Values: The State of Our Understanding.* American Water Resources Association, Minneapolis, MN.

Boulding, K. 1969. Economics as a moral science. *Amer. Econ. Rev.* 59(2):1-12.

Environmental Law Institute. 1979. State wetland and floodplain regulatory programs. *National Wetlands Newsletter* 1:5-8.

Freeman, A. M. 1979. *The Benefits of Environmental Improvement: Theory and Practice.* Johns Hopkins Press, Baltimore, MD. 272 pp.

Gosselink, J. G., E. P. Odum and R. M. Pope. 1974. The Value of the Tidal Marsh. Publication No. LSU-56-74-03, Center for Wetland Resources, Louisiana State U., Baton Rouge, LA. 30 pp.

Howe, C. W. 1971. *Benefit-Cost Analysis for Water System Planning.* Water Resources Monograph No. 2, American Geophysical Union, Washington, D.C. 144 pp.

Kneese, A. V. and B. T. Bower. 1973. *Managing Water Quality, Economics, Technology, Institutions.* Johns Hopkins Press, Baltimore, MD. 328 pp.

Margolis, J. 1969. Shadow prices for incorrect or nonexistent market prices, pp. 553-546. *In:* U.S. Congress, Joint Economic Committee. *The Analysis and Evaluation of Public Expenditures: The PPB System.* U.S. Government Printing Office, Washington, D.C.

Midwest Research Institute. 1979. *Economic Aspects of Wildlife Habitat and Wetlands.* MRI Report No. 4626-D, Kansas City, MO. 108 pp.

Mills, E. S. 1978. *The Economics of Environmental Quality.* W. W. Norton and Company, New York. 304 pp.

Odum, E. P. 1979. Rebuttal of "Economic value of natural coastal wetlands: A critique." *Coastal Zone Mgmt. J.* 5:231-237.

Odum, E. P. and H. T. Odum. 1972. Natural areas as necessary components of man's total environment. *Trans. North Amer. Wildl. Nat. Res. Conf.*, pp. 178-189.

Odum, H. T. 1979. "Principle of environmental energy matching for estimating potential economic value: a rebuttal." *Coastal Zone Mgmt. J.* 5:239-241.

Odum, H. T. and E. Odum. 1976. *Energy Basis for Man and Nature.* John Wiley & Sons, Inc., New York 288 pp.

Park, W. and S. S. Batie. 1978. Alternative Management Strategies for Virginia's Coastal Wetlands: A Program of Study. Sea Grant Project Report, Virginia Tech, Blacksburg, VA. 15 pp.

Shabman, L. A. and S. S. Batie. 1978. Economic value of natural coastal wetlands: A critique. *Coastal Zone Mgmt. J.* 4:231-247.

Shabman, L. A., S. S. Batie and C. C. Mabbs-Zeno. 1980. The economics of wetlands preservation in Virginia. *J. Northeast. Agric. Econ. Council* (in press).

Shabman, L. A. and M. A. Bertelsen. 1979. The use of development value estimates for coastal wetland permit decisions. *Land Economics* 55:213-222.

Stevenson, J. C., D. R. Heinle, D. A. Flemer, R. J. Small, R. A. Rowland and J. F. Ustach. 1976. Nutrient exchanges between brackish water marshes and the estuary, pp. 219-240. *In:* M. L. Wiley (ed.), *Estuarine Processes, Vol. II,* Academic Press, New York.

Valiela, I., J. M. Teal and W. Sass. 1973. Nutrient retention in salt marsh plots experimentally fertilized with sewage sludge. *Est. Coastal Mar. Sci.* 1:234-253.

Virginia Code Annotated. Sec. 62.1-13.1, *et.seq.* (1972), as amended (Supp. 1976).

Walker, R. A. 1973. Wetlands preservation and management on Chesapeake Bay: the role of science in natural resource policy. *Coastal Zone Mgmt. J.* 1:75-101.

Wass, M. L. and T. D. Wright. 1969. Coastal Wetlands of Virginia. Interim Report No. 1, SRAMSOE No. 10, Virginia Institute of Marine Science, Gloucester Point, VA. 154 pp.

Westman, W. E. 1977. How much are nature's services worth? *Science* 197:960-964.

WETLANDS' VALUES—CONTRIBUTIONS TO ENVIRONMENTAL QUALITY OR TO NATIONAL ECONOMIC DEVELOPMENT?

William J. Hansen, Sue E. Richardson

Environmental Laboratory
U.S. Army Engineer Waterways Experiment Station
Vicksburg, Mississippi

Richard T. Reppert

U.S. Army Engineer Institute for Water Resources
Kingman Building
Fort Belvoir, Virginia

and

G. E. Galloway, Jr.

Department of Earth, Space and Graphic Sciences
U.S. Military Academy
West Point, New York

Abstract: The ability of decisionmakers to evaluate impacts of actions on estuarine areas is impaired by the absence of procedures to compare these impacts to the benefits and costs of the action. Many of the goods and services provided by marine wetland areas are done so outside the market system and therefore do not have monetary values attached. For some of the goods and services, techniques are not available for establishing an acceptable proxy value. Although absolute quantification and, ultimately, monetization of impacts on estuarine areas may be desirable, current technology does not allow us to be that definitive. There are, however, methods which allow for the relative quantification of impacts. An interim technique developed by the U.S. Army Institute for Water Resources (IWR) provides for the establishment of relative indices of value based on a case by case analysis of the biological, physical and human use characteristics of wetlands. The recently developed Wetland Evaluation System is somewhat more analytical than the methods in the IWR manual and utilizes a weighting-scaling system to incorporate value and technical judgments. Additional methods include computer tools such as policy capturing which may also be used to incorporate key factors systematically into and allow the relative quantification of the evaluation of impacts on estuarine areas.

Introduction

Recently there has been an increasing public awareness of the value of our Nation's wetland areas and the need for their preservation. President Jimmy Carter, in his first environmental message to Congress on 23

May 1977, focused national attention on the importance of wetland areas when he stated:

"The important ecological function of coastal and inland wetlands is well known to natural scientists. The lasting benefits that society derives from these areas often far exceeds the immediate advantage their owners might get from draining or filling them."

On 24 May 1977, the President reaffirmed this message with the issuance of Executive Order 11990. This Order directed all Federal agencies to ensure the proper protection of wetlands in all actions undertaken under their jurisdiction.

The U.S. Army Corps of Engineers has significant responsibility in decisions affecting the development of wetland areas. In the Federal Water Pollution Control Act Amendments of 1972 (Public Law 92-500, 86 Stat. 884), Congress specified that placement of dredged or fill material in wetlands would require a Federal permit. A subsequent Federal court ruling *(NRDC v. Callaway)* extended the Federal jurisdiction of this Act from just navigable waters to the "waters of the United States." As a result, the Corps has the responsibility for review of permits concerning development in most wetland areas of the United States.

Applications for Department of the Army permits are subject to a public interest review that involves consideration of a broad set of engineering, economic, social, and environmental factors. Corps of Engineers general regulatory policies (U.S. Army Corps of Engineers 1977) state in part that:

Effect on wetlands. (1) Wetlands are vital areas that constitute a productive and valuable resource, the necessary alteration or destruction of which should be discouraged as contrary to the public interest.

The Corps' responsibilities in its regulatory as well as its Civil Works programs frequently require detailed study and evaluation of wetland areas. For years the Corps, like most other Federal agencies, has used benefit/cost analysis as one of its primary tools for evaluating its water resources development projects. However, exclusive use of benefit/cost analysis requires that all impacts on society of alternative management actions be monetized.

Although the absolute quantification and ultimately monetization of impacts on wetland areas may be desirable, current technology does not allow us to be that definitive. This problem is not unique to wetland areas, but is typical in the evaluation of most water resource development projects.

Recognizing the limitations of the exclusive use of benefit/cost analysis, the Water Resources Council (WRC) in 1973 established the *Principles and Standards for Planning Water and Related Land Resources* (Water Resources Council 1973). In *Principles and Standards,* the WRC acknowledges that not all impacts of water resources development projects can be monetized and that "plans for the use of the Nation's water and land resources will be directed to improvement in the quality of life through contributions to the objectives of national economic development and environmental quality." These objectives are defined respectively as "to enhance national economic development by increasing the value of the Nation's output of goods and services and improving national economic efficiency" and "to enchance the quality of the environment by the management, conservation, preservation, creation, restoration, or improvement of the quality of certain natural and cultural resources and ecological systems."

Although the Corps' permit responsibilities under PL 92-500 are not subject to the administrative requirements of *Principles and Standards,* the evaluative framework provided therein is still applicable for projects that have an impact on wetland areas.

Evaluating Estuarine Areas

Economic Evaluation

Placing a monetary value on many of the services provided by wetlands (e.g., nursery and feeding habitat for wildlife) is constrained because the services are provided outside the normal market system and, therefore, do not have monetary values attached to them. Several recent research efforts have attempted to derive dollar benefits for some of these services. Two serious problems, however, preclude decisionmakers from utilizing the results of most of the research to date.

The first problem involves conceptual errors in the research design that negate the validity of the results for use in traditional benefit/cost analysis. An example is the use of expenditures by fishermen to evaluate sports fisheries. The values obtained do not reflect contributions to national economic development since they only reflect a substitute for some other consumer expenditures. The use of techniques that result in the determination of the gross value rather than the value added have long been recognized as invalid for measuring alternative uses of a particular resource (Clawson and Knetsch 1966).

The second problem associated with the use of research results to date is that, even if correctly derived, they generally reflect the average value of wetlands in a specific area. Average values are not appropriate for evaluating the incremental impacts that would result from most modification plans. In addition, they have little relevancy even as average values to other wetland areas without adjustment for the appropriateness of the benefits considered (wetland services provided will vary from place to place), variations in wetlands' productivity, and regional variations in the availability of wetland services (a region with an abundance of wetland areas, other things being equal, would have lower average values than a region in which wetlands are scarce).

Monetization of the value of wetland services is desirable and additional research in this area is encouraged. Optimal resource allocations will not result, however, unless conceptually sound benefit evaluation techniques are utilized. Additional knowledge of the production functions of the natural systems of wetlands is required as well as of the probable incremental effects that alternative modification plans will have on such systems. It is here that the greatest need exists for joint research efforts from the economic, social, and biological communities to determine the incremental impacts of modifications on wetland natural systems and to evaluate the marginal impacts on societal values.

In the interim, decisions will continue to be made affecting the preservation or loss of wetland areas. If the decision process is systematic, analytical, and open, it can provide a framework for explicitly addressing the conceptual problems described above. Such a framework could expedite the incorporation of new scientific knowledge into the decision process as well as assist in the eventual monetization of values associated with wetland services.

Environmental Quality Evaluation

The evaluation of wetlands on the basis of their contribution to environmental quality presently comes closer to being feasible, given the state of the art, than does economic evaluation. More is known about the ecological aspects or characteristics of estuarine areas than is known about the economic value of such areas. It is only logical to expect that a complete and accurate economic evaluation is dependent on a scientifically accurate knowledge of the processes and components that make up an estuary.

Although the ecological understandings of wetlands are improving each day, a systematic evaluation of their worth or of the "goodness" or "badness" of potential impacts is still difficult. The primary reason behind the difficulty is that environmental quality is measured in a wide variety of ways that are not necessarily comparable. An effective evaluation technique must overcome this problem, as well as the problem that the value of an estuarine area is highly dependent on the context in which it is found.

Judgment in Evaluation

Two differing kinds of judgment must be integrated to accomplish adequate valuation of estuarine areas—technical judgment and value judgment. Technical judgment includes the projections or predictions of potential impacts on wetland areas, these forecasts for the most likely future being accomplished by technical experts. In our society, however, the experts may not impose their value judgments on the public without the public's knowledge and consent. It is important, therefore, that the experts educate the public about the potential impacts and their ramifications so that the public can make enlightened choices about future options. Therefore, once the experts have made their technical judgments known to the public, the public is then in the position of determining the importance of the estuarine resource, given their knowledge and feelings about the environmental quality of their area.

Evaluation Techniques

Presently, there is no single best technique for the evaluation of wetlands. Numerous environmental impact models are available. Some have been developed specifically for evaluating estuarine areas while others are more general in nature. Consideration of the attributes described above should be included in the selection of a particular approach to be used, as well as the resources and time available to conduct the analysis.

Resource and time considerations are especially critical when considering the range in scale of projects to be evaluated and the magnitude of the workload, especially in the Corps' regulatory program. In fiscal year 1977, for example, the Corps received nearly 11,000 applications for permits involving discharge of dredged or fill materials in waterways and wetlands. Some of these applications involved wetland areas of miniscule size and importance, while others involved major alterations. The same level of specificity of data collection and analysis and public involvement cannot be expected for all decisions.

Briefly described below are three alternative approaches that could be used for evaluating estuarine areas. They are similar in as much as each provides a systematic and analytic approach to the evaluation process, recognizes the need for site-specific applications, addresses the differences between technical and value judgments, and emphasizes the need for an open and carefully documented decision process. Yet each differs significantly in levels of specificity, public participation, and resource requirements.

IWR Manual

The Institute for Water Resources has recently prepared a manual for wetlands' evaluation (U.S. Army Engineer Institute for Water Resources 1979). A list of important wetland characteristics was synthesized for consideration and divided into two main categories: primary functions and cultural values. The primary functions closely correspond to the list of wetland characteristics contained in the Corps' permit regulations, while the cultural values incorporate socioeconomic and other socially perceived considerations.

Specifically addressed under primary functions are: food chain production; general and specialized habitat for land and aquatic species; aquatic study areas; sanctuaries and refuges; hydraulic support function; shoreline protection; storm and flood water storage; natural groundwater recharge; and water purification. Cultural values include commercial fisheries; renewable resources and agriculture; recreation; aesthetics; and other special values.

The IWR manual presents a general description of these functions and values and provides techniques for the assessment of individual evaluation criteria. Also described in the manual are two suggested frameworks for the overall evaluation of wetlands, a deductive and a comparative analysis. The manual provides a base for the evaluation of wetlands through consideration of the functions and values listed above. Each function or value is described, including discussion of the relative efficiency with which they are carried out in a particular wetland system. Evaluation criteria are given for each function and value, delineating how to assign a numerical rating of 1, 2, or 3 based on relative efficiency or importance. For example, for the wetland function of water purification through natural water filtration, the nature of wetland water quality improvement is discussed and three types of evaluative criteria are considered. These encompass some of the key environmental conditions relevant to the wetland function of water quality improvement: type of wetland, areal and waste loading relationships, and geographic and other locational factors. Then evaluation guidance is provided; e.g., if total wetland size is greater than 100 acres, it is rated 3 (of high value for water quality improvement); if between 10-100 acres, rated 2 (moderate); and less than 10 acres, rated 1 (low).

The deductive analysis is intended as a nonquantitative approach and is based on a systematic evaluation of the degree to which the wetlands under examination satisfy each of the functional characteristics and cultural values. It is suggested for use in such applications as the Corps' permit program where site alternatives are not normally presented. It is also recommended that a nonquantitative summary of the conclusions, as well as the reasons for these conclusions, be included as an integral part of the method. Adequate documentation is essential to any method where public attitudes are to be incorporated into the decisionmaking process. An example of the quality and quantity of information necessary for a typical deductive analysis evaluation is provided in the manual. It includes a table with each type of function or value described for the wetland being analyzed, and is primarily a narrative/descriptive approach.

The comparative analysis is intended as a quantitative analysis that involves the systematic evaluation of the degree or efficiency with which two or more wetland areas under study satisfy criteria pertaining to functional characteristics and cultural values. The manual suggests that the comparative analysis provides a framework for evaluating proposed water resources development projects since the quantitatively ranked comparison of wetlands' values and functions makes it possible to assess the relative environmental importance of site alternatives. An example of a comparative analysis is provided which is aimed at determining the relative value of alternative locations for a proposed marina site. The comparison is accomplished by summing the numerical rankings for each function or value, in addition to a strong narrative presentation. The strength and validity of this approach rest on the foundation of the function and value scheme described in the manual.

Wetland Evaluation System

Another recently developed method is the Wetland Evaluation System (WES) (Galloway 1978). It is somewhat more analytical than the methods in the IWR manual and utilizes a weighting-scaling system to incorporate value and technical judgments. The WES also incorporates a computer graphics element to display the impacts of the various actions being considered to the decisionmaker and analyst. The stated purpose of WES ". . . is to produce information concerning the change in value of the environmental quality of a wetland area (or areas) as a result of the intrusion of man into the area(s)."

Galloway reviews several approaches that have been used in the assessment of man's impact on the environment in general and on wetlands in particular. These approaches include attempts to monetize the value of wetlands, and are divided into two categories: macro and micro. Choosing to focus on the macro perspective, Galloway classifies and discusses models as graphic, computer-assisted graphic, quantitative, and matrix.

Three types of evaluations are conducted in WES: the determination of the relative value of a wetland indicator, an assessment of the percentage change in this base value that will occur under various conditions, and the determination of the relative weight or importance of each indicator being used. It is recommended that a local team of experts, representing the social and natural sciences, provide the first two evaluations, while the latter is accomplished by a team of local and regional elected officials. Thus, the WES provides for the separation of technical and value judgments and explicitly incorporates public values into the analysis.

The WES incorporates nine Environmental Quality Indicators into the analysis. It is recognized that, although the nine factors may not represent 100% of a given wetland's quality, they do represent a most substantial amount for modeling purposes. The nine indicators are: endangered species; fish and other aquatic ecosystems; wildlife and other terrestrial ecosystems; waterfowl; uniqueness; appearance; natural protection; life-cycle support; and historical/cultural (Galloway 1978).

The Environmental Quality Indicators represent the principal features of a wetland and the weighted sum of their values provides a measure of the quality of a designated wetland. To provide a degree of focus, evaluators must determine which six indicators (of the nine) best represent the wetland area under study. To permit the wetland to be evaluated with some degree of specificity, the total wetland is divided into relatively homogeneous areas. The WES also incorporates a probability-of-impact factor to bring the overall ratings in closer touch with reality. Two well-developed examples are provided of how WES might work in an actual situation. Detailed information is presented about one subbasin/estuary in each case, while information on the other subbasins/estuaries is provided without explanation. Examples of final displays, both graphic and numeric, are given.

Judgment Analysis

A third approach to estuarine evaluation is represented by methods developed for judgment analysis. These generic tools seek to provide a framework for integrating scientific information and public values in the formation or accomplishment of public policy. This approach is scientifically and socially defensible for wetlands' evaluation and is ably represented by policy capturing, developed at the Center for Research on Judgment and Policy at the University of Colorado (Hammond and Adelman 1976).

The underlying assumptions of policy capturing are as follows:

> "Basic to any policy involving scientific information are objectively measurable variables. Scientific judgments regarding the potential effects of technological alternatives are also required. And social value judgments by policy makers or community representatives are required. The overall acceptability of an alternative is determined by how closely its potential effects satisfy the social values of the community." (Hammond and Adelman 1976, p. 18)

> "Every decision involves the selection among an agenda of alternative images of the future, a selection that is guided by some system of values. The values are traditionally supposed to be the cherished preserve of the political decisionmaker, but the agenda, which involves fact, or at least a projection into the future of what are presumably factual systems, should be very much in the domain of science." (Boulding 1975, p. 679)

Policy capturing is a technique for building a model that describes how an individual combines multiple pieces of information into a single judgment or forecast. Specifically, the technique uses multiple regression analysis to derive a mathematical description of (a) the relative weights (importance) that an individual places on each type of data entering into an evaluation judgment, and (b) the functional relationship between data values and the individual's evaluation. Individuals are required to input information about the desirability of several different types of possible impacts on an estuarine area. The program statistically analyzes the relationship between the values of the variables used to describe the impact and the impact itself. For example, suppose the task was to evaluate the impact on number of shrimp in an estuary from information about pH level, temperature, salinity, and turbidity. Then, policy capturing could specify for each individual making judgments (a) the relative importance of pH level, temperature, salinity, and turbidity in evaluating impacts on shrimp, and (b) the functional relationship between values of each type of data and the level of impact—e.g., the higher the salinity, the greater the number of shrimp; the greater the turbidity, the lower the number of shrimp. The judgment policy, i.e., weights and functional relations, can be used to evaluate the level of impact on the estuarine resource for any new set of data values for these variables.

Commonalities

Each of the techniques described above is scientifically defensible because the procedures are systematic and available for public inspection; therefore, the process provides the opportunity for cumulative knowledge, as scientific efforts should. Each of the methods is socially responsible because it provides a public framework for separating technical judgments from social value judgments and integrating them analytically, not judgmentally. The separation phase permits elected representatives to function exclusively as policy makers, and scientists to function exclusively as scientists. The integrative phase provides an overt, rather than covert, process for combining facts and values. Because the social values of the community are identified before the decision is implemented, the decision process is not seen to be a mere defense of a predetermined choice; rather it can be evaluated in terms of its rational basis before the final choice is made.

Conclusions

The purpose of this paper is to describe several key factors essential to the evaluation of impacts on wetland areas, as well as some of the available techniques that can assist the decisionmaker in this process. As noted earlier, there is no single best or universal technique available for the evaluation of estuarine areas, and there may never be, given the rapidly expanding knowledge of the natural systems of wetland areas. In selecting between acceptable techniques, the analyst should be aware of data and resource requirements and form of results relevant to his study resources and needs.

Additional research towards improving techniques for monetizing wetlands values is encouraged. No matter how analytical the evaluation process, it will always be difficult to analyze the trade-offs between impacts measured in dollars relative to those measured in some other units.

The use of techniques such as those described herein can provide an interim step towards the eventual monetization of many of the values associated with wetland areas. These techniques utilize analytical approaches that are also essential to proper economic analysis.

Note

The views expressed in this paper are those of the authors and do not necessarily represent official policy of the Department of the Army or any of its agencies.

References Cited

Boulding, K. E. 1975. Truth or power? *Science* 190:423.

Clawson, M. and J. L. Knetsch. 1966. *Economics of Outdoor Recreation.* Johns Hopkins Press. Baltimore and London.

Executive Order 11990, 24 May 1977, Protection of Wetlands, Office of the White House Secretary.

Federal Water Pollution Control Act Amendments, Section 404 (86 Stat. 884), 1972.

Galloway, G. E. 1978. Assessing Man's Impact on Wetlands. University of North Carolina, Sea Grant Publication UNC-SG-78-17 or University of North Carolina Water Resources Research Institute Publication UNC-WRRI-78-136, Raleigh, NC.

Hammond, K. R. and L. Adelman. 1976. Science, values, and human judgment. University of Colorado, Institute of Behavioral Science, Program for Research on Human Judgment and Social Interaction, Report No. 182.

NRDC v. Callaway (7ERC1784).

U.S. Army Engineer Institute for Water Resources. 1979. Wetland Values: Concepts and Methods for Wetlands Evaluation. Fort Belvoir, VA.

U.S. Army Corps of Engineers. 1977. Regulatory Program of the Corps of Engineer—General Regulatory Policies. *Federal Register,* (33 CFT 320), 42 FR 138, Tuesday, 19 July 1977, pp. 37133-37138.

Water Resources Council. 1973. Water and Related Land Resources; Establishment of Principles and Standards for Planning. *Federal Register,* 38(174), pp. 24779-24869.

QUANTIFICATION OF ENVIRONMENTAL IMPACTS IN THE COASTAL ZONE

Peter N. Klose

Environmental Resources Management, Inc.
999 West Chester Pike
West Chester, Pennsylvania

Abstract: Protection of coastal zone ecosystems historically has been hampered by an inability to quantify environmental impacts of various types of development. Also, the completion of environmental impact assessment studies has been an involved, subjective process hindered by a lack of quantitative techniques which could aid in establishing cumulative impacts of a project. However, several studies have established some of the economic values of wetlands or other ecological systems and the biological functions they support. This has allowed for the direct comparison of economic with ecological impacts of certain environmental alterations. A study funded by the National Science Foundation allowed further investigation of the feasibility of developing a methodology for assessing onshore impacts for outer continental shelf oil and gas development. Results of this study indicated that no impact assessment methodologies were available which could be used to predict ecosystems' changes quantitatively from most types of expected coastal zone developments. For coastal ecosystems to be adequately considered in policy decisions affecting their fate, increased use of valuational techniques of an ecosystem's natural functions must be made.

Introduction

The National Environmental Policy Act (NEPA) of 1970 established the inclusion of ecological considerations in the planning and development of major Federal projects by demanding the development of an Environmental Impact Statement (EIS) for each project. Estuarine scientists had a direct interest in NEPA since this legislation called for biological involvement in environmental planning. The intent of the Act is described in Section 102, which calls for inclusion of ecological information, via:

a) the use of an interdisciplinary approach to decision making which affects man's environment;

b) developing procedures to include presently unquantified environmental values in environmental decision making;

c) producing an environmental impact statement for major legislation and Federal actions;

d) initiating and utilizing ecological information in the planning and development of resource oriented projects.

It was soon after NEPA's passage that scientists became actively involved in the impact assessment of natural resource decisions in the coastal

27

zone. Much of this early involvement revolved around the assessment of environmental impacts of proposed projects, such as ports, refineries, or power plants. Although impact assessment was a new concept and not often based on available ecological literature, estuarine researchers were asked to determine the impact, or effect, of such diverse activities as: dredging of barge channels and marshes, construction of pipeline rights-of-way, and entraining ichthyoplankton in once-through power plant cooling systems.

Assessing the ecological impact of these types of activities proved difficult. Little biological or ecological research had been conducted prior to this period which was concerned with "impact analysis". Few scientists were interested in the applied ecological aspects of new industrial facility siting, dredging and filling activities, or housing development in the coastal zone. In addition, little money was available to pay for such studies during the first few years in which environmental impact assessment (EIA) was required.

To aid in overcoming expected problems for this new field of environmental impact assessment, Congress had the foresight, under NEPA, to establish the Council on Environmental Quality (CEQ). This agency produced guidelines for the preparation of EIS reports (Council on Environmental Quality 1973). These guidelines were helpful in directing environmental information in the planning and development of resource oriented projects. Despite the CEQ guidelines, several problems with regard to environmental impact assessment remained, including:

1. ***Definition of an impact:*** Although NEPA declared that EIS's should include a statement of "the environmental impact of the proposed action, and any adverse environmental effects which cannot be avoided should the proposal be implemented", no clear definitions of impacts or effects (which are treated as synonyms in the 1978 Regulations) were given in NEPA. CEQ (1978) only went so far as to state, "Effects includes (sic) ecological (such as the effects on natural resources and on the components, structures, and functioning of affected ecosystems), aesthetic, historic, cultural, economic, social, or health, whether direct, indirect, or cumulative". No precise definitions of "impact" or "effect" were given, nor was any guidance given as to how severe an effect would have to be in order to be considered important for the EIA process.

2. ***Development of data bases.*** The intent of NEPA was negated to a great degree in the early years due to insufficient ecological data which were available for the establishment of a baseline from which environmental effects could be predicted. In many cases, impacts of proposed projects could only be approximated since few data were available as to what the actual impacts of the pro-

ject would be. That is, the environmental effect of past activities had not been monitored.

3. ***Establishment of ecological effects on a cost-benefit basis.*** The greatest hindrance to effective use of ecological impact data was the inability to compare ecological effects with economic effects. That is, it was difficult for estuarine scientists to convey the economic loss of a 25% reduction in primary productivity in a specific estuary and compare such data to a cost/benefit ratio of 1:2 and the annual generation of $10 million in revenues to the local community where the project would be located. The unfortunate response of decision-makers to an ecologist's statement on the severity of an ecological impact was all too often, "So what? What does it mean in terms of economic loss to the fishing or recreation industry?" Unfortunately, primarily due to the complex nature of ecological systems, establishing direct links between ecological effects and economic values proved extremely difficult.

Due to the above difficulties in properly applying the EIA process, many EIS's became lengthy collections of poorly linked and inadequately analyzed data. Furthermore, the process of determining the types and severity of impacts of a particular project commonly was based on the experience and preconception of the document's preparer. Such lack of specific guidance led to much variation in the content and length of EIS's.

The great variety of projects which required environmental assessment studies further increased the difficulty of assessing the effects, particularly quantitative effects, of those projects. Estuarine scientists had to deal with a wide variety of coastal zone projects and activities for which the impacts were elusive. Examples of such assessment scenarios included:

1. quantifying the difference in effect on water quality, fish, and shrimp populations of discharging the effluents of a large industrial facility into a tidal river as opposed to direct discharge to the ocean at the mouth of the river;
2. calculating the impact of reducing the volume of flow of estuarine water through an enclosed bay on oyster growth and reproduction;
3. determining the impact of reducing freshwater inflow and decreasing salinity of an entire estuary on its biological "health and vitality".

Many of the problems of EIS preparation were dealt with in CEQ's "Regulations on Implementing National Environmental Policy Act Procedures", (Council on Environmental Quality 1978). With this Act, EIS guidelines became regulations. The objective of the regulations was "to reduce paperwork and the accumulation of extraneous background data, in order to emphasize the need to focus on real environmental issues and alternatives". Thus, administrative and content issues regarding impact assessment were resolved, but the task of predicting, much less quantifying, impacts which might accrue to an ecosystem was still a source of scientific guesswork.

Scientists working in the environmental assessment field attempted, over the years, to overcome problems inherent in the environmental assessment process by developing standardized EIA methodologies which would convert subjective assessments into objective, numerical results (Warner and Preston 1974). Such studies attempted to allow for comparison between similar projects, as well as explain in detail how a particular environmental assessment decision had been reached so that readers of the EIS could understand the decision making process. Also, environmental assessment methodologies attempted to remove the inherent bias of the preparer by subjecting the process to a systematic sequence of interpretive steps (Coastal Engineering Research Laboratory 1975). Analytical methodologies available in the mid-1970's consisted chiefly of check lists, map overlays, matrices, or quantitative models of limited expanse, such as for water quality only (Warner and Preston 1974).

In 1976, as the search for America's offshore oil and gas escalated, the Department of the Interior and Department of Commerce, coordinated by the National Science Foundation, supported a study entitled, "Methodology for Assessing Onshore Impacts for OCS Oil and Gas Development". The objective of this study was to develop a set of methodologies which could be applied to all U.S. coastal regions to assess and predict environmental impacts associated with onshore development of OSC activities. The product of this two-year study was a set of EIA methodologies capable of determining the environmental impacts of the various stages of OCS oil and gas activities, namely, exploration, development, production, and shut-down (Weston 1978a). The outcome of the environmental analysis section of this study is valuable in providing guidance for applied ecological studies undertaken in support of environmental impact statements.

Methods

The study was divided into six separate elements: industry requirements, location analysis, economic analysis, fiscal analysis, demographic analysis, and environmental impact analysis. For this study, the word "environmental" referred to the natural environment, that is, ecosystems. Environmental impact analysis was among the last steps to be completed since project location, population changes, water demands, land requirements, and required infrastructure all influenced the final environmental situation.

The specific study tasks included:
   a) review existing impact assessment methodologies;
   b) determine which methodologies were applicable to coastal zone requirements;

c) develop a composite methodology which could be applied to various sized projects and coastal zone regions;

d) determine the extent to which impacts could be quantified; and

e) establish data bases, information sources and provide descriptive material wherever a specific impact assessment methodology could not be applied.

Study tasks were completed via detailed literature searches, meetings with Federal agencies involved in the EIS process, and peer review via advisory and oversight committees. Impact assessment methodologies were developed on an interdisciplinary team-task basis in which each of the principal disciplines (engineering, economics, sociology, ecology, planning, and government) was represented. For the environmental analysis, it was necessary to explore the use of assessment methods of varied sophistication for different levels of expected use, such as by municipal, township, county, or state agencies. All concerned Federal agencies, state agencies, and many university groups were consulted during the impact methodology evaluation and development steps. The resultant methodology was tested and revised on the basis of a hypothetical Baltimore Canyon OCS region test case (Weston 1978b).

Results and Discussion

The analytical and review process associated with this project produced the following results regarding impact assessment:

a) Environmental impact assessment of the natural environment could not be approached with accuracy until specific sites were identified and known environmental alterations and effects quantified.

b) Impact assessment could not deal accurately with long-term, low-rate impacts, such as the inexorable but continued loss of wetlands, in which each parcel is not a measurable loss, but the total loss might be significant in a given region over a long period of time.

c) The most flexible and easy-to-use impact analysis technique was a basic question and answer technique. This method of analysis allowed for increased complexity and sophistication of study as the expected effects became more complex; it is the common "technique" employed in making most types of decisions and is widely practiced in the preparation of environmental impact statements.

d) A matrix analysis technique was useful when impact-related effects were not clearly defined (Leopold et al. 1971). This technique was able to provide specific guidance, especially if prepared for a generic series of projects, such as marine terminals. Matrix techniques became excessively burdensome, however, if they were constructed with too large an analytical field.

e) The Optimum Pathway Matrix technique, developed by the University of Georgia (Institute of Ecology 1971) was one of the most useful numerical techniques applicable to comparison of the environmental suitability of two or more alternative projects. This technique was noteworthy in that it reduced the inevitable variability of subjective analysis via a normalization process, by conducting several passes through the analysis, and by incorporating randomly-generated error variation in its impact measurements.

f) There were no assessment techniques available which could quantitatively relate general physical habitat alterations, such as dredging of an estuarine pipeline channel, to future biological impacts, such as the percentage reduction in blue crab populations.

g) Quantitative impact assessment was possible only in situations where the action was well understood and the estuarine environment to be affected was limited in size and well characterized as to biological structure and function.

It was further learned that the greatest effort towards impact assessment methodologies, and the mathematical modeling of such impacts, had been directed toward primary impacts, those generally associated with a project's construction or operation. The results of these types of impacts are often ameliorated quickly in the natural environment, as exemplified by the re-establishment of benthic communities after channel deepening. Also, impacts which had been the focus of massive, region-wide research efforts, such as the entrainment of striped bass eggs in power plant cooling systems, were quantifiable. In this case, the impact-producing process (cooling water entrainment) and the biological population affected (striped bass eggs near the power plant cooling water intake) were well understood. In addition, the economic value of the striped bass fishery was well documented. Secondary impacts, those which occur over a long period of time and only indirectly from the project in question, were included in impact assessment techniques via guidelines, descriptive data, or Delphi panels (Dee et al. 1973). Little quantitative information could be developed by available methodologies on ecological impact, e.g., on the long-term effects on fish of increasing the petroleum-derived hydrocarbon concentration of a bay's waters by 0.5 mg/l. The background field data, as well as supportive laboratory data, simply were not available.

Two results of the NSF study described above were: a) the importance of using the proper definitions of terms involved in EIA studies, and b) the necessity for evaluating environmental impacts in terms comparable with economic considerations so that the total environmental impact, positive or negative, could be determined.

For the first result, the definition for "impact" was taken from Clark and Terrell (1978) as, "a degradation of the ecosystem which has an adverse effect on human society". This definition of impact fits with NEPA's use of impact assessment to weigh the pros and cons of a project in relation to man's economic and environmental requirements. Such assessment requires an equal base. One cannot compare apples and oranges. Determination of impacts, not biological effects, is what policy makers and natural resources agencies are looking for to base their environmental management decisions on.

Although the CEQ Regulations (1978) use "effects" and "impacts" synonymously, the determination of environmental value of a project is enhanced by applying "impact" only to effects directly affecting man. In forcing this definition, we force ecologists, wherever possible, to evaluate biological effects in economic (or "human") terms, that is quantifying effects as they relate to man's use of the environment. A simplified example serves to clarify this concept.

If a wetlands' dredging project produces environmental disturbances (loss of benthic habitat; increased turbidity; hydraulic changes; filling of wetlands), which result in biological effects (20% local reduction in benthos; decrease in local fish populations; reduction in wetlands acreage), can these biological effects be described on an economic (quantitative) basis? If such quantification is not feasible, one is left with a summary similar to, "Dredging of the proposed shipping channel will result in some loss of benthic organisms, local fish populations, and 100 acres of wetlands". In an EIS, such a conclusion could be weighed against such possible facts as, "The new shipping channel will generate $25 million of new business for the local port city and adjacent counties. Commercial fishing boats will have easier access to the ocean and save $1.1 million per year in fuel costs. Also the deeper channel will decrease grounding and shoaling by 80%, which, last year, resulted in $1.9 million in lost revenues and damages to shippers and recreational boaters".

The fact remains that where biological effects are compared to economic impacts on an unequal basis, the outcome favors economics. Most people, especially government and industry leaders who will be making the environmental decisions, can more readily grasp the meaning of a $10 million increase in business than a 25% reduction in primary productivity.

Much work has been completed toward resolving the three major problems associated with environmental impact assessment stated earlier. *First*, the definitions of impact (and effects) have been clarified in the CEQ Regulations. The evaluation of the importance (degree) of impacts has been addressed by the many public hearings and review procedures to which EIS's are subjected. That is, the democratic process, even though primarily involving scientists, decides on the importance of the severity of impacts. In addition, much specific research on the impact of various alterations and actions on ecosystems has been completed over the past ten years. *Second*, extensive data bases have been developed on important environmental activities, as well as on whole ecosystems. For instance, the Bureau of Land Management and NOAA have completed extensive oceanographic baseline surveys for most OCS regions where oil and gas activities are expected. The U.S. Fish and Wildlife Service has been active in completing a National Wetlands' Inventory, as well as a Coastal Ecosystems Characterization for many coastal regions (Fish and Wildlife Service 1979); the Environmental Protection Agency has conducted numerous water quality and biological studies throughout the United States; while the Corps of Engineers has been in the forefront of compiling data on America's major river systems and studying the impacts of dredging activities. In addition, many Federal agencies have been involved in countless laboratory and field studies which have investigated the responses of species, populations, communities, and ecosystems to various environmental pollutants or alterations. *Third,* where ecologists once defined quantification of biological/environmental data in terms of pounds of fish caught, number of recreational user days, or numbers of waterfowl harvested (Government Accounting Office 1979), more recent emphasis has been placed on the quantification of an ecosystem's natural services.

Initial work done by Gosselink et al. (1974) focused attention on the economic value of the natural functions of ecosystems. Among such functions, particularly of wetlands, are: treatment of pollutants, floodwater retention, ground water storage, nursery and food sources for commercial species, erosion control, minimization of storm damage, cycling of nutrients, and fixation of solar energy.

Studies by Westman (1977), and others, have carried forward the concept that natural ecosystem functions can be evaluated on economic terms. Westman describes the many natural functions which provide clean air, pure water, a green earth, and a balance of organisms. When such functions are greatly disrupted, society may be burdened with previously unrecognized costs. Such costs become particularly apparent as the natural resource in question, such as wetlands, is significantly diminished.

Although there are technical problems associated with imputing values to natural functions (Shabman and Batie 1978; this volume), this avenue of research needs to be pursued diligently. Quantification of coastal zone impacts depends to a great degree on the ability and willingness of estuarine scientists to establish the value of natural ecosystem functions. These values then can be compared with economic effects and a reasonable approximation of the overall environmental impact established. The outcome will result in policy and legislative decisions which will reflect the true value of America's coastal systems.

The Federal government has hypothesized that, "there is a question of whether the Nation may be paying more to provide flood control, ground water recharge, and control of pollution and sediment in Federal public works projects than would have been paid to obtain these same benefits by protecting wetland from drainage", (General Accounting Office 1979). Continued studies on the natural values and quantitative environmental costs of coastal and wetland ecosystems will be critical in substantiating this hypothesis and, ultimately, in protecting America's more vulnerable natural systems.

Acknowledgments

This study was funded by the U.S. Department of Interior, U.S. Department of Commerce, and the National Science Foundation under NSF contract: ENV 76-22611-A03. It was conducted by Roy F. Weston, Inc., Frederic R. Harris, Inc., and the University of Delaware—Center of Policy Studies. The author was employed with Weston at the time of the study.

References Cited

Clark, J. and C. Terrell. 1978. Environmental planning for offshore oil and gas. Vol. III: Effects on living resources and habitats. The Conservation Foundation, Washington, D.C., U.S. Fish and Wildlife Service, Biological Services Program, FWS/OBS-77/14. 220 pp.

Coastal Engineering Research Laboratory. 1975. Environmental impact analysis—current methodologies, future directions. U.S. Army—CERL, Fort Belvoir, VA. and Department of Architecture, University of Illinois, Champaign-Urbana, IL. 192 pp.

Council on Environmental Quality. 1973. Guidelines on preparation of environmental impact statements. *Federal Register,* August 1, 1973. pp. 20549-20562.

Council on Environmental Quality. 1978. Regulations on implementing national environmental policy act procedures. *Federal Register,* November 28, 1978. pp. 55978-56006.

Dee, N., J. Baker, N. Drobney, K. Duke, I. Whitman and D. Fahringer. 1973. An environmental evaluation system for water resource planning. *Water Resources Research* 9:523-535.

Fish and Wildlife Service. 1979. Coastal ecosystems characterizations—a summary of activities FY75 through FY79. U.S. Department of the Interior, NSTL Station, MS. 21 pp.

General Accounting Office. 1979. Better understanding of wetland benefits will help water bank and other federal programs achieve wetland preservation objectives. GAO Report PAD-79-10. Office of the Comptroller General of the United States, Washington, D.C. 20548. 58 pp.

Gosselink, J. G., E. P. Odum and R. M. Pope. 1974. The value of the tidal marsh. Pub. No. LSU-SG-74-03. Center for Wetland Resources, Louisiana State University, Baton Rouge, LA. 30 pp.

Institute of Ecology. 1971. Optimum pathway matrix analysis approach to the environmental decision making process. University of Georgia, Athens, GA. 13 pp. and Appendix.

Leopold, L. B., F. E. Clarke, B. B. Hanshaw and J. R. Balsley. 1971. A procedure for evaluating environmental impact. *Geol. Surv. Circ.* 645. 13 pp.

Shabman, L. A. and S. S. Batie. 1978. Economic value of natural coastal wetlands: A critique. *Coastal Zone Management J.* 4:231-247.

Warner, M. L. and E. H. Preston. 1974. A review of environmental impact assessment methodologies. U.S. Environmental Protection Agency—Office of Research and Development, Washington, D.C. 26 pp.

Westman, W. E. 1977. How much are nature's services worth? *Science* 197:960-964.

Weston, Roy F., Inc. 1978a. Methodology for assessing onshore impacts for outer continental shelf oil and gas development; Methodology—Vol. II. West Chester, PA 19380. multi-paged.

Weston, Roy F., Inc. 1978b. Methodology for assessing onshore impacts for outer continental shelf oil and gas development; Baltimore Canyon Test Case, Vol. III. West Chester, PA 19380. multi-paged.

WETLANDS AND WILDLIFE VALUES: A PRACTICAL FIELD APPROACH TO QUANTIFYING HABITAT VALUES

Melvin L. Schamberger

U.S. Fish and Wildlife Service
Office of Biological Services
Western Energy and Land Use Team
Fort Collins, Colorado

and

Herman E. Kumpf

National Marine Fisheries Service
Southeast Fisheries Center
Office of Fishery Management
Miami, Florida

Abstract: The U.S. Fish and Wildlife Service (FWS) is developing a methodology to evaluate quantitatively the value of habitat to fish and wildlife. The system of Habitat Evaluation Procedures (HEP) will be a standardized system designed to provide indices of habitat value which are based on the ability of the habitat to support fish and wildlife. HEP, first developed for terrestrial habitats, has recently been refined for use in inland aquatic areas. A software package is being developed to implement the Procedures, and evaluation criteria will be developed as species "Handbooks." An effort is currently underway to expand the Procedures for use in estuarine and other nearshore saltwater ecosystems. The establishment of habitat values for estuarine habitats must occur within the perspective of three general value systems: (1) the ability of the habitat to provide a resource, such as fish, wildlife, etc.; or (2) the relative importance society places on these resources; and (3) the dollar value of these resources to man. The HEP methodology will provide a mechanism to establish habitat values for fish and wildlife, establish societal weighting factors to adjust habitat and dollar values, and place dollar values on these resources. This paper addresses only the first two topics.

Introduction

The problem of quantifying fish and wildlife values has plagued scientists for many years. Yet the quantification of habitat values is essential if we are to evaluate or manage biotic systems effectively. In some cases, estuarine development projects are a fact of life and the only option available to biologists is to determine the area where the least impact will occur. The decision must be made to locate the impact in a type of habitat that is "less valuable" to either man or to fish and wildlife species. Another use of quantification occurs when mitigation of project impacts is possible or required. Unless we can quantify the impacts, we cannot determine when they have been mitigated. In terms of Federal projects, such deter-

37

minations become extremely important not only from the legal perspective but also from the perspective of the bill-payer.

Scientists have recognized that important resources were being undervalued and lost primarily because the dollar value did not represent the ecological worth of a piece of habitat adequately, particularly estuarine habitats. Some turned to "intrinsic" value assessments . . . "this piece of land is highly valuable because there is not much left" or "the ecological value is self-evident" and other emotional pleas to consider and conserve habitat. This approach was effective in a few instances, but failed to satisfy the need for a method to determine the real worth of a piece of land for both economic and ecological assessments.

The problem of adequately expressing the economic and environmental worth of ecological systems has been made acute as a result of recent environmental legislation, implementing regulations for the Fish and Wildlife Coordination Act, National Environmental Protection Act (NEPA), and the Water Resources Planning Act. These acts and their implementing regulations require a meaningful display of values for both environmental and economic parameters that are being affected as a result of some land use change or management alternative.

Several methodologies have been used in recent years in an attempt to place a value on fish and wildlife habitat (Gosselink, Odum and Pope 1974; Tihansky and Meade 1976). Early efforts were designed primarily to provide the traditional dollar value of a piece of land or habitat; this has been used with only limited success. It does provide a dollar value of current resource utilization but does not account adequately for nonconsumptive uses nor does it consider biological supply.

Gosselink et al. (1974), Shabman and Batie (1978), and others have demonstrated in different ways that economic values are present in estuarine systems, even though the authors may differ on the method of determining and expressing those values. The free mobility of species in the estuarine system usually does not permit the landowner to realize the economic return generated from that estuary when fish, shellfish, or waterfowl may be harvested many miles from his holding. Nonetheless, the economic and ecological value of estuarine systems is indeed enormous when we consider the various life-support functions, productivity, and storm protection values involved. Pure economic assessments do have a role to play in evaluations, particularly where data and time are available to develop such values. However, in many situations time is short, funds are limited, and decisions are needed in a short period. We need a system that is based on basic biological productivity, yet can be translated into some economic standard.

Habitat Evaluation Procedures

The Fish and Wildlife Service (FWS) is responding to this situation by developing a quantitative assessment methodology that relies on the habitat and its biological productivity as a measure of value. This procedure differs from other evaluation methods by integrating habitat quality and quantity into a single index number. A subjective system developed by Daniel and Lamaire (1974), based on the philosophy that all land has wildlife values that can be expressed numerically, was selected as the system most likely to fit FWS needs. This system was modified by the FWS and published as the Ecological Planning and Evaluation Procedures (Hickman 1974), and later simplified and published as the Habitat Evaluation Procedures (U.S. Fish and Wildlife Service 1976). These procedures have recently been updated and revised and are now being reviewed prior to publication in early 1980.

The basic concepts of the Habitat Evaluation Procedures (HEP) have been presented in more detail in other publications (e.g., Schamberger and Farmer 1978) and are beyond the scope of this presentation. Basically, HEP provides a methodology to supply a quantitative evaluation of baseline habitat conditions from which various changes through time can be predicted and compared. The Procedures utilize environmental data concerning baseline and potential future conditions and use index values to display impacts. This system assumes that: 1) habitat value can be quantified; 2) habitat suitability for a particular animal species can be determined by evaluating the physical and chemical parameters of the habitat; and 3) habitat quantity and quality are directly related to animal populations.

An index of habitat suitability is determined for individual species, selected species groups, or selected habitat types on a scale of 0-1.0 with 1.0 representing optimal habitat. Suitability is determined by inventorying physical and biological characteristics of the habitat and determining how well those characteristics fulfill the life requisites of the species being evaluated. The Habitat Suitability Index (HSI) can be obtained by a variety of methods depending upon the amount and quality of data available for the species of interest. The first level is simply the subjective opinion of an experienced ecologist. Other levels include several types of verbal or mathematical models that integrate individual suitability indices for habitat parameters or characteristics. In some cases HSI values can be derived from biomass or population models. The user must select the most appropriate method for developing HSI values for each species and document the method of use. Once the HSI value is obtained, it is multiplied by area to provide the "Habitat Unit" that is used to represent both habitat quality and quantity: HSI and AREA = HU. Habitat Units are calculated for existing or baseline conditions for various habitats or species. This index serves as a unit of comparison between sites, or can be used for comparisons over time for the same site (i.e., with as opposed to without project conditions). This methodology assumes that optimal habitat conditions for a species can be characterized and that any habitat can be compared to the optimum to provide an index of habitat suitability for that species.

The Procedures are designed to be accurate and detailed enough to be applied in all phases of resource planning at all levels of decisionmaking (Schamberger and Farmer 1978). They provide supporting data necessary for practical use at the field level. The Habitat Evaluation Procedures do not provide a "bottom line" for decisionmakers; they do, however, provide the types of environmental data that are necessary in the decisionmaking process and can display probable development impacts on habitat, fish and wildlife species, or both. The Procedures can help assess the degree to which mitigative measures can be used to offset productivity lost as a result of a particular project plan or land use change.

An Applied Example

An example will serve to illustrate the application of the Habitat Evaluation Procedures to estuarine situations. Harbor Island, adjacent to Port Aransas, Texas (Fig. 1), has been under consideration as a deepwater port since the early 1970's. Data for this example were provided by Ruebsamen and Moore (pers. comm., August 1979). The island of 3,232 hectares (ha) already supports petroleum storage tanks, dock facilities, a plant for fabricating drilling rigs, bait stands, boat ramps, and campsites. The island is in close proximity to several major bays. Nearby are two barrier islands, and three major channels either cross or are adjacent to the island.

Vegetation includes black mangrove marsh (*Avicennia* sp.), smooth cord grass (*Spartina alterniflora*), and extensive submerged grass flats (the

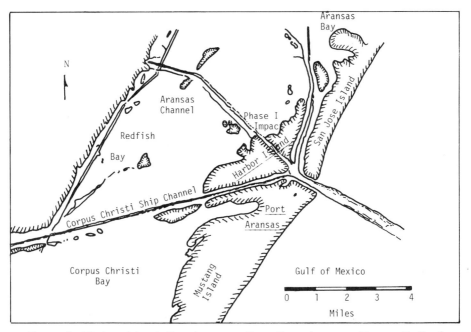

Figure 1. Harbor Island Development Site, Port Aransas, Texas, Gulf of Mexico, showing area of Phase 1 impact.

largest aggregation of these types on the Texas coast). The Harbor Island area supports about 300 species of fish and about the same number of benthic species. It supports a productive fishery and serves as a nursery ground for red drum *(Sciaenops ocellata)*, spotted seatrout *(Cynoscion nebulosus)*, black drum *(Pogonias cromis)*, flounder *(Paralichthys lethostigma)*, sheepshead *(Archosargus probatocephalus)*, brown *(Penaeus aztecus)*, white *(P. setiferus)*, and pink shrimp *(P. duorarum)*, and blue crabs *(Callinectes sapidus)*. Over 300 species of birds are found in the project area. Large and diverse populations of fish-eating birds use the area for feeding, nesting and breeding. Waterfowl such as pintail *(Anas acuta)*, baldpate *(Mareca americana)*, redhead *(Aythya americana)*, shoveler *(Anas clypeata)*, teal *(Anas* spp.), merganser *(Mergus* spp.), bufflehead *(Bucephala albeola)*, and lesser scaup *(Aythya affinis)* inhabit the area each winter.

There are seven alternative plans to accomplish the development goals and three phases in the project proposal. This example only evaluates the first phase of one alternative and focuses on the alteration of submerged grassflats by dredging or filling. The Harbor Island Multi-Purpose Alternative evaluated here provides for jetty relocation and the construction of docking and turning basins for Very Large Crude Carriers. Three channels would be modified by deepening and widening, and the size of existing turning basins would be substantially increased.

Habitat mapping based on aerial photography was carried out by the FWS. Six sample locations were selected by stratified random sampling techniques in the submerged grassflat areas to evaluate habitat values for eight indicator species (Table 1). The habitat evaluation team consisted of biologists from the Corps of Engineers, FWS, National Marine Fisheries Service, Texas Parks and Wildlife Department, and the State of Texas General Land Office. The entire team visited each sample site to determine suitability index values for each species subjectively. The team viewed the sample site and, based on their knowledge and experience, assigned HSI values (1-10) to the species; this process involved discussion among the team members to reach a mutually agreed upon value.

Recent modifications of HEP now use a 0-1.0 scale, and data in this example have been adjusted to this modification. The HSI provides a relative value of the habitat type for the species evaluated. The individual values for each species were aggregated into one value per biotype. For example, the Habitat Unit Value for the submerged grassflat area under Alternative 4, Phase 1, of the project was 0.89/ha (Table 2). The other biotypes affected by the Harbor Island development project were also evaluated in a similar manner, and included: open Gulf, open bay, deep-water channels, basins, high marsh, *Spartina alterniflora* assemblages, sand and mud flats, mangroves, dune and barrier island, upland vegetated and upland non-vegetated areas.

Table 1. Habitat Suitability Index values for eight animal species at six submerged grassflat sample sites at Harbor Island Development site. (Raw data converted from base 10 to base 1.0).

| | Sample Sites | | | | | | | |
Species Evaluated	1	2	3	4	5	6	Total	Mean
Great blue heron (Ardea herodias)	0.9	0.9	0.9	0.9	0.9	0.9	5.4	0.9
Pintail (Anas acuta)	1.0	1.0	0.8	1.0	1.0	1.0	5.8	0.97
Brown shrimp (Penaeus aztecus)	1.0	1.0	1.0	1.0	1.0	1.0	6.0	1.0
Blue crab (Callinectes sapidus)	1.0	1.0	1.0	1.0	1.0	1.0	6.0	1.0
Mullet (Mugil sp.)	1.0	1.0	1.0	1.0	1.0	0.9	5.9	0.98
Pinfish (Lagodon rhomboides)	1.0	0.9	1.0	0.9	1.0	0.9	5.7	0.95
Red drum (Sciaenops ocellata)	1.0	0.8	1.0	1.0	1.0	1.0	5.8	0.97
Oyster (Crassostrea virginica)	0.4	0.4	0.1	0.4	0.4	0.6	2.3	0.38
Site Totals	7.3	7.0	6.8	7.2	7.3	7.3	42.9	
Site Mean	0.91	0.88	0.85	0.9	0.91	0.91		
Habitat Type Mean								0.89

In this study, the alternative calls for the total loss of 311 ha of submerged grassflats with a value of 0.89/ha for a total loss of 277 HU's (311 ha × 0.89 = 277 HU's) (Table 2). This loss represents an annualized loss for the life of the project. Recovery of this habitat type is not anticipated because of construction and permanent maintenance of facilities. These basic biological data can now be used for further interpretations, and can be displayed in various ways to demonstrate impacts and to make recommendations or decisions relative to the proposed alternative. These data should be used in conjunction with other environmental data to provide composite information on the project and its impacts. For example, grassflats in this area are extremely important wintering areas for waterfowl. Because the facility will be developed, the resource manager must set resource objectives, and rank in order of priority species or habitats so that the various alternatives can be evaluated. Therefore, the evaluation team

Table 2. Summary of existing habitat conditions and expected habitat changes under conditions of Phase 1, Alternative 4, at the Harbor Island Development Site. Habitat Unit value multiplier has been adjusted to preset 0-1.0 FWS ratings.

Habitat Type	Existing Conditions			Area Loss-gain	Phase I	
	Area (ha)	Habitat Unit Value	Total Habitat Units		Annualized HU Value of Loss-gain[1]	Relative Importance[2]
Submerged grassflats	311	0.89	277	-311	-277	10
Open gulf	3,309	0.71	2,349	0	-247	7
Dune & barrier islands	441	0.38	168	+23	-2	8
Upland-vegetated	268	0.58	156	+101	-62	5
Upland-nonvegetated	140	0.20	28	+611	+58	1
Sand and mudflats	330	0.52	171	-330	-171	7
Spartina alterniflora	128	0.65	83	-128	-83	9
Mangrove	30	0.65	19	-30	-19	10
High marsh	73	0.52	38	-73	-38	9

[1] Values reflect changes in HSI values (i.e., habitat suitability) not shown in table.

[2] Hypothetical value derived from biological and societal values using pair by pair comparison technique.

must determine which habitats or species are more important before further decisions can be made. This step in resource evaluation is the initial blending of biological with political and societal values.

Endangered species, for example, represent political (legal) considerations. If one habitat type is critical for an endangered species, then all alternatives that negatively impact that type must be eliminated. In this study, there were no endangered species, so this was not a consideration in the ranking of habitat types. Biological productivity was the main factor used to assign relative weights subjectively to the various habitat types, although other biological factors were considered. Since this is an ongoing project, analysis has not proceeded beyond this point. However, for purposes of completing this example, the remaining assumptions and recommendations are hypothetical and are not related to the actual assessment and recommendations of this particular project.

Relative weights or importance values were assigned hypothetically (Table 2) based on the consideration of biological productivity, other biological factors, scarcity, vulnerability, and local importance. The method used to assign relative importance values was that of pair by pair comparison techniques in which each biotype was compared against each of the others to derive relative ranking values. With the combined use of Habitat Unit values, number of hectares, and relative importance values, we are better prepared to evaluate impacts and make resource decisions and project recommendations.

Submerged grassflats were determined to be one of the most important resources in the project area because of their value to wildlife and their scarcity in the area. The development option presented results in a total loss of all grassflats, an option less than ideal. An attempt would be made to search for options with less impact on the important habitat or options utilizing areas of lower habitat unit value. For example, the recommendation could be made to locate the facility on 100 hectares of upland vegetated habitat (100 ha of upland at 0.5 HU/ha = 50 HU lost) rather than to locate on 100 hectares of submerged grassflat (100 ha grassflat at 0.8 HU/ha = 80 HU lost).

Discussion

The HEP approach does not provide data that are not already available. In most cases HEP is simply used to document the basis for a recommendation or decision, and is used to provide a standardized, systematic approach to handling data. HEP does not provide a decision, but rather provides a display of data in such a manner that the data can be readily understood and relative comparisons made.

HEP has been used effectively on a number of other saltwater-estuarine studies from the evaluation of permit applications to large projects. Experiences with HEP during the past several years have indicated that results are more replicable and supportable when evaluation criteria are available for the species and habitats being evaluated. The FWS is currently preparing standards from which species evaluation criteria can be developed on a project by project basis. The FWS concurrently is developing evaluation criteria for a limited number of species for inland aquatic and terrestrial habitats. For better known species and habitats we have been able to develop sets of curvilinear relationships between habitat parameters and habitat suitability. Examples of this approach are available for terrestrial and inland aquatic species (Schamberger et al. 1979).

The FWS recently conducted a joint workshop with the National Marine Fisheries Service to determine the feasibility of the adaptation of HEP to saltwater systems. The conclusion was that the basic approach of HEP is applicable to estuarine habitats; however, development of evaluation criteria appears to be the single most limiting factor in such applications. At the workshop it was further determined that, although it is desirable to have evaluation criteria for species, in many cases such criteria do not exist and are difficult to obtain. In some studies it may be appropriate to develop criteria for habitat types rather than for species, but the ultimate goal should be to develop evaluation criteria for species of importance in the project area. Such an approach would be facilitated by the grouping of species into guilds. Evaluation criteria could then be developed for a small number of carefully selected species that would represent a large segment of those species present. Further research and development is needed to identify evaluation criteria; the National Coastal Ecosystems Team of the FWS will play a role in this effort.

The economic value of fish and wildlife resources can be calculated by a number of different ways. The problem of supply is complicated when evaluating biological systems that do not respond to demand projections in the same manner as in normal marketplace conditions. Overutilization of the resource results in depleted brood stock and ever-decreasing supply. Further, fish and wildlife resources may be used both consumptively as well as non-consumptively—one individual animal may contribute at different times to both uses. The Water Resources Council has published several conceptual approaches to developing dollar values for fish and wildlife resources (U.S. Water Resources Council 1979). The major conceptual addition that we have made to those procedures is that biological supply is the driving force that limits projected demand for wildlife use. In other words, all use demands that involve consumptive uses of wildlife are constrained by the ability of the resource to provide a sustainable yield for consumption.

In summary, it is clear that quantification of environmental values is essential and there is a need for several approaches to habitat evaluation. The habitat approach proposed in this paper is not advocated in lieu of other systems that provide monetary or societal values; rather the approach is advocated because it can provide important ecological data that should be used in impact assessment. There are three general considerations that enter into the resource decision-making process: (1) resource related data such as the size of the area, resources present, and probable change in the resources; (2) economic data that includes some measure of the dollar value of the resource; and (3) societal values that include legal protection of resources perceived as important by society.

The Habitat Evaluation Procedures can accommodate all of these considerations because they consider biological components, the human uses affected by a proposed development action, and a method to develop societal weighting values. The Procedures integrate societal values as well as scientific data so that both can be utilized in resource evaluation and the decision-making process.

References Cited

Daniel, C. and R. Lamaire. 1974. Evaluating effects of water resource developments on wildlife habitat. *Wildl. Soc. Bull.* 2:114-118.

Gosselink, J. G., E. P. Odum and R. M. Pope. 1974. The value of the tidal marsh. Publication No. LSU-SG-74-03, Center for Wetland Resources, Louisiana State U., Baton Rouge, LA.

Hickman, G. 1974. Ecological planning and evaluation procedures. Joint Federal-State-Private Conservation Organization Committee, U.S. Dept. of Interior, Fish and Wildlife Service Mimeo. Report. 269 pp.

Schamberger, M. and A. Farmer. 1978. The habitat evaluation procedures: Their application in project planning and impact evaluation. *Trans. N. Amer. Wildl. Nat. Res. Conference* 43:274-283.

Schamberger, M., C. Short and A. Farmer. 1979. Evaluating wetlands as wildlife habitat, pp. 74-83. *In:* P. E. Greeson, J. R. Clark and J. E. Clark (eds.), *Wetlands Functions and Values: The State of Our Understanding.* American Water Resources Association, Minneapolis, MN.

Shabman, L. A. and S. S. Batie. 1978. Economic value of natural coastal wetlands: A critique. *Coastal Zone Mgmt. J.* 4:231-247.

Tihansky, D. P. and N. F. Meade. 1976. Economic contribution of commercial fisheries in valuing U.S. estuaries. *Coastal Zone Mgmt. J.* 2:411-421.

U.S. Fish and Wildlife Service. 1976. Habitat evaluation procedures: For use by the Division of Ecological Services in evaluation of water and related land resource development projects. U.S. Dept. of Interior, Fish and Wildlife Service Mimeo. Report. 30 pp.

U.S. Water Resources Council. 18 CFR Part 713, 1979. *In:* Procedures for Evaluation of National Economic Development (NED) Benefits and Costs in Water Resources Planning (Level C). In Fed. Reg. Vol. 44, No. 242, Friday, Dec. 14, 1979 Rules and Regulations. Final Rule Effective Date: January 14, 1980.

EVALUATING THE EFFECTIVENESS OF COASTAL ZONE MANAGEMENT

Virginia K. Tippie

Center for Ocean Management Studies
University of Rhode Island
Kingston, Rhode Island

Abstract: Since the passage of the Coastal Zone Management Act of 1972, 35 coastal states and territories have participated in the federal coastal zone effort, and 19 have established approved management programs. Most of the coastal planning efforts focus on resolution of use conflicts and amelioration of impacts. As a result of this emphasis, decisions are frequently made on the basis of permissibility of use within the conflict resolution framework rather than on the best or priority use of an area. In order to move from this "responsive" mode to a more "comprehensive" planning mode, I maintain that states should implement ongoing evaluation programs that monitor their efforts. This can only be achieved if measurable criteria are established for program goals. The evaluation process and suggested criteria for goals in the area of environmental protection, economic development, and management are outlined.

Introduction

In 1972 the U.S. Congress recognized in the passage of the Coastal Zone Management Act that there is a national interest in the effective management, beneficial use, protection, and development of the coastal zone. The Act encouraged states through grants and incentives to develop management programs for the coastal zone in cooperation with federal and local governments and other vitally affected interests. To date, 19 states and territories have implemented management programs that provide for protection of coastal resources such as wetlands, beaches, and dunes; management of coastal development; increased access and recreational use of the coast; and coordination of permit requirements. These states and the federal government are now faced with the task of evaluating coastal zone management (CZM) in terms of both substantive and process goals. However, since many government agencies are involved in coastal zone management, such an evaluation must recognize that progress toward stated goals cannot be attributed to the designated state CZM agency *per se*.

The evaluation of the effectiveness of coastal management efforts is critical to the viability of the CZM concept. Such review provides the feedback that permits state management agencies to adapt and thereby survive in an uncertain social-political environment. Although there are many forms of feedback in day-to-day operations and in the budgetary process, there is really no systematic evaluation of the effectiveness of state coastal management efforts. All approved state coastal plans have monitoring/enforcement programs but none have established comprehensive evaluation efforts. It should be noted, however, that the "312" review process required by the federal coastal zone management act does measure state progress under "306" implementation against broad national goals (R. W. Knecht, Office of Coastal Zone Management, NOAA, Washington, DC, unpublished memorandum, 1979). While this review process is useful in terms of the national perspective, it cannot really evaluate progress against standards specific to each state.

Evaluation Research

The benefits of a systematic program review have been described extensively in the literature (see bibliography by Hoole and Friedheim 1978). There have also been several papers which have discussed the value of applying such techniques to coastal zone management (Englander et al. 1977; Feldmann and McCrea 1978). In addition, several researchers have used evaluation methods to assess specific programs (Healy 1974; Rosentraub and Warren 1976; McCrea and Feldman 1977; Sabatier 1977; Warren et al. 1977). According to Hoole (1978), evaluation research and other policy analysis techniques permit a "modeling through" as opposed to a "muddling through" policy-making process. It is obvious that systematic evaluations of coastal management programs would encourage improvements in the decision-making process and help agencies demonstrate program and policy accomplishments to their constituents and funding sources.

Essentially, the purpose of evaluation research is to measure the effects of a program against the goals it set out to accomplish as a means of contributing to subsequent decision-making about the program and improving future programming (Weiss 1972). The evaluation process itself is fourfold: 1) define program goals; 2) formulate evaluation criteria which are measurable; 3) collect and organize data; and 4) evaluate the results. Once the results are evaluated, one can often recommend constructive changes in program operations.

Evaluation and Coastal Zone Management

It is difficult to assess the effectiveness or success of a program as complex as "coastal zone management." One has a general sense that it has improved the quality of our life. However, several studies specific to the California Coastal Plan have argued that coastal management has resulted in no real improvement and that society has incurred a great social cost due to economic losses (Institute for Contemporary Studies 1976; Warren et al. 1977). This contradiction is the result of different perceptions of the Act and the federal program. The differences become apparent when one compares statements by representatives of the Natural Resources Defense Council (NRDC) and the oil industry (Coastal States Organization 1979). NRDC and other environmental groups argue that the CZM program is not effectively protecting the coastal resources; on the other hand, industry argues that CZM is restrictive and can cause excessive delays.

Needless to say, it is difficult to meet or interpret the mandate of the CZM Act to "preserve, protect, develop, and, where possible, to restore or enhance the resources of the nation's coastal zone" (U.S. Congress 1972). States generally try to achieve these objectives by establishing a review process which attempts to balance different interests and resolve conflicts. This approach generally addresses the permissibility rather than the priority of a proposed activity. As a result of the "review" nature of CZM, Hershman (1980) concludes that CZM is not a planning effort but rather an impact amelioration and conflict resolution process. Since CZM is basically a review process and many of its decisions can be controversial and costly to society, it is important that the judging process be fair and convincing. An ongoing evaluation effort would indicate trends and thereby assure more comprehensive and objective decisions.

Some form of evaluation should follow the planning and implementation stages of a coastal zone program at any level of government. As indicated in Fig. 1, the planning process includes gathering of information on resources and problems, definition of goals and objectives, and formulation of policies and regulations. Once a plan is established, the permit process, designation of sites for specific uses (i.e., sanctuaries, public access, zoning) and enforcement procedures are the mechanisms for implementation. These actions result in what I call "process outcomes" (i.e., management impacts, such as public awareness, coordination among public agencies, etc.) and "resource or substantive outcomes" (i.e., economic development

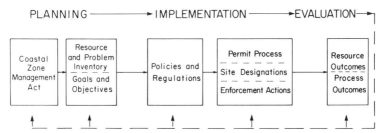

Figure 1. Coastal Zone Management Model. The solid line indicates the sequence of activities in the coastal zone management process. The dashed line indicates the feedback to subsequent CZM activities after outcomes are evaluated.

and environmental impacts as indicated by public access, water quality, etc.). The evaluation process measures these outcomes or effects against the goals and objectives. Depending on the results, one may wish to do more planning (i.e., change the goals and objectives) or revise the implementation process. If absolutely necessary, it may be appropriate to change the mandate (i.e., the federal or state coastal management act). In the best of all worlds, the feedback process described in the model would lead to improvements and more effective management.

The Federal Coastal Zone Program

In order to evaluate the federal program, it is useful to examine the intent of the Coastal Zone Management Act. The act was passed in response to a growing dissatisfaction with the state of natural resources and a concern that existing organizational processes were preventing effective management. Englander et al. (1977) identified major coastal zone problems articulated in congressional hearings, reports, etc. Table 1 outlines problems which received primary emphasis. It can be assumed that it was

Table 1. Resource outcome problems and process outcome problems (modified from Englander et al. 1977).

Resource Outcome Problems
1. Intense-use conflicts among competing uses
2. Population growth with residential, commercial, and industrial development pressures
3. Extensive environmental pollution
4. Destructive dredging, filling, and bulkheading
5. Destruction of coastal habitat and degradation of fish and wildlife resources
6. Limited public access and recreational opportunities

Process Outcome Problems
1. Lack of coordination among public agencies
2. Insufficient planning and regulatory authority
3. Lack of clearly stated goals
4. Insufficient data base and lack of information for decision-making
5. Little understanding or knowledge about coastal ecosystems
6. Primitive analytical tools and predictive methodologies
7. Lack of state and local government funds to manage the coastal zone adequately
8. Dominance of short-term management over long-range planning
9. Resource decisions made primarily on the basis of economic considerations to the exclusion of ecological considerations

the intent of the Coastal Zone Management Act to alleviate these expressed problems.

To achieve this, the federal act explicitly states that: ". . . the key to more effective protection and use of the land and water resources of the coastal zone is to encourage states to exercise their full authority over the lands and waters in the coastal zone by assisting the states, in cooperation with federal and local governments and other vitally affected interests, in developing land and water use programs for the coastal zone, including unified policies, criteria, standards, methods, and processes for dealing with land and water use decisions of more than local significance." (U.S. Congress 1972). Thus, the focus of the act is to encourage states through planning grants ("305") to develop a management program which will lead to substantive outcomes. If their programs are approved, the states will receive additional funds to administer the program ("306").

In a summary of the first five years of CZM, the federal Office of Coastal Zone Management (1979) identified four broad goals of the act which it felt states should address:

(1) protection of significant natural coastal resources and areas, such as wetlands, fisheries, beaches, dunes, and barrier islands;

(2) management of coastal development to minimize loss of life and property because of improper development in areas subject to coastal hazards, giving priority to coastal dependent development, including energy facility siting, and identifying sites for dredge spoil disposal;

(3) increased access to the coast for recreation purposes, including revitalization of urban waterfronts, and protection and restoration of cultural, historic, and aesthetic coastal resources; and

(4) improved predictability and efficiency in public decision-making, including increased intergovernmental cooperation and coordination.

The Office of Coastal Zone Management (OCZM) recognized that these goals can only be achieved if states first establish procedures and institutions that address these objectives. For this reason, institution-building has been the primary emphasis of the federal effort. Therefore, any evaluation of the federal program should be focused on OCZM's ability to encourage the 35 participating coastal states and territories to implement some effective management procedures and institutions ("process outcomes") regardless of whether they develop an approved management program.

For discussion purposes, let us assume that, to achieve the goals stated above, the federal program intended to encourage the establishment of new or expanded laws or programs in all states that did not have adequate legislation. Using legislative or process criteria in each goal area, one can determine the percentage of states that implemented appropriate legislation since receiving federal CZM monies. For example, prior to the federal coastal zone management program, 12 of the 31 states with significant wetlands had either inadequate or no legislation concerning wetlands.

Since then, 8 of the 12 states have established laws or programs. In short, the federal program has achieved two-thirds of its specific goal of institutional protection of wetlands. It should be noted, however, that this accomplishment cannot be solely credited to OCZM. The point is that there is progress towards a national objective of institution-building for protection of wetlands regardless of the factors that influenced states to implement legislation. This fact does not necessarily indicate that more wetlands are being preserved; it simply reflects the fact that more states are now actively managing wetlands.

Table 2 summarizes the application of this evaluation process to specific criteria in the four national goal areas. As indicated in the table, the federal program has managed to convince approximately half of the "negligent" states to establish institutions or procedures to meet the goals of protecting natural resources and managing coastal development. It appears, however, that OCZM has a long way to go to convince states to

Table 2. Establishment of state institutions or procedures in national goal areas. Percentage goal achievement figures are based on the number of states that have implemented legislation, institutions, or procedures since receiving federal CZM monies, divided by the total number that had inadequate laws or programs when OCZM was established. Data from the Office of Coastal Zone Management (1979).

National Goal Area	Percentage Goal Achievement
1. Protection of significant natural coastal resources	
Wetlands	67
Floral and faunal habitats	36
2. Management of coastal development	
Erosion-prone areas	44
Flood plains	42
3. Increased recreational access/protection of historic/cultural resources	
Protection of historic and cultural sites	46
Requirement for coastal access	7
4. Efficiency in public decision-making	
Decrease in permit review time	18
Consolidation of state permits	10

streamline government decision-making. Obviously, these conclusions would be different if other criteria were selected. However, if these criteria are acceptable, then the exercise provides a simplistic but useful tool.

Another obvious goal of the federal program is to convince all states to prepare comprehensive plans that meet the specified requirements. Of the 35 coastal states and territories that have participated in the program, 19 have received federal approval. This affects approximately 75% of the nation's shoreline. Considering that the federal coastal zone program only has a "carrot" and not a "stick," I think it is doing fairly well. However, there has been some criticism of the inconsistencies between approved programs, thus bringing into question the criteria used to evaluate plans. Basically, critics are demanding measurable and acceptable national standards (Coastal States Organization 1979).

State Coastal Management Goals

As previously discussed, the federal coastal zone management program encourages states to develop and implement management programs by offering a financial incentive and the promise of federal consistency. They are given planning monies ("305" grants) to develop a program. If their program is approved they will receive management funds ("306" grants). In order to receive federal approval of their plan, states are required to:

(1) identify important resources, areas, and uses within a state's coastal zone which require management or protection;

(2) establish a policy framework to guide decisions about appropriate resource use and protection;

(3) identify the landward and seaward boundaries of the coastal zone;

(4) provide for the consideration of the national interest in the planning for and siting of facilities that meet more than local requirements; and

(5) include sufficient legal authorities and organizational arrangements to implement the program and ensure conformance to it (Office of Coastal Zone Management 1979).

The federal program essentially encourages states to develop a management process that addresses multiple objectives from a comprehensive perspective. Once state programs are approved, they undergo a yearly federal review. This review is useful in ascertaining each state's progress in meeting broad national institutional goals. However, since there are no national standards for substantive policies, the "312" review can only evaluate general process outcomes.

In order for a state to develop an internal substantive evaluation process, it must establish "yardsticks" for measuring progress. Since the federal requirements are rather broad and flexible, states have not established specific objectives. The goal statements of approved state programs are general. Phrases such as "preservation", "protection", "restoration of coastal resources", "maintain balance between conservation and development", and "involve the public", appear in most plans. These statements can be grouped into three broad goal areas: economic development, management process, and environmental protection. The categories reflect the intent to have a "comprehensive" and "balanced" approach to Coastal Zone Management. If some sort of "balance" between goal areas is not maintained, the state CZM agency can lose its credibility and be perceived as "a bunch of environmentalists", "a rubber stamping pro development agency", or "a bureaucratic mess".

In order to evaluate progress toward the broad goal areas it is necessary to establish specific objectives and criteria or monitoring indices. The parameters selected for study should reflect the comprehensiveness and emphasis of the program. It is also essential that they be measurable and acceptable to the public. In some cases, if the indices cannot be measured quantitatively, a state may wish to have an unbiased panel of judges assess progress on a scale. Regardless of the techniques used to measure or represent progress towards the objectives, the effort should give the state a sense of trends. Table 3 lists a set of suggested criteria or monitoring indices in each goal area.

The establishment of criteria to measure *economic development* may be critical to justifying the continued existence of a management program.

Table 3. Criteria for evaluation of state programs.

Economic Development
  1. Value of undeveloped land
  2. Applications for permits received
  3. Permits issued/denied
  4. Rights of way/public access established

Management Process
  5. Permit time lag
  6. Headlines in papers
  7. Submitted/funded local programs
  8. Enforcement actions

Environmental Protection
  9. Heavy metals
  10. Shellfish closure areas
  11. Acres of wetlands
  12. Recreational fishing catch/unit effort

The value of undeveloped land is a useful index of the effect the program has on development. In a critique of the California Coastal Plan, Frech and Lafferty showed that between 1973 and 1976 the coastal commission regulations lowered the value of undeveloped land in the permit zone by 15% (Institute for Contemporary Studies 1976). Presumably, this reflected the perception that the California Commission was biased against development during that time. The number of applications for permits and the permits issued/denied in all categories are important to monitor for they indicate whether development projects are being proposed and how the management authority is responding. An assessment of the rights of way/ public access areas measures use of the coastal area.

The coastal *management process* itself must be monitored if "process outcome" problems are to be resolved. A useful index for measuring the efficiency of the process is the permit time lag. Another important aspect of the process is public awareness. Newspaper headlines of the major state papers can provide a good monitor of public awareness and sentiment toward CZM. If coastal zone problems are not being discussed on the front page or editorial section of the paper, something is wrong with public involvement and citizen participation efforts. On the other hand, if discussion focuses on public criticism of the management authority *per se*, then it may be time for administrative changes. Effective communication between state and local government is another important component of the process. To evaluate how well the Michigan state program was communicating "306" goals and priorities to local governments, applications for funds were categorized, and a ranking by category for submitted/funded was established (Michigan Division of Land Resource Programs 1978). The study indicated that several communities were not aware of state priorities. Last, but perhaps most important, the process does not work unless it is enforceable and people recognize this and begin to respect it. Thus, number of enforcement actions per effort is a useful index.

In the goal area of *environmental protection*, it is important to select indices that reflect the complexity of the environment. Also, for certain criteria, long-term time series data are necessary to insure that the changes observed in the time frame being evaluated are not just noise. Most states have good time series data on heavy metals and shellfish closure areas, both of which are good indices for monitoring water quality. To assess the protection of valuable floral resources, it is useful to monitor the acreage of wetlands or other important environments such as seagrass beds. I personally feel that in many cases it may be appropriate to fill a degraded salt marsh for the public good and, in turn, one could create a new marsh on dredge spoil. If data are available, the index, recreational fish catch per unit effort, can give a state an indication of whether it is protecting the faunal resources in its coastal waters.

The above criteria are useful indicators of the "health" of the coastal zone. However, as previously noted, many of the changes in selected in-

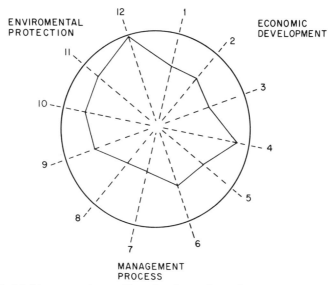

ENVIROMENTAL
PROTECTION

ECONOMIC
DEVELOPMENT

MANAGEMENT
PROCESS

Figure 2. Multiparametric profile of a hypothetical management program. The spokes 1-12 represent the criteria indicated in Table 3. The circle represents 100% achievement of goals. The polygon represents the extent to which goals have been satisfied.

dices cannot be directly attributed to coastal zone management *per se.* For example, improvements in water quality in the coastal area would probably be the result of EPA regulations and the 1972 Federal Water Pollution Control Act rather than the Coastal Zone Management Act. Nonetheless, it can be argued that the state coastal program should be the guardian for the entire coastal area, for in concept coastal zone management should be comprehensive. Most states achieve this through interagency agreements or "networking". For this reason, I feel a state CZM program has the obligation to monitor environmental, economic, and management impacts in the coastal area regardless of who gets the credit.

In order to assess the balance between goal areas and get a better sense of the overall picture, it is useful to compare progress towards different objectives. There is a variety of analytical techniques that a state can use to get a sense of its progress. For example, Fig. 2 is a multi-parametric profile of a hypothetical program. The circle represents 100% achievement of goals. Its center is the time of program implementation, the spokes are the indices grouped by goal areas, and the polygon represents goal achievement at the time of evaluation. In this case, the program appears to be protecting the environment fairly well, but it needs to improve results in the management process and economic development areas.

Conclusion

To assure the continuing viability of coastal zone management at the federal and state levels, it is essential that established programs periodically evaluate progress towards substantive goals. The federal "312" review process does provide a feedback mechanism for the states; however, it is specific to national goals and objectives. I strongly recommend that states establish their own evaluation process which would assess their progress toward meeting specified goals in the broad areas of environmental protection, economic development, and management process. An effective monitoring program will help a state program redirect its efforts and assure its credibility.

References Cited

Coastal States Organization. 1979. Coastal Management - "Options for the 80's." NOAA, U.S. Dept. of Commerce, Washington, DC. 45 pp.

Englander, E., J. Feldmann and M. Hershman. 1977. Coastal zone problems: a basis for evaluation. *Coastal Zone Mgmt. J.* 3:217-236.

Feldman, J. H. and M. McCrea. 1978. Evaluating the effectiveness of CZM programs, pp 117-128. *In: Coastal Zone '78.* American Society of Civil Engineers. New York, NY.

Healy, R. A. 1974. Saving California's coast: The coastal zone initiative and its aftermath. *Coastal Zone Mgmt. J.* 1:365-394.

Hershman, M. J. 1980. Coastal Zone Management in the United States. *In: Comparative Marine Policy.* Center for Ocean Management Studies, U. Rhode Island. Bergin Publishers, New York, NY (in press).

Hoole, F. W. 1978. From muddling through to modeling through. *In: Formulating Marine Policy: Limitations to Rationale Decision-Making.* Center for Ocean Management Studies, U. Rhode Island, Kingston, RI. 174 pp.

Hoole, F. W. and R. L. Freidheim. 1978. A selected bibliography of evaluation research for marine and coastal specialists. Occasional Paper Number 5, Institute for Marine and Coastal Studies, U. Southern California, Los Angeles, CA.

Institute for Contemporary Studies. 1976. *The California Coastal Plan: A Critique.* Institute for Contemporary Studies, San Francisco, CA.

McCrea, M. and J. H. Feldmann. 1977. Interim assessment of Washington State shoreline management. *Coastal Zone Mgmt. J.* 3:119-150.

Michigan Division of Land Resource Programs. 1978. An assessment of Michigan's 306 grant-review process. The Coastal Zone Laboratory, U. Michigan. Technical Report No. 114. Lansing, MI. 49 pp.

Office of Coastal Zone Management. 1979. *The First Five Years of Coastal Zone Management: An Initial Assessment.* NOAA, U.S. Dept. of Commerce, Washington, DC. 60 pp.

Rosentraub, M. S. and R. Warren. 1976. Information utilization and self-evaluating capacities for coastal zone management agencies. *Coastal Zone Mgmt. J.* 2:193-222.

Sabatier, P. 1977. State review of local land use decisions: The California Coastal Commissions. *Coastal Zone Mgmt. J.* 3:255-291.

U.S. Congress. 1972. Coastal Zone Management Act of 1972. Public Law 92-583, 92nd Congress, S. 3507. October 27 (86 Stat. 1280).

Warren, R., L. F. Wechsler and M. S. Rosentraub. 1977. Local-regional interaction in the development of coastal land use policies: A case study of metropolitan Los Angeles. *Coastal Zone Mgmt. J.* 3:331-363.

Weiss, C. H. 1972. *Evaluation Research: Methods of Assessing Program Effectiveness.* Prentice-Hall, Inc., Englewood Cliffs, NJ. 160 pp.

A NEW ZEALAND RESEARCH PROGRAMME
TO ASSIST COMMUNITY MANAGEMENT OF AN ESTUARY

W. Bernard Healy

*Soil Bureau, Department of Scientific
and Industrial Research, Private Bag,
Lower Hutt, New Zealand*

Abstract: Community concern at the effects of accelerated land development on Pauatahanui Inlet, an estuary close to Wellington, New Zealand, led to a 3-year multidisciplinary study. Its objectives were to evaluate the ecosystem, define its sensitive features, establish data baselines against which to monitor change, and provide information to planners, developers and the community itself. In the catchment, studies have been carried out on geology, surficial deposits, land classes, vegetation, and hydrology including flood events and water quality. In the estuary, studies have been carried out on morphology, bottom sediments, salinity and temperature characteristics, residence time, subtidal and intertidal biology, fish and bird populations, salt marsh vegetation, sedimentation and effects of dredging. Scientific results will be published in appropriate journals. In addition a more popular integrated account is being published in book form to inform the community of the results of the study and its planning implications. An informed community will contribute to and evaluate development decisions and ensure wise management of a valuable resource.

Introduction

Estuaries provide maximum contact between land and sea. Man selected them as sites for settlement because of the attraction of sheltered waters, access to sea, freshwater habitats, varied landscape, choice of food sources and so on. Initially it was possible to live in harmony with this environment, but as populations increased, man's activities became more varied and pressures increased. Many estuaries are now centres of population and under these pressures. Although New Zealand has a long coastline (approx. 5400 km) and relatively small population (approx. 3 million), population tends to be concentrated around harbours and estuaries and there is increased concern at the effects of development.

Pauatahanui Inlet (Pauatahanui is a Polynesian word meaning "side of a large shell") is an estuary near Wellington, the capital city of New Zealand. Apart from providing the only sheltered harbour between Wellington Harbour itself and Kawhia Harbour, some 400 km north, it is of considerable aesthetic and recreational value to the Wellington Region (population 300,000). It has changed little since man removed the forest and established grassland.

Initially Pauatahanui was a weekend retreat with houses built singly with minimum disturbance to the environment. In recent years large scale development has taken place with extensive changes to the landscape. The resultant deposition of silt in a local popular bay became the focus for community concern, and resulted in a demand for a study to evaluate the effects of development on the future of the estuary.

The Department of Scientific and Industrial Research was asked to accept responsibility for the study and I was asked to organise and direct it. The Department responded not only because of community requests but also because it was seen as a study in its own right which would have application in other parts of New Zealand. I invited other appropriate agencies to co-operate and so the Pauatahanui Environmental Programme (PEP), a 3-year study, was set up with the following objectives:

To study the estuary and its catchment in order to understand the present ecosystem;

To define its sensitive features;

To establish data baselines against which to monitor change;

To provide data of assistance to developers and planners;

To present the results to the community in a suitable form so as to encourage wise use of the resource.

Resources

Once the objectives were set, specific project areas were defined and scientists "selected" to work on them. Researchers came mainly from the pool of scientists already employed by government agencies or from universities. Some 35 scientists were involved, some for a short time, others over extended periods. My preference was for workers already experienced in specialist fields who could, with minimum delay, design and carry out specific projects appropriate to the objectives of PEP. It was my task to persuade their Directors that such projects were worthwhile scientifically and appropriate to the role of their own organisations. Only a small number were graduate students working towards a Ph.D. degree.

Every effort was made to encourage co-operative studies and ensure contact between scientists of different disciplines. The need to consider the catchment and estuary as an integral unit was continually emphasised.

There was no overall budget for PEP and in the main it was supported from within the budgets of the participating organisations. Some funds were made available for Ph.D. students. However some attempt was made to assess the monetary cost of PEP. It was considered important that the community appreciate that environmental studies are expensive and cannot be undertaken casually. Even where scientists work from within existing budgets there is a national cost in relation to deployment of science resources. It is estimated that PEP cost about $400,000 (about 15 scientist-years in New Zealand terms), a relatively modest sum compared to a suggested monetary value of the resource, estuary and catchment, of $1000-2000 million. Hopefully the results will also have wider application.

Research Projects

The studies carried out can be grouped as follows:

Catchment:

> Geology of the Region
> Surficial Deposits
> Soils and Land Classification
> Climate
> Vegetation
> Hydrology
> Water Quality

Estuary:

> Bathymetry
> Nature of Bottom Sediments
> Tides
> Temperature and Salinity
> Water Quality
> Intertidal Biology
> Subtidal Biology
> Plankton
> The bivalve, *Chione stutchburyi*
> Fish
> Birds
> Vegetation
> Heavy Metals and Pesticides

Estuary Processes:

> Residence Time
> Seiches
> Heating and Cooling
> Sedimentation
> Movement of Bottom Sediments by Waves
> Effects of Dredging and Reclamation

Results

It is not possible to consider in detail the results of PEP but some of the more important findings that should influence planning and management of the resource are:

Catchment:

- Land classified of high value for food production, and at the other extreme, land of potential erosion risk, has been defined.
- Loess (surficial deposits) mantle all but the steepest land. The nature of this material was assessed in relation to soil engineering requirements.

- Small but significant forest remnants occur in the mainly grassland cover. Successional diagrams indicate the benefits of conservation.
- Discharge of water and sediment is a function of catchment area. Higher specific discharges were identified with the catchment under development. A permanent monitoring station in the largest stream enables integrated data to be calculated for all streams, and is providing valuable data baselines.

Estuary:

- A bathymetric chart (0.2 m contour) defines the present features of the estuary.
- Analysis of aerial photographs shows channels and banks have been relatively stable for the last 40 years.
- Sedimentation studies define sites and amounts of erosion and deposition, and indicate a mean deposition rate of 2.9 mm per year.
- Radiocarbon dating of undisturbed sediment cores, together with pollen and diatom studies, allows a reconstruction of the history of the estuary. Over the last 3500 years, a mean annual deposition rate of 2.4 mm is indicated.
- A residence time of approx. 3 days indicates a strong flushing action in the estuary.
- Intertidal and subtidal biology shows over 40 species present in diverse and stable communities. Macrofaunal populations were significantly reduced by excessive local deposition of sediment.
- The N. Z. Cockle, *Chione stutchburyi* (a bivalve), makes up most of the biomass, excluding birds and fish. It was studied in detail because it provided a useful data baseline. The volume of water filtered by the cockle is about 40% of the tidal exchange.
- The diversity of fish and bird populations confirms that the estuary is in a healthy condition.
- Vegetation surveys emphasise the importance of salt marsh and other vegetation to food chains and the viability of the estuary.
- Heavy metal levels in plant, bird and fish tissue were low. Pesticides in polychaetes, birds and fish were also low.
- Water quality, based on N, P, and *Escherichia coli* levels, in both stream and estuarine waters, was good. Increases in nutrients and *E. coli* from streams during storm events were monitored in the estuary. Some 3 days after stream levels dropped, the levels of nutrients and *E. coli* in the estuary also returned to normal.
- The importance of storm events on sediment inputs to the estuary and contribution to annual sediment inputs was established. Inputs from individual streams allowed the input from the small catchment under development (the initial cause of community concern) to be seen in perspective. Sediment budgets were attempted over two storm events.

Presentation of Results

Significant scientific results have been obtained from the various projects that make up PEP. Scientists are expected to publish their findings in appropriate scientific journals or reports. These are not read by the community nor do they provide an overall appreciation of the study. A more popular, integrated account has been prepared and will shortly be published in book form (Healy 1979). The contents list is given as an appendix. It is aimed at the intelligent layman and tries to present data in a relatively simplified form and to interpret the results so that they can be related back to the local issues. The objective is to report back to the community since it was from here that the stimulus for the study came. Since not all results are presented and many data are simplified, references to more detailed technical publications are included.

Planners have been involved in PEP from the start so that they can present the planning implications resulting from the study. Scientists are not prepared to project their results beyond a certain point especially outside their own field. Liaison between scientists and planners has been encouraged so that appropriate scientific findings are used in planning recommendations.

The aim of the popular account is to inform the community on the nature of the resource, especially the need to consider the estuary and its catchment as one environmental unit. Hopefully, an informed community, aware of the planning implications, will contribute to and evaluate development decisions and ensure wise management of a valuable resource.

References Cited

Healy, W. B. 1979. *Pauatahanui Inlet—An Environmental Study*. DSIR Information Series 141. Department of Scientific and Industrial Research, Wellington, New Zealand. 195 pp.

Appendix

Pauatahanui Inlet—An Environmental Study

W. B. Healy

Table of Contents

Why the study was done
    Scientists involved
    This book
An early history of Pauatahanui
    The landscape as the Europeans found it
    The Maori
    The whalers
    The military period
    The development of roads
    Sawmilling
    Amenities
    Gold mining

TERRITORIAL SEA FISHERIES MANAGEMENT AND ESTUARINE DEPENDENCE

Kenneth L. Beal

Fisheries Management Operations Branch
National Marine Fisheries Service
State Fish Pier
Gloucester, Massachusetts

Abstract: The management of marine fisheries is an imprecise science, based on biological information and an understanding of ecological inter-relationships, tempered with socio-economic and political considerations. Each coastal State in the United States has management jurisdiction over the resources within its waters, yet most fish species are migratory or extend beyond the control of a single State. Interstate fishery conflicts can usually be traced to divergent management goals. Commercial and recreational harvests along the Atlantic and Gulf coasts of the United States are dominated by estuarine-dependent species. These species exist in a dynamic environment; consequently, their abundance reflects the many variables affecting survival. Recent technological advances in fish-finding and harvesting, coupled with increased fishing pressure, have resulted in greater total landings, and often in declining stocks. The impacts of developmental pressures on fish habitats and other man-induced changes magnify the socio-political conflicts which wax and wane along our coasts. Although our understanding of the complex life histories of estuarine and marine species is improving, effective management will only result from inter-State cooperation, patience, and a coordination of research and management programs. In the United States, the groundwork has been laid for this objective in the efforts of the State—Federal Fisheries Management Program and the Fishery Conservation and Management Act of 1976.

Introduction

At least 50% of the United States domestic commercial harvest and about 80% of the recreational catch is taken within three nautical miles of the coastline (the Territorial Sea). Coastal estuaries have been described as among the most highly productive biological systems on earth, with an average annual gross production of 10,000 to 25,000 kilocalories per m^2 (Odum 1971). The distribution and extent of estuaries varies considerably; about 80% of the Atlantic and Gulf coasts and 10-20% of the Pacific coast is represented by estuaries and lagoons (Emery 1967). Saila (1975) described an increase in diversity of fish species on the east coast of the United States as one proceeds from north to south. McHugh (1966, 1976) has shown the importance of estuarine-dependent species to the total landings. Using 1970 statistics, McHugh (1976) estimated that 69% by weight of all U.S. landings are estuarine-dependent, ranging from a low of slightly over 3% in California to 98% in the Gulf of Mexico. Consequently, the National Marine Fisheries Service and many of the State marine fishery

67

agencies are very concerned with estuaries, wetlands, habitat protection, and effective measures to assure the conservation of these resources. Lack of a compatible management system in the past has led to user conflicts and resource reduction, and created inequities and economic hardships for both commercial and recreational fishermen. The responsibility of the Federal Government and the concerned States is to assure effective conservation and management of interjurisdictional fishery resources throughout their range.

In this paper, I will a) examine the estuary in terms of its value to fishery resources, b) review factors affecting fishery management, and c) discuss the existing institutional structures created to provide for conservation and management of the fishery resources of the Territorial Sea.

Estuaries: The "Commons" for Marine Fisheries

In his essay, "The tragedy of the commons," Garrett Hardin (1968) refers to the exploitation of natural resources as a race by the exploiters to derive the greatest immediate benefits at the expense of future harvests. The ocean has historically been regarded by maritime nations as "inexhaustible". We have recently learned that concentrated exploitation by fishing vessels can result in severe overharvest of stocks to the point that restrictive measures must be imposed to restore the stocks to a level where the yield is optimum for biological, ecological, sociological, and economic interests. In many parts of the world, as in the United States, estuaries and the wetlands which fringe and nourish them can be linked to the diversity and abundance of fish and shellfish. The estuary is often the spawning area, nursery ground, or feeding area for many species, and in the concept of a common property resource, or "commons," an area set aside for use by all. The estuary unfortunately experiences the same pressures which led to the "tragedy" Hardin wrote about.

I propose to borrow a concept from wildlife management and apply it to estuaries. Leopold (1933) refers to the "edge-effect" of two different habitat types. Game of low mobility, such as quail, rabbit, snipe, deer, pheasants, partridge, and wild turkeys are frequently found in greater abundance where the meadow and the woods meet, where the field and the hedgerow join, or where the hardwood stand rubs shoulders with the pine thicket or the cedar swamp. The potential density of organisms of low radius or mobility which require two or more habitat types is proportional to the sum of the type peripheries i.e., the edges of the types. Likewise, the edge-effect can be observed in the estuary, where land and water meet. The wetlands serve as the marine equivalent to a hedgerow, and the population density of estuarine-dependent species appears proportional to the wetlands found in the estuary. Characteristically, the wetlands provide benefits to prey and predator, and to land and aquatic species. The "edge" is habitat for nesting, spawning, feeding, and rearing young. This area also provides protection from predators, and eventually, a place for dying, decomposing, and nutrient recycling.

The estuary is characterized as dynamic, and consequently, those organisms which successfully dwell there must also be tolerant of change. Haedrich and Hall (1976) feel that estuaries are more important as nursery areas than as spawning grounds. Perhaps this distinction has some evolutionary and biogeographical significance and may account for Saila's (1975) observation that faunal diversity is greater in tropical rather than temperate latitudes. Nikolsky (1963) states that in northern latitudes where the vegetative period is short, there are no herbivorous fishes. But in lower latitudes with longer growing seasons, plants play a greater role as food, with some fishes feeding entirely on vegetation. This is not to say that vegetation is unimportant in northern latitudes; submerged and emergent marshes are convenient hiding places for larval and juvenile fishes, and are also used as substrate for eggs spawned in the estuary. Walker (1973) questioned the importance of vascular plant detritus to estuarine consumers, but this was refuted by Odum and Heald (1975), based on their work in Florida on estuarine mangrove communities. Nixon and Oviatt (1973) in their study of a New England saltmarsh found the embayment was dependent on the marsh for organic matter to balance the annual energy budget. The effective transport to the estuary was between 10% and 30% of the total emergent saltmarsh cordgrass (*Spartina alterniflora*) production. Teal (1962) suggested that as much as 45% of the grass may be exported to the estuary.

Estuaries and their associated flora and fauna are transient, and in a geological sense, ephemeral. Furthermore, Schubel and Hirschberg (1978) believe the geomorphological record does not reveal the presence of estuaries as a persistent feature. Nevertheless, it seems unlikely that periodic changes in sea level resulting from major long-term climatic events would effectively prevent the relocation and re-establishment of a given estuarine ecosystem up-river (when sea livel is rising), or down-river (when sea level is falling). Pritchard (1967) defined an estuary as a semi-enclosed coastal body of water which is freely connected with the ocean and within which seawater is measurably diluted with freshwater runoff. If those characteristics prevail, even after dramatic sea level changes, the coastal area will remain an estuary.

The process of ecological succession is first accommodated in an estuary by sedimentation at the mouth of the rivers and creeks providing the freshwater runoff. As sediments accumulate, seawater is displaced from the semi-enclosed area until the estuary is taken over by riverine sediments and the river itself. As physical conditions change, so too will the species diversity and distribution associated with the estuary. Some species will be displaced; others may be exposed to intolerable conditions and become the subject of heated debate in scientific and political forums concerning the "desirability" of their imminent demise. It is to be hoped that our descendants will be better able to deal with the snail darters of the future. Whether a species is estuarine-dependent or ocean-dependent is in reality a matter of semantics. It may require both habitat types at different life stages. If the species possesses the ability to adapt to a changing environment, it will survive. However, the expanding populations of man have brought other consumptive uses of the estuary, such as highways, airports, residential and industrial development, ports, marinas, power plants, and parking lots. At this point we must ask ourselves if these uses are necessary and appropriate for a common property resource over which we must act as trustees. In one respect, the "succession" which takes place when an estuary is consumed by development is an ecological dead end. We seldom hear of a parking lot or an industrial site being converted back into a wetland.

Factors Affecting Fisheries Management

Man has been using the ocean for centuries for fishing, transportation, communication, and for conducting warfare. Boundaries were established by the sovereign nations for various reasons. In the United States, we have four principal zones which relate to fishing activities; 1) internal waters—lakes and rivers within the State, including some estuaries; 2) the Territorial Sea—coastal waters extending 3 nautical miles (5.5 km) from shore (originally established by the range of cannon fire), and including some estuaries; 3) the Fishery Conservation Zone—oceanic waters beginning at the outer edge of the Territorial Sea and extending to 200 nautical miles (370 km) from shore; and 4) international waters—those oceanic waters beyond 200 miles. Fisheries management in the United States is administered differently for each zone and these management measures will be discussed later. Interjurisdictional fishery resources are migratory or have a natural distribution which extends beyond the boundaries of one State. Although control of these resources within the Territorial Sea is normally a matter for interstate cooperation, the range of the species may also extend into the internal waters of one or more States, or into the Fishery Conservation Zone. In such cases, more than one management regime may be involved. On the east coast of the United States, commercial landings from the Territorial Sea of these interjurisdictional species are shown in Table 1 as a percentage of total landings (recreational landings are not reflected). Table 2 shows the range of these species in terms of overlapping jurisdiction between adjacent States and Canada. These data are derived from NMFS fishery statistics, 1971-1978. The fisheries for migratory interjurisdictional stocks have historically been characterized by a tendency toward over-exploitation. Since the oceans have been viewed as a common property resource, where no individual has ownership, a fisherman has little incentive to reduce fishing effort on declining stocks because he has no guarantee the fish he saves will not be caught by another fisherman (Crutchfield 1973).

Table 1. Percent of U.S. commercial landings occurring in the Northwest Atlantic from 0-3 miles by species for the years 1971 - 1978.

SPECIES	1971	1972	1973	1974	1975	1976	1977	1978
Alewife, Atlantic & Gulf	100	100	99	100	100	100	100	100
Bluefish	81	80	88	90	88	90	83	78
Butterfish	17	50	35	47	37	50	32	10
Cod	8	6	5	6	8	8	6	12
Cusk	7	8	4	5	3	11	9	9
Winter Flounder	21	14	16	12	9	17	13	26
Summer Flounder	19	20	49	38	41	34	21	19
Yellowtail Flounder	7	7	6	3	4	5	4	17
Other Flounder	12	32	50	60	40	33	23	19
Haddock	1	1	1	2	2	1	1	3
Red Hake	6	18	47	17	23	15	12	14
White Hake	17	20	12	11	17	10	8	27
Halibut	10	11	11	41	13	13	13	46
Herring	14	62	71	59	58	83	89	85
Mackerel	32	46	59	51	23	36	55	58
Atlantic Menhaden	76	83	92	100	100	95	96	99
Ocean Perch	<1	<1	<1	<1	<1	<1	<1	<1
Pollock	2	3	12	5	9	4	4	8
Scup	32	13	46	44	46	38	48	45
Black Sea Bass	5	6	6	7	9	14	7	9
Weakfish	60	84	86	85	64	63	72	50
Striped Bass	91	91	92	91	94	97	97	94
Whiting	4	13	7	13	13	12	9	11
Wolffish	25	3	1	13	17	10	3	16
American Shad	100	100	100	100	100	100	100	100
Hard Clam	100	100	100	100	100	100	100	100
Soft Clam	100	100	100	100	100	100	100	100
Surf Clam	5	13	12	23	49	14	16	20
Ocean Quahog	-	-	-	-	-	-	13	12
Oysters	100	100	100	100	100	100	100	100
Squid	54	85	78	84	83	79	33	50
Bay Scallop	100	100	100	100	100	100	100	100
Calico Scallop	<1	<1	<1	<1	<1	<1	<1	1
Sea Scallop	11	17	15	8	17	4	2	5
Blue Crab	100	100	100	100	100	100	100	100
Northern Shrimp	28	29	35	24	16	40	7	-
South Atlantic Shrimp	95	100	88	86	89	88	84	68
Gulf Shrimp	31	22	29	29	37	35	38	26
Lobster	67	72	80	72	74	73	79	79

Table 2. Approximate geographical range of species landed commercially in the Northwest Atlantic. Abbreviations used are: Maine - ME, New Hampshire - NH, Massachusetts - MA, Rhode Island - RI, Connecticut - CT, New York - NY, New Jersey - NJ, Delaware - DE, Maryland - MD, Virginia - VA, SE Region - North Carolina, South Carolina, Georgia, Florida.

Range of Landings

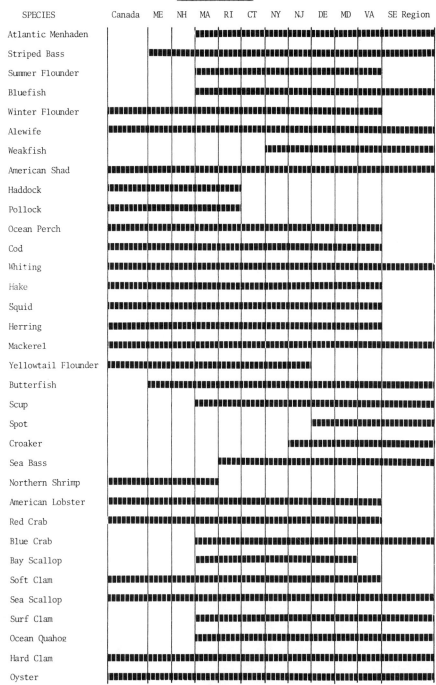

| SPECIES | Canada | ME | NH | MA | RI | CT | NY | NJ | DE | MD | VA | SE Region |

The regulation of commercial fisheries is really a matter of economics. Fishing is a risky business, and expected profits have to be worth high risk. Conservation and management are designed to assure continued fishing for market-sized fish, rather than simply to protect fish. Some of the measures used to reduce overfishing are catch limits, closed areas, and gear restriction. All of these restrictions adversely affect a fisherman's ability to make a successful trip at the outset and increase the chance of not making a profit. It is understandable that a fisherman may regard regulations as intolerable meddling. If he observes large numbers of the regulated species, he may trust his conclusion that the stock as a whole is healthy, and doubt the view of fishery scientists that the stock is in trouble (Tait and DeSanto 1975). Unless regulations can be easily and equitably enforced on all fishermen, overfishing will continue, and the regulations will be unsuccessful in their intent. Consequently, regulations are likely to be opposed by those people they were designed to benefit. In the absence of property rights, regulations which affect all are inevitable following a resource disaster or when a resource is threatened.

Property rights do exist in at least one fishery, and regulations have been advocated by these fishermen, primarily as a means of excluding "outsiders," though conservation has also been achieved. The lobster fishermen in the coastal area of central Maine have extended ownership of shore property seaward to include "fishing rights". These rights are traditionally held, and inherited patrilineally, as are land rights (Acheson 1975). This unique practice is slowly breaking down as technology has provided fishermen with more efficient gear and a greater fishing range. Yet tradition is hard to change in Maine, particularly in the offshore islands where the fishing community exerts social and political pressure to restrict nonconformists. For instance, the residents of Monhegan Island have persuaded the State of Maine Legislature to pass a law forbidding lobster fishing in Monhegan waters from June 25 to January 1. The closure prevents fishing during the lobster molting season (July and August) when high mortality from cannibalism by other lobsters caught and confined in the traps would normally occur. The season then opens at a time when prices and catches are normally high and competition from "outsiders" will be low. Such "conservation measures" will only work in areas where everyone agrees to them. Acheson (1975) found unmistakable evidence that reduction in fishing effort in this area has resulted in both biological and economic benefits to the fishery. However, it has not been possible to institute similar closed seasons in the lobster fishery throughout the Maine coast or in any other State (although Canada has done so).

Coordinated interstate conservation efforts have not been successful until recently. The stumbling block is not wholly one of disagreement on conservation, but rather a realistic appraisal of the political possibilities of implementing certain restrictions as a State regulation or law. Many State Legislatures jealously guard "States' rights," and historically have shown reluctance to address problems of a regional nature. I call this the Kaleidoscope Theory of Fishery Management, which may be defined as tunnel vision reflecting the multi-colored and multi-faceted prisms of economic, social, conservation and political viewpoints. The President's Commission on Marine Science, Engineering and Research (the Stratton Commission) recognized the lack of a single focus of management responsibility in marine fisheries in its 1969 report. Citing the common property nature of the resource and the fact that fisheries are regulated (or not regulated) under split or multiple jurisdictions, the Commission recommended a definitive review and restructuring of fisheries laws and regulations, and the creation of a new framework based on national objectives for fisheries development and on the best scientific information.

Territorial Sea Fisheries Management

There are several scenarios of management in the Territorial Sea, based primarily on whether a species is confined to one State or is interjurisdictional in its distribution. One facet important to managers is the location of the preponderance of landings. Species which are caught primarily within the Territorial Sea are usually managed under the State-Federal Fisheries Management Program. This program was formally established in 1971 in response to the Stratton Commission report. The Regional Director of the National Marine Fisheries Service is the federal participant, and the heads of each State's marine fisheries agency are the State participants. Federal funds are used to defray the costs of meetings, to encourage coordination and cooperation, and to develop management plans. The greatest weakness to date is that the program has no authority to require States to implement approved management plans within their respective territorial waters. Once a management plan is adopted, it is up to the individual marine fishery administrators to attempt to have the management measures adopted in their respective State Legislatures.

If the species concerned is caught in the Fishery Conservation Zone, as well as in the Territorial Sea, the possibility exists that a management plan may be prepared under the Fishery Conservation and Management Act. This Act provides for regional management councils and includes a provision for preempting State authority for managing a fishery when it has been determined that the implementing provisions of a plan are threatened by an action taken (or not taken) by the State. The majority of those species shown in Table 1 have an extensive fishery in both the Territorial Sea and in the Fishery Conservation Zone. The Table does not reflect the recreational fishery which catches about 80% of its fish within the Territorial Sea, including numerous estuaries. The commercial catch of some species, such as summer flounder (Paralichthys dentatus), which one may suspect from reviewing Table 1 are taken primarily in the Fishery Conservation Zone, are in fact taken primarily by recreational fishermen in great numbers within three nautical miles of the shore in the Territorial Sea and the coastal estuaries. The extent of the recreational fishery is a serious consideration in selecting species in greatest need of management. It should be noted that commercial fishery statistics are available on an annual basis, but recreational surveys are conducted only periodically. The results of the 1978 recreational survey are not yet available in final form. Nevertheless, we do know that the recreational fishery, including that within estuaries, exceeds that of the commercial fishery for many species.

To date, the following species have been selected and studied under the State-Federal Program in the Northeast: summer flounder, Atlantic menhaden *(Brevoortia tyrannus)*, striped bass *(Morone saxatilis)*, surf clam *(Spisula solidissima)*, ocean quahog *(Arctica islandica),* Northern shrimp *(Pandalus borealis),* and American lobster *(Homarus americanus).* All but the surf clam, ocean quahog, and American lobster are estuarine-dependent at some stage of their life history. Both the surf clam/ocean quahog plan and the lobster plan were presented to the Fishery Management Councils for implementation in the Fishery Conservation Zone. It is anticipated that the Northern shrimp plan and the summer flounder plan will also be presented to the Councils because the commercial fishery is primarily conducted in the Fishery Conservation Zone. Other species will be studied as soon as funding is available.

Gulland (1974) stated that the absence of disaster does not imply that management is not desirable. He cites the collapse of the California sardine fishery as a monument both to the failure to act in time and to the insistence of having conclusive scientific evidence before acting. Management of all fisheries depends on biological and other information. Unless there is a good analysis of the likely consequences of alternative management strategies, no rational decision can be made. Fishery scientists today must be more willing to make judgments in the face of incomplete knowledge. This fact is now obvious to the State directors of marine fisheries and the members of the regional fishery management councils, and they must be willing to make management decisions on this basis. Fisheries management plans must be tailored to specific needs of regional fisheries problems and prepared cooperatively with State and local input. Management beyond State control is a federal responsibility, but this must be carried out in concert with the States through the regional fishery management council framework.

Conclusions

Concurrent with great increases in world landings of fish and shellfish, oceans and estuaries have reacted to man's progressive encroachment, and these productive "commons" are now less capable of recycling raw materials and replenishing the resources upon which our fisheries depend. Diking, filling, pollution, and consumptive uses have handicapped the ecological balance in estuaries. Hardin (1968) made an accurate assessment in his statement that "Ruin is the destination toward which all men rush, each pursuing his own best interest in a society that believes in the freedom of the commons." Education can counteract the tendency to do the wrong thing, but the basis for our temperance must be constantly refreshed. We have reached a point when the use or misuse of all wetlands must be guarded, even those considered "unproductive" (Horwitz 1978). We have also learned enough to know that there is much more to learn. Management in the Fishery Conservation Zone, in the Territorial Sea, and in the estuaries is complicated by many factors. The challenge of today's fisheries managers is to recognize the kaleidoscope nature of the problem and to reach beyond basic biology to develop a multi-disciplinary strategy which can be presented with diplomacy to decision-makers. The State-Federal Fisheries Management Program and the Fishery Conservation and Management Act are two vehicles with which to meet this challenge. And like the experienced hunter who walks the hedgerow, we are optimistic.

Acknowledgments

I gratefully acknowledge Peter D. Colosi, Jr. for his help in designing and drafting the tables, and for his constructive comments on the text. Richard G. Seamans, Jr. and Bruce G. Nicholls also critically read the manuscript and provided several valuable suggestions.

References Cited

Acheson, J. M. 1975. The lobster fiefs: economic and ecological effects of territoriality in the Maine lobster industry. *Human Ecology* 3:185-207.

Crutchfield, J. 1973. Resources from the sea, pp. 105-133. In: T. S. English (ed.), *Ocean Resources and Public Policy*, Univ. of Washington Press, Seattle.

Emery, K. O. 1967. Estuaries and lagoons in relation to continental shelves, pp. 9-11. In: G. H. Lauff (ed.), *Estuaries*. Amer. Assoc. Adv. Sci. Pub. 83, Washington, D.C.

Gulland, J. A. 1974. *The Management of Marine Fisheries*. Univ. of Washington Press, Seattle. 198 pp.

Haedrich, R. L. and C. A. S. Hall. 1976. Fishes and estuaries. *Oceanus* 19:55-63.

Hardin, G. 1968. The tragedy of the commons. *Science* 162:1243-1248.

Horwitz, E. L. 1978. *Our nation's wetlands. An interagency task force report.* Council on Environmental Quality, Washington, D.C. 70 pp.

Leopold, A. 1933. *Game Management.* Charles Scribner's Sons, New York, N.Y. 481 pp.

McHugh, J. L. 1966. Management of estuarine fisheries, pp. 134-154. In: R. F. Smith, A. H. Swartz and W. H. Massmann (eds.), *A Symposium on Estuarine Fisheries*. Amer. Fish. Soc. Spec. Pub. 3.

McHugh, J. L. 1976. Estuarine fisheries: Are they doomed? pp. 15-27. In: M. L. Wiley (ed.), *Estuarine Processes, Vol. 1.* Academic Press, New York.

Nikolsky, G. V. 1963. *The Ecology of Fishes.* Academic Press, New York. 352 pp.

Nixon, S. W. and C. A. Oviatt. 1973. Ecology of a New England salt marsh. *Ecol. Monogr.* 43:463-498.

Odum, E. P. 1971. *Fundamentals of Ecology.* W. B. Saunders Co., Philadelphia. 574 pp.

Odum, W. E. and E. J. Heald. 1975. The detritus-based food web of an estuarine mangrove community, pp. 265-286. *In:* L. E. Cronin (ed.), *Estuarine Research, Vol. 1.* Academic Press, New York.

Pritchard, D. W. 1967. What is an estuary: Physical viewpoint, pp. 3-5. *In:* G. H. Lauff (ed.), *Estuaries.* Amer. Assoc. Adv. Sci. Pub. No. 83, Washington, D.C.

Saila, S. B. 1975. Some aspects of fish production and cropping in estuarine systems, pp. 473-493. *In:* L. E. Cronin (ed.), *Estuarine Research, Vol. 1.* Academic Press, New York.

Schubel, J. R. and D. J. Hirschberg. 1978. Estuarine graveyards, climatic change, and the importance of the estuarine environment, pp. 285-303. *In:* M. L. Wiley (ed.), *Estuarine Interactions.* Academic Press, New York.

Tait, R. V. and R. S. DeSanto. 1975. *Elements of Marine Ecology.* Springer-Verlag, New York. 327 pp.

Teal, J. M. 1962. Energy flow in the salt marsh ecosystem of Georgia. *Ecology* 43:614-624.

Walker, R. A. 1973. Wetlands preservation and management on Chesapeake Bay. *Coastal Zone Management J.* 1:75-101.

SOCIO-CULTURAL VALUES OF WETLANDS

Robert J. Reimold, Jeannette H. Phillips

Georgia Department of Natural Resources
Coastal Resources Division
1200 Glynn Ave.
Brunswick, Georgia
and
Michael A. Hardisky

University of Delaware
College of Marine Studies
Newark, Delaware

Abstract: If the general public is to appreciate and protect wetlands, they must learn these areas have values other than strictly economic ones. Nonconsumptive uses that neither deplete wetlands themselves, nor rob them of their wealth, can be divided into aesthetic, recreational and educational categories. Wetland areas provide unique sensory experiences. Examples are used to show how these have been documented in the graphic arts, music, and literature. Early prototype models have been developed for assessing visual-cultural values of freshwater wetlands; however, these values usually reflect the bias of the evaluator. Recreational use of wetlands is often most appropriate for an individual or small group. Although wetland playground or park activities may not in themselves destroy or deplete the resource, they are geared toward mass recreation which obscures the isolation and subtle mood changes of swamp, marsh or beach. Educational values of wetlands range from the simple joys of discovery to complex scientific research. Marshes provide a good place for ecosystem study because the physical stresses on the environment limit diversity and the number of metabolic pathways in the system. Educating the non-science public about wetland ecology is essential if these fragile environments are to escape development. A conceptual approach is developed in this paper to assess the educational, aesthetic, recreational, anthropological and/or theological functions and values of wetlands. The importance of these considerations is assessed, as are long term rewards that wetlands afford to man and biosphere.

Introduction

As the world's population grows and becomes more demanding of the planet's natural resources, there is increasing pressure to use those resources to provide the maximum or optimum benefit for man. Without being wasteful, we must look for new ideas about uses of existing resources in order to support human life in the future.

Thus, justification for preservation or conservation of resources becomes very important. The most persuasive line of argument for preventing alteration or destruction (consumption) of natural resources is to evaluate the monetary benefits of leaving resources in their present state.

For instance, as more offshore areas are surveyed for their potential in gas and oil, we have an economic basis to compare the ocean's returns in fisheries products against the possible benefits of energy development. Even when looking at some wetlands, we now begin to assign monetary values, e.g. the role of salt marshes in nurturing commercially important species, or in protecting high ground property from storm damage.

Is a resource worth preserving when its benefits cannot be measured in dollars? What about non-economic benefits provided to mankind by wetland systems? How can we attempt to determine how best these resources can serve the public good? This discussion centers on the major socio-cultural contributions of wetlands to human society and suggests some ways in which we might determine values for these benefits to the whole planet and to man.

We must first define wetlands descriptively and functionally. For the purposes of this paper, wetlands are areas inundated or recharged with water frequently enough that they can support herbaceous vegetation that requires or tolerates saturated soils for growth and reproduction under normal circumstances.

By "value", we mean the worth of something. Perhaps if we were seeking to define worth in strictly economic terms, we might ask, "What products can we extract from wetland areas that are useful to mankind?" We would list the role of the salt marsh in feeding and sheltering commercially important marine organisms, or the amounts of peat that can be extracted from bogs, or the tons per year per acre of cranberries harvested. However, in evaluating the often less tangible socio-cultural values of wetland areas, we need to ask the previously suggested broader question: "What benefits do wetland systems offer the biosphere?"

Before going any further into an exploration of what these socio-cultural values are, it is prudent to consider the definitions for "social" and "cultural" as given in a standard dictionary. The pertinent definition for "social" is: ". . . of or relating to human society, the interaction of the individual and the group, or the welfare of human beings as members of society." Definitions of "culture" or "cultural" have a common element: they all mention acquisition of certain qualities or tastes through training or transmitting of knowledge; "The integrated pattern of human behavior that includes thought, speech, action, and artifacts and depends upon man's capacity for learning and transmitting knowledge to succeeding generations."

Socio-cultural values, then, are generally removed from the economic plane and involve mankind's higher aspirations: philosophy, beauty, learning, spiritual and humanitarian concerns—those elusive elements that make up the equally elusive thing known as "quality of life." Not only are they hard to define; they are very difficult to evaluate in any tangible terms. In this paper, we divide socio-cultural values into three categories: aesthetic, recreational, and educational. The importance of these uses and values of the wetlands of our nation was underscored when President Carter (1977) issued his Executive Order 11990 to

> ". . . minimize the destruction, loss or degradation of wetlands, and preserve and enhance the natural and beneficial values of wetlands . . ."

Among the factors to be considered by federal agencies when evaluating possible effects on wetlands of proposed activities were

> ". . . maintenance of natural systems, including conservation and long term productivity of existing flora and fauna, species and habitat diversity and stability, hydrologic utility, fish, wildlife, timber, and food and fiber resources; and other uses of wetlands in the public interest, including recreational, scientific, and cultural uses."

Aesthetic Values

Mankind has always been drawn to water. He has built his settlements near water sources for sustenance, for transport, for irrigation. Yet wetland areas, attractive as they were in many ways, posed frustrations for man because they were not always reliable sources of water; they often bred disease-carrying insects; and the land could not be developed for use in traditional ways. Occasionally, the changeable nature of the land was used to the advantage of the owner. The Abbey of Mont St. Michel, off the coast of Normandy, stands atop a craggy island separated from the mainland by wide strands of tidal sands. These days, a causeway connects the rock to the shore for the convenience of visitors, and the sands have accreted, blocking much water access. But for centuries the wetland barrier at the rock's base protected the monastery, since the unwary could be easily swept away by incoming waves if careful entrance or exit plans had not been made.

An aesthetic appreciation of marshes, swamps, and other wetlands is essentially a sensual one. Wetlands stimulate the vision, hearing, sense of smell, touch and taste in ways that have been recorded by painters, musicians, and writers of many ages. As our experience expands and our senses become increasingly attuned to wetland stimuli, we are able to appreciate and sense more. Wisconsin naturalist Aldo Leopold (1949) summarized this developmental process:

> "Our ability to perceive quality in nature begins, as in art, with the pretty. It expands through successive stages of the beautiful to values as yet uncaptured by language."

Wetlands have captured the imagination of mankind for centuries because of the fantastic changes that take place in the landscape as a result of weather conditions. Horwitz (1978) mentions the severe extremes to which these areas are subject, and notes ". . . floods, drought, winter ice, high winds, waves, violent storms, and hurricanes are important factors which help shape these ecosystems. In short, the wetlands have evolved with natural catastrophe as a partner."

In order to really evaluate and understand wetlands and their functions, wetlands must be observed over time. Their very nature changes with the hour, the day, and the season. Different animal populations utilize the resource depending on the stage of the tide. Even the soil color and texture reflect the system's recent history; recording algal activity, the presence of grazing creatures, or the path of an animal across the newly-drained creek bank.

Western European painters since the Middle Ages have used every excuse to incorporate water landscapes into their paintings. Academies of Art, except in Holland, frowned on the practice of landscape painting. If a painter wanted to represent outdoor scenes, he incorporated them around more acceptable subjects like religious stories, portraits, or allegories. Look at the "Mona Lisa" and you will see the floodplain of an imaginary river behind her. St. Sebastian, rendered in painstaking detail by Pollaiuolo (1475), hangs with arrows piercing his chest, but the painter gave exquisite attention also to the river landscape behind him.

Painters used a variety of media to capture the fluid essence of wetland subjects. At one end of the graphic spectrum, the 16th century Dutch and Flemish masters meticulously reproduced scenes from daily life in their low country environment which they carefully guarded from intrusion by the sea. Pieter Brueghel the Elder (1525-1569) created a number of wetland "genre" or everyday scenes in oil paint, of which "Hunters in the Snow" is a good example. His "Temptation of St. Anthony" supposedly ilustrates a religious subject, but in the background we see a typically "Netherlandish" water landscape, so central to the lives of Brueghel and his countrymen.

Gradually, graphic craftsmen began noticing that the wetland landscape's appearance changed depending on the time of day, the clouds, the quality of light, and the season. Until the 18th Century, painters made only their rough sketches from nature, preferring to work at the actual painting in a static studio atmosphere. John Constable (1776-1837) began a new tradition. He set up his canvas and palette in the English countryside and studied the transient effects of light and cloud formations on the rural landscapes he knew so well. Two of his wetland scenes, "View at Hampstead Heath" and "Flatford Mill" combine realism with the nuances produced by light and water.

Constable's departure from a strict adherence to total clarity of subject matter led the way for new ideas. Later in the 19th Century, a group of painters moved further down the graphic spectrum, and were scornfully called "Impressionists" because their work did not try to reproduce the actual appearance of objects, but the impression the objects created when flooded with light. Water and wetlands provided a never-ending source of inspiration to these French painters, who perched their easels on the river-bank, the pond verge, and at the seashore.

Claude Monet (1840-1926) was probably the master of wetland impressionism. He celebrated pond life in his "Water Lilies" paintings, and also recorded many river scenes, such as those in his "Argenteuil" series, and captured the look of urbanized wetlands in "Palazzo da Mula, Venice". Many others in the Impressionist school, such as Renoir (1841-1919), Sisley (1839-1899), Morisot (1841-1895), and Boudin (1824-1898), showed a special affinity for and a demonstrated ability to depict tidal rivers, beach scenes, and wetland vegetation in a painterly fashion.

Vincent Van Gogh (1853-1890) was an expressionist; that is, he distorted color and form to convey his emotions and feelings about the subjects of his paintings. His work, done in oil, was produced primarily in the south of France, where Mediterranean wetlands served many times as his landscapes. Tidal wetlands figure in his "Drawbridge at Arles" and "The Cypress Road".

Many American painters too have delighted in displaying the picturesque and aesthetic qualities of wetlands. John James Audubon (1785-1851), illustrator as much as painter, recorded the images of many marsh and swamp birds of the southeast. His "Snowy Heron" is displayed against a backdrop of flooded rice fields in South Carolina. Thomas Eakins (1844-1916), Winslow Homer (1836-1910), Frederic Church (1826-1900), and Thomas Hart Benton (1889-   ) were a few of the painters who also found subjects for their canvas among the swamps, marshes and rivers of the southern United States. Still other painters portrayed the wetlands of the northeast. Thomas Cole (1801-1848), Martin Heade (1819-1904), and James McNeil Whistler (1834-1903) recorded the sights of nearby rivers, lakes and marshes in this country's youthful days; Andrew Wyeth (1917-   ) takes a contemporary view of the bleached wetlands of Maine.

During the last century, photography has documented the visual aesthetics of wetlands. Striking images are also produced from side-scanning radar and infrared photography, technologies developed for use in research to assess wetland primary productivity, plant species, habitat and other ecological characteristics (Gosselink et al. 1974; Odum and Skjei 1974; Reimold and Linthurst 1975).

Although man finds it easiest to create and transmit impressions visually, our literature abounds in references to the beauty of wetlands. William Byrd (1674-1744), representing the Virginia colony at the British court, praised his country's coastal marshes. William Cullen Bryant (1794-1878) described New England wetlands. Ralph Waldo Emerson (1803-1882), James Russell Lowell (1819-1891), and Henry Wadsworth Longfellow (1807-1882) recorded in prose and poetry images of the valuable wetlands and water life of our emerging nation. John and William Bartram, father and son, used their skills as naturalists to make an inventory of the plant and animal life of the southeastern United States' wetlands, observing the land, the climate, and the people they met.

In his *Travels,* published in Philadelphia in 1791, William Bartram observed the coastline from Cowford (Jacksonville, Florida) northward:

> "There is a large space betwixt this chain of seacoast islands and the main land . . . but all this space is not covered with water; I estimate nearly two-thirds of it to consist of low salt plains, which produce Barilla, Sedge, Rushes, etc."

The changing and sometimes unpredictable moods of wetlands give these areas a feeling of mystery which makes them ideal settings for stories. Robert Louis Stevenson (1850-1894) placed the action for many of his pirate stories in tidal wetlands. Dorothy Sayers' (1893-1957) *The Nine Tailors,* set in England's fen country, depends for much of its celebrated suspense on the natural processes of the marsh and the canals that drain the area. Elizabeth Goudge, another English writer, drew on the same boggy countryside as the backdrop for her book, *The Dean's Watch.* Non-fiction works dealing with wetlands include Sally Carrighar's *One Day at Teton Marsh,* Margaret Craven's *I Heard the Owl Call My Name,* and Olive M. Anderson's *Seekers at Cassandra Marsh.*

In the United States, Sidney Lanier (1842-1881) is the man of letters most closely associated with our coastal marshes. In 1880, he wrote:

> "Reverend of marsh, low couched along the sea, old chemist, wrapt in alchemy, distilling silence, — low, that which our Father-age had died to know — the menstruum that dissolves all matter — thou has found it; for this silence, filling now the globed charity of the seething space, this solves us all; man, matter, doubt, disgrace." (Lanier, M. 1944)

In James Michener's recent book, *Chesapeake,* the young boy who is the story's central figure finds reinforcement for his own values when he chances upon the works of Lanier. The poet's admiration for the expanses of marsh and the life within them leads the boy to realize that others, too, love the environment which the boy's parents see only as a place from which to eke out a living.

Wetlands offer a variety of sounds to the careful listener, which composers have often tried to capture in music. Historically, life at the water's edge has been colorful and exotic, furnishing plenty of material for song as well as story. Stephen Foster (1826-1864) wrote many songs about life near the river deltas and along the levees of the South; Louis Moreau Gottschalk (1829-1869), a brilliant concert pianist and composer from Louisiana, used many rice plantation tunes as the basis for his melodies. The seafaring traditions of the British Isles and the New World gave birth to sea shanteys and other nautical tunes such as "Lowlands Away", "The Eddystone Light", and "The Greenland Whale Fishery."

More recently, Florida composer Dale Crider has made wetland ecology the focus of his music. His songs, such as "Sea Oats", "O Kissimmee River", "A Swamp is a Natural Systems Machine", and "Natural Cycles", acquaint the public with the delicate balance of wetland systems and the potential threats to this balance.

Claude Debussy (1862-1918) often used water or wetland themes for his music, employing the wide intervals of the whole-tone scale to advantage in portraying the fluid nature of his subject. In "La Cathedral Engloutie" (The Sunken Cathedral), "Clair de Lune" (Moonlight) and "Beau Soir" (Beautiful Evening) he created moods of contemplation and mystery.

Debussy's "La Mer" is a well-known tonal seascape. The most famous of the genre, however, is doubtless Richard Wagner's overture to "Der Fliegende Holländer" (The Flying Dutchman). In his opera, Wagner protrays not only the ocean's dramatic changes, but, more significantly, the effect produced on the protagonists by the tumultuous sea and illusory shoreline.

Man's quest to establish values is usually based on the gold standard. Value determinations for wetlands often fail to evaluate intangible wetland attributes such as beauty and purity. Martin, Hotchkiss, Uhler and Bourn (1953), Pope and Gosselink (1973), and Gosselink, Odum and Pope (1974) recognized the importance of non-consumptive values (that is, uses that do not expend or diminish the resources' value), but did not attempt to place a value on them. Messman, Reppert and Stakhiv (1977) incorporated both cultural and aesthetic values of wetlands into an evaluation methodology. The recognition of visual, cultural and aesthetic values of wetlands does not require a landscape architect. However, comparing these values to those of basic life support of fisheries is inherently difficult due to the absence of a convenient common denominator.

Smardon and Fabos (1976) developed a model for assessing visual-cultural values of freshwater wetlands. The three-level model evaluates wetlands in terms of landform contrast, landform diversity, wetland edge complexity, wetland type diversity, educational and recreational quality and possible outstanding elements. These attributes were weighed and the resultant rating proved useful in comparing visual-cultural values of different freshwater wetlands (Fabos 1971; Smardon 1972; Larson 1976). Their objective attempt to describe subjective values certainly reflects the perspective of the evaluator.

For example, a large expanse of *Spartina alterniflora* in Georgia would not fare well in terms of plant diversity, wildlife diversity, wetland types diversity or visual diversity when compared to a southern river swamp. However, to the people who visit or live in the vicinity of the marsh, the luxuriant green or golden brown coloration of these meadows undulating in golden waves in the gentle southeast breezes may be much more pleasurable than the sight of cypress trees festooned with Spanish moss. From afar, a viewer can easily admire homogenous expanses of *S. alterniflora* and the dendritic pattern incised in the marsh by creeks and rivulets. At close range, however, the viewer might demand much diversity of vegetation, coloration, or movement in order to find beauty in the scene. An observer's mental and physical perspectives dictate the final value determination.

All of these aesthetic components, and others, underscore the utility of having wetlands without doing anything to them. All of these aesthetic values are attributable to the inherent presence of the wetlands. Whether they be a prairie pothole in North Dakota, a coastal *S. alterniflora* marsh, or an inland bog, the wetlands all provide a variety of aesthetic experiences to mankind. All of these have value and yet none of the values can be adequately priced on the present economic marketing structure.

Recreational Values

Recreational activities in wetland areas form a continuum from those that are consumptive or destructive to those that are purely non-consumptive. Somewhere along this scale there are activities which, depending on one's view of the full cycle of nature, may or may not be consumptive. The notion of consumptive as opposed to non-consumptive is very important when we consider use of the wetland resources for the public pleasure. It often happens that an area has such aesthetic appeal that people flock to it, over-use or abuse it, thereby reducing the aesthetic qualities that attracted visitors in the first place.

Swimming or diving in waters near the wetland edge is an example of a use that can leave the resource without human trace. Hiking through marsh, bog, or beach, when responsibly done, similarly maintains quality of the resource while providing a pleasurable pastime for many.

Picnic areas, playgrounds, and local, state and national parks within wetland bounds are geared to mass recreation activities that are not generally considered consumptive. However, it requires good management of the resource to maintain its quality while providing pleasure for large numbers of people. Loosely-controlled use of off-road vehicles, popular in those very environments which are the least resilient to the impact of these vehicles, threatens animals, vegetation, and terrain within the wetland realm. Human litter quickly accumulates to mar the landscape, leaving man's unmistakable imprint on a fragile system. Even when well-managed and maintained, sites of mass recreation are by definition not for those who value the solitude and subtle mood changes of wetland areas. Consequently, this type of socio-economic value attributed to the wetland for recreational purposes often results in consumption and diminishment of the overall value over time. Between the individual bird-watcher on the marsh verge and the reckless jeep or dirt-bike enthusiast, there is a large number of people who appreciate wetland resources and—within their perspective—use these resources wisely. Hunters, trappers, or fisher folk remove animal life from the environment, and can therefore be considered to benefit from the socio-economic values of wetlands. However, their harvests, when limited, can be seen also as a portion of the wetland cycle.

Sometimes the hunter or fisherman acts as a resource advocate. He or she recognizes the need for conservation measures and provides financial support to maintain the recreational resource for future use.

One way of estimating the recreational resource value of wetlands is to compute the dollars spent by participants in these pleasurable activities. People spend increasing amounts of money on marine recreational fishing, an industry whose existence derives directly from wetlands. A recent study by National Marine Fisheries Service (1977) estimated a total national economic benefit of $2.9 billion in 1975 from marine recreational fisheries. This represents a significant economic value from a resource that relies upon wetlands to provide nursery grounds for the fish later harvested in open waters. Wetlands are not depleted during this activity; therefore, it may be considered non-consumptive.

Another economic value of wetlands is provided by expenditures on waterfowl. A recent national survey reports that the average duck hunter spends over $730 per year on hunting. The value of the marsh to duck hunters, based on our traditional economic structure, can then be computed by the equation: number of waterfowl hunters per area × average expenditure per hunter ÷ the number of wetlands in the same area. For example, if 1000 duck hunters solely use a 7300 acre wetland wildlife management area, then the average value of each acre of wetland would be $100/year just for duck hunting.

Similar equations have been used to assess the economic impact of other recreational activities such as fishing, boating, and camping. Such computations, however, are probably misleading and inaccurate in terms of establishing a value system. Since these are not directly cause-and-effect related expenditures, the economic value is not indicative of the true total value. The peace of mind derived from a day of marsh hunting, birding or daydreaming in the sun beside a tidal creek is a definite value that cannot be measured in any terms presently available.

As stated earlier, aesthetic considerations are an important element in a locale's recreational appeal. The open space associated with natural wetlands, when protected from exploitation, attracts many people from their crowded metropolitan habitat. These visitors are not only inland inhabitants, spending vacation dollars near a wetland wilderness, but local residents enjoying wetland spots close to home. Over 50% of the population in the United States now lives within the coastal counties (Ketchum 1972) and this coastal migration appears unstemmed. During the remainder of this century, this coastward move by the U.S. population will doubtless continue. Consequently the money people spend to enjoy themselves within or adjacent to wetland locales to some degree measures the recreational value of marshes, swamps, beaches and bogs and can be expected to increase significantly during the next decades. At the same time, if wetlands' socio-cultural benefits to the human populace are not recognized, intensified population pressures could deplete or destroy the resource.

Recreational experiences of wetlands are just now being evaluated in a quantitative, complex way—as accurately as the ways in which we quantify energy flow or productivity. There is however at this time no matrix within which to equate recreational use data and yet derive a direct statement of value. This is true because man has not yet identified a universal unit of measure that describes an increase in man's emotional, intellectual, and physical well-being as a result of wetland recreation activities.

Educational Values

The educational values obtained from wetlands have not been well studied or documented but do enter into the overall socio-economic equation for value. Educational and research values of wetlands include studies in botany, natural history, ornithology, and environmental metabolism. Not only are wetlands functional educational settings for elementary, secondary, and undergraduate students, but they are also of significant value in basic and applied scientific research.

The physical stresses of inundation, salinity and anaerobic sediments typical in wetlands reduce the diversity and the number of metabolic pathways within a system. Marshes have provided an invaluable tool for ecosystem study. The basic principles developed and refined in these environments have been used to study more complex systems masked by multiple layering of energy pathways.

Wetlands have also become natural laboratories for non-scientists. Education of citizens is essential for continued conservation of wetlands. Marsh walks conducted in coastal Georgia have attracted people of all ages in large numbers. A firm attitude of wetland conservation is quickly developed in citizens who have never before walked in a marsh and previously knew little of its organismal composition or environmental values. This attitude can easily be expanded and reinforced with continued education about wetland values, and we must educate people if seemingly "useless" wetlands are to escape development.

An analysis of federal and state funding of wetland programs provides another economic handle on the dollar value of wetlands. Private interest groups such as the Conservation Foundation and federal agencies such as the National Science Foundation, U.S. Army Corps of Engineers, U.S. Environmental Protection Agency, U.S. Department of Interior and U.S. Department of Commerce all spend billions of dollars annually on research aimed at better understanding the complexities of the wetland continuum. Considering the emphasis made by President Carter on better understanding of wetlands, probably no system is funded at as high a dollar per acre value as is the study of the interaction of wetlands and their surrounding environments.

In summary, educational, aesthetic, recreational, anthropological and/or theological functions and values of wetlands are important. It is time we develop an awareness of these values, for they promise long-term rewards for man and the biosphere.

References Cited

Bartram, W. 1791. *Travels through North and South Carolina, Georgia . . .* James and Johnson, Philadelphia, PA. 534 pp.

Carter, J. E. 1977. Executive Order 11990. *U.S. Federal Register,* Vol. 42:26961.

Fabos, J. G. 1971. An analysis of environmental quality ranking systems. *Proceedings Forest Recreation Symposium,* U.S. Dept. Agriculture, Forest Service, Northeast Forest Experiment Station, p. 40-55.

Gosselink, J. G., E. P. Odum and R. M. Pope. 1974. The value of the tidal marsh. Center for Wetland Resources, Louisiana State University, Baton Rouge, LSU-SG-74-03. 30 pp.

Horwitz, E. L. 1978. Our nation's wetlands; an interagency task force report. U.S. Gov't. Printing Office, Washington, D.C. 70 pp.

Ketchum, B. H. 1972. *The Water's Edge: Critical Problems of the Coastal Zone.* MIT Press, Cambridge, MA. 393 pp.

Lanier, M. (Editor). 1944. *The Poems of Sidney Lanier.* U. Georgia Press, Athens, GA. 262 pp.

Larson, J. S. (Editor). 1976. *Models for Assessment of Freshwater Wetlands.* Water Resources Research Center, U. Massachusetts, Publ. No. 32, 91 pp.

Leopold, A. 1949. *A Sand County Almanac.* Oxford U. Press, New York, NY. 269 pp.

Martin, A. C., N. Hotchkiss, F. M. Uhler and W. S. Bourn. 1953. Classification of wetlands of the United States. U. S. Fish and Wildlife Service, Washington, D.C., *Special Scientific Report—Wildlife* No. 20. 14 pp.

Messman, L., R. Reppert and E. Stakhiv. 1977. Wetland values; Interim assessment and evaluation methodology [review draft]. U.S. Army Corps of Engineers, Ft. Belvoir, VA. 263 pp.

National Marine Fisheries Service. 1977. Economic activity associated with marine recreational fishing. U.S. Dept. Commerce, Washington, D.C. 63 pp.

Odum, W. E. and E. Skjei. 1974. The issue of wetlands preservation and management: A second view. *Coastal Zone Management J.* 1:151-163.

Pope, R. M. and J. G. Gosselink. 1973. A tool for use in making land management decisions involving tidal marshland. *Coastal Zone Management J.* 1:65-74.

Reimold, R. J. and R. A. Linthurst. 1975. Use of remote sensing for mapping wetlands. *Transp. Eng. J., Proc. Paper* 11293, ASCE, 1(TE2):189-198.

Smardon, R. C. 1972. Assessing visual-cultural values of inland wetlands in Massachusetts. MLA Thesis, U. Massachusetts, Amherst, MA. 295 pp.

Smardon, R. C. and J. G. Fabos. 1976. Visual-cultural sub-model, pp. 35-51. *In:* J. S. Larson (ed.), *Models for Assessment of Freshwater Wetlands.* U. Massachusetts, Water Resources Center Publ. No. 32.

CHEMICAL CYCLES AND FLUXES

NUTRIENT FLUXES ACROSS THE SEDIMENT-WATER INTERFACE IN THE TURBID ZONE OF A COASTAL PLAIN ESTUARY

Walter R. Boynton

Chesapeake Biological Laboratory
University of Maryland
Solomons, Maryland

W. Michael Kemp

Horn Point Environmental Laboratories
University of Maryland
Cambridge, Maryland

and

Carl G. Osborne

Chesapeake Biological Laboratory
University of Maryland
Solomons, Maryland

Abstract: Oxygen and nutrient fluxes across the sediment-water interface were measured over an annual cycle in the turbid portion of the Patuxent Estuary. Benthic respiration rates ranged from 0.5 to 4.1 g O_2 m^{-2} d^{-1} and were positively correlated with temperature and primary production. Net fluxes of ammonium (NH_4^+) and dissolved inorganic phosphorus (DIP) ranged from -105 to 1584 μg-at N m^{-2} h^{-1} and 1 to 295 μg-at P m^{-2} h^{-1}, respectively. These rates, which were positively correlated with temperature, are among the highest yet reported in the literature. Fluxes of nitrate plus nitrite were small during summer when water column concentrations were low, but high and directed into the sediments during winter when water column concentrations were high. In general it appears that nutrient fluxes across the sediment-water interface represent an important source to the water column in summer when photosynthetic demand is high and water column stocks are low and, conversely, serve as a sink in winter when demand is low and water column stocks high, thereby serving a "buffering" function between supply and demand. A simple budget of sediment-water exchanges and storages of nitrogen indicated that, of the total particulate nitrogen deposited annually onto the sediments, about 34% was returned to the water column as NH_4^+, 41% was stored as particulate nitrogen in the sediments and, by difference, we estimated that the remaining 24% was denitrified. We also observed considerable uptake of nitrate by the sediments during winter months (1.1 g-at m^{-2} y^{-1}), suggesting an additional source of annual denitrification, since this nitrate uptake was not accompanied by ammonium release back to the water column. The ecological implications of these large nutrient fluxes are discussed in terms of sources and sinks of nutrients, as well as couplings with carbon productivity.

Introduction

In recent years we have witnessed a rapid expansion of available information concerning nutrient dynamics of coastal and estuarine eco-systems. While a comprehensive synthesis of nutrient patterns and strategies for estuarine nutrient cycling is still lacking, quantitative data on major pathways of nutrient flux continues to be amassed. As Nixon (1979) has pointed out, our perspective on the importance of various components and mechanisms in the nutrient cycling scheme has changed remarkably over the last several decades. One of the most striking of these changes has been the relatively widespread documentation of benthic remineraliza-tion as a principle source of recycled nutrients available for photosynthesis in several types of aquatic ecosystems (Davies 1975; Hale 1975; Rowe et al. 1977; Fisher and Carlson 1979). This runs counter to the previously held view that water-column processes dominate estuarine remineralization (e.g. McCarthy et al. 1974; Carpenter and McCarthy 1978). Although geochemists have understood for some time the chemistry of estuarine sedimentology (Berner 1974) as part of early diagenesis, ecologists are just now beginning to understand its importance to the dynamics of estuarine ecosystems. Nixon and his co-workers (Nixon 1979; Nixon et al. 1980) have even suggested that benthic remineralization may be the primary fac-tor controlling relative availability of nitrogen and phosphorus for photosyn-thesis.

Despite the large number of nutrient-related measurements made in estuarine systems, surprisingly few efforts have been made to develop mass balances of nutrient fluxes for these areas. Nutrient budgets have been developed to various levels of detail and completeness for a few estuaries including the Hudson River Estuary (Simpson et al. 1975), San Francisco Bay (Peterson et al. 1975), Long Island Sound (Bowman 1977), Narragan-sett Bay (Nixon 1979), and portions of Chesapeake Bay (Taft et al. 1978; Kemp and Boynton 1979). Placing nutrient cycling pathways in a budget is essential for judging the relative importance of these fluxes in the context of overall estuarine nutrient dynamics (Kemp and Boynton 1979; Nixon 1979).

One of the most extensively studied estuaries on the east coast of the United States is that of the Patuxent River, a tributary of Chesapeake Bay. Measurements of water column nutrient concentrations, which were initiated prior to the 1940's (e.g. Newcomb and Brust 1940; Nash 1947), have continued into the 1970's (e.g. Cory and Nauman 1967; Flemer et al. 1970). A few estimates of nutrient fluxes have been made for watershed inputs from point (FWPCA 1968) and diffuse (Correll 1977) sources and for exchanges with brackish marshes (Heinle et al. 1977), while net biochemical fluxes have been estimated as the residual term in a one-dimensional model of longitudinal mass-transport along the estuary (Ulanowicz and Flemer 1977). For the past several years, we have been conducting nutrient studies in the Patuxent Estuary both at the scale of the entire watershed for purposes of constructing nutrient budgets (Kemp and Boynton 1979) and at distinctive locations in the estuary to investigate nutrient cycling between the water column and the benthos. The purpose of this communication is to report the results of benthic nutrient flux measurements made in the turbid portion of the Patuxent Estuary and to relate these measurements to certain aspects of our preliminary nutrient budget. Since most, if not all, coastal plain estuaries are characterized by zones of maximum turbidity (Postma 1967; Schubel 1968) we would expect that our findings in this study might have broad applicability for that portion of these estuaries.

Study Area

The Patuxent River has a drainage basin of 2230 km², ranks sixth in volume (7.6×10^8 m³) among the primary tributaries of Chesapeake Bay, has a total length of 175 km, and an average annual stream flow of 45 × 10^7 m³ y^{-1} (Mihursky and Boynton 1979). The estuary can be characterized as a typical partially-stratified, coastal plain system. Our study area was in the middle portion of the estuarine system, approximately 33 km upstream of the mouth and 81 km downstream of the head of tide. Throughout the year our stations were located close to the downstream edge of the turbidity maximum and just upstream of the estuarine areas characterized by two-layer circulation. The turbidity maximum zone expands in areal extent, primarily in an upriver direction, during periods of high flow (spring) and encompasses all of the upper 81 km of the estuary. During periods of lower flow (fall), this zone contracts and occasionally exhibits bimodal turbidity maxima (Keefe et al. 1976; Roberts and Pierce 1976). Sediments are largely silts and clays (90-95%) and organic matter ranges from 1.5 to 2.6% by weight. On an annual basis, water temperature and salinity range between 0 and 30C and 6 and 13°/oo, respectively. The euphotic zone (1% light) generally extends only 1.5 to 2.0 m beneath the water surface depending on season and, because the mean depth of the study area is about 3 m, the sediment surface receives very little light energy. Sedimentation rates range from 5 to 37 mm y^{-1} (Fox 1974; Roberts and Pierce 1976), with substantial resuspension of bottom sediments during periods of high wind and wave action.

During the last 25 years the river has been receiving increasing amounts of sewage treatment plant effluent as well as materials associated with non-point sources (FWPCA 1968; Correll 1977). Nitrogen and phosphorus concentrations in the water have increased significantly, dissolved oxygen concentrations have decreased in deeper waters during the summer, and turbidity has significantly increased for most times of the year (Mihursky and Boynton 1979). The river also receives the cooling water discharge from an electric generating station (Chalk Point) located at the upper end of our study area.

We selected two sampling sites for this study, one in the vicinity of the power plant discharge (CP) approximately 38.6 km upstream of the estuary's mouth and the other 5.6 km downstream of the discharge, beyond direct power plant influence (R). At each site, one station was established on each side of the river (CP1 and CP2; R1 and R2) in about 3 m of water, a depth characteristic of this region of the estuary. Benthic respiration measurements were taken at all four stations while nutrient flux measurements were made at stations R2 (0.5 km upstream of Rt. 231 bridge) and CP2 (near Potts Point), both of which are on the eastern side of the estuary.

Methods and Materials

Benthic community respiration was measured *in situ* at the four stations in May, July, August, October and December 1978, and March 1979. Four replicate measurements of benthic respiration were made at each station using four opaque plexiglass chambers. Each chamber was cylindrical in shape with a height of 18 cm and a circular base of 62 cm diameter. The volume in the chambers was 38 liters and surface area covered was 0.3 m². A 10 cm wide flange was attached to the edge of the chamber, 6 cm above the bottom opening, to provide a base on which the chamber could rest on the bottom. A submersible pump and dissolved oxygen probe were attached to the inside of each chamber and connected to shipboard by cables. These pumps discharged recirculated water under each chamber through a manifold and were capable of maintaining mean current speeds within the chamber of 1 to 20 cm s^{-1}, typical of those encountered in bottom waters in the study area. All measurements reported in this paper were performed under conditions where a mean speed of water movement was maintained at about 10 cm s^{-1}. Prior to a measurement, the chambers were filled with water to remove all air and slowly lowered to the bottom. Placement was accomplished with the assistance of SCUBA divers to ensure that the chambers were securely placed on the bottom and to note penetration into the bottom. After a wait of about 15 min for equilibration, the pumps were started. Temperature and dissolved oxygen readings were taken at intervals ranging from 10-20 min over a period lasting from 1-3 h. The spacing and duration of measurements depended on the observed rates of dissolved oxygen change. In no case did measurements extend for such a period that dissolved oxygen concentration decreased by more than 2 mg l^{-1} or to less than 2.5 mg l^{-1}. Chemical oxygen demand (COD) was estimated as the residual oxygen consumption under a chamber following formalin treatment. Duplicate dark bottles were also incubated contemporaneously with the chamber experiments to provide a blank (plankton oxygen demand).

Nutrient fluxes across the sediment-water interface were also measured at stations R2 and CP2, with a single experiment performed at each station during each cruise. In subsequent experimental work during the summer of 1979, we have made three replicate measurements for which variability was relatively low, with coefficients-of-variation ranging from 10 to 30% (Boynton et al. unpublished data). Concentrations of the following nutrient species were measured for water samples from the chambers, taken at the beginning, mid-point and end of incubation periods (1-3 h): nitrate (NO_3^-); nitrite (NO_2^-); ammonium (NH_4^+); dissolved inorganic phosphorus (DIP); dissolved organic phosphorus (DOP); total phosphorus (TP); and dissolved organic nitrogen (DON). Measurements of particulate carbon, particulate nitrogen, chlorophyll *a* and seston were also made for the water column at each station. The analytical methods for nutrient analyses are described in detail elsewhere (Osborne et al. 1979) but in general they were as follows: nitrate, nitrite, ammonium and dissolved inorganic phosphate were analyzed using the automated method of EPA (1979); total nitrogen was estimated according to D'Elia et al. (1977); total dissolved phosphate (filtered) and total phosphate (unfiltered) analyses used the digestion and neutralization procedure of D'Elia et al. (1977) followed by standard phosphate analysis (EPA 1979); particulate organic carbon and nitrogen samples were analyzed using a Model 240B Perkin Elmer Elemental Analyser. Methods given in Strickland and Parsons (1968) were used for chlorophyll *a* and seston determinations. Nutrient flux across the sediment-water interface was estimated by following the concentration changes within a chamber throughout a measurement using the mean rate as calculated from first to second and from second to third sampling times. Nutrient concentration changes generally appeared linear over the sampling period. Nutrient changes in triplicate dark BOD bottles incubated at ambient temperature were used as an estimate of water column nutrient changes.

Estimates of water column photosynthesis and respiration were made at each station using the oxygen light-dark bottle method. Depending on the turbidity conditions encountered on the day of the measurement, triplicate light bottles were suspended for a daylight period at depths approximating 90%, 60% and 10% of surface insolation. A single group of three dark bottles was suspended at the 10% light level.

Results

Over a temperature range from 3C in March to 29C in August, total oxygen utilization by the sediment surface ranged from 4.1 g O_2 m^{-2} d^{-1} in late June to 0.50 g O_2 m^{-2} d^{-1} in March (Fig. 1). There were no consistent differences among the four stations at which oxygen uptake was measured. Linear regression analysis indicated a good correlation between total oxygen uptake and water temperature ($r^2 = 0.68$) and the positive intercept value (0.56 g O_2 m^{-2} d^{-1}) was representative of the lowest uptake rates observed during the winter (Table 1). A reasonably good correlation was also observed between benthic respiration and primary production ($r^2 = 0.59$). From May through October, COD was estimated at stations R2 and CP2 and averaged 32% of total oxygen uptake.

Net release of NH_4^+ from the sediment surface dominated nitrogen fluxes during the summer period. Ammonium fluxes ranged from 1584

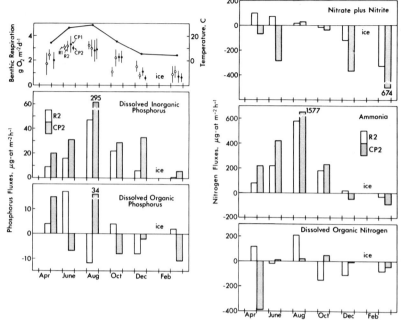

Figure 1. Seasonal estimates of benthic respiration rate and flux rates of dissolved phosphorus and nitrogen in the Patuxent Estuary, 1978-1979. Daily means and standard errors (vertical bars) of benthic respiration rate are shown (left to right) for stations R1 and R2 (open circles) and CP1 and CP2 (closed circles).

μg-at m^{-2} h^{-1} at station CP2 in August to -105 μg-at m^{-2} h^{-1} at the same station in March (Fig. 1). The annual average (time weighted) NH$_4$$^+$ flux rate was 403 μg-at m^{-2} h^{-1} and 186 μg-at m^{-2} h^{-1} for stations CP2 and R2, respectively. The high average flux rate for station CP2 resulted, in part, from an extremely high flux rate recorded in August. Measurements repeated in triplicate at CP2 a year later yielded NH$_4$$^+$ flux rates of about 750 μg-at m^{-2} h^{-1}. If the latter values are used, the annual average flux for this station is 259 μg-at m^{-2} h^{-1}. Ammonium fluxes at both stations were significantly correlated with oxygen uptake rates and temperature (Table 1). Both positive and negative fluxes of NO$_3$$^-$ plus NO$_2$$^-$ were observed during the year and, in the winter, these fluxes into the sediments dominated all other nitrogen fluxes (Fig. 1). The average annual flux (time weighted) was -141 μg-at m^{-2} h^{-1}. Nitrate fluxes were always higher than nitrite fluxes and, in all but one instance, both fluxes were in the same direction. These fluxes were significantly correlated with oxygen uptake rate and the concentration of NO$_3$$^-$ in the overlying water (Table 1). In general, large NO$_3$$^-$ fluxes into the sediments occurred when water column concentrations of NO_3^- were high, whereas small fluxes of NO_3^- occurred in both directions when water column concentrations were low. Finally, during the winter, large $NO_3^- + NO_2^-$ fluxes into the sediments were not accompanied by large fluxes of NH_4^+ or dissolved organic nitrogen either going into or coming from the sediments. Fluxes of dissolved organic nitrogen (DON) ranged from -428 μg-at m^{-2} h^{-1} at station CP2 in May to 216 μg-at m^{-2} h^{-1} at station R2 in August (Fig. 1). Fluxes were variable both in magnitude and direction throughout the year and were often quite different between stations on the same date. Fluxes of DON did not appear to be related to temperature, oxygen uptake rate, or the concentration of DON in the water column. Except for one occasion, the flux of DON was not the major component of dissolved nitrogen fluxes between the sediment surface and the water column.

Dissolved inorganic phosphorus (DIP) fluxes ranged from 1 to 295 μg-at m^{-2} h^{-1} and were generally highest during the warm seasons (Fig. 1). Fluxes at station CP2 were significantly higher ($p < 0.05$) than those at station R2. The average annual DIP flux was 16.7 and 70.9 μg-at m^{-2} h^{-1} at stations R2 and CP2, respectively. Fluxes at R2 were significantly correlated with both temperature and respiration (Table 1) but significant correlations were not found for DIP fluxes at station CP2. Dissolved organic phosphorus fluxes ranged from 34 μg-at m^{-2} h^{-1} at CP2 in August to -11 μg-at m^{-2} h^{-1} at the same station in March. Time-weighted average annual DOP flux at stations R2 and CP2 were 1.2 and 3.5 μg-at m^{-2} h^{-1}, respectively. Fluxes were variable both in magnitude and direction throughout the year and consequently no significant correlations between DOP flux and temperature, respiration or water column DOP concentration were found. In general, DOP flux was a small fraction of the total dissolved phosphorus flux, particularly when annual average fluxes were considered.

Discussion

Benthic Respiration

In overview, benthic respiration rates in the Patuxent were comparable to the highest previously reported measurements from estuarine areas (Smith 1973; Hall et al. 1979) with summer values ranging from 2.1-4.1 g O_2 m^{-2} d^{-1}. Mean daily respiration (time-weighted, annual average) amounted to about 2.0 g O_2 m^{-2} d^{-1} or about 300 g C m^{-2} y^{-1}. Our respiration measurements were well correlated with both temperature and primary production (Table 1). The relationship between temperature and benthic respiration has been consistently observed in many aquatic systems (Hargrave 1969), presumably being attributable to temperature effects on chemical kinetics. More recently, Hargrave (1973) reported that the estimates of benthic respiration in various ecosystems could be related to an index of the amount of plankton production reaching the bottom, a conclusion which suggests that organic matter supply, rather than

Table 1. Summary of linear regression analyses relating temperature, benthic respiration, primary production, and nutrient concentrations to benthic nutrient fluxes. Entries in the table are b = y-intercept; m = slope; r^2 = coefficient of determination; n = sample size.

Independent Variables	Regression Parameters	Benthic Respiration ($g\ O_2\ m^{-2}\ d^{-1}$)	Fluxes, $\mu g\text{-}at\ m^{-2}\ h^{-1}$				
			NO_x	NH_4^+	DON	DIP^a	DOP
Temperature	b	0.564^b	−343.29	$−183.48^b$	−111.32	$−4.10^b$	−6.08
(C)	m	0.092	12.84	25.10	4.74	1.14	0.53
	r^2	0.67	0.29	0.83	0.08	0.73	0.14
	n	100	12	12	12	6	12
Benthic Respiration	b	—	$−562.46^b$	$−294.94^b$	−107.71	$−10.35^b$	−11.49
($g\ O_2\ m^{-2}\ d^{-1}$)	m	—	203.19	244.37	34.25	13.13	6.66
	r^2	—	0.56	0.61	0.03	0.50	0.17
	n	—	12	12	12	6	12
Primary Production	b	1.31^b	—	—	—	—	—
($g\ O_2\ m^{-2}\ d^{-1}$)	m	0.17	—	—	—	—	—
	r^2	0.59	—	—	—	—	—
	n	23	—	—	—	—	—

Dependent Variables

[a] Data from station R2.
[b] Indicates significant coefficients at $p < 0.05$.

temperature, regulates annual benthic respiration. Nixon (1979) extended this notion and found what appears to be a very strong correlation between benthic respiration and the amount of organic matter produced and imported to the area of interest. Our data from the Patuxent fit this relationship reasonably well. Deciphering of the individual effects of primary production, organic matter flux to the bottom, temperature and other factors is confounded by the fact that many of these factors interact with each other. As mechanistic information on causal relationships emerges, non-linear ecological simulation models may be useful as analytical tools in gaining a clearer understanding of the interactions of these factors.

Nutrient Fluxes

As with benthic respiration values, relative to other reported measurements of nutrient fluxes from estuarine and coastal sediments, our values from the Patuxent were also among the highest (Fig. 2). For instance, the annual mean flux (average of all stations) of NH_4^+ and DIP was 295 and 43 μg-at m^{-2} h^{-1}, respectively. Fluxes of NH_4^+ and DIP in Narrangansett Bay were estimated to be 100 and 20 μg-at m^{-2} h^{-1} (Nixon et al. 1976) whereas in Buzzards Bay (Rowe et al. 1975) these fluxes were 68 and -3 μg-at m^{-2} h^{-1}, respectively. Rowe et al. (1977) reported average NH_4^+ and DIP fluxes of 250 and 29 μg-at m^{-2} h^{-1}, respectively, for measurements taken in the Sahara upwelling zone. Ammonium and DIP fluxes reported from North Carolina estuaries (Fisher and Carlson 1979), the California coast (Hartwig 1974), and a Scottish sea loch (Davies 1975) are all considerably lower than those observed in the Patuxent estuary. The generally large ranges evident in Fig. 2 also indicate that there are large differences in flux rates probably related to some function of season.

The high flux rates we observed may be related to seasonal organic matter supply to the benthos and physical dynamics of the sediment-water interface. Inputs of organic matter are primarily derived from *in situ* phytoplankton production during summer and fall and from upland runoff and adjacent marshlands during winter and early spring. Numerous investigations have shown this region of the Patuxent to be characterized by high rates of primary production and deposition (Mihursky and Boynton 1979). In addition, carbon-nitrogen-phosphorus ratios of suspended particulate material (Flemer et al. 1970) are close to those observed in healthy phytoplankton, indicating that ample nitrogen and phosphorus are reaching the bottom to support high regeneration rates. The second feature which probably contributes to high flux rates concerns the dynamic behavior of the sediment surface. Preliminary results from sediment trap studies suggest very high rates of sedimentation and resuspension (Boynton et al. unpublished data). Moreover, it appears that sediments are extensively mixed via the burrowing activities of benthic infauna (Holland et al. 1979). The combination of these processes would tend to enhance the exchange of dissolved materials between the sediment and water column. Hale (1975)

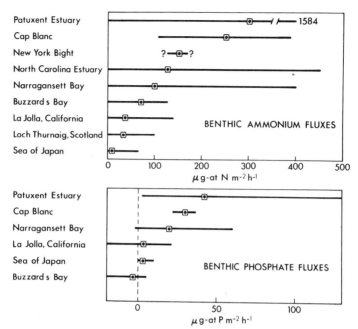

Figure 2. Comparative benthic fluxes (annual mean and range) of ammonium and dissolved inorganic phosphate from selected estuarine and coastal ecosystems. Data sources by area were as follows: Patuxent (this study); Cap Blanc (Rowe et al. 1977); New York Bight (Rowe et al. 1975); North Carolina Estuary (Fisher and Carlson 1979); Narrangansett Bay (Nixon et al. 1976); Buzzards Bay (Rowe et al. 1975); La Jolla, California (Hartwig 1974); Loch Thurnaig (Davies 1975); Sea of Japan (Propp et al. 1980).

and Nixon et al. (1976) reported strong correlations between NH_4^+ and DIP fluxes and benthic respiration in Narrangansett Bay. They concluded that such a relationship was expected because of the coupling between aerobic remineralization mediated mainly by bacteria and the release of early regeneration products such as NH_4^+. We observed a similar correlation of NH_4^+ with temperature and respiration at both stations and of phosphorus fluxes at station R2 (Table 1).

Nitrate fluxes were significantly correlated with benthic respiration and particularly with the concentration of nitrate in the water column (Table 1). The positive correlation with respiration resulted because the largest fluxes of NO_3^- were directed into the sediments during winter when benthic respiration was low. Nitrate fluxes were small during summer when benthic respiration was high, suggesting that NO_3^- fluxes were coupled to processes other than aerobic metabolism. Nitrate fluxes were erratic and small during summer when NO_3^- concentrations were low, but large and directed into the sediments when NO_3^- concentrations in the water were large. The strong correlation of NO_3^- concentration with the negative flux of NO_3^- suggests a nitrate consumptive process in the sediments which, in some seasons, may be limited by diffusion across the sediment-water interface, and apparently is not closely linked to aerobic respiration. These relationships suggest that NO_3^- fluxes into the sediments represent the first stages of either denitrification ($NO_3^- \rightarrow N_2O$ and N_2) or nitrate respiration $NO_3^- \rightarrow NH_4^+$). Van Kessel (1977) has shown that denitrification rates are dependent on NO_3^- concentration in the overlying water. This, coupled with the fact that we did not observe NH_4^+ fluxes out of the sediments when NO_3^- was entering sediments (Fig. 1), strongly suggests that this NO_3^- uptake from water to sediments was associated with denitrification.

Annual fluxes of DON and DOP were less important features of the overall exchange of nitrogen and phosphorus between the sediments and water column. However, on occasion these fluxes were reasonably large as at CP2 in May when DON dominated nitrogen fluxes. It appears that both components require additional study. At present DON and DOP are treated as if they were defined forms of nitrogen and phosphorus; in fact, they represent unknown and probably complex mixtures of dissolved organic compounds. As defined, we presently find no seasonal or consistent patterns in these fluxes. Interpretable patterns might emerge if more effort was expended in better defining the specific components which comprise DON and DOP.

Coupling of Benthic Nutrient Fluxes with Plankton Productivity

We can make a preliminary assessment as to the relative importance of benthic releases of nitrogen and phosphorus by comparing these fluxes to the nutrient demands of planktonic net production in the water column. In the Patuxent we have found that benthic fluxes of NH_4^+ and DIP can satisfy from 0-190% and 52-330% of the estimated daily demand for these nutrients, respectively, where atomic ratios of O:N:P observed in phytoplankton were used to estimate nutrient uptake (Redfield 1934). These percentages are similar to those reported by Hale (1975) for Narragansett Bay and Davies (1975) for a Scottish sea loch. There are, however, strong seasonal changes in both benthic nutrient fluxes, photosynthetic demand and the concentration of nutrients available in the water column. In August, for instance, primary production (8.3 g O_2 m^{-2} d^{-1}), and presumably nutrient demand, were at seasonal maxima. At that time, available nitrogen in the water column was low (4 μg-at N l^{-1}) and was sufficient to satisfy about 30% of the estimated daily demand. Ammonium fluxes from the sediments were sufficient at that time to satisfy 65% of the estimated daily phytoplankton demand and to replace the water column supply twice in a single diel period.

A similar pattern was evident for inorganic phosphorus. However, in winter, phytoplankton nutrient demand was low and water column supplies of NO_3^-, which appear to be derived primarily from terrestrial sources, were high (40-60 μg-at N l^{-1}) and sufficient to maintain observed levels of primary production for about two weeks. There was a net uptake of nitrogen by sediments during winter at rates sufficient to turn over the water column stock of NO_3^- in about one week. Thus, during warm periods, net releases of NH_4^+ from the sediments constituted a substantial nutrient source. However, winter net nitrogen fluxes into the sediments occurred at rates which had a considerably larger effect on water column nutrient stocks than did phytoplankton demand and constituted a substantial seasonal nitrogen sink. Thus, it appears that the flux of nitrogen across the sediment-water interface represents an important source to the water column in summer when demand is high and water column stocks low and conversely it constitutes a sink in winter when demand is low and water column stocks are high, thereby serving a "buffering" function between supply and demand.

Benthic Nutrient Flux Ratios

Over the annual cycle, approximately 2.5 g-at m^{-2} of NH_4^+ was released by the benthic community. During the same period some 0.4 g-at m^{-2} of phosphorus was released and about 46 g-at m^{-2} of oxygen was consumed. In terms of the amount of oxygen consumed, the NH_4^+ released was low, yielding an O/N atomic ratio of 18.4 which is substantially different from the O/N ratio of 13:1 associated with aerobic decomposition of organic matter (Redfield 1934). The O/N ratio departs even further if annual dissolved inorganic nitrogen ($NO_3^- + NO_2^- + NH_4^+$) flux (1.2 g-at m^{-2} y^{-1}) is considered rather than just NH_4^+ flux (O/N = 38.3). On an annual basis, N/P ratios of benthic fluxes were also considerably lower (3.0) than the expected 16:1, indicating that nitrogen flux back to the water column was only about 19% of that expected considering phosphorus release. Benthic flux ratios in summer were very similar to N:P concentration ratios observed in the water column. From studies conducted by Flemer et al. (1970), we calculated that particulate material in the water column was rich in phosphorus relative to carbon and nitrogen. The annual average particulate N/P ratio (11:1) was in the same general range of values observed for phytoplankton cell composition (Parsons et al. 1961) but was considerably higher than dissolved N/P ratios associated with benthic fluxes.

Since input of organic matter to the sediment surface is not strikingly deficient in nitrogen, some other process must have been operative in producing the low N:P flux ratios we observed. Nixon et al. (1976) reported preliminary results from Narragansett Bay which indicated that nitrogen flux from sediments to water was a mixture of both NH_4^+ and DON. Thus, they suggested that partitioning of the flux into inorganic and organic fractions may have been responsible for low N/P benthic flux ratios and, to a lesser extent, the low N/P ratios observed in the water column. In our study, DON flux was generally small and, on an annual basis, was directed into the sediments. Hartwig (1974) also reported that DON flux was a small portion of total nitrogen flux in a sandy system in California.

Alternative explanations for low N/P flux ratios are that (1) nitrogen is being stored in the sediments as particulate nitrogen or as one or more dissolved forms and/or (2) that denitrification is taking place during all seasons and is particularly noticeable during winter. Inspection of a simple nitrogen budget (Fig. 3) which considers dissolved and particulate storages in sediment, as well as fluxes to and from the sediments, has led us to conclude tentatively that low N/P benthic flux ratios resulted from both incomplete remineralization of particulate nitrogen and denitrification. Annual sedimentation rates (2 cm y^{-1}) coupled with the concentration of particulate nitrogen in the water column yielded an estimate of particulate nitrogen input to the sediments of about 7 g-at N m^2 y^{-1}. Vertical profiles of particulate nitrogen concentration in the sediments (Flemer et al. 1970) were quite constant, suggesting that after some initial remineralization a considerable amount of nitrogen remained (2.9 g-at N m^{-2} to a depth of 2 cm). Storage of dissolved nitrogen in interstitial water was small, amounting to only 0.014 g-at N m^{-2} in a 2 cm sediment column. Flux of NH_4^+ from the sediments accounted for virtually all dissolved nitrogen exchange from the sediments to the water column and amounted to 2.4 g-at N m^{-2} y^{-1}.

Thus, of the annual input of particulate nitrogen to the sediments, it appears that about 41% remains as particulate nitrogen, 34% is returned to the water column as NH_4^+, and 24% is not accounted for and may represent that portion of the annual particulate input which undergoes

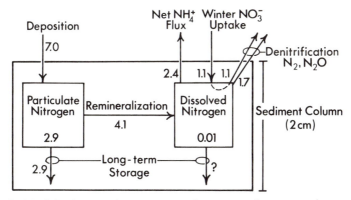

Figure 3. Model of annual nitrogen exchanges and storages between the water column and sediments in the turbid portion of the Patuxent Estuary. Units are g-at N m^{-2} y^{-1} for fluxes and g-at N m^{-2} for storages.

denitrification. This is consistent with observations summarized by Nixon (1979). If one were also to consider NO_3^- flux into the sediments during winter (Dec.-March) which amounted to 1.1 g-at m^{-2} and to assume further, as we have earlier argued, that most of this nitrate is denitrified, then about 35% of total nitrogen input to the sediments is lost as gaseous nitrogen. Moreover, since NO_3^- is the form of nitrogen which first enters the denitrification process (Brezonik 1977), it appears that seasonally there must be different mechanisms which supply the necessary NO_3^-. In winter, NO_3^- probably comes from the water column where the supply is abundant. In summer, however, NO_3^- concentration in the water column is very low and during this period, nitrification in the aerobic zone of sediments probably supplies the needed NO_3^- (Patrick and Reddy 1976; Billen 1978).

In summary, rates of benthic respiration and nutrient flux across the sediment-water interface measured over an annual cycle in the Patuxent estuary were found to be among the highest values yet reported in the literature. It is unclear as to whether these high rates are generally characteristic of the turbid portion of coastal plain estuaries, or whether they are particular to the Patuxent which receives large nutrient inputs from point and non-point sources. Comparable measurements for other actively depositional regions of estuaries are not presently available in the literature. The magnitude of sediment nutrient fluxes was found to be sufficient to satisfy calculated photosynthetic demand during summer periods of peak productivity. Preliminary nitrogen budget calculations suggested that denitrification may be a significant process in both summer and winter with the required NO_3^- being supplied via nitrification during summer while coming directly from the water column in winter. Denitrification and sediment storage of particulate nitrogen may contribute to the low dissolved N/P ratios associated with benthic nutrient fluxes, and, moreover, may strongly influence water column concentrations of dissolved N and P during summer. This observation has implication in terms of factors limiting primary production as well as management of nutrient waste discharges to this and other estuaries. In order to resolve some of the questions raised by this study, further research is needed to obtain direct measurements of denitrification, nitrification and nitrogen fixation rates, and to describe patterns which these and other nutrient-cycling processes exhibit along the longitudinal axis of such temperate, coastal plain estuaries.

Acknowledgments

We are indebted to a number of people whose assistance greatly enhanced the conduct of this project. Kenneth Kaumeyer and Mark Jenkins participated in all phases of the study and Ms. Carolyn Keefe conducted all nutrient analyses with remarkable efficiency. Ms. Kathryn Wood assisted in review of the manuscript. Capt. William Keefe and Mate John Crane operated the R. V. Orion. This work was supported in part by Grants P2-72-02(78) Mod. 4 & 5, and P2-72-02(79) Mod. 6, from the

Maryland Department of Natural Resources, Power Plant Siting Program, Dr. J. A. Mihursky, Coordinating Principal Investigator. Contribution No. 986, Center for Environmental and Estuarine Studies of the University of Maryland.

References Cited

Berner, R. A. 1974. *Principles of Chemical Sedimentology.* McGraw-Hill, New York. 175 pp.

Billen, G. 1978. A budget of nitrogen recycling in North Sea sediments off the Belgian coast. *Est. Coastal Mar. Sci.* 7:127-146.

Bowman, M. J. 1977. Nutrient distributions and transport in Long Island Sound. *Est. Coastal Mar. Sci.* 5:531-548.

Brezonik, P. L. 1977. Denitrification in natural waters. *Prog. Wat. Tech.* 8:373-392.

Carpenter, E. J. and J. J. McCarthy. 1978. Benthic nutrient regeneration and high rate of primary production in continental shelf waters. *Nature* 274:188-189.

Correll, D. (ed.). 1977. *Watershed Research in North America.* Chesapeake Bay Center for Environmental Studies. Smithsonian Institution, Washington, D.C. 857 pp.

Cory, R. L. and J. W. Nauman. 1967. Temperature and water quality conditions for the period July 1963 to December 1965. Patuxent River Estuary, Maryland. U.S. Dept. Interior, Geol. Survey, Washington, D.C. Open file report. 176 pp.

Davies, J. M. 1975. Energy flow through the benthos in a Scottish sea loch. *Mar. Biol.* 31:353-362.

D'Elia, C. F., P. A. Steudler and N. Corwin. 1977. Determination of total nitrogen in aqueous samples using persulfate digestion. *Limnol. Oceanogr.* 22:760-764.

Environmental Protection Agency (EPA). 1979. Methods for Chemical Analysis of Water and Wastes. USEPA-600/4-79-020. Environmental Monitoring and Support Laboratory, Cincinnati, OH.

Federal Water Pollution Control Administration (FWPCA). 1968. A comprehensive analysis and program for water quality management in the Patuxent River Basin. Chesapeake Field Station. U.S. Dept. of Interior, Annapolis, MD. 312 pp.

Fisher, T. and P. Carlson. 1979. The importance of sediments in the nitrogen cycle of estuarine systems. Amer. Soc. Limnol. Oceanogr., Abstracts, 42 Annual Meeting, Stony Brook, NY.

Flemer, D. A., D. H. Hamilton, C. W. Keefe and J. A. Mihursky. 1970. The effects of thermal loading and water quality on estuarine primary production. Chesapeake Biological Laboratory, Solomons, MD. NRI Ref. No. 71-6. 217 pp.

Fox, H. L. 1974. The urbanizing river: A case study in the Maryland Piedmont, pp. 245-271. In: D. R. Coats (ed.), *Geomorphology and Engineering.* Dowden, Hutchison and Ross. Stroudsburg, PA.

Hale, S. S. 1975. The role of benthic communities in the nitrogen and phosphorus cycles of an estuary. pp. 291-308. In: F. G. Howell, J. B. Gentry and M. H. Smith (eds.), *Mineral Cycling in Southeastern Ecosystems.* Published by Tech. Info. Center, U.S. Energy Research and Development Administration, Washington, D.C.

Hall, C. A. S., N. Tempel and B. J. Peterson. 1979. A benthic chamber for intensely metabolic lotic systems. *Estuaries* 2:178-183.

Hargrave, B. T. 1969. Similarity of oxygen uptake by benthic communities. *Limnol. Oceanogr.* 14:801-805.

Hargrave, B. T. 1973. Coupling carbon flow through some pelagic and benthic communities. *J. Fish. Res. Board Canada.* 30:1317-1326.

Hartwig, E. O. 1974. Physical, chemical, and biological aspects of nutrient exchange between the marine benthos and the overlying water. Ph.D. Thesis, U. California, San Diego, CA. 189 pp.

Heinle, D. R., D. A. Flemer and J. F. Ustach. 1977. Contributions of tidal marshlands to mid-Atlantic estuarine food chains, pp. 309-320. In: M. Wiley (ed.), Estuarine Processes, Vol. II. Academic Press, New York.

Holland, A. F., M. H. Hiegel, D. G. Cargo, R. V. Lacouture, N. K. Mountford and J. A. Mihursky. 1979. Interim Report on Benthic Community Studies at Chalk Point: January 1978-July 1978 data. Chesapeake Biological Laboratory, Solomons, MD. U. Maryland, Center for Estuarine and Environmental Studies Ref. No. 78-220-CBL. 125 pp.

Keefe, C. W., D. A. Flemer and D. H. Hamilton. 1976. Seston distribution in the Patuxent River Estuary. Chesapeake Sci. 17:56-59.

Kemp, W. M. and W. R. Boynton. 1979. Nutrient budgets in a coastal plain estuary: Sources, sinks and internal dynamics. Amer. Soc. Limnol. Oceanogr., Abstracts, 42 Annual Meeting, Stony Brook, NY.

McCarthy, J. J., W. R. Taylor and M. E. Loftus. 1974. Significance of nanoplankton in the Chesapeake Bay estuary and problems associated with the measurement of nanoplankton productivity. Mar. Biol. 24:7-16.

Mihursky, J. A. and W. R. Boynton. 1979. Review of the Patuxent Estuary data base. Chesapeake Biological Laboratory, Solomons, MD. U. Maryland, Center for Environmental and Estuarine Studies Ref. No. 78-157-CBL. 327 pp.

Nash, C. B. 1947. Environmental characteristics of a river estuary. J. Mar. Res. 6:147-174.

Newcombe, C. L. and H. F. Brust. 1940. Variations in the phosphorus content of estuarine waters of the Chesapeake Bay near Solomons Island, MD. J. Mar. Res. 3:76-88.

Nixon, S. W. 1979. Remineralization and nutrient cycling in coastal marine ecosystems. In: B. Neilson and L. E. Cronin (eds.), Nutrient Enrichment in Estuaries. Humana Press, Clifton, NJ (in press).

Nixon, S. W., C. A. Oviatt and S. S. Hale. 1976. Nitrogen regeneration and the metabolism of coastal marine bottom communities, pp. 269-283. In: J. M. Anderson and A. Macfayden (eds.), The Role of Terrestrial and Aquatic Organisms in Decomposition Processes. Blackwell Sci. Publ., Oxford.

Nixon, S. W., J. R. Kelly, B. N. Furnas and C. A. Oviatt. 1980. Phosphorus regeneration and the metabolism of coastal marine communities. In: K. R. Tenore and B. C. Coull (eds.), Marine Benthic Dynamics. U. South Carolina Press, Columbia, SC (in press).

Osborne, C. G., K. R. Kaumeyer, C. W. Keefe, W. R. Boynton and W. M. Kemp. 1979. Community metabolism and nutrient dynamics of the Patuxent Estuary interacting with the Chalk Point Power Plant. Chesapeake Biological Laboratory, Solomons, MD. U. Maryland, Center for Environmental and Estuarine Studies Ref. No. 79-8-CBL.

Parsons, T. R., K. Stephens and J. D. H. Strickland. 1961. On the chemical composition of eleven species of marine phytoplankton. J. Fish. Res. Bd. Canada 18:1001-1016.

Patrick, W. H., Jr. and K. R. Reddy. 1976. Nitrification-denitrification reactions in flooded soils and water bottoms: Dependence on oxygen supply and ammonium diffusion. Environmental Quality 5:469-472.

Peterson, D. H., T. J. Conomos, W. W. Broenkon and E. P. Scrivani. 1975. Processes controlling the dissolved silica distribution in San Francisco Bay, pp. 153-157. In: L. E. Cronin (ed.), Estuarine Research, Vol. I. Academic Press, New York.

Postma, H. 1967. Sediment transport and sedimentation in the estuarine environment, pp. 158-179. In: G. H. Lauff (ed.), Estuaries. Amer. Assoc. Adv. Sci., Publ. No. 83, Washington, D.C.

Propp, M. V., V. G. Tarasoff, I. I. Gherbadgi and N. V. Lootzik. 1980. Benthic pelagic oxygen and nutrient exchange in a coastal region of the Sea of Japan. *In:* K. R. Tenore and B. C. Coull (eds.), *Marine Benthic Dynamics.* U. South Carolina Press, Columbia, SC (in press).

Redfield, A. C. 1934. On the proportions of organic derivatives in seawater and their relation to the composition of the plankton, pp. 176-192. *In:* James Johnstone Memorial Volume. Liverpool U. Press, Liverpool.

Roberts, W. P. and J. W. Pierce. 1976. Deposition in upper Patuxent Estuary, Maryland, 1968-1969. *Est. Coastal Mar. Sci.* 4:267-280.

Rowe, G. T., C. H. Clifford and K. L. Smith, Jr. 1977. Nutrient regeneration in sediments off Cap Blanc, Spanish Sahara. *Deep-Sea Res.* 24:57-63.

Rowe, G. T., C. H. Clifford, K. L. Smith, Jr. and P. L. Hamilton. 1975. Benthic nutrient regeneration and its coupling to primary productivity in coastal waters. *Nature* 255:215-217.

Schubel, J. R. 1968. Turbidity maximum of the Northern Chesapeake Bay. *Science* 161:1013-1015.

Simpson, H. J., D. E. Hammond, B. L. Deck and S. C. Williams. 1975. Nutrient budgets in the Hudson River estuary, pp. 618-635. *In:* T. M. Church (ed.), *Marine Chemistry in the Coastal Environment.* Amer. Chem. Soc., New York.

Smith, K. L. 1973. Respiration of a sublittoral community. *Ecology* 54:1065-1075.

Strickland, J. D. H. and T. R. Parsons. 1968. *A Practical Handbook of Sea-water Analysis.* Fish. Res. Bd. Canada Bull. No. 167. 311 pp.

Taft, J. L., A. J. Elliott and W. R. Taylor. 1978. Box model analyses of Chesapeake Bay ammonium and nitrate fluxes, pp. 113-130. *In:* M. L. Wiley (ed.), *Estuarine Interactions.* Academic Press, New York.

Ulanowicz, R. E. and D. A. Flemer. 1977. A synoptic view of a coastal plain estuary, pp. 1-26. *In:* J. C. J. Nihoul (ed.), *Hydrodynamics of Estuaries and Fjords.* Elsevier, Amsterdam.

Van Kessel, J. F. 1977. Factors affecting the denitrification rate in two water sediment systems. *Water Research* 11:259-267.

NUTRIENT VARIABILITY AND FLUXES IN AN ESTUARINE SYSTEM

Theodore C. Loder

Department of Earth Sciences and
Ocean Process Analysis Laboratory
University of New Hampshire
Durham, New Hampshire

and

Patricia M. Glibert

The Biological Laboratories
Harvard University
Cambridge, Massachusetts

Abstract: Nutrient data from Great Bay Estuary, New Hampshire, were analyzed using a conservative mixing model to assess nutrient distribution and variability with time, and to determine a nutrient budget. Analytical precision was estimated as was the magnitude of short term (5 min intervals) environmental variability. Finally, monthly data were analyzed using this mixing model and a phosphate budget was calculated for a one year period. In well mixed waters, the short-term sampling showed that short-term mixing processes can mix estuarine waters so well that variability due to small-scale turbulence or imperfect mixing is less than the analytical and sampling variability. However, when the system is not well mixed then environmental variability can be distinct from analytical variability and may be attributed to some other process or processes. The conservative mixing model was then used to predict apparent average river concentration of phosphate based on the zero salinity intercept calculated with salinity and phosphate data from near the mouth of the estuary. Predicted average river concentrations agreed well with measured concentrations during winter and early spring, but were much higher during summer and fall. Total flux of dissolved phosphate into this estuary was calculated to be 70.7×10^5 moles/yr during 1976 with 78% of the input from municipal sewage and the rest from rivers (13%), sediment flux (7%) and rainfall (2%). Only about 12% of this dissolved phosphate was exported to coastal waters, while the rest of the phosphate apparently remained trapped in estuarine sediments.

Introduction

Quantitative assessment of the distribution and variability with time of biological nutrients, in both estuarine and oceanic systems, has become essential in interpreting many oceanographic processes. In estuarine systems this information is important for understanding the flux of a component through the system as well as sources and sinks of the component within the system, and for constructing overall budgets of chemical components. In both estuarine and oceanic systems, information on short term variations of nutrient concentrations may be helpful in determining the rate

111

of biological turnover of these parameters in the euphotic zone (Beers and Kelly 1965), or in indicating areas of turbulence or incomplete mixing. In addition, information on short-term variability (as determined by high-resolution sampling) will also help to define the limits of analytical precision necessary to discern both small scale and large scale phenomena in a particular area (Kelley 1975).

One common way of estimating nutrient variability in an estuarine system is to plot the measured concentration of a nutrient against a conservative index of mixing, such as salinity. When all data points fall on the straight line joining the end members (water masses), within the resolution of the data, then the only mixing process occurring is simple dilution or "conservative mixing". When the points lie above or below the line then there is evidence for a source or a sink within the system (Peterson et al. 1975; Simpson et al. 1975; Liss and Burton 1976). In addition, deviations from linearity may be caused by the mixing of multiple sources such as several rivers (Boyle et al. 1974) and changing concentrations in otherwise conservatively mixing end members (Loder and Reichard 1980).

Although a nutrient-salinity relationship is not always linear over a wide range of salinities, it will be essentially linear over a very narrow range. By applying this assumption, the amount of sampling variability and short-term environmental variability can be calculated and will be equal to the standard error of the y-estimate (SEE) for the nutrient data relative to salinity. Thus,

$$SEE = (\sum_{i=1}^{n} (C_{oi} - C_{ei})^2 \div (n-2))^{1/2}$$

where C_{oi} equals an observed nutrient concentration, C_{ei} equals a nutrient concentration estimated from the sample salinity for C_{oi} and computed regression, and n equals the number of samples. This equation is analogous to the equation for standard deviation. An assumption in using this approach is that the analytical error of the salinity determination is small relative to the analytical error of the nutrient determination and will not significantly affect the SEE of the nutrient being studied (Loder 1978).

In this paper we apply this conservative mixing model to nutrient and salinity data from Great Bay Estuary, New Hampshire for 1975 and 1976. Estimates of analytical precision were first defined, as was the magnitude of short-term environmental variability. Finally, monthly data were analyzed using this mixing model and a total budget for phosphate was calculated for a one year period.

Materials and Methods

Sampling sites for the Great Bay Estuary are shown in Fig. 1. Sampling was conducted on several time scales: short-term sampling in which water samples were collected from one site every 5 min for 1 h periods; semidiurnal sampling in which samples were collected from one site every

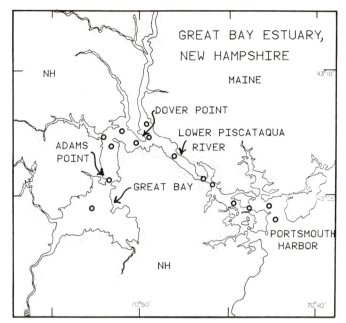

Figure 1. Water column station locations in Great Bay Estuary sampled during 1976.

h throughout a tidal cycle; and monthly sampling in which samples were collected near low tide from throughout the estuary.

A submersible pumping system was used for sample collection. Unfiltered nutrient samples were collected in acid-washed, sample-rinsed linear polyethylene bottles and preserved with mercuric chloride (final concentration ~ 100 ppm). They were immediately cooled on ice and remained refrigerated until analysis (within 2-3 d). The suspended matter was allowed to settle during storage and the analyses were made on the supernatant using a Technicon AutoAnalyzer II System (Glibert and Loder 1977). Samples for salinity were collected in aged linear polyethylene bottles and analyzed on a Guildline Autosal Salinometer (Model 8400).

Results

Analytical variability was determined for each nutrient method as the average standard deviations of 9-10 replicate analyses of the same samples. These results are shown in Table 1.

Variability was determined from a set of samples collected every 5 min for 1 h near high tide at a center channel location off Adams Point (Fig. 1). Phosphate, nitrate and silicate data were plotted against salinity (Fig. 2) and a significant correlation was found even though range of salinities was very narrow (0.16°/₀₀). The SEE values for these data are

Table 1. Analytical and replicate sampling variability of various chemical parameters for samples collected in Great Bay Estuary, New Hampshire during summer (Glibert 1976). The sampling and analytical variability is approximately the sum of a and b or less if samples are analyzed all together.

Parameter (units)	Salinity ($^o/_{oo}$)	NO_2 ($\mu m/l$)	NO_3 ($\mu m/l$)	PO_4 ($\mu m/l$)	SiO_4 ($\mu m/l$)
Range of method	0-40	0-2	0-5	0-5	0-10
Approx. sample concentration	28-30	0.1-0.4	0.1-1.0	1.0-2.0	4.0-7.0
Analytical variability[a]	—	0.009 (2.6%)	0.05 (1.2%)	0.02 (.17%)	0.08 (1.4%)
Replicate sampling variability[b]	0.003 (0.01%)	0.002 (0.7%)	0.04 (0.8%)	0.01 (1.0%)	0.43 (6.4%)

[a]Based on the average standard deviations of 9-10 replicate analyses of the same samples.

[b]Based on the average standard deviations of several sets of replicate samples run at the same time.

Table 2. Observed variability for nutrients relative to salinity using a conservative mixing model. The values given are standard error of y-estimate (SEE).

Locations and Study Type	Nutrient SEE in $\mu m/l$			
	NO_3	NO_2	PO_4	SiO_4
Adams Point time series (5 min) for 1 h in July (Glibert 1976) Salinity range (28.95-29.13$^o/_{oo}$)	0.03	—	0.02	0.12
Lower Piscataqua time series (1 h) for 12 h in August (Glibert 1976) Salinity range (30.2-31.0$^o/_{oo}$)	0.06	0.01	0.02	0.21
Great Bay No. 2, Survey sampling 5 h in February (Loder 1978) Salinity range (6-31$^o/_{oo}$)				
Lower estuary	—	—	0.02	0.90
Upper estuary	—	—	0.05	5.07

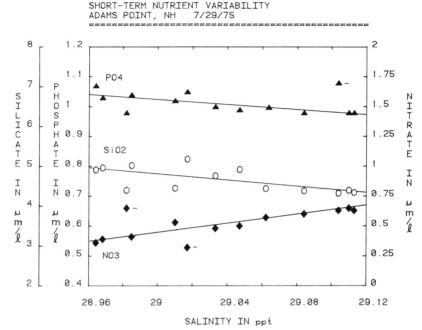

Figure 2. *Salinity-nutrient plots for surface samples collected at 5-min intervals for 1 h near high tide. Dashed data points were not used in the regression calculation. The regressions for these data are: Phosphate = −0.4357 (S°/₀₀) + 13.66, r = −0.724; Silicate = −3.682 (S°/₀₀) + 111.6, r = −0.646; and Nitrate = 1.776 (S°/₀₀) − 51.06, r = 0.967.*

given in Table 2 and are very similar to the values of analytical variability (Table 1). This suggests that short-term mixing processes can mix estuarine waters so well that variability due to small-scale turbulence or imperfect mixing can be less than the analytical and sampling variability.

Nutrient variability determined for time series samples collected every h for 12 h in the center of the well-mixed Lower Piscataqua River (Fig. 1) was similar to nutrient variability data from Adams Point (Table 2). Thus, even over longer time scales, mixing processes can be thorough enough to remove localized inhomogeneities caused by longer period events such as biological processes or sediment-water interactions.

Variability was also determined for the upper estuary above Dover Point. Unlike the Lower Piscataqua River, the Upper Estuary has several rivers with differing nutrient concentrations flowing into it. Figures 3 and 4 illustrate the variability in phosphate and silicate respectively for samples collected at a series of stations located between Portsmouth Harbor and the center of Great Bay (Fig. 1). Significantly more scatter was observed

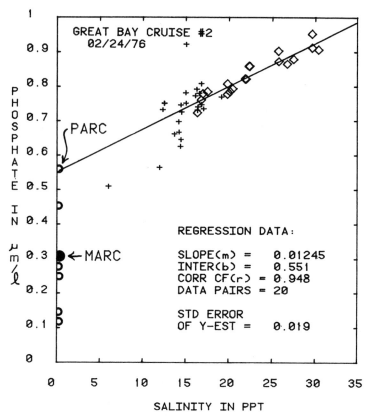

Figure 3. Salinity-dissolved phosphate plot for samples collected in Great Bay Estuary. Only data from the lower Piscataqua River (diamonds) were used to calculate the regression line. Predicted average river concentration (PARC) or intercept is higher than measured average river concentration (MARC) which is equal to 0.31 μm/l. Crosses represent upper estuary samples. Open circles indicate individual river phosphate concentrations.

for samples collected in the upper estuary (crosses) than those from the better mixed lower section (diamonds). The upper estuary apparently had not had sufficient time for mixing to smooth out the initial input variability. Thus, where there are multiple input sources (such as several rivers) with differing nutrient concentrations, and where conservative mixing is the major process affecting nutrient variability, there will be a reduction in the variability of the data relative to salinity near the lower end of the estuary.

Discussion

The analytical variability and short-term environmental variability data are similar when the system is well mixed and events on the scale of turbulence cannot be determined. If the system is not well mixed, then the environmental variability (observed variability minus analytical and sampling variability) can be determined to be distinct from analytical noise and can be attributed to some process, such as the mixing of several end members or biological processes. Having established the sources of variability, a conservative mixing model may then be applied to variations with season and may be used to estimate annual budgets for nutrients into and out of the estuary.

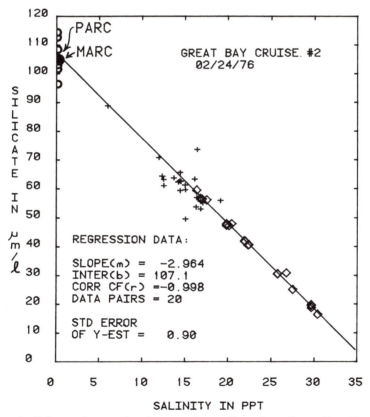

Figure 4. Salinity-silicate plot for samples collected in Great Bay Estuary. Only data from the lower Piscataqua River (diamonds) were used to calculate the regression line. Predicted average river concentration (PARC) or intercept is not significantly different from measured average river concentration (MARC) which is equal to 105.6 μm/l. Crosses represent upper estuary samples. Open circles indicate individual river silicate concentrations.

In an estuarine system with several fresh water sources, the zero salinity intercept value on a nutrient-salinity plot should reflect the volume-weighted average concentration of a nutrient in the fresh water if there are no other significant sources of the nutrient within the estuary. When there are several fresh water inputs or other nutrient sources at different locations along an estuary, then only samples collected between the lowest input and the ocean will reflect a true average input to the estuary. In Great Bay Estuary, samples collected between Dover Point and Portsmouth Harbor would fit this criterion (Fig. 1). Figure 4 illustrates this principle for a time of year (February) when silicate input from sediments is very low, biological uptake is minimal, and rivers are the major source of silicate. The predicted average river concentration based on the lower estuary data was 107.1 μm/l, while the measured average river concentration on a volume-weighted basis was 105.6 μm/l. However, the predicted average river concentration for phosphate (0.6 μm/l) is slightly higher than the measured average river concentration (0.3 μm/l) at the same time of year indicating a phosphate source within the estuary below the river input (Fig. 3). Figure 5 shows that the predicted average phosphate concentration values are

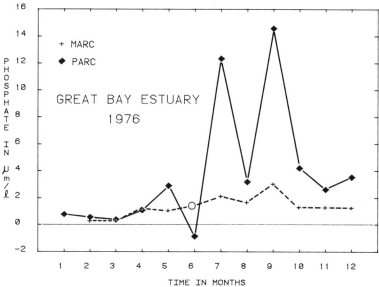

Figure 5. Plot of monthly measured average river concentrations (MARC) and predicted average river concentrations (PARC) for dissolved phosphate. Higher predicted than measured values indicate that a source of phosphate exists within the estuary. Point indicated as a circle represents the MARC value from 1977 as rivers were not sampled in June, 1976.

similar to the measured average river concentration values for the early spring months of high river flow and low biological activity. In May, predicted values were significantly higher than measured values and remained higher for the rest of the year (except for June). In June, the predicted value became negative, indicating that the upper estuary was a sink for phosphate. This "sink" was most likely a large plankton bloom as indicated by (unpublished) chlorophyll data.

The predicted average values of phosphate then increased significantly during the late summer months relative to the river concentrations, indicating a significant non-river source to the estuary. This phosphate increase occurs annually in Great Bay Estuary (unpublished data) and results from both benthic regeneration of phosphate and non-river sewage input. Benthic regeneration of organic matter deposited during the spring bloom is partly a function of temperature, and thus is at a maximum when water temperature is also at a maximum. During late summer, non-river sewage input (relatively high in phosphate) makes up a larger-than-normal fraction of the fresh-water phosphate input since river flow is at a minimum, and thus its dilution effect is minimal.

Thus, over the course of a year the measured concentrations of phosphate did not always agree well with the predicted concentrations (Fig. 5). The difference between the measured and predicted concentration indicates the magnitude of the non-river sources or sinks in the estuary. The product of this difference and river flow indicates the average quantity of non-river carried nutrients added to or removed from the estuary. A predicted average concentration value multiplied by net estuary outflow (essentially equal to river flow into the estuary) gives a good estimate of the net transport of a nutrient through an estuary. In other words, this transport value will equal the total input of a nutrient into an estuary, minus the internal sources and sinks.

We have used the above application of a conservative mixing model to compute an annual budget of dissolved phosphate for Great Bay Estuary during 1976. First, the processes that can act as sources to the estuary were identified and the importance of each assessed. The major sources were identified as: sediment-water column exchange processes, river flow exclusive of sewage wastes, sewage inputs, and rainfall. The phosphate added to the water column from sediment processes was estimated using measured sediment flux data, determined with benthic bottom chambers (Glibert 1976 and unpublished data). It was calculated for an average annual water temperature of 10C (Norell and Mathieson 1976) and one half the area of the estuary since part of the estuary has a sand or gravel bottom where flux rates are significantly lower. The river input was estimated by multiplying the measured concentration value times the appropriate river flow (obtained for gauged rivers from the United States Geological Survey (1977 and 1978) and for ungauged rivers from estimates based on watershed area ratios) and integrated over the year

(Loder et al. 1979). The average flow rates for the sewage treatment plants emptying into the upper estuary were multiplied by the average concentrations of dissolved phosphate (P. Bishop, U. New Hampshire, pers. comm.). The rainfall flux was calculated using phosphate concentrations from the literature (Valiela et al. 1978) and the average rainfall over the estuary. These inputs are summarized in Table 3.

Second, annual export of phosphate from the estuary was estimated by multiplying the predicted average concentration by the total average daily river flow for a 4 d period at the time of sampling and integrating over the year. These flux calculations show that the quantity of phosphate entering the estuary from municipal treatment plants is many times that from other sources (Table 3). However, at one time of the year (June) biological uptake was able to remove much of this input. The amount of phosphate exported in the dissolved form was a small percentage (12%) of the phosphate entering the estuary. Thus, most phosphate must be either exported in the particulate form or remain trapped in the sediments by iron phosphate reactions such as described by Taft and Taylor (1976).

Supporting evidence for the sediments as a phosphate sink comes from W. B. Lyons (U. New Hampshire, pers. comm.) who found a large increase of total phosphate in the surface layers of Great Bay Estuary sediments. His calculations indicate that the excess phosphate in these sediments can be accounted for by the difference between imported and exported phosphate (Table 3) over the past 20 to 30 years.

Conclusions

Conservative mixing models were applied to data from Great Bay Estuary, New Hampshire, over time scales ranging from minutes to months. In well mixed waters, high resolution or short-term sampling (5 min intervals) showed that short-term mixing processes can mix estuarine

Table 3. Dissolved phosphate input and export for Great Bay Estuary, New Hampshire during 1976.

Dissolved Phosphate Input (10^5 moles/year)		
Rivers (measured)		9.1
Sediment (1/2 area, 10C)		4.9
Rainfall (estimated)		1.7
Municipal sewage (estimated)		55.0
	Total	70.7

Dissolved Phosphate Export (10^5 moles/year)		
Based on monthly predicted average river values concentrations and river flow	Total	8.8

waters so well that variability due to small-scale turbulence or imperfect mixing is less than the analytical and sampling variability. In other words, the reproducibility of numerous replicate samples was on the order of the SEE values for short-term samples. Even over slightly longer periods of time (tidal cycles), processes such as turbulance cannot be resolved in the data from the lower estuary since mixing is thorough enough to remove localized inhomogeneities. However, when the system is not well mixed then environmental variability can be determined to be distinct from analytical variability and can be attributed to some process, such as the mixing of several end-members.

However, for the upper estuary where there are numerous rivers flowing into the estuary, the conservative mixing model can be used to isolate sources and sinks within the estuary based on the difference between predicted concentration values and actual measured values. For Great Bay Estuary it was shown that, for the majority of the year, municipal sewage is the major source of phosphate to the estuary, and that most of it remains trapped in sediments, whereas only a small percentage is exported to the coastal waters in the dissolved form.

Acknowledgments

We thank Jane Hislop, Gordon Smith and James Love for technical assistance, and W. Berry Lyons for discussions of the data. This research was funded by the Office of Sea Grant, National Oceanic and Atmospheric Administration, U.S. Department of Commerce through a grant to the University of New Hampshire (4-20237-R/EM-2) and by the University of New Hampshire Leslie S. Hubbard Marine Program Fund. Contribution No. UNH-SG-JR-122 from the University of New Hampshire.

References Cited

Beers, J. R. and A. C. Kelly. 1965. Short-term variation of ammonia in the Sargasso Sea off Bermuda. *Deep-Sea Res.* 12:21-25.

Boyle, E., R. Collier, A. T. Dengler, J. M. Edward, A. C. Ng and R. F. Stallard. 1974. On the chemical mass-balance in estuaries. *Geochim. Cosmochim. Acta* 38:1719-1728.

Glibert, P. M. 1976. Nutrient flux studies in the Great Bay Estuary, New Hampshire. M.S. Dissertation. U. New Hampshire. 89 p.

Glibert, P. M. and T. C. Loder. 1977. Automated analysis of nutrients in seawater: A manual of techniques. WHOI Tech. Rept. No. 77-47, Woods Hole, MA. 46 p.

Kelley, J. C. 1975. Time-varying distributions of biologically significant variables in the ocean. *Deep-Sea Res.* 22:679-688.

Liss, P. S. and J. D. Burton. 1976. *Estuarine Chemistry.* Academic Press, New York. 229 p.

Loder, T. C. 1978. AutoAnalyzer methodology and its role in estuarine and oceanic ecosystem modeling, pp. 443-450. *In: Environmental Changes and Biological Response.* Proceedings of 6th Technicon International Symposium, Tokyo, Japan.

Loder, T. C., J. E. Hislop, J. P. Kim and G. M. Smith. 1979. Hydrographic and chemical data for rivers flowing into the Great Bay Estuary, New Hampshire. U. New Hampshire, Sea Grant Rep. No. 161. 46 p.

Loder, T. C. and R. P. Reichard. 1980. The dynamics of conservative mixing in estuaries. *Estuaries* (in press).

Norell, T. L. and A. C. Mathieson. 1976. Nutrient and hydrographic data for the Great Bay Estuarine System and the adjacent open coast of New Hampshire-Maine. U. New Hampshire, Jackson Estuarine Laboratory Rep. 88 p.

Peterson, D. H., T. J. Conomos, W. W. Broenkow and E. P. Scrivani. 1975. Processes controlling the dissolved silica distribution in San Francisco Bay, pp. 153-187. *In:* L. E. Cronin (ed.), *Estuarine Research, Vol. I.* Academic Press, New York.

Simpson, H. D., D. E. Hammond, B. L. Deck and S. C. Williams. 1975. Nutrient budgets in the Hudson River Estuary, p. 618-635. *In:* T. Church (ed.), *Marine Chemistry in the Coastal Environment.* Amer. Chem. Soc., Symp. Series, Vol. 18.

Taft, J. L. and W. R. Taylor. 1976. Phosphorus dynamics in some coastal plain estuaries, pp. 79-89. *In:* M. L. Wiley (ed.), *Estuarine Processes, Vol. I.* Academic Press, New York.

United States Geological Survey. 1977. Water resources data for New Hampshire and Vermont, water year 1976. USGS Water-Data Report NH-VT-76-1, 208 p.

United States Geological Survey. 1978. Water resources data for New Hampshire and Vermont, water year 1977. USGS Water-Data Report NH-VT-77-1, 200 p.

Valiela, I., J. M. Teal, S. Volkmann, D. Shafer and E. J. Carpenter. 1978. Nutrient and particulate fluxes in a salt marsh ecosystem: Tidal exchanges and inputs by precipitation and ground water. *Limnol. Oceanogr.* 23:798-812.

NUTRIENT INTERACTIONS BETWEEN SALT MARSH, MUDFLATS, AND ESTUARINE WATER

T. G. Wolaver

Department of Environmental Sciences
University of Virginia
Charlottesville, Virginia

R. L. Wetzel

Virginia Institute of Marine Science
College of William and Mary
Gloucester Point, Virginia

J. C. Zieman

Department of Environmental Sciences
University of Virginia
Charlottesville, Virginia

and

K. L. Webb

Virginia Institute of Marine Science
College of William and Mary
Gloucester Point, Virginia

Abstract: A salt marsh, mudflat and contiguous estuarine water body form a set of systems linked by tidal flushing, with each system sequentially adding to or removing available ammonia. The salt marsh acts both as a source and sink for the nutrient depending on the time of year. During winter, the marsh removes a small amount of the nutrient load made available by tidal flushing. With the onset of spring the marsh system is an active site for uptake of ammonia, such removal becoming especially efficient during the early summer months. In late summer, if biologic activity is sufficient to lower the water oxygen level (especially during nighttime slack water), the marsh becomes an exporter of ammonia. Also during this period the export of ammonia can be quite large due to seepage of interstitial water from the reduced sediments after the tidal water has receded from the marsh surface. During fall, the marsh again acts as an active sink for this constituent.

Introduction

There has been a variety of studies conducted during the past several years investigating the following questions: (1) Is the salt marsh a source or sink for nutrients with respect to the adjacent estuary? and (2) What are the processes controlling nutrient exchange? Research conducted to address these questions has centered around three basic approaches or

123

methodologies: (1) Specific studies on various marsh components such as vegetation and sediments to determine if they act as a nutrient source or sink, (2) routine or periodic sampling efforts to determine nutrient concentrations in the water flowing past various fixed sampling locations in tidal creeks and marshes, and (3) nutrient mass balance studies at the entrances of tidal creeks connecting marsh complexes with the surrounding estuarine water body.

These approaches have produced useful information but have failed to provide a clear solution to the proposed questions. Studies conducted in the first category have investigated specific processes which affect the flux of nutrients between components of the marsh and the surrounding estuary but have contributed little in relation to understanding the importance of these processes within the total marsh ecosystem. Examples are work with *Spartina alterniflora* as a shunt for orthophosphate to tidal water (Reimold 1972) and investigations of estuarine sediments controlling orthophosphate concentrations in the overlying water column (Pomeroy et al. 1965). Aurand and Daiber's (1973) work on nitrite and nitrate concentrations in the tidal water of marsh creeks and vegetated marsh areas is a good example of the second approach. Investigators using this methodology can infer that one area may act as a source or sink for nutrients, but cannot specify either the direction or magnitude of nutrient flux between adjacent systems. The last approach *a priori* appears to be the one most capable of providing a solution to the above questions. To date it has been used by several investigators (Axelrad 1974; Heinle and Flemer 1976; Stevenson et al. 1976; Woodwell and Whitney 1977; Valiela et al. 1978). A synthesis of these nutrient mass balance studies with respect to the various nitrogen species (V. Lee, Grad. School of Oceanography, U. Rhode Island, unpub. table) indicates there is little consistency with respect to either direction or magnitude of flux, seasonal trends often being obscured due to the variability in the flux data.

There appear to be several reasons why this last approach has generally failed to provide the necessary information to address the initial questions:

1. Water mass balances for large systems are hard if not impossible to quantify (Boone 1975; Kjerfve and Proehl 1979; Pritchard and Shubel 1979). The ability to determine the flux of the medium containing the nutrients of interest, including spatial and temporal variability, sets the lower limit on the accuracy of the nutrient flux calculations.

2. Water samples, properly preserved and adequate in number, must be taken so that the spatial and temporal variability with respect to nutrient concentrations is well documented. Characteristically, studies have addressed one or more of these problems inadequately.

3. Nutrients contained in the flooding water are made available to a series of systems including the marsh vegetation and soil, mudflats,

and the particulate material within the water column. Each of these systems in different estuarine environments may act similarly with regard to the direction of nutrient flux, but vary in the magnitude of interaction. If this is the case, the overall behavior of the coupled systems with respect to nutrient exchange as measured in marsh creeks may vary both as a function of season and locale, thus explaining why the results of many studies are inconsistent.

A more refined approach should be adopted if we are to answer the questions proposed. Initially a solution may be in coupling nutrient mass balance studies of the various systems, such as the vegetated marsh and mudflats, with experimental work. The latter would help specify which components are responsible for the observed nutrient fluxes found using the mass balance approach and would aid in elucidating which processes are controlling nutrient movement.

For the past 18 months we have employed such an approach to investigate nutrient exchange in several marsh-estuarine systems. Initially a nutrient mass balance study of a *Spartina alterniflora* marsh was completed. A nutrient mass balance study of the mudflats contiguous with another marsh is now being conducted. Concurrently, chamber experiments have been initiated to investigate specific subsystems and processes responsible for the observed nutrient fluxes.

Methods and Materials

A nutrient mass balance study of a *Spartina alterniflora* marsh was conducted on Carters Creek in the lower York River estuary (Fig. 1). The marsh was chosen for study because of a fairly uniform elevation gradient between the low and high marsh zones. This allowed for a uniform flow of tidal water onto and off the marsh surface. A flume, 1.8 m wide and 23 m long, was constructed with fiberglass fencing and placed in the marsh to prevent lateral water movement during tidal water exchange. The structure was aligned perpendicular to the creek flow.

Water samples for nutrient analysis were collected over complete tidal cycles from two marsh locations within the flume; one, in front of the tall *S. alterniflora* zone and the second between the tall and short *S. alterniflora* zones. Water samples (nutrient) collected from the leading edge of the tidal water as it traversed the marsh indicated that the high and low *S. alterniflora* zones differed with respect to their nutrient processing capabilities. To characterize the depth profile of nutrients in the water column, four depth samples were taken from the first station (1, 10, 25 and 45 cm above the sediment) and three from the second (1, 10, and 25 cm above the sediment). Samples were taken every 0.5 or 1 h at each station over the tidal cycle using a peristaltic pump. Water samples were split in the field and frozen immediately in a dry-ice acetone bath. One fraction was left unfiltered for total phosphorus (Gales et al. 1966) and nitrogen (Kammerer et al. 1967) determinations; the second fraction was filtered in-

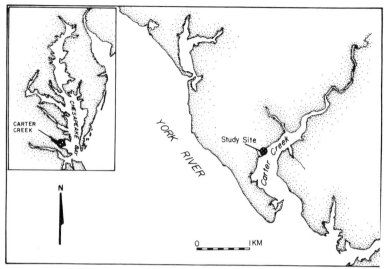

Figure 1. Site map.

line (Gelman glass-fiber, GF/C) and the filtrate analyzed on a Technicon AutoAnalyzer II system for dissolved ammonia (O'Brien and Fiore 1962), nitrite (F.W.P.C.A. 1969), nitrate (F.W.P.C.A. 1969) and orthophosphate (Murphy and Riley 1962). The sampling procedure was repeated every two weeks for the period January 1978-January 1979.

The flux of nutrients onto and off the marsh surface was determined using the nutrient data and tidal discharge information. Discharge was determined over each tidal cycle from relative tide height measurements and marsh topography data. These data were then used to calculate the volume of water on the marsh at any time over the tidal cycle. By subtracting the volumes of water on the marsh at successive time intervals, volume discharge was determined. A computer program was written to couple the discharge and nutrient data to estimate the direction and magnitude of nutrient flux between the marsh and surrounding estuarine water.

Experiments have been conducted in the two marsh zones to segregate the water column and sediment contribution to the observed overall nutrient flux. Experimental chambers were constructed from 15 cm diameter plexiglas tubing cut into 122 cm sections. The chamber design allows flooding water to enter at the sediment-water interface through a single inlet and exit the same route on the ebb tide (Fig. 2). This design is adequate since the tidal water adjacent to the sediment is often representative (nutrient concentration) of that in the rest of the water column. Chambers used to determine sediment exchange were open at the bottom so that water flooding the chamber contacted the marsh sediments. The control chambers were identical in design except that a plexiglas plate was attached to the bottom to prevent contact with the sediment.

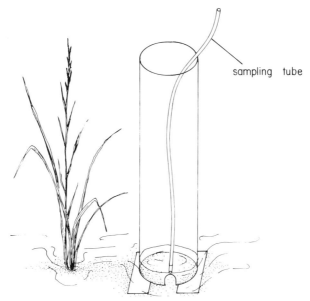

sampling tube

Figure 2. Chamber design for studying nutrient exchange between tidal water and salt marsh sediment surface.

Results

To demonstrate the validity of this nutrient mass balance approach as applied to the marsh, we are presenting the dissolved NH_4 data obtained from the first year of study. Figure 3 depicts the NH_4 concentrations in the tidal water as a function of time during a tidal cycle at Station 1, located at the intersection of the mudflat and marsh. The four curves represent NH_4 concentrations at four depths within the tidal water: curve 1 is the concentration at the sediment-water interface and curves 2 through 4 are concentrations respectively higher in the water column. These data indicate the need to quantify vertical nutrient heterogeneity.

The marsh flux data determined by coupling the nutrient data with discharge information indicates NH_4 was removed on an annual basis from tidal water inundating the marsh (Fig. 4). As illustrated, there were two periods when large amounts of NH_4 were removed; late spring and early fall. There is some indication that, for a short interval before the late summer removal of NH_4, this constituent was released from the marsh to the surrounding water. Comparison of the NH_4 flux data and the NH_4 concentration in the flooding water (Fig. 5) indicates that, during every period when high concentrations of NH_4 are available, the total marsh system responds by efficiently removing a fairly large percentage of this constituent. An r value of 0.643 was obtained when correlating the flux of ammonia onto or off the marsh with the value of ammonia on the incoming

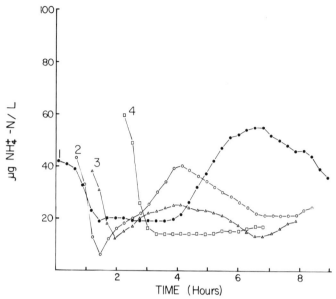

Figure 3. Ammonia concentrations at four depths within the water column versus time over a tidal cycle. Depth 1: sediment surface; Depth 2: 10 cm above sediment surface; Depth 3: 25 cm above sediment surface; Depth 4: 40 cm above sediment surface (June 29, 1978).

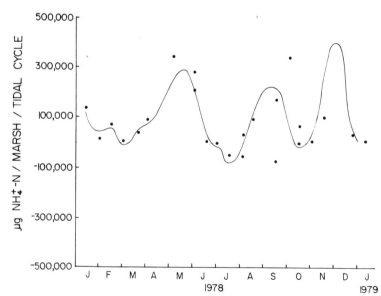

Figure 4. Annual flux of ammonia between the marsh surface and the surrounding estuary (negative flux to estuary, positive flux onto marsh; the curve is a fitted 10th order polynomial).

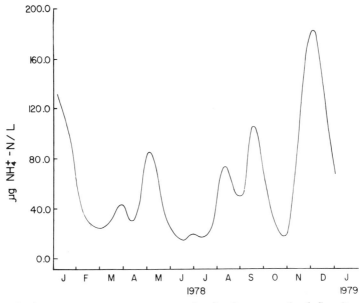

Figure 5. Ammonia concentration in the leading parcel of flooding tidal water entering the vegetated marsh versus time of year.

tide. This is significant at the 0.001 level. Figures 6 and 7 illustrate the flux of ammonia within the high and low marsh areas, respectively. It is evident that the high marsh (short *Spartina alterniflora*) acts as a sink for NH_4 throughout the year whereas the low marsh acts as a source or sink depending on the season.

In general, the ammonia mass balance data indicate that ammonia was removed from the tidal water as it passed over the marsh surface. Figure 8 illustrates data obtained from three experimental chambers in the low marsh; two were open to the sediment and one was closed isolating the water column. It is evident that water column processes are responsible for most of the observed removal of ammonia from the tidal water over the low marsh during this tidal cycle, because the shapes of the nutrient curves for the sediment chambers are similar to that of the control (all curves indicating an uptake of ammonia). This also implies that there was no net ammonia exchange between the sediment and the tidal water. The complete set of chamber data for the low marsh shows that water column processes result in either an uptake of ammonia or no net exchange while the sediment is either neutral with respect to nutrient exchange or is in fact a source of ammonia.

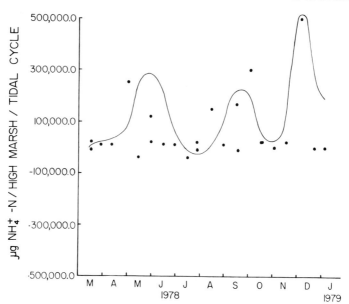

Figure 6. Annual flux of ammonia between the tidal water and the low marsh (negative flux into tidal water, positive flux out of tidal water; the curve is an 8th order polynomial).

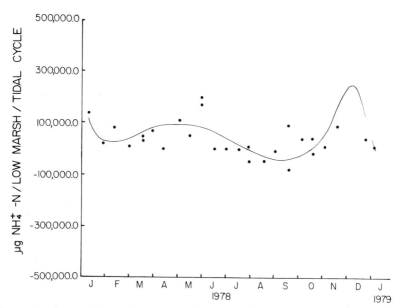

Figure 7. Annual flux of ammonia between the tidal water and the high marsh (negative flux into tidal water, positive flux out of tidal water; the curve is an 11th order polynomial).

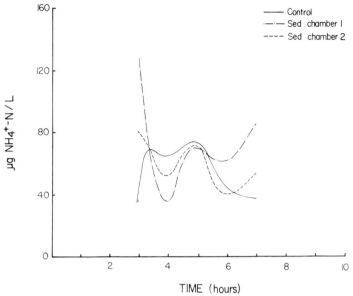

Figure 8. Ammonia concentrations in the water within the experimental chambers placed on the low marsh versus time over the tidal cycle (Aug. 22, 1979).

Figure 9 depicts a similar data set obtained from chambers on the high marsh. In this case there was no utilization of ammonia within the water column on August 22, 1979, the sediment surface removing most of this constituent. This result is also consistent with the other complete data set collected using the high marsh chambers.

Discussion

Synthesis of the nutrient mass balance data collected to date indicates that the marsh system should be divided spatially into three subsystems; mudflat, low marsh and high marsh. As tidal water traverses the mudflats, the nutrients are made available for processing within the water column and sediment. In addition, flooding water can be enriched by interstitial marsh seepage flowing over the mudflats. Flooding water next interacts with the low marsh components including plants and the sediments. Finally, at or near high tide water often covers the high marsh, its nutrient concentration determined by its previous interaction with mudflat and low marsh environments. The sequence is reversed on ebb tide.

To expand upon this conceptual model we will discuss the ammonia data collected to date. As tidal water progresses across the mudflat, its ammonia concentration is affected by biological and chemical processes within the water column and/or the sediment subsystem. Data collected by Nixon et al. (1976) indicated no consistent exchange of ammonia between inter-

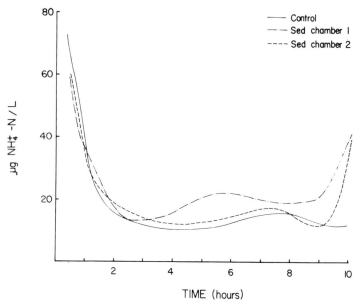

Figure 9. Ammonia concentrations in the water within the experimental chambers placed on the high marsh versus time over the tidal cycle (Aug. 22, 1979).

tidal sediments and estuarine water. Our data from Carters Creek indicate that the first parcel of water entering the low marsh is already enriched with ammonia. This enrichment of the tidal water as it traverses the mudflats could arise either from its mixing with ammonia-rich interstitial water which has seeped from the high marsh sediment during low tide exposure or from direct sediment release (diffusion and/or microbial mineralization). With the exception of late summer, the low marsh is a site of ammonia removal from the water column. The ammonia not removed at this site is subsequently removed in the high marsh, this zone consistently acting as a sink for this constituent. The original data also indicate that as the tidal water traverses the low marsh it is often enriched with ammonia seepage from high marsh sediments, this ammonia being removed by both the low and high marsh subsystems. As the water recedes from the marsh, it is usually depleted in ammonia except during the later summer months. During this period low marsh sediments show a net release of ammonia. The data illustrate a fair degree of internal ammonia cycling between the various subsystems with little ammonia escaping to the surrounding estuary.

Once this general picture was developed concerning the processing of ammonia within the marsh complex we next attempted to determine which components within the mudflats or vegetated marsh systems were responsible for the observed nutrient dynamics. The chamber experiments indicate that in the low marsh, water column processes are primarily responsible for the observed removal of ammonia during the summer, the sediment possibly acting as a source for this constituent depending on the season. Within the high marsh the sediment appears to dominate the exchange.

The next stage in the experimental design is to investigate which specific processes control the observed nutrient exchange. If the experimental design described herein is adopted, a foundation for valid comparisons between systems with respect to how the nutrients are processed will be established.

References Cited

Aurand, D. and F. C. Daiber. 1973. Nitrate and nitrite in the surface waters of two Delaware salt marshes. *Chesapeake Sci.* 14:105-111.

Axelrad, D. M. 1974. Nutrient flux through the salt marsh ecosystem. Dissertation, College of William and Mary, Williamsburg, VA. 134 pp.

Boone, J. D., III. 1975. Tidal discharge asymmetry in a salt marsh drainage system. *Limnol. Oceanogr.* 20:71-80.

Federal Water Pollution Control Administration (F.W.P.C.A.) 1969. Methods for Chemical Analysis of Water and Wastes. U.S. Government Printing Office, Washington, D.C.

Gales, M., E. Julian and R. Kroner. 1966. Method for quantitative determination of total phosphorus in water. *J. Amer. Waterways Assoc.* 58:1363.

Heinle, D. R. and D. A. Flemer. 1976. Flows of materials between poorly flooded tidal marshes and an estuary. *Mar. Biol.* 35:359-373.

Kammerer, P. A., M. G. Rochel, R. A. Hughes and G. F. Lee. 1967. Low level Kjeldahl nitrogen determination on the Technicon AutoAnalyzer. *Environ. Sci. Tech.* 1:4-340.

Kjerfve, B. and J. A. Proehl. 1979. Velocity variability in a cross-section of a well-mixed estuary. *J. Mar. Res.* 37:409-418.

Murphy, J. and J. P. Riley. 1962. A modified single solution method for the determination of phosphate in natural waters. *Anal. Chim. Acta,* 27:30.

Nixon, S., C. Oviatt, J. Garber and V. Lee. 1976. Diel metabolism and nutrient dynamics in a salt marsh embayment. *Ecology* 57:740-750.

O'Brien, J. and J. Fiore. 1962. Ammonia determination by automated analysis. *Wastes Eng.* 33:352.

Pomeroy, L. R., E. F. Smith and C. M. Grant. 1965. The exchange of phosphate between estuarine water and sediments. *Limnol. Oceanogr.* 10:167-172.

Pritchard, D. W. and J. R. Schubel. 1979. Physical and geological processes controlling nutrient levels in estuaries, pp. 000-000. *In:* B. J. Nielsen (ed.), International Symposium on Nutrient Enrichment in Estuaries. The H.U.M.A.N.A. Press, Inc., Clifton, NJ.

Reimold, R. J. 1972. The movement of phosphorus through the salt marsh cord grass *Spartina alterniflora* Loisel. *Limnol. Oceanogr.* 17:606-611.

Stevenson, J. C., D. R. Heinle, D. A. Flemer, R. J. Small, R. A. Rowland and J. F. Ustach. 1976. Nutrient exchanges between brackish water marshes and the estuary, pp. 219-240. *In:* M. L. Wiley (ed.), *Estuarine Processes, Vol. II.* Academic Press, New York.

Valiela, I., J. Teal, S. Volkmann, D. Shafer and E. Carpenter. 1978. Nutrient and particulate fluxes in a salt marsh ecosystem: Tidal exchanges and inputs by precipitation and groundwater. *Limnol. Oceanogr.* 23:798-812.

Woodwell, G. M. and D. E. Whitney. 1977. Flax Pond ecosystem study: exchanges of phosphorus between a salt marsh and the coastal waters of Long Island Sound. *Mar. Biol.* 41:1-6.

AN EVALUATION OF MARSH NITROGEN FIXATION

David W. Smith

School of Life and Health Sciences
University of Delaware
Newark, Delaware

Abstract: Reported rates of nitrogen fixation in different salt marshes vary widely. Although some of this difference results from geographical considerations, there is inherent variability in the methods used for measurement of nitrogen fixation which may contribute significantly to the observed differences. This variability may be seen in two broad areas: experimental procedure (mainly sampling) and data analysis. In sampling, the effects of spatial heterogeneity are not always considered. In a Delaware marsh system, spatial effects are pronounced in all three dimensions in the marsh sediment. Normalized fixation rates increase as sample size decreases, presumably reflecting differences in gas diffusion. Therefore, extrapolations from small samples are unlikely to generate accurate quantitative results at the system (marsh) level. Concerning data analysis, it must be kept in mind that most field studies of nitrogen fixation predominantly involve the acetylene reduction technique. Different rates of acetylene reduction can be derived from many data sets collected in fixation studies, often varying by 5- to 10-fold. There is uncertainty surrounding the conversion of these rates to equivalent nitrogen fixation rates. Many conversion factors are used, varying by 10-fold. In spite of the difficulties in quantitative measurement, marsh nitrogen fixation has emerged as a significant process. It is possible to measure relative rates in small samples accurately. It follows that studies of the physiological ecology of marsh nitrogen fixation should prove informative and useful.

Introduction

The ecology of nitrogen fixation has received tremendous attention in the past 20 years in many systems, especially those in which the legume-*Rhizobium* symbiosis is significant. Detailed studies of this process in marshes began more recently, but are quite numerous (Jones 1974; Whitney et al. 1975; Hanson 1977a,b; Patriquin and McClung 1978; Teal et al. 1979; Dicker and Smith 1980a,b). Marshes are involved in significant nutrient fluxes (Heinle and Flemer 1976; Valiela et al. 1978; Woodwell et al. 1979). It is a natural extension from such data to determine individual components in the cycles, and nitrogen fixation is an obvious and important process.

However, extreme care must be used in presenting quantitative statements concerning system-level processes. Quantitative extrapolation of data in nitrogen fixation studies is often an overextension which cannot be justified. This problem is greatest in those studies which use acetylene reduction activity (ARA) to estimate nitrogen fixation, i.e. the vast majority

135

of nitrogen fixation studies. I will present data from my laboratory to support a series of conclusions which I hope will be viewed as constructive criticisms of the procedures used in marsh nitrogen fixation studies. My data were gathered with the acetylene reduction method. One of my main conclusions will be that acetylene reduction with natural samples may not reflect nitrogen fixation quantitatively. Therefore I will present results in the most conservative manner, namely in terms of the parameter actually measured, ARA. The purpose of the present report is to examine in detail some of the factors involved in system-level conclusions and to urge considerable caution in the derivation of quantitative expressions of marsh nitrogen fixation activity.

Methods and Materials

Detailed characteristics of the Canary Creek marsh in Lewes, Delaware have been presented elsewhere (Dicker and Smith 1980a). Samples were collected from the top 1 cm of sediment by surface scraping with a sterile spatula in a zone of short-form *Spartina alterniflora*. Samples were immediately placed in sterile plastic bags (Whirl-Pak) for transportation to the laboratory. Samples of four sizes were prepared: 0.05 g; 0.10 g; 5.0 g; 10.0 g. The 5 and 10 g samples were prepared by subdividing the sediment with a spatula. The 0.05 and 0.10 g samples were prepared by creating a slurry of sediment and mineral salts medium adjusted to marsh salinity with NaCl and applying the slurry to membrane filters (Millipore) to remove the excess liquid. This procedure ensured uniform geometry for these very small samples. Samples were incubated in 60 ml serum bottles and were pre-incubated under an argon atmosphere for 8 h. Incubations were begun by removing the argon by evacuation and replacing it with an atmosphere of 10% acetylene, 5% oxygen, and 85% argon (Patriquin 1978). Rate of acetylene reduction was determined by daily withdrawals of atmosphere samples (50 μl) over 5 days for gas chromatographic analysis. Ethylene production displayed no lag and was linear over this period.

Results and Discussion

Table 1 presents a summary of methods from several laboratories which shows that the incubation times and sample sizes used in field nitrogen fixation studies vary greatly. I examined in some detail the effect on ARA of varying sample size from 0.05 g to 10 g (Fig. 1). Normalized ARA rates are clearly an exponential function of sample size, with small samples leading to high rates. This effect is related to the geometry of the samples, since a very similar curve is obtained by plotting the ratio of sample surface area to sample volume as a function of sample size (data not shown). Gas diffusion is a function of surface area and the acetylene reduction procedure relies on the diffusion of two gases: acetylene (into the sample) and ethylene (out of the sample). It is therefore quite likely that the results of Fig. 1 represent differential gas exchange by samples with different surface

Table 1. Summary of nitrogen fixation procedures from several marsh studies.

Studies	Sample Type	Length of Incubation	Sample Size (g)
Carpenter et al. 1978	Cores - blue-greens	1 h	0.1
VanRaalte et al. 1974	Cores - blue-greens	2 h	0.1
Rice et al. 1967	Cores	24 h	0.6
Teal et al. 1979	Cores	48 h	4
Hanson 1977a	Cores - slurry	30 h	10
Whitney et al. 1975	Cores - slurry	2 h	20
Dicker and Smith 1980a	Slices of sod	5 d	20
Patriquin and McClung 1978	Slices of sod	1 to 3 d	1 to 60
Herbert 1975	Cores	36 h	95

to volume ratios. Patriquin (1978) has also expressed concern with gas diffusion through large samples in ARA studies. It would seem that the smaller samples are to be preferred because diffusion is accelerated. However, the large surface to volume ratio of the small samples also encourages desiccation which damages the microbial community. This size factor is clearly important in quantitative calculations, but is difficult to address in a field situation.

Another diffusion-related problem arises when sediment samples are mixed with aqueous solutions to form slurries as is done by some workers (Table 1). Flett et al. (1976) have examined ARA by aqueous samples in physical-chemical detail. Based on considerations of the greatly different solubilities of acetylene and ethylene and the geometry of the incubation vessel, these workers conclude that accurate, quantitative nitrogen fixation data are difficult to obtain from such incubations.

The choice of incubation period is also highly variable in marsh ARA studies (Table 1). The major effect of different incubation times is seen when a lag occurs, i.e., when there is a delay (often up to 24 h) after acetylene addition before ARA commences. Lags have been observed by several workers in different systems (Dobereiner et al. 1972; Herbert 1975; Hanson 1977b; Patriquin and Keddy 1978; Teal et al. 1979; Dicker and

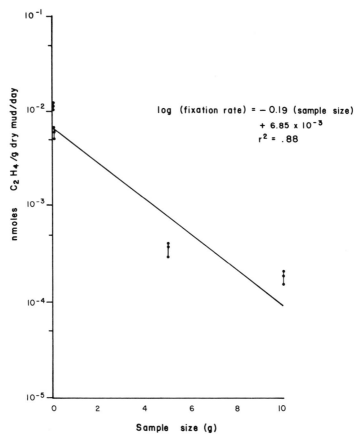

Figure 1. *Effect of sample size on measured ARA. The mean and 95%
confidence intervals are presented for each sample size. The total
number of samples was 32.*

Smith 1980a). In these samples, short incubation times yield low rates and
long incubation times which neglect the lag period yield much higher rates.
The question of which procedure is more correct depends on the assumptions made about the nature of the lag. Nitrogen-fixing bacteria may be
damaged during sampling and the lag may therefore represent recovery to
in situ conditions. In this interpretation, the lag period should be ignored
and long incubations used. Herbert (1975) and Teal et al. (1979) present
this argument in analyzing their data. Alternatively, the lag may represent a
physiological change or "bottle effect" in the community. In this interpretation, the lag represents the more natural conditions and very short incubations should be used. Patriquin (1978) and David and Fay (1977) are in
sympathy with this position and indeed the latter authors argue that prolonged incubations with acetylene lead to gross overestimations of *in situ*

nitrogen fixation rates. It is difficult to determine precisely the cause of these lag periods, although we have data which suggest that they result from community-mediated ammonia interactions (Dicker and Smith 1980b). I believe that there is no clear answer to the question of incubation length. Marsh systems differ from each other in many physical, chemical, and vegetational characteristics. It is extremely unlikely that any single incubation period is the "correct" one. If an attempt is made to determine total system rates, the variation from this source will add substantial error.

The large variation in surface ARA over very short distances casts more doubt on quantitative extrapolations. The standard procedure is to examine small portions of representative marsh areas (differing usually in vegetational type), determine the ARA in each, and multiply by a factor relating that area to the total marsh system (Whitney et al. 1975; Carpenter et al. 1978; Teal et al. 1979). A key assumption in this procedure is that each major vegetational area has been accurately delimited and that the samples taken from each area are indeed representative. The question of spatial heterogeneity on a small scale (1 m^2) has been addressed (Dicker and Smith 1980a). We examined the natural variability of a nitrogen-fixing population and the extent to which this variability could be reduced by mechanical homogenization. We found that the coefficient of variation (n = 6) of the native sediment ranged from 12 to 31% and that of the homogenized sediment ranged from 9 to 21%. The means of the two groups of samples were not different (p = 0.05). Acetylene reduction rates reported by others contain coefficients of variation of similar magnitude to ours: 10 to 15% (Whitney et al. 1975; Patriquin 1978); 25 to 30% (Carpenter et al. 1978; Teal et al. 1979). We also found that similar areas of short-form *Spartina alterniflora*, separated by only 100 m, differed in rates of ARA by up to 3-fold (Dicker and Smith 1980a). This considerable variability in nitrogen-fixing populations at a single sample site and within one vegetational zone make it difficult to know if a given series of samples is representative and if large scale extrapolations can be meaningfully made from the data. It is possible to calculate the number of samples needed to obtain confidence intervals of a desired probability for a population with a given natural variability. The calculated value for a particular study will depend on details of assay techniques and exact sampling procedures. The general formula relevant here is easy to apply, but the number of samples needed is exceptionally sensitive to the true standard deviation of the natural population (Sokal and Rohlf 1969). Therefore, proper use of this statistical technique requires extensive measurements of population variance before detailed system surveys are performed. There is a practical limit to the number of samples which can be analyzed in a given project, therefore accurate determination of population variance in one marsh area will limit the number of different areas which can be sampled. The result is that spatial variations can easily go undetected.

Many different patterns of ARA with depth have been reported in marshes. Jones (1974) found ARA in the surface, but not at 5 cm; Whitney et al. (1975) and Teal et al. (1979) found maximum activity at or near the surface; Haines et al. (1975), Hanson (1977b), Patriquin and Mc-Clung (1978), and Dicker and Smith (1980a) found the bulk of ARA at depths ranging from 3 to 20 cm. These differences may represent geographical variability as well as interactions with physical (tide, temperature), chemical (O_2, interstitial ammonia or nitrate), or biological (plant root extent) factors. Teal et al. (1979) noted considerable variation in depth profiles during their sampling year and suggested the involvement of temperature and ammonia especially. Whatever the depth profile in a given marsh, it is necessary for quantitative studies to design sampling schemes sufficient to represent the natural populations. This problem is not insoluble but, again, effort directed toward this aspect of a project decreases what can be done in other areas.

The conversion of ARA to equivalent nitrogen fixation activity must also be examined. The basic theoretical approach is based on the number of electrons transferred during the two reduction processes. The values used are generally in the range of 3 to 6 moles acetylene reduced per mole of N_2 reduced, with a maximum reported value of 25 (Hardy et al. 1973). There is recent evidence that the *in situ* relation between these two activities may not necessarily be what is predicted in theory (Oremland and Taylor 1975; Potts et al. 1978). A possible cause of these differences is that nitrogen gas is much less soluble in water than is acetylene or ethylene (Flett et al. 1976). Samples of different water content or different water holding capacity may therefore display quite different ARA rates even if they had similar levels of nitrogen fixation activity. Variation in the value chosen for the conversion of ARA to nitrogen fixation of course has significant impact on the results of quantitative studies.

A final concern is that essentially all investigators report their ARA and nitrogen fixation results in different terms. Rates are normalized on any of three bases: per unit surface area; per unit volume; or per unit mass of sediment. Each investigator has valid reasons for the units chosen. However, this diversity makes interstudy comparisons quite difficult. Conversions are of course possible, but depend on assumptions concerning sediment density and water content which are not always presented. It would be valuable if investigators studying marsh nitrogen fixation and ARA could agree on a standard set of units. Individual research reports could then be presented in two forms: the standard units and the units preferred by the author.

In summary, I have pointed out many ways in which field nitrogen fixation studies are imprecise, especially when done with the acetylene reduction technique. My purpose is to emphasize the many possible sources of error in creating "absolute" system-level estimates. The problems I have identified can combine to produce an exceptionally high or exceptionally low value of system nitrogen fixation. Each individual feature can be addressed separately to some degree. Difficulty arises when all of these factors must be considered simultaneously in a quantitative fashion. Nitrogen fixation is an important ecological process and the acetylene reduction technique has great value. This procedure, performed in a standard fashion, allows precise determination of relative levels of ARA. I conclude that the physiological ecology of nitrogen fixation is well studied with ARA. Difficulties arise when one attempts to draw system-level conclusions from ARA data. The implication of my presentation is that the values obtained in system studies must be evaluated with the greatest possible caution and with a knowledge of the shortcomings of the procedures used to obtain the final answers.

Acknowledgments

I am grateful to my coworker Howard Dicker. Guerard Byrne provided technical assistance. This work was supported by the Delaware Sea Grant College program.

References Cited

Carpenter, E. J., C. D. VanRaalte and I. Valiela. 1978. Nitrogen fixation by algae in a Massachusetts salt marsh. *Limnol. Oceanogr.* 23:318-327.

David, K. A. V. and P. Fay. 1977. Effects of long-term treatment with acetylene on nitrogen-fixing microorganisms. *Appl. Environ. Microbiol.* 34:640-646.

Dicker, H. J. and D. W. Smith. 1980a. Acetylene reduction (nitrogen fixation) in a Delaware salt marsh. *Mar. Biol.* (in press).

Dicker, H. J. and D. W. Smith. 1980b. Physiological ecology of acetylene reduction (nitrogen fixation) in a Delaware salt marsh. *Microbial Ecol.* (in press).

Dobereiner, J., J. M. Day and P. J. Dart. 1972. Nitrogenase activity and oxygen sensitivity of the *Paspalum notatum-Azotobacter paspali* association. *J. Gen. Microbiol.* 71:103-116.

Flett, R. J., R. D. Hamilton and N. E. R. Campbell. 1976. Aquatic acetylene-reduction techniques: Solutions to several problems. *Can. J. Microbiol.* 22:43-51.

Haines, E. A., A. Chalmers, R. Hanson and B. Sherr. 1975. Nitrogen pools and fluxes in a Georgia salt marsh, pp. 241-254. *In:* M. L. Wiley (ed.), *Estuarine Processes, Vol. II.* Academic Press, New York.

Hanson, R. B. 1977a. Comparison of nitrogen fixation activity in tall and short *Spartina alterniflora* salt marsh soils. *Appl. Environ. Microbiol.* 33:596-602.

Hanson, R. B. 1977b. Nitrogen fixation (acetylene reduction) in a salt marsh amended with sewage sludge and organic carbon and nitrogen compounds. *Appl. Environ. Microbiol.* 33:846-852.

Hardy, R. W. F., R. C. Burns and R. D. Holsten. 1973. Applications of the acetylene-ethylene assay for measurement of nitrogen fixation. *Soil Biol. Biochem.* 5:47-81.

Heinle, D. R. and D. A. Flemer. 1976. Flows of material between poorly flooded tidal marshes and an estuary. *Mar. Biol.* 35:359-373.

Herbert, R. A. 1975. Heterotrophic nitrogen fixation in shallow estuarine sediments. *J. Exp. Mar. Biol. Ecol.* 18:215-225.

Jones, K. 1974. Nitrogen fixation in a salt marsh. *J. Ecol.* 62:553-565.

Oremland, R. S. and B. F. Taylor. 1975. Inhibition of methanogenesis in marine sediments by acetylene and ethylene: validity of the acetylene reduction assay for anaerobic microcosms. *Appl. Microbiol.* 30:707-709.

Patriquin, D. G. 1978. Factors affecting nitrogenase activity (acetylene reducing activity) associated with excised roots of the emergent halophyte *Spartina alterniflora* Loisel. *Aquatic Bot.* 4:193-210.

Patriquin, D. G. and C. Keddy. 1978. Nitrogenase activity (acetylene reduction) in a Nova Scotian salt marsh: its association with angiosperms and the influence of some edaphic factors. *Aquatic Bot.* 4:227-244.

Patriquin, D. G. and C. R. McClung. 1978. Nitrogen accretion and the nature and possible significance of N_2 fixation (acetylene reduction) in a Nova Scotian *Spartina alterniflora* stand. *Mar. Biol.* 47:227-242.

Potts, M., W. E. Krumbein and J. Metzger. 1978. Nitrogen fixation rates in anaerobic sediments determined by acetylene reduction, a new ^{15}N field assay, and simultaneous total N ^{15}N determination, pp. 753-769. *In:* W. E. Krumbein (ed.), *Environmental Biogeochemistry and Geomicrobiology. Vol. 3: Methods, Metals and Assessment.* Ann Arbor Science Publishers, Inc., Ann Arbor, MI.

Rice, W. A., E. A. Paul and L. R. Wetter. 1967. The role of anaerobiosis in asymbiotic nitrogen fixation. *Can. J. Microbiol.* 13:829-836.

Sokal, R. R. and F. J. Rohlf. 1969. *Biometry.* W. H. Freeman and Co., San Francisco, CA. 776 pp.

Teal, J. M., I. Valiela and D. Berlo. 1979. Nitrogen fixation by rhizosphere and free-living bacteria in salt marsh sediments. *Limnol. Oceanogr.* 24:126-132.

Valiela, I., J. M. Teal, S. Volkman, D. Shafer and E. Carpenter. 1978. Nutrient and particulate fluxes in a salt marsh ecosystem: tidal exchanges and inputs by precipitation and ground water. *Limnol. Oceanogr.* 23:798-812.

VanRaalte, C., I. Valiela, E. J. Carpenter and J. M. Teal. 1974. Inhibition of nitrogen fixation in salt marshes measured by acetylene reduction. *Est. Coastal Mar. Sci.* 2:301-305.

Whitney, D. E., G. M. Woodwell and R. W. Howarth. 1975. Nitrogen fixation in Flax Pond: a Long Island salt marsh. *Limnol. Oceanogr.* 20:640-643.

Woodwell, G. M., C. A. S. Hall, D. E. Whitney and R. A. Houghton. 1979. The Flax Pond ecosystem study: exchanges of inorganic nitrogen between an estuarine marsh and Long Island Sound. *Ecology* 60:695-702.

NITROGEN AND PHOSPHORUS CYCLING IN A GULF COAST SALT MARSH

R. D. DeLaune *and* **W. H. Patrick, Jr.**

Laboratory for Wetland Soils & Sediments
Center for Wetland Resources
Louisiana State University
Baton Rouge, Louisiana

Abstract: The Gulf Coast salt marshes in the deltaic plain of the Mississippi River are in a rapidly subsiding zone. Accretion processes are essential for maintenance of the marsh surface within the intertidal range and for supplying essential nutrients for plant growth. Incoming sediment is the major source of plant nutrients for the *Spartina alterniflora* marsh grass with an input of 23 and 2.3 g $m^{-2}yr^{-1}$ of nitrogen and phosphorus, respectively. Nitrogen fixation also provides a significant source of nitrogen to the marsh. Mineralization of nutrients in the sediment from these sources provides a significant portion of the plants' requirements, but the marsh is still limited in nitrogen, as nitrogen fertilizer experiments show. Although there was no increase in plant biomass from added phosphorus, soil profile evidenced depletion of phosphorus with depth. Detrital export probably accounts for the largest losses of nitrogen and phosphorus from these salt marshes.

Introduction

Louisiana contains a vast and very productive wetland ecosystem. Features of its coastal wetland are related to the geological history of the Mississippi River. Over the past several thousand years, frequent flooding of the distributaries over their natural levees resulted in the formation of broad expanses of marshland near the coast. The salt marshes have a lower tidal range than their counterparts on the Atlantic coast and marsh inundation is generally influenced more by wind than tides. Marsh relief is extremely low and is characterized by slightly elevated natural levees adjacent to streams and water-bodies which gradually slope into inland depressions where the dominant marsh grass *Spartina alterniflora* is sparse or completely absent.

Over historic time, land building has exceeded erosion and subsidence and has resulted in the formation of a large deltaic plain. However, leveeing of the Mississippi River over the past century has prevented distributary switching and annual overbank flooding. Consequently, Louisiana salt marshes do not receive a direct supply of fluvial sediment sufficient to counteract rapid subsidence (Swanson and Thurlow 1973). Louisiana is currently losing (primarily to subsidence) approximately 16 square miles per year of its coastal wetlands (Gagliano and Van Beek 1970). The

143

marsh surface maintains an approximate elevation with respect to some
sea level datum by continual accumulation of dead plant material and by
entrapment and stabilization of organic detritus and some nutrient enriched
mineral sediment.

Nutrient cycling in these salt marshes is closely interrelated with these
geological processes. Cycling of nitrogen is of special interest since it has
been found to be the limiting nutrient for growth of *Spartina alterniflora*
(Patrick and DeLaune 1976; Buresh et al. 1979). We present in this paper
the major aspects of nitrogen and phosphorus cycling of a streamside salt
marsh ecosystem in Louisiana's Barataria Basin (29°13'N, 90°7'W). The
site is the same study area used by DeLaune et al. (1978).

Nitrogen and Phosphorus Pools

Aboveground plant biomass of the streamside marsh is on the order
of 1800 g m^{-2} containing approximately 0.7% N and 0.07% P. Separation
of live turgor roots shows that there is approximately an equal amount of
belowground biomass. Approximately 13 g N m^{-2} is incorporated in
aboveground plant biomass and an equal amount is in belowground plant
material. The soil contains 450 g N m^{-2} in the root zone (0-30 cm). Almost
all of this nitrogen is in the organic form which is not readily available to
plants. There is generally less than 2 g m^{-2} soil inorganic nitrogen of which
practically all is in the ammonium form (DeLaune et al. 1976). There is 1.3
g m^{-2} phosphorus in both the aboveground and belowground biomass. The
streamside soil contains 20 g m^{-2} of phosphorus.

Sedimentation

Cesium-137 dating has been used to document sedimentation rates
in these rapidly accreting marshes (DeLaune et al. 1978). Cesium-137 is a
fallout product of nuclear testing and does not occur naturally in the environment. Rates were calculated from peak ^{137}Cs concentrations found in
the profile which corresponds to 1963, the year of peak ^{137}Cs fallout, and
1954, the first year of significant ^{137}Cs fallout.

The streamside marsh is accreting at the rate of 1.35 cm yr^{-1}
(DeLaune et al. 1978). Density and carbon content of the marsh soil show
that the marsh grows vertically through organic detritus accumulation and
sediment input. The equivalent of 23 g m^{-2}yr^{-1} nitrogen and 2.3 g m^{-2}yr^{-1}
phosphorus is being supplied through sedimentation processes to the
streamside marsh (DeLaune et al. 1980). The nitrogen added is not all in a
form which can be taken up by plants. Most of the nitrogen is organic
nitrogen which must first be mineralized. Due to low mineralization of
organic nitrogen under anaerobic conditions only a small amount will be
available during the first year. However, in each succeeding year additional
nitrogen will be mineralized. These figures were calculated from the
nitrogen and phosphorus content of the incoming sediment caught in sediment traps and the bulk density of the marsh soil at the 0-3 cm depth

along with the vertical accretion rate of 1.35 cm yr^{-1} obtained from ^{137}Cs dating.

The most probable source of the sediment being deposited on the marsh surface is resuspended sediment from other environments within Barataria Basin, because the basin receives no significant riverborne sediment supply. The resuspended sediment is rich in plant nutrients which supplies nutrients for marsh plants which in turn enhance further sediment entrapment and stabilization. Productivity of *S. alterniflora* is greater in soils with greater bulk density, a result of greater mineral sediment input (DeLaune et al. 1979). A portion of the inorganic nitrogen and phosphorus incorporated in plant tissue is recycled and made available to succeeding years' plant growth. A large portion of the organic matter from primary production remains on the marsh, contributing to the aggradation processes of vertical marsh accretion. This helps compensate for a rapid subsidence rate due to compaction and dewatering of a several hundred-foot layer of recent Mississippi River alluvial sediment underlying the marsh surface.

Accumulation of Nitrogen and Phosphorus

The marsh is undoubtedly a sink for both nitrogen and phosphorus. Nitrogen and phosphorus is accumulating at the rate of 21 g m^{-2}yr^{-1} and 1.7 g m^{-2}yr^{-1} respectively (DeLaune et al. 1980). These rates were calculated from an accretion rate of 1.35 cm yr^{-1}, bulk density of the sediment profile and the average total nitrogen and phosphorus content of the sediment. The bulk of active roots of *S. alterniflora* is found above a depth of 30 cm. Thus, as the marsh grows vertically, the nitrogen and phosphorus below 30 cm depth become unavailable to marsh plants. This loss of potential nutrients to the root zone is being compensated for by the addition of nitrogen- and phosphorus-enriched sediment to the marsh surface.

Nitrogen Fixation

Nitrogen fixation provides the second largest input of nitrogen to the marsh. Casselman (1979) found fixation to occur in all habitats of the marsh except in the water column where no significant fixation was measured. A greater mean fixation was measured on dead plant material as compared to live plants. However, the combined rate of nitrogen fixation on both live and dead *S. alterniflora* was less than 0.2 g m^{-2}yr^{-1}. Seasonally, the greatest measured fixation was in the streamside marsh soil where fixation rates equivalent to 15.4 g m^{-2}yr^{-1} were measured. Fixation was considerably less (4.5 g m^{-2}yr^{-1}) in the adjoining inland marsh. Here fixation closely paralleled root distribution in the soil profile. A highly significant negative relationship between nitrogenase activity and 2*N* KCl extractable ammonium nitrogen in the soil profile was observed (Fig. 1).

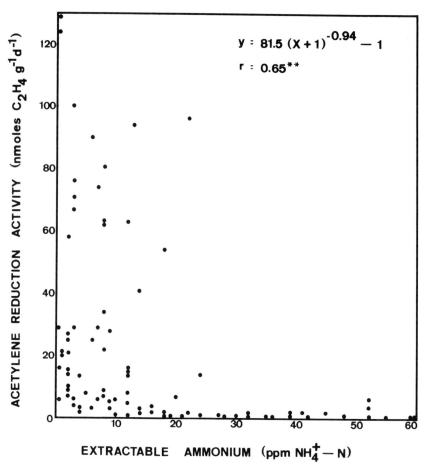

Figure 1. Relationship between acetylene reduction activity and extractable ammonium nitrogen in a Louisiana Spartina alterniflora salt marsh.

There was clearly a lack of nitrogenase activity with soil containing more than 20 μg/g ammonium nitrogen. The inhibiting effect of added fertilizer ammonium nitrogen on nitrogenase activity in wetland soil has been documented by Teal et al. (1979). The work of Casselman (1979) is the first on the inhibiting effect of native soil ammonium concentrations on nitrogenase activity in marsh soils.

Nitrogen Mineralization

Mineralization of organic nitrogen to the ammonium form is the primary source of inorganic nitrogen for S. alterniflora. Very little nitrate is present in these flooded marsh soils because there is a lack of oxygen for nitrification and also any nitrate which may be formed is rapidly denitrified.

Laboratory studies have shown that streamside soils are capable of mineralizing approximately 9.0 μg nitrogen per gram of soil per week through mineralization (Fig. 2). Ammonium nitrogen released by mineralization from the streamside marsh soil was continuously extracted with a slow flow of oxygen-free 2% NaCl solution at 30 C. Using an active rooting depth of 30 cm and a bulk density of 0.30 g cm^{-3} for the streamside marsh, the equivalent of 40 g m^{-2} of nitrogen would become available over a one year period. True mineralization rates in the field would be somewhat less since mineralization is slower during winter months. We thus estimate that the marsh is supplying 25 g m^{-2}yr^{-1} of ammonium nitrogen through the mineralization process. Most of this is quickly taken up by plants.

Nitrogen and Phosphorus Fertilization Experiments

Even though the mineralization of these nutrients in the sediment provides a significant portion of the plant requirement, the marsh is still limiting in nitrogen as nitrogen fertilization experiments have shown. Supplemental labelled inorganic nitrogen applied in the spring increased the aboveground biomass *S. alterniflora* at the streamside marsh by 15% (Patrick and DeLaune 1976). The added nitrogen caused a yield increase equivalent to 250 g m^{-2}. Ammonium nitrogen was used in this experiment. Practically all of the inorganic nitrogen in a reduced soil is found in the

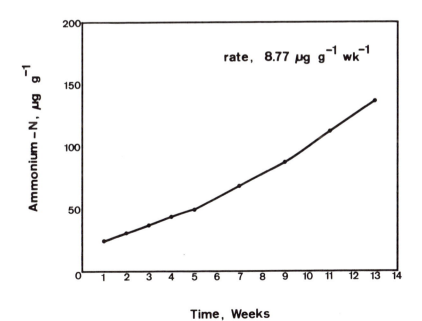

Figure 2. *Nitrogen mineralization rate of streamside marsh soil.*

ammonium form. Nitrate is ineffective as a nutrient because it is subject to rapid denitrification under reduced soil conditions. Not only was productivity increased by the added nitrogen, but the content of nitrogen in the plant material was also increased. The use of labelled nitrogen made it possible to distinguish between plant nitrogen derived from the sediment and the added fertilizer nitrogen. The amount of plant nitrogen derived from the sediment was about 59% during June and July and increased to about 69% by September. The supply of added nitrogen was apparently diminished toward the end of the growing season, probably through nitrogen losses from nitrification-denitrification reactions as well as prior plant uptake. Only 29% of the 20 g m^{-2} of added nitrogen was recovered in the aboveground portion of the plants in September. Assuming an equal amount was incorporated in the belowground biomass, we can account for 60% of the added labelled nitrogen.

A similar experiment in which supplemental nitrogen was applied to the adjacent inland marsh which is receiving less mineral sediment showed a greater response to added nitrogen (Buresh et al. 1979). The addition of 20 g m^{-2} of labelled nitrogen in May significantly increased total aboveground plant biomass and plant height by 28% and 25%, respectively, during the growing season. The increase in plant biomass was almost twice the increase observed from the addition of an equal amount of nitrogen at the streamside location. The inland marsh sediment supplied approximately 50% of the nitrogen taken up by the plants over the growing season as compared to 69% for sediments from the streamside marsh. These fertilization experiments using labelled nitrogen show that the inland marsh soils are apparently supplying less nitrogen for plant growth than streamside marsh soils. In this study 57% of the added nitrogen was recovered in the aboveground and belowground biomass.

The relatively large recovery from a single addition of a large amount of added ammonium nitrogen in these field studies indicates that S. alterniflora has a high capacity to assimilate supplemental nitrogen in this nitrogen-limiting environment.

There was no response of S. alterniflora to the addition of phosphorus at the rate of 20 g m^{-2} at either the streamside or inland marsh (Patrick and DeLaune 1976; Buresh et al. 1979). However, the concentration of phosphorus in the plant tissue increased about 20% from these additions of inorganic phosphorus. The increase in phosphorus uptake represented a very small fraction (about 1%) of the added phosphate. Even though no plant response to added phosphorus has been observed and sedimentation is apparently supplying adequate phosphorus, a statistically significant decrease in phosphorus with depth has been observed (Fig. 3) (DeLaune et al. 1980). The decrease was not related to any changes in soil mineral content or organic matter. Sediment at the lower depth of the soil profile has had more time for its phosphorus to be removed by S. alterniflora than more recently deposited sediment. Other

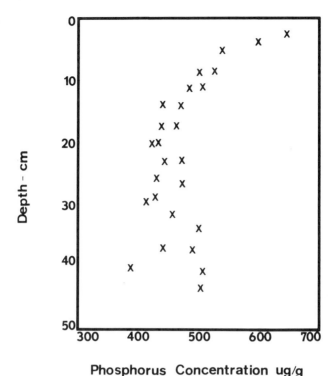

Phosphorus Concentration ug/g

Figure 3. Profile distribution of phosphorus in a Louisiana Spartina alterniflora *salt marsh soil.*

researchers have shown that the dominant mechanism of nutrient removal from marsh soil is the uptake by marsh grasses through their root system followed by release of some of these nutrients to the water column through excretion from stem and leaves of living marsh plants and through detrital export of dead plant material (Pomeroy et al. 1969; Reimold 1972).

Denitrification

Greenhouse studies in which labelled ammonium nitrogen equivalent to 100 μg nitrogen per gram of soil was added in 10 μg increments to marsh soil cores containing *S. alterniflora* have shown that less than 10% of the added nitrogen is lost from the system over a 21 week period (Buresh 1978). The ammonium nitrogen was added in increments so as to approximate the release of nitrogen to plants by soil mineralization processes. Nitrification-denitrification reactions were presumably the mechanism of the ammonium nitrogen loss. These results indicate that the ammonium nitrogen mineralized in these marsh soil systems is rapidly assimilated by *S. alterniflora* before it can undergo nitrification-denitrification reactions and be lost as gas. Current preliminary research is showing no N_2O flux from these salt marshes which may also indicate little denitrification. We estimated that only 10-15% (~3 g $m^{-2}yr^{-1}$) of the mineralization ammonium nitrogen is lost through nitrification-denitrification reactions.

Our estimates of denitrification are lower than those reported for Atlantic coast salt marshes. Valiela and Teal (1979) reported large losses of nitrogen through denitrification in the Great Sippewissett marsh. High denitrification rates would be expected in these marshes since they receive large inputs of nitrate from ground water. Louisiana Gulf Coast salt marshes receive no inorganic nitrogen from ground water sources and tidal flushing brings in only small quantities of nitrate. Haines et al. (1977) estimated denitrification to be 12 g N $m^{-2}yr^{-1}$ in a Georgia salt marsh.

Tidal Exchanges

Tidal exchanges are also major processes in these salt marshes. Ho and Lane (1973) hypothesized that tidal water moving onto the marsh will have a higher nitrate content while water departing the marsh will have more ammonium, primarily as a result of exchanges with interstitial water in the marsh soil profile. Any nitrate in tidal water entering the marsh will be rapidly denitrified. There is probably a small net export of dissolved inorganic nitrogen from the marsh.

We estimated that the streamside marsh exports 10-12 g N $m^{-2}yr^{-1}$ of particulate nitrogen or 40% of the net aboveground production of *S. alterniflora*. An estimated 1.0 g $m^{-2}yr^{-1}$ of phosphorus is lost through the same process. Peak aboveground plant biomass of the streamside marsh as measured by clip plot techniques are on the order of 1800 g m^{-2}. Calculated net production is even greater (Hopkinson et al. 1978).

Conclusions

In summary, the total nitrogen input to the saltwater marsh by sedimentation, fixation, and rainfall is about 38 to 40 g N $m^{-2}yr^{-1}$. Losses of nitrogen through detrital export, denitrification, export of dissolved inorganic nitrogen etc. are estimated to be approximately 18 g N $m^{-2}yr^{-1}$, leaving a net accumulation of 21 g N $m^{-2}yr^{-1}$. Most of the nitrogen is organic nitrogen associated with the large amounts of organic matter accumulating in this rapidly accreting marsh. Sedimentation is also the major phosphorus input to the marsh, supplying 2.3 g $m^{-2}yr^{-1}$. Appreciable soil phosphorus is apparently being removed through plant uptake and detrital export as evidenced by depletion of phosphorus in the soil profile.

Acknowledgments

This research was sponsored by the Louisiana Sea Grant Program, part of the National Sea Grant Program maintained by the National Oceanic and Atmospheric Administration, U.S. Department of Commerce.

References Cited

Buresh, R. J. 1978. Nitrogen transformation and utilization by *Spartina alterniflora* in a Louisiana salt marsh. Ph.D. Thesis, Louisiana State University, Baton Rouge, La. 117 p.

Buresh, R. J., R. D. DeLaune and W. H. Patrick, Jr. 1980. Nitrogen and phosphorus distribution and utilization in a Louisiana Gulf Coast marsh (accepted for publication in *Estuaries*).

Casselman, M. E. 1979. Biological nitrogen fixation in a Louisiana Gulf Coast salt marsh. M.S. Thesis, Louisiana State University, Baton Rouge, La.

DeLaune, R. D., W. H. Patrick, Jr. and J. M. Brannon. 1976. Nutrient transformation in Louisiana salt marsh soils. Sea Grant Publication No. LSU-T-76-009. Louisiana State University, Baton Rouge, La.

DeLaune, R. D., W. H. Patrick, Jr. and R. J. Buresh. 1978. Sedimentation rates determined by ^{137}Cs. *Nature* 275:532-533.

DeLaune, R. D., R. J. Buresh and W. H. Patrick, Jr. 1979. Relationship of soil properties to standing crop biomass of *Spartina alterniflora* in a Louisiana marsh. *Est. Coastal Mar. Sci.* 8:477-487.

DeLaune, R. D., C. N. Reddy and W. H. Patrick, Jr. 1980. Accumulation of plant nutrients and heavy metals through sedimentation processes and accretion in a Louisiana salt marsh (accepted for publication in *Estuaries*).

Gagliano, S. M. and J. L. Van Beek. 1970. Hydrologic and geologic studies of coastal Louisiana. Rep. No. 1, Center for Wetland Resources, Louisiana State University, Baton Rouge, La.

Haines, E., A. Chalmers, B. Hanson and B. Sherr. 1977. Nitrogen pools and fluxes in a Georgia salt marsh, pp. 241-254. *In*: M. Wiley (ed.), *Estuarine Processes. Vol. II.* Academic Press, New York.

Ho, C. L. and J. Lane. 1973. Interstitial water composition in Barataria Bay Louisiana sediment. *Est. Coastal Mar. Sci.* 1:125-135.

Hopkinson, C. S., J. G. Gosselink and R. T. Parrondo. 1978. Aboveground production of seven marsh plant species in coastal Louisiana. *Ecology* 59:760-769.

Patrick, W. H., Jr. and R. D. DeLaune. 1976. Nitrogen and phosphorus utilization by *Spartina alterniflora* in a salt marsh in Barataria Bay, Louisiana. *Est. Coastal Mar. Sci.* 4:59-64.

Pomeroy, L. R., R. E. Johannes, E. P. Odum and B. Roffman. 1969. The phosphorus and zinc cycles and productivity of a salt marsh. Proc. Nat. Symp. Radioecol., pp. 412-419.

Reimold, R. J. 1972. The movement of phosphorus through the salt marsh cordgrass *Spartina alterniflora*. *Limnol. Oceanogr.* 17:606-611.

Swanson, R. L. and C. I. Thurlow. 1973. Recent subsidence rates along the Texas and Louisiana coast as determined by tide measurements. *J. Geophys. Res.* 78:2665-2671.

Teal, J. M., I. Valiela and D. Berlo. 1979. Nitrogen fixation by rhizosphere and free-living bacteria in salt marsh sediments. *Limnol. Oceanogr.* 24:126-132.

Valiela, I. and J. M. Teal. 1979. The nitrogen budget of a salt marsh ecosystem. *Nature* 280:652-656.

MICROBIAL NITROGEN CYCLING IN A SEAGRASS COMMUNITY

Douglas G. Capone
Marine Sciences Research Center
State University of New York
Stony Brook, New York
and
Barrie F. Taylor
University of Miami
Rosenstiel School of Marine and Atmospheric Science
Miami, Florida

Abstract: The seagrass, *Thalassia testudinum,* is a highly productive marine plant which grows in shallow waters of the tropical and sub-tropical Atlantic Ocean. Its production in these nutrient poor environments is regulated by availability of nitrogen. Studies of $N_2(C_2H_2)$ fixation associated with *T. testudinum* communities indicate that bacterial activity in the rhizosphere contributes a substantial fraction of the plant's nitrogen requirement. In the phyllosphere, N_2 fixation by heterocystous cyanobacteria may promote development of other leaf epiphytes. Nitrogenase activity is also associated with detrital leaves, and this provides another locus of nitrogen input to the seagrass community. Nitrogen fixation and denitrification (measured by C_2H_2 blockage/N_2O production) occur simultaneously in anaerobic assays of rhizosphere sediments, but the latter process requires the addition of NO_3^-. The significance of the presence of denitrifying bacteria therefore depends on the rate of *in situ* NO_3^- production. Nitrification would be promoted both by the high levels of NH_4^+ in these sediments and by the O_2 conveyed into the rhizosphere by the extensive rhizome system of *T. testudinum.* A comprehension of the interplay of bacterial activities and their relevance to the nitrogen cycle of this seagrass, particularly with respect to nitrification, requires further scrutiny.

Introduction

Seagrasses are highly productive plants which thrive over broad expanses of shallow bottoms and form the essential framework of an important marine community (McRoy and McMillan 1977). *Thalassia testudinum* is the predominant seagrass in the tropical and subtropical Atlantic Ocean (Den Hartog 1970) and its growth in these generally nutrient poor waters has been correlated with the concentration of NH_4^+ in its sediments (Patriquin 1972). Undoubtedly nitrogen often plays a crucial role in the functioning of marine ecosystems (Ryther and Dunstan 1971; Webb et al. 1975; Valiela and Teal 1979) but the value of N_2 fixation, a *de novo* source of nitrogen, in seagrass communities has been in dispute (Goering and Parker 1972; Patriquin and Knowles 1972; McRoy et al. 1973). We have examined this question in detail over the past few years (Capone et al. 1977, 1979; Capone and Taylor 1977, 1980) and, more recently, another aspect of the microbial nitrogen cycle associated with *T. testudinum*, namely denitrification (Capone and Taylor in preparation). This paper briefly synthesizes our data and current interpretation of the nitrogen cycle in these communities.

Methods and Materials

The stands of *T. testudinum* under study are located in Biscayne Bay, Florida (Capone and Taylor 1977), and in Bimini Harbor, Bahamas (Capone et al. 1979). Site 2 is on the bay side of Soldier Key (Biscayne Bay) and has standing stocks of leaves of about 250 g dry wt m^{-2} in water 2-3 m deep. Site 2A, on the ocean side of Soldier Key, has leaf densities of about 50 g dry wt m^{-2} with water depths of 0.3-0.5 m. Leaf standing stocks are of the order of 200 g dry wt m^{-2} at sites A and B in Bimini Harbor. Mean water depth at these sites is approximately 1 m and strong tidal currents in the harbor assure an exchange with the nearby ocean waters.

Nitrogen fixation was assayed by the C_2H_2 reduction technique (Stewart et al. 1967; Hardy et al. 1968) and denitrification by the C_2H_2 blockage method (Balderston et al. 1976; Yoshinari et al. 1977). Nitrogen fixation and denitrification results were calculated from short-term linear rates of C_2H_4 or N_2O production, respectively; detailed procedures have been previously described (Capone et al. 1977, 1979; Capone and Taylor 1977, 1980, in preparation). A theoretical molar ratio of 3:1 was used to convert C_2H_4 production into N_2 fixation (Patriquin and Knowles 1972). Primary productivity was measured by a $^{14}CO_2$ method (Wetzel 1964) and the procedures of Zieman (1968, 1974) and Patriquin (1973); these techniques were described in detail in earlier publications (Penhale 1977; Capone et al. 1979).

Results

Phyllosphere

N_2 fixation associated with leaves of *T. testudinum* was generally correlated with the presence of heterocystous cyanobacteria, particularly *Calothrix* sp. (Capone and Taylor 1977). Accordingly the activity was substantially stimulated by light (Table 1), but not by organic compounds, and proceeded equally well under either aerobic or anaerobic conditions. Phyllosphere N_2 fixation showed diurnal, seasonal and spatial variations. Table 2 illustrates the seasonal and spatial effects and complements earlier data (Capone and Taylor 1977). Diurnal fluctuations reflected light stimulation of the process and seasonal changes were likely determined by water temperature, although maximal activities occurred in early summer rather than later in the year (Capone and Taylor 1977). Factors controlling spatial variation probably included water depth, and its accompanying changes in

Table 1. Nitrogen fixation associated with whole shoots and detrital leaves of *Thalassia testudinum* from Bimini Harbor. Samples were incubated aerobically *in situ* in Roux bottles (vol. 900 ml) for 3-6 h. Results are means ± standard error with number of determinations in parentheses. N.D. = not determined. C = water temperature.

		C_2H_4 Production (nmoles g dry wt^{-1}h^{-1})			
		Leaves		*Detrital Leaves*	
Date	*(C)*	*Light*	*Dark*	*Light*	*Dark*
Feb 1975	25	83 (2)	N.D.	N.D.	N.D.
Jul 1975	30	142 ± 10 (4)	N.D.	N.D.	N.D.
Aug 1976	30	125 ± 7 (12)	17 ± 4(4)	222 ± 27 (3)	9 (2)
Jul 1977	30	210 ± 28 (3)	13 (1)	307 ± 69 (3)	16 (1)
Aug 1978	30	77 ± 13 (3)	17 (1)	113 ± 20 (30)	25 (1)

light intensity and quality, and nutrient levels, especially of nitrogen compounds. The latter parameter almost certainly accounts for higher nitrogenase activities consistently observed on the ocean side relative to the Biscayne Bay side of Soldier Key (Table 1). In another study of a grass bed located near Fowey Rock just outside Biscayne Bay, spatial variation was inversely correlated to areal density of the plants (Capone and Taylor 1977). N_2 fixation rates (per dry weight of leaf and epiphyte material), were highest in the sparser developing regions of the bed where recycling of detrital material might be less efficient.

Table 2. Variation in N_2 fixation in the phyllosphere of *Thalassia testudinum* at two stations near Soldier Key, Florida. Whole shoots (5-6) were assayed aerobically in the light in Roux bottles (vol. 900 ml) *in situ* between 1000 and 1400 hrs. Values are means ± standard error with number of determinations in parentheses.

		C_2H_4 Production (nmoles g dry wt^{-1} h^{-1})		
Date	*Water Temp. (C)*	*Bay Side*	*Ocean Side*	*Ratio (Ocean:Bay)*
March 1976	21	43 ± 4 (3)	248 ± 60 (3)	5.8
June 1976	28	200 ± 10 (3)	519 ± 42 (3)	2.6
July 1976	31	195 ± 49 (3)	887 ± 25 (3)	5.0
April 1977	23	56 ± 2 (4)	422 ± 81 (4)	7.5

Light-enhanced N_2 fixation was also associated with detrital leaves (Table 1) and, since the short-term activities were neither stimulated by organic compounds nor inhibited by aerobic conditions, the principal microbes involved were most likely heterocystous cyanobacteria.

Rhizosphere

Nitrogen fixation was associated with rhizosphere cores (containing roots, rhizomes and sediments) from *T. testudinum* beds (Capone et al. 1977, 1979; Capone and Taylor 1980). The activity occurred at depths of at least 40 cm but was predominantly located in the upper 20 cm of the sediments. Rates were an order of magnitude higher for material from grassy regions relative to cores collected from nearby areas of bare bottom. Also, the activity with rhizosphere cores was correlated to the mass of roots and rhizomes rather than sediment weight. Undoubtedly the fixation of N_2 in the root zone of *T. testudinum* is affected by a variety of factors such as root exudates, gas transport, and decay of plant tissue. Diurnal fluctuations in activity support this idea, with initially increasing rates probably related to root exudation, and depressed rates in the afternoon attributable to a partial inactivation of N_2 fixing populations by elevated O_2 levels in the rhizosphere (Gotto and Taylor 1976). Seasonal variations in N_2 fixation, like those for the phyllosphere, were most likely in response to changes in temperature (Capone and Taylor 1980).

Even though N_2 fixation provides an input of nitrogen to *T. testudinum* communities, possible losses need evaluating, especially since denitrification occurs in marine sediments (Koike and Hattori 1978; Sorensen 1978; Oren and Blackburn 1979). Therefore, the rate of N_2 fixation and the potential for denitrification associated with rhizosphere cores were compared (Capone and Taylor in preparation). Denitrification was undetectable unless NO_3^- was added to the assay systems and its rate increased in accord with the concentration of NO_3^-. Nitrogen fixation rates, in contrast, were inversely related to NO_3^- levels (Fig. 1). Recoveries of N_2O from NO_3^- were substantially less than 100% thereby suggesting, as observed in other marine sediments (Koike and Hattori 1978; Sorensen 1978), a dissimilative reduction of NO_3^- to NH_4^+ operating simultaneously with denitrification.

Discussion

Previous investigations of N_2 fixation in seagrass communities have been limited in scope and this might partly account for the conflicting data obtained. A very high estimate of N_2 fixation in the phyllosphere (300 mgN m^{-2} day^{-1}) was derived by Goering and Parker (1972) for *T. testudinum* beds in Redfish Bay, Texas. When small pieces of leaves heavily covered with *Calothrix* sp. colonies were selected, and assayed in small serum bottles (volume = 21 ml), we obtained specific rates similar to Goering and Parker's average value of 11 μmoles C_2H_4 h^{-1} g dry wt^{-1} of plant material. Thus the inadvertent use of such experimental material might have biased

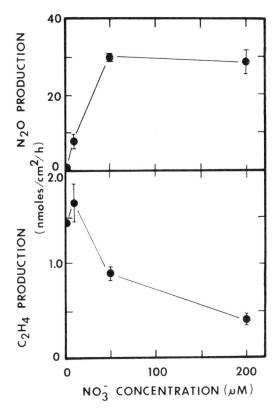

Figure 1. Effect of NO_3^- concentration on N_2 fixation (C_2H_4 production) and denitrification rates (N_2O production) associated with rhizosphere samples from Soldier Key, Florida. Results from Capone and Taylor (in preparation).

their extrapolations, especially since they were based on a small number of assays (6 light, 6 dark). Our assays using large Roux bottles (volume = 900 ml) containing whole shoots from several plants probably gives more realistic estimates for mean rates of N_2 fixation in the phyllosphere (Capone and Taylor 1977).

Patriquin and Knowles (1972) reported N_2 fixation rates of from 27 to 137 mgN m^{-2} day^{-1} for the rhizosphere based on a single sampling at one site in Barbados, West Indies. Furthermore, as noted by McRoy et al. (1973), their delay before assaying (2 days) and the duration of the assays (2-3 days) might have greatly enhanced activities. However, the highest N_2 fixation rates we noted for rhizosphere cores were with samples from seagrass stands near Bath, Barbados (Table 3). This result was obtained in short-term assays with freshly collected material and it falls into the lower part of the range estimated by Patriquin and Knowles (1972). Finally, the negligible or minimal N_2 fixation rates detected by McRoy et al. (1973) associated with the phyllosphere and rhizosphere of *T. testudinum* might be partly attributable to sampling during winter months (January and March).

A summary of our estimates for N_2 fixation in the phyllosphere and rhizosphere during the summer months at several sites is presented in Table 3. Activities in the rhizosphere were quantitatively the most important and this observation supports our previous data obtained in a comparative study of N_2 fixation and productivity of *T. testudinum* communities in Bimini Harbor (Capone et al. 1979). It was calculated from the Bimini data that N_2 fixation in the rhizosphere might provide between 20 to 50% of nitrogen required for leaf production. A similar conclusion is reached using our more recent data from Biscayne Bay. These data and those of the Bimini investigation are summarized and compared in Table 4.

An empirical relationship observed in the Bimini study (Capone et al. 1979) between CO_2 and N_2 fixation suggests that N_2 fixation is important to the productivity of *T. testudinum*. The observed relationship also suggests that leaves and their epiphytes are relatively independent of each other with respect to nitrogen and carbon metabolism. Also, plant growth as indicated in the earlier work of Patriquin (1972) probably depends mainly on nitrogen in the sediments. We have now determined the

Table 3. Comparison of N_2 fixation in the phyllosphere and rhizosphere of *Thalassia testudinum*.

Sample	Station	Date	N_2 Fixation $(mgN\ m^{-2}\ d^{-1})$
Leaves	Soldier Key, Site 2	1976	5.3
	Soldier Key, Site 2A	1976	4.7
	Bimini, Site B	1976	3.2
		1977	5.0
		1978	2.1
Detritus	Bimini, Site A	1976	0.6
	Bimini, Site B	1976	0.8
		1977	1.1
		1978	0.5
Rhizomes, roots and sediments	Soldier Key, Site 2	1977	7.6
	Soldier Key, Site 2	1977	11.4
	Bimini, Site A	1976	5.1
	Bimini, Site B	1976	5.3
		1977	13.9
		1978	12.1
	Bath, Barbados	1978	37.6

Table 4. Estimates of nitrogen demand and supply for the production of *Thalassia testudinum* calculated as described in Capone et al. (1979).

Site	Production Method	Productivity ($gC\,m^{-2}\,d^{-1}$)	N-Demand ($mgN\,m^{-2}\,d^{-1}$)	Sediment N-Fixation	
				($mgN\,m^{-2}\,d^{-1}$)	(% Demand)
Bimini A	^{14}C	2.36	34-44	5.14	12-15
Bimini A	Z	1.28	18-24	5.14	21-28
Bimini B	^{14}C	1.84	24-32	5.32	17-22
Bimini B	Z	0.85	11-15	5.32	36-47
Biscayne Bay	^{14}C[a]	1.92[b]	28-37	7.6	21-27
Soldier Key, Site 2	Z[c]	1.47[d]	22-29	7.6	26-36

[a] From Penhale and Thorhaug (1977)

[b] From yearly average

[c] From Zieman (1968)

[d] Average of determinations made April through October

magnitude and distribution of N_2 fixation associated with seagrass communities at our study sites. Several other processes need quantifying before nitrogen budgets can be developed for these communities. Clearly denitrification represents a possible loss, especially since it has been shown to outweigh inputs by N_2 fixation in a salt marsh ecosystem (Valiela and Teal 1979). The lack of endogenous N_2O production in our denitrification assays indicates a shortage of NO_3^- in the root zone of *T. testudinum*. In fact, dissolved NO_3^- levels in seagrass sediments have been reported as low or undetectable whereas NH_4^+ concentrations can be up to 0.1 mM (Patriquin 1972; Curry 1975). The presence of NH_4^+, together with the probable transport of O_2 into the sediments via the plant (Oremland and Taylor 1977), suggest a potential for nitrification. Nitrification rates must be determined for all components of the system but with an emphasis on sediment-O_2 interfaces (i. e. rhizosphere, sediment surface). Conceivably, as can happen in salt marshes (Valiela and Teal 1979), a high degree of nutrient conservation and recycling occurs in seagrass communities and this aspect of the nitrogen cycle also deserves attention.

Acknowledgments

We thank D. Fishbein, J. Gotto, T. Jereb, P. Keyser, C. Lee, J. Martin, R. Oremland, P. Penhale and M. Roman for their various assistances in this work. The RSMAS Marine Department provided logistic support and the Bahamian Government gave operational clearance. Financial support was supplied by the Oceanography Section of the National Science Foundation (GA-41135, OCE 74-01986, OCE 78-08618).

References Cited

Balderston, W. L., B. Sherr and W. J. Payne. 1976. Blockage by acetylene of nitrous oxide reduction in *Pseudomonas perfectomarinus*. Appl. Environ. Microbiol. 31:504-508.

Capone, D. G., R. S. Oremland and B. F. Taylor. 1977. Significance of N_2 fixation to the production of *Thalassia testudinum* communities, pp. 71-85. In: H. B. Stewart, Jr. (ed.), Cooperative Investigations of the Caribbean and Adjacent Regions, Vol. 2. FAO, Rome.

Capone, D. G., P. A. Penhale, R. S. Oremland and B. F. Taylor. 1979. Relationship between productivity and N_2 (C_2H_2) fixation in a *Thalassia testudinum* community. Limnol. Oceanogr. 24:117-125.

Capone, D. G. and B. F. Taylor. 1977. Nitrogen fixation (acetylene reduction) in the phyllosphere of *Thalassia testudinum*. Mar. Biol. 40:19-28.

Capone, D. G. and B. F. Taylor. 1980. $N_2(C_2H_2)$ fixation in the rhizosphere of *Thalassia testudinum*. Can. J. Microbiol. 26 (in press).

Curry, R. W. 1975. The concentration and distribution of nitrate-nitrogen and nitrite-nitrogen in the sediments of Biscayne Bay. 129 pp. M. S. thesis, University of Miami, Coral Gables.

Den Hartog, C. 1970. *The Sea Grasses of the World*. North Holland, Amsterdam and London. 275 pp.

Goering, J. J. and P. L. Parker. 1972. Nitrogen fixation by epiphytes on sea grasses. Limnol. Oceanogr. 17:320-323.

Gotto, J. W. and B. F. Taylor. 1976. N_2 fixation associated with decaying leaves of the Red Mangrove (*Rhizophora mangle*). Appl. Environ. Microbiol. 31:781-783.

Hardy, R. W., R. D. Holsten, E. K. Jackson and R. C. Burns. 1968. The acetylene-ethylene assay for N₂-fixation: Laboratory and field evaluation. *Plant Physiol.*43: 1158-1207.

Koike, I. and A. Hattori. 1978. Denitrification and ammonia formation in anaerobic coastal sediments. *Appl. Environ. Microbiol.* 35:278-282.

McRoy, C. P., J. J. Goering and B. Chaney. 1973. Nitrogen fixation associated with sea grasses. *Limnol. Oceanogr.* 18:998-1002.

McRoy, C. P. and C. McMillan. 1977. Production ecology and physiology of sea grasses, pp. 53-87. *In:* C. P. McRoy and C. Hefferich (ed.), *Seagrass Ecosystems.* Dekker, New York.

Oremland, R. S. and B. F. Taylor. 1977. Diurnal fluctuations of O_2, N_2 and CH_4 in the rhizosphere of *Thalassia testudinum. Limnol. Oceanogr.* 22:566-570.

Oren, A. and T. H. Blackburn. 1979. Estimation of sediment denitrification rates at *in situ* nitrate concentrations. *Appl. Environ. Microbiol.* 37:174-176.

Patriquin, D. 1972. The origin of nitrogen and phosphorus for growth of the marine angiosperm *Thalassia testudinum. Mar. Biol.* 15:35-46.

Patriquin, D. 1973. Estimation of growth rate, production and age of the marine angiosperm *Thalassia testudinum* Konig. *Carib. J. Sci.* 13:111-123.

Patriquin, D. and R. Knowles. 1972. Nitrogen fixation in the rhizosphere of marine angiosperms. *Mar. Biol.* 16:49-58.

Penhale, P. A. 1977. Macrophyte-epiphyte biomass and productivity in an eelgrass (*Zostera marina* L.) community. *J. Exp. Mar. Biol. Ecol.* 26:211-224.

Penhale, P. A. and A. Thorhaug. 1977. Macrophyte-epiphyte productivity in a subtropical *Thalassia* community. *Bot. Soc. Amer.* 63(5):24.

Ryther, J. H. and W. M. Dunstan. 1971. Nitrogen, phosphorus and eutrophication in the coastal marine environment. *Science* 171:1008-1012.

Sorensen, J. 1978. Denitrification rates in a marine sediment as measured by the acetylene inhibition technique. *Appl. Environ. Microbiol.* 36:139-143.

Stewart, W. D., G. P. Fitzgerald and R. H. Burris. 1967. *In situ* studies on N₂-fixation using the acetylene reduction technique. *Proc. Natl. Acad. Sci. U.S.* 58:2071-2078.

Valiela, I. and J. M. Teal. 1979. The nitrogen budget of a salt marsh ecosystem. *Nature* 280:652-656.

Webb, K. L., W. D. DuPaul, W. Wiebe, W. Sottile and R. E. Johannes. 1975. Enewetak (Eniwetok) Atoll: aspects of the nitrogen cycle on a coral reef. *Limnol. Oceanogr.* 20:198-210.

Wetzel, R. G. 1964. Primary production of aquatic macrophytes. *Int. Ver. Theor. Angew. Limnol. Verh.* 15:426-436.

Yoshinari, T., R. Hynes and R. Knowles. 1977. Acetylene inhibition of nitrous oxide reduction and measurement of denitrification and nitrogen-fixation in soil. *Soil Biol. Biochem.* 9:177-183.

Zieman, J. C. 1968. A study of the growth and decomposition of the sea grass, *Thalassia testudinum.* 50 pp. M.S. thesis, University of Miami, Coral Gables.

Zieman, J. C. 1974. Methods for the study of the growth and production of turtle grass, *Thalassia testudinum. Aquaculture* 4:139-143.

EFFECTS OF BIOTURBATION AND PREDATION BY *MELLITA QUINQUIESPERFORATA* ON SEDIMENTARY MICROBIAL COMMUNITY STRUCTURE

David C. White, Robert H. Findlay, Steven D. Fazio,
Ronald J. Bobbie, Janet S. Nickels, William M. Davis,
Glen A. Smith and Robert F. Martz

Department of Biological Science
Florida State University
Tallahassee, Florida

Abstract: Processing of sand by sand dollars (Echinodermata: *Mellita quinquiesperforata*) resulted in modification of the benthic microbial community without a significant effect on gross nutrient balances. Measures of cellular and membrane biomass (total adenosine nucleotides, lipid phosphate and chlorophyll *a*) were essentially unchanged. Muramic acid concentration and thymidine incorporation into DNA, which are measures of prokaryotic biomass and activity, remained unchanged. Total metabolic activity, measured as acetate incorporation into lipid, was unchanged. Lipid glycerol and the inositol and glucosamine remaining in the extracted residue were reduced in the processed sediment, as was sulpholipid biosynthesis. Fatty acids characteristic of prokaryotes were enriched whereas fatty acids characteristic of microeukaryotes decreased in processed sands. The same was true for the lipid neutral carbohydrates. Examination of meiofauna showed significant reduction in foraminifera, suggesting that sand dollars are selective predators for a portion of the non-photosynthetic microeukaryotes, having little effect on the biomass or metabolic activity of benthic prokaryotes.

Introduction

The benthic microbial community is a most important component in estuarine nutrient recycling (Cosgrove 1977; Fenchel and Jorgensen 1977; Focht and Verstraete 1977). This complex assemblage has been difficult to study because classical plate count or methods of enumeration by epifluorescent microscopy require quantitative recovery of the microbes from the sediments. The release of microbes from sediments by various techniques has been shown not to be reproducible when the uniquely prokaryotic wall component, muramic acid, is measured after the microbial stripping procedure (Moriarty 1980).

Measurements of lipids extracted from the sediments offer a method free from the selectivity of methods requiring growth or quantitative release from sediments. Measurements of lipid phosphate give data that correlate with sedimentary adenosine triphosphate (ATP) and the rate of ^3H methyl thymidine incorporation into DNA (White et al. 1979b). Simple hydrolytic procedures when applied to lipids can provide indications of the microbial community structure (White et al. 1979a).

Use of more defined methods of lipid analysis of detrital microflora has shown that lipid changes correspond to the morphology observed by scanning electron microscopy of members of estuarine detrital microbial communities, whose structure was altered by changing the water column chemistry (White et al. 1980). The success of these experiments prompted their application to the sedimentary microbial community.

Mechanical bioturbation changes microbial biomass, activity and nutritional state of benthic microbes (Nickels et al. 1979; White et al. 1979b). Deposit feeding invertebrates have a marked effect on sediment structure (Rhoads and Young 1970) and are attracted to high concentrations of microbes (Tunnicliffe and Risk 1977). Consequently, bioturbation and predation by deposit feeders should have marked effects on benthic microbial activity and community structure which in turn could affect estuarine nutrient recycling. In this study, the effect of the deposit feeding invertebrate *Mellita quinquiesperforata* on benthic microbiota was examined and shown to affect the structure of the benthic microbial community.

Methods and Materials

In the summer of 1979, experimental glass tanks 21.5 × 42 × 26 cm (w × l × h) were established and filled with 16 cm of sand sediments (ϕ = 2.5, mean grain size 0.177 mm) collected from a sand bar in Alligator Harbor near the Florida State University marine station. The sands were sieved through 0.5 mm mesh screens and allowed to equilibrate with filtered sea water at a flow rate of 400 ml/min (80% turnover/day) for 2 weeks. Experimental conditions included 26.5%$_{oo}$ salinity, pH 7.35, 28.5C, 3.5 ppm dissolved oxygen and a light intensity of 875 lux.

Four sand dollars (*Mellita quinquiesperforata*, Leske 1778) averaging 10 cm in diameter were added to each of 3 tanks. After 3 weeks, surface sediments above the redox discontinuity were recovered with a scoop from the 3 tanks with the sand dollars and 3 control tanks to which no sand dollars were added. Meiofaunal composition was determined for recovered sediment using four 10 g subsamples of each treatment after staining with Rose Bengal. Carbon, hydrogen and nitrogen (CHN) analyses were performed on the sediments with a Carlo Erba elemental analyzer using triplicate 15-20 mg subsamples.

Rates of incorporation of the isotopes ^{3}H-methyl thymidine into DNA, $H_2^{35}SO_4$ and Na acetate-1-^{14}C into lipids were measured in separate experiments using 10-20 g of sediment shaken in 25 ml of sea water with each flask containing 5 μCi of isotope. ^{3}H-thymidine was recovered in sedimentary microbial DNA using the mild alkaline extraction technique of Tobin and Anthony (1975). Lipid synthetic rates were measured as described by White et al. (1977).

Three sediment subsamples from control and sand dollar tanks were extracted with a modified one-phase chloroform-methanol extraction system (0.8:2:1, V/V) (White et al. 1979b) in the dark at 4C. The monophasic solution was partitioned into an organic and an aqueous phase by addition of equal volumes of chloroform and water for a final solvent ratio of 0.9:1:1(V/V). The aqueous phase contained the adenosine nucleotides. These nucleotides were converted to the 1-^{6}N-ethenoadenosine derivatives which were separated by high pressure liquid chromatography and the nucleotide content was determined fluorometrically (White et al. 1980). Acid methanolysis of the organic phase resulted in chloroform and water-soluble fragments of the lipids. A portion of the water-soluble fragment was analyzed for lipid phosphate (White et al. 1979b). The remaining portion was reduced in the presence of sodium borohydride, separated by column chromatography into neutral carbohydrates and amines, and analyzed by glass capillary chromatography (GLC) after formation of volatile derivatives (White et al. 1980). The chloroform-soluble lipid components were separated by thin layer chromatography and the fatty acid methyl esters were analyzed by capillary GLC (Bobbie and White 1980). Sediment residue remaining after lipid extraction was analyzed for muramic acid and other cell wall components as described by Fazio et al. (1979).

Fatty acids are defined as: number of carbon atoms: number of double bonds: number of carbon atoms to first double bond from the omega (ω) end of the molecule. The prefix a, i, Δ, indicates anteiso, iso or cyclopropane.

Results and Discussion

Sand dollar density

The four sand dollars covered 0.314 m², or 34% of the 0.904 m² surface area of each tank. This is comparable to the 5.9 − 489 animals/m² of various size classes reported for Tampa Bay, Florida (Lane 1977). The level of the redox discontinuity was not different between the control and sand dollar-containing tanks. The similarity in the appearance of the sediments was reflected in measurements of organic carbon and nitrogen of the two sediments (Table 1). Sand dollars in the tanks did not produce much visible bioturbation in agreement with the findings of Lane (1977). Consequently, sand dollars are probably not the most satisfactory animal to demonstrate effects of bioturbation on nutrient cycling.

Microbial biomass

Microbial biomass of sediments can be estimated by measurements of extractable lipids (White et al. 1979b). Extractable lipid phosphate represents a measure of the membrane content, as phospholipids are not normally endogenous storage materials, are localized in membranes, and their synthesis parallels bacterial growth (White and Tucker 1969). Lipid

Table 1. Effects of *Mellita quinquiesperforata* on sediment composition. Values for carbon and nitrogen are mean ± standard deviation (n = 9) and refer to % of the sediment dry weight.

	Control tanks	Sand dollar tanks
Total organic carbon	0.053 ± 0.008	0.057 ± 0.004
Total nitrogen	0.012 ± 0.005	0.013 ± 0.007
C/N ratio	4.6	4.5

glycerol measures both phospholipids and neutral lipids, especially in the microeukaryotes. The total adenosine nucleotide represents a preferable measure of the viable biomass over ATP. Rapid enzymatic hydrolysis of ATP can complicate the interpretation unless special precautions are taken (White et al. 1980).

The glucosamine and inositol recovered in hydrolysates of the residue from which the lipids have been extracted tend to be higher in microbial assemblages richer in microeukaryotes (White et al. 1980). Muramic acid constitutes a component unique to the cyanophyte and bacterial peptidoglycan. The decreased residue inositol, glucosamine and lipid glycerol in sand processed by sand dollars suggest predation on microeukaryotes (Table 2).

Microbial activities

Effects of bioturbation by sand dollars on synthesis of lipids, sulpholipids and DNA are shown as differences in the incorporation of ^{3}H-

Table 2. Effects of *M. quinquiesperforata* on microbial biomass determination. Values are means ± one standard deviation. * = significant at the 90% level (t test). n = 9.

μmoles/g dry wt	Control tanks	Sand dollar tanks
Total lipid phosphate	0.022 ± 0.001	0.021 ± 0.005
Total lipid glycerol	0.026 ± 0.013	0.012 ± 0.004*
Chlorophyll *a* (nmoles/g dry wt)	0.020 ± 0.016	0.012 ± 0.003
Total adenosine nucleotide	1.04 ± 0.32	0.96 ± 0.28
Wall inositol	0.78 ± 0.26	0.29 ± 0.16*
Wall glucosamine	0.74 ± 0.32	0.28 ± 0.16*
Muramic acid	0.09 ± 0.06	0.03 ± 0.02

Table 3. Effects of *M. quinquiesperforata* on microbial activity measured by hours taken to double the isotope incorporation (a relative measure dependent upon the specific activities of the labeled precursors). Values are mean ± one standard deviation. ** = significant at the 95% level (t test). n = 9.

	Control tanks	Sand dollar tanks
^{3}H-DNA formation	1.0 ± 0.64	1.2 ± 0.59
^{14}C-lipid formation	1.0 ± 0.41	1.2 ± 0.55
^{35}S-lipid formation	0.8 ± 0.28	1.7 ± 0.80**

methyl thymidine into DNA and ^{14}C-acetate incorporation into lipids (Table 3). ^{3}H-methyl thymidine incorporation into DNA represents a measure of prokaryotic growth as microeukaryotes lack the salvage pathway. ^{14}C-acetate incorporation into lipids has been shown to correlate with microbial mass and activity in estuarine detritus (Morrison et al. 1977; White et al. 1977). Incorporation of ^{14}C-acetate into lipids occurred in both prokaryotes and microeukaryotes. Incorporation of ^{35}S into sulpholipids was greater and more rapid in microeukaryotes than in prokaryotic microorganisms present in the detrital microflora (White et al. 1980). Incorporation of ^{35}S into sulpholipids was depressed in the sands processed by the sand dollars (Table 3).

Fatty acid methyl esters

Fatty acids characteristic of prokaryotes (iso + anteiso 15:0, cyclopropane fatty acids Δ17:0, Δ19:0 and cis-vaccenic acid [18:1ω7]) were elevated in sands processed by sand dollars (Table 4). Fatty acids characteristic of microeukaryotes (the polyenoic 18:2ω6, 22:4ω6, 22:5ω3, 22:6ω3, and saturated long chain 22:0, 24:0) were lower in sands processed by sand dollars. Fatty acid methyl esters derived from lipids have proven useful in defining the community structure of detrital microbiota (White et al. 1979a; Bobbie and White 1980; White et al. 1980). Eukaryotic-enriched populations showed increased relative amounts of 18:0, 22:0, 24:0, and polyenoic fatty acids of both the *a* linolenic (ω3) and γ linolenic series (ω6). Prokaryotic-enriched populations showed high levels of branched 15:0, Δ17, Δ19, and the cis-vaccenic fatty acid (18:1ω7) formed by the bacterial anaerobic pathway.

Lipid neutral carbohydrates

Sand dollar processing of sediments resulted in a reduction in glycerol, I and M in the neutral carbohydrate derived from the lipids (Table 5), whereas I, L, arabinose and glucose increased in the sand reworked by

Table 4. Effects of *M. quinquiesperforata* on fatty acid methyl ester composition. Esters are abbreviated as number of carbon atoms: number of double bonds. i, a, br, indicate iso, anteiso and branched; Δ indicates cyclopropane; ω indicates the position of the first double bond from the ω end of the ester. Values are the mean ratios of esters $\pm$ standard deviation (n = 9). Statistical significance (t test) *** 99%; ** 95%; * 90%.

Ratios of fatty acid esters	Control tanks	Sand dollar tanks
i + a 15:0/15:0	0.56 ± 0.11	2.00 ± 0.07***
Δ17:0/16:0	0.012 ± 0.003	0.023 ± 0.003***
Δ19:0/16:0	0.009 ± 0.0006	0.015 ± 0.0000***
18:1ω7/18:1ω9	1.42 ± 0.14	2.12 ± 0.1***
18:0/19:0	651 ± 134	607 ± 168
16:0/19:0	9300 ± 3600	4200 ± 1200*
18:2ω6/19:0	370 ± 93	224 ± 79*
22:0/19:0	73 ± 6	46 ± 15**
24:0/19:0	170 ± 6	108 ± 8***
22:4ω6/19:0	110 ± 10	63 ± 15***
22:5ω3/19:0	120 ± 60	16 ± 17***
22:6ω3/19:0	197 ± 28	82 ± 32***

the sand dollars. These neutral carbohydrates are associated with a prokaryote-enriched detrital microflora (White et al. 1980).

Lipid amines

The lipid-derived amines, ethanolamine, serine, components B, C, D, were similar in sand dollar-processed and control sands. Monohydroxy steroids, which offer great potential use in differentiating the various micro-eukaryotes (White et al. 1980), were not detected by GLC.

Differences in meiofaunal populations

Activity of sand dollars resulted in changes in the meiofauna after 3 weeks (Table 6). There were more harpacticoid copepods and juvenile polychaetes and significantly less vital staining foraminiferans in the presence of the sand dollars.

From these studies it was possible to make inferences as to the effects of sand dollars on the benthic microbes. It appeared that the sand dollars did not affect the benthic prokaryotes. Prokaryotic biomass measured as muramic acid was not changed. The unchanged bacterial biomass probably did not represent sand dollar feeding followed by rapid recovery, as ^{3}H thymidine incorporation into DNA was not increased.

Table 5. Effect of *M. quinquiesperforata* on the ratios of neutral carbohydrates and primary amines derived from the lipids. A through M refer to undefined carbohydrate components with progressively increasing retention volumes eluted from a polar SP-2330 column. Values are ratios relative to a component eluting just before rhamnose given as the mean ± standard deviation (n = 9). Statistical significance (t test) *** 99%; ** 95%; * 90%.

	Control tanks	Sand dollar tanks
Glycerol	168.0 ± 34.0	81.0 ± 3.0***
A	2.0 ± 1.4	2.6 ± 1.2
B	2.2 ± 0.8	3.1 ± 1.1
C	1.2 ± 1.0	1.4 ± 0.2
D	0.7 ± 0.7	1.9 ± 0.8***
E	0.9 ± 0.5	0.8 ± 0.3
F	1.1 ± 0.5	1.4 ± 0.4
G	1.8 ± 0.6	0.8 ± 0.7
H	1.6 ± 1.0	1.2 ± 1.1
I	1.5 ± 0.3	0.6 ± 0.5**
J	0.7 ± 0.7	2.7 ± 2.4
Rhamnose	3.3 ± 1.4	3.2 ± 0.5
Fucose	1.0 ± 1.0	1.8 ± 0.5
Ribose	1.0 ± 0.2	1.2 ± 0.1
Arabinose	5.2 ± 0.5	8.3 ± 2.2**
K	4.4 ± 1.9	6.2 ± 0.3
L	2.6 ± 2.6	9.2 ± 2.7**
Xylose	5.7 ± 2.7	5.4 ± 0.1
M	2.2 ± 0.6	1.0 ± 0.3***
Mannose	6.9 ± 0.7	7.3 ± 4.0
Galactose	153.0 ± 67.0	95.0 ± 13.0*
Glucose	5.1 ± 1.4	7.9 ± 1.6**
Inositol	2.7 ± 0.7	3.4 ± 1.0

Table 6. Effect of *M. quinquiesperforata* on meiofaunal animal populations. Values are means ± one standard deviation. *** = significant at the 99% level (t test). n = 3.

Number per 10 g dry sediment	Control tanks	Sand dollar tanks
Harpacticoid sp.	1.8 ± 1.26	8.0 ± 3.4***
Nematode sp.	17.0 ± 15.3	12.3 ± 2.8
Foraminifera sp.	14.3 ± 4.9	2.8 ± 1.3***
Polychaeta (juvenile) sp.	0.3 ± 0.5	1.8 ± 1.0***

There was little change in ratios of cyclopropane and short-branched fatty acids. Since the fatty acids are each reasonably specific to portions of the bacterial population (Bobbie et al. 1980), the community structure of the bacterial population did not shift markedly. Cyclopropane fatty acids accumulate in older and stressed bacterial monocultures of some bacteria (Knivett and Cullen 1965). Sand dollar feeding provoked no evidence of stress on that portion of the bacterial community. Cyanophytes and microalgae were not affected by sand dollar predation as the total chlorophyll *a* and the lipid galactose were similar in disturbed and control sediments. Sand dollar predation depressed microeukaryotic activity and biomass. Depressed activity was reflected in the decreased rate of sulpholipid synthesis. Decreased biomass was seen in decreased lipid glycerol, lipid long chain and polyenoic fatty acids, and residue inositol and glucosamine. Losses in microeukaryotic lipid components in sand dollar-processed sands were reflected by increases in fatty acids and lipid carbohydrates associated with prokaryotes. Lipid-derived amines were not affected by sand dollar bioturbation. Foraminifera disappeared in the disturbed tanks.

Lane (1977) found that sand dollars selectively concentrate particles with diameters less than 64 µm. She found living phytoplankton in the feces, indicating that sand dollars selectively digest organisms that pass through the gut.

Lipid analysis provides a number of components whose identity and relevance to microbial community structure are unknown. Structural confirmation by capillary gas chromatography-electron impact mass spectrometry has thus far only been applied to the major fatty acid methyl esters (Bobbie and White 1980, unpublished data). Further refinements of these analyses and experimental techniques will allow for the development of biochemical methods to characterize the benthic microbial community more fully.

Acknowledgments

This work was supported by National Oceanic and Atmospheric Administration Office of Sea Grant, Department of Commerce, under grant 04-7-158-4406. The U.S. government is authorized to produce and distribute reprints of this article for governmental purposes regardless of any copyright notation that may appear in this volume. The work was also supported by grant OCE 78-21174 from the Biological Oceanography program of the National Science Foundation, grant R-0806143010 from the U.S. Environmental Protection Agency and contract 31-109-38-4502 from the Department of Energy (Argonne National Laboratory). RJB received a grant DEB 7807498 from the National Science Foundation in support of Doctoral Dissertation Research.

References Cited

Bobbie, R. J. and D. C. White. 1980. Characterization of benthic microbial community structure by high resolution gas chromatography. *Appl. Environ. Microbiol.* 39 (in press).

Cosgrove, D. J. 1977. Microbial transformations in the phosphorous cycle. *Adv. Microb. Ecol.* 1:95-134.

Fazio, S. D., W. R. Mayberry and D. C. White. 1979. Muramic acid assay in sediments. *Appl. Environ. Microbiol.* 38:349-350.

Fenchel, T. M. and B. B. Jorgensen. 1977. Detritus food chains of aquatic ecosystems: The role of bacteria. *Adv. Microb. Ecol.* 1:1-58.

Focht, D. D. and W. Verstraete. 1977. Biochemical ecology of nitrification and denitrification. *Adv. Microb. Ecol.* 1:135-214.

Knivett, V. A. and J. Cullen. 1965. Some factors affecting cyclopropane acid formation in *Escherichia coli. Biochem. J.* 96:771-776.

Lane, J. M. 1977. Bioenergetics of the sand dollar *Mellita quinquiesperforata* (Leske 1778). Ph.D Thesis, U. South Florida, Tampa, FL.

Moriarty, D. J. W. 1980. Problems in the measurement of bacterial biomass in sandy sediments. *In:* M. R. Walter, P. A. Trudinger and B. J. Ralph (eds.), Proc. 4th Internat. Symp. Environ. Biogeochem., Australian Academy of Science, Canberra (in press).

Morrison, S. J., J. D. King, R. J. Bobbie, R. E. Bechtold and D. C. White. 1977. Evidence for microfloral succession on allochthonous plant litter in Apalachicola Bay, Florida, USA. *Mar. Biol.* 41:229-240.

Nickels, J. S., J. D. King and D. C. White. 1979. Poly-β-hydroxybutyrate accumulation as a measure of unbalanced growth of the estuarine detrital microbiota. *Appl. Environ. Microbiol.* 37:459-465.

Rhoads, D. C. and D. K. Young. 1970. The influence of deposit-feeding organisms on bottom sediment stability and community trophic structure. *J. Mar. Res.* 28:150-178.

Tobin, R. S. and D. H. J. Anthony. 1978. Tritiated thymidine incorporation as a measure of microbial activity in lake sediments. *Limnol. Oceanogr.* 23:161-165.

Tunnicliffe, V. and M. J. Risk. 1977. Relationships between the bivalve *Macoma balthica* and bacteria in intertidal sediments, Minas Basin, Bay of Fundy. *J. Mar. Res.* 35:499-507.

White, D. C., R. J. Bobbie, S. J. Morrison. D. K. Oosterhof, C. W. Taylor and D. A. Meeter. 1977. Determination of microbial activity of estuarine detritus by relative rates of lipid biosynthesis. *Limnol. Oceanogr.* 22:1089-1099.

White, D. C., R. J. Bobbie, J. D. King, J. Nickels and P. Amoe. 1979a. Lipid analysis of sediments for microbial biomass and community structure, pp. 87-103. *In:* C. D. Litchfield and P. L. Seyfried (eds.), *Methodology for Biomass Determinations and Microbial Activities in Sediments.* ASTM STP 673, American Society for Testing and Materials, Philadelphia.

White, D. C., W. M. Davis, J. S. Nickels, J. D. King and R. J. Bobbie. 1979b. Determination of the sedimentary microbial biomass by extractible lipid phosphate. *Oecologia* 40:51-62.

White, D. C., R. J. Bobbie, J. S. Nickels, S. D. Fazio and W. M. Davis. 1980. Nonselective biochemical methods for the determination of fungal mass and community structure in estuarine detrital microflora. *Bot. Marina* 23:239-250.

White, D. C. and A. N. Tucker. 1969. Phospholipid metabolism during bacterial growth. *J. Lipid Res.* 10:220-233.

A BIOENERGETIC MODEL OF ANAEROBIC DECOMPOSITION: SULFATE REDUCTION

James P. Reed

The Ecosystems Center
Marine Biological Laboratory
Woods Hole, Massachusetts

Abstract: A bioenergetic model of anaerobic decomposition has been developed in which rates of sulfate reduction were comparable to published experimental results for a detritus-sand system when the soluble organic matter pool in the detritus was the sole substrate source. These results were based on a model in which the proportion of organic substrate which was used to reduce sulfate was determined from Gibbs' free energy of reactions, cell maintenance requirements, and population turnover time. The actual rates of reduction in the model were predicted from a Monod-type equation. The model suggests that release of substrate for sulfate reducers by cellulose degradation is important only after the initial soluble organic matter pool has been released from detritus. From sensitivity analyses of the model, it was concluded that 1) sulfate reduction is limited primarily by the rate of supply of organic substrate; 2) the form of substrate, i.e., carbohydrate or fatty acid, has only a slight effect on the rate of sulfate reduction; and 3) knowing the fraction of the total microbial biomass that is active does not help to determine the sulfate reducing activity.

Introduction

Decomposition releases nutrients from organic matter to the environment. However, factors which regulate decomposition are not yet fully understood. Mathematical models of decomposition have been constructed which elucidate some of the regulatory factors. Most decomposition models which are discussed in the literature are based on aerobic processes (e.g., Bunnel et al. 1977; Hunt 1977). There are some models of anaerobic decomposition in marine sediments but these lack an active microbial component (Berner 1977). Models of anaerobic sewage treatment processes incorporate both microbial dynamics and decomposition of a variety of substrates (Christensen and McCarty 1975).

The model discussed here was developed in order to determine the factors that regulate anaerobic decomposition. It is based on the theory of bioenergetics as developed by Christensen and McCarty (1975) and McCarty (1975). Because sulfate reduction is the predominant respiratory process in anoxic marine sediments (Fenchel 1978), this model predicts rates of decomposition as mediated by sulfate reducing microbes. Results were compared to data reported by Jorgensen and Fenchel (1974).

The model was used to investigate the following questions about the sulfate reduction process:

1) How many microbes are required to reduce sulfate at a given rate?
2) To what level of accuracy do we need to know cellular respiration rate, rate of supply of organic matter to microbes, or exact composition of substrate used by sulfate reducers in order to predict sulfate reduction rates?
3) Because sulfate reducers use only dissolved organic material (DOM), is the release of DOM from detritus mediated solely by microbes or is physical leaching of importance?

Description of the sulfate reduction model

In the model, formation of microbial biomass and end products is predicted from chemical half reactions and Gibbs' free energy of reaction. These free energies of reaction have been adjusted for conditions where the activity of hydrogen ions is at pH 7 and where all other compounds are assumed to be at unit activity.

The rate of formation of microbial biomass is a function of rate of uptake of substrates, cell yield for the substrate, and cell decay rate:

$$dM/dt = Y \cdot dS/dt - Q \cdot M \tag{1}$$

where:

dM/dt is the change in microbial biomass

Y is the microbial yield (total biomass produced/substrate taken up)

dS/dt is the rate of uptake of the limiting substrate, S

Q is the cell decay rate due to endogenous respiration and death

M is the microbial biomass.

The organic substrate which is taken up is either synthesized into cellular material or converted to end products. The ratios of concentrations of reactants (such as sulfate, lactate, and ammonium) to products (such as new cells, carbon dioxide, and hydrogen sulfide) are determined from the combination of half reactions for the electron acceptor, for the electron donor, and for cell production:

$$OR = Rd - fe \cdot Ra - fs \cdot Rc \tag{2}$$

where:

OR is the overall reaction

Rd, Ra, Rc are the half reactions for the electron donor, acceptor, and cells, respectively. For example, the half reaction for lactate is:

$$-1.0 \, lactate - 4.0 \, H_2O + 2.0 \, CO_2 + 1.0 \, HCO_3 + 12.0 \, H^+ + 12 \, e^-$$

All half reactions are written so that the free electrons have positive coefficients. When multiplied by the factors, fe and fs, which are described below, the reactants have negative coefficients and the products have positive coefficients.

fs, the fraction of the substrate used to produce new cells, defined as $(dM/dt)/(dS/dt)$ when M and S are expressed in equivalent units (see McCarty 1975), is derived from equation (1) and the equation for the turnover time, T:

$$T = M/(dM/dt)$$

$$fs = Y/(1 + Q \cdot T) \tag{3}$$

Q is cell decay rate

Y, yield, $= 1/(1 + A)$ (McCarty 1975)

$$A = - \frac{\text{amt. of energy required to make a unit of cell material}}{\text{amt. of energy obtained from the reaction of a unit of substrate with the electron acceptor}}$$

These calculations are based on Gibbs' free energies and conversion efficiencies (see McCarty 1975).

fe is the fraction of the substrate that provides electrons for the reduction of the electron acceptor:

$$fe = 1 - fs. \tag{4}$$

The rate at which the overall reaction takes place is determined by the lowest rate of uptake of any of the reactants. The rate of uptake of each reactant is based on the Monod-type of equation:

$$dS/dt = Vmax \cdot M \cdot S/(S + h) \tag{5}$$

where:

dS/dt is the rate of uptake of substance S

Vmax is the maximum rate of uptake per unit microbial biomass

M is microbial biomass

S is the concentration of the substance

h is the half-saturation coefficient

Change in concentration of soluble ions in the sediment due to diffusion is estimated by a finite difference approximation to a simple Fickian model, where it is assumed that the material is uniformly distributed within discrete layers in the sediment:

$$dS/dt = Ds \cdot dS/dz \tag{6}$$

where:

Ds is the coefficient of diffusion in the sediment

dS/dz is the concentration gradient of substance S over the distance, dz

dz is the distance between the midpoints of the layers.

The rate of release of dissolved organic matter (DOM) from eelgrass was modeled for two different cases (equations 7 and 8). The amount of DOM released per hour as calculated by equation 7 exponentially decreases with time. The amount of DOM released is constant over time when calculated by equation 8.

$$dS/dt = k1 \cdot S \tag{7}$$

or

$$dS/dt = k2 \tag{8}$$

where:

dS/dt is the release rate of DOM

S is the concentration of DOM in the eelgrass

k1 and k2 are constants.

Parameter evaluation

The values for most of the physiological parameters and all of the chemical half reactions were taken from Christensen and McCarty (1975) and McCarty (1975). Some of the parameters defined by McCarty can take on a range of values. For the model, therefore, I set the following conditions: cell decay equaled 0.056% h^{-1}, turnover time was 180 h, the sulfate reducing microbial biomass was 10^8 cells cm^{-3} sediment, ammonia was available for cell growth, Vmax was 8.2 mmol lactate (10^8 bacteria•h)$^{-1}$, and the half saturation coefficient for lactate was 8.3 μmole liter $^{-1}$.

Most of the values for the physical components of the model were derived from Jorgensen and Fenchel (1974). They used an aquarium which was filled with sand to a depth of 20 cm. Fresh dead eelgrass was mixed into the sand so that the final dry weight concentration of eelgrass was 5.5 mg cm^{-3}. Porosity of the sand was 32%. Sulfate concentration in water overlying the sand was maintained at 18.4 mmol SO_4 liter^{-1} by regularly passing fresh seawater ($23^o/_{oo}$) over the sediment surface. Temperature in the aquarium, even though it ranged from below freezing to 22C, was a constant 12C in the model. Sediment diffusion coefficients which include the effects of tortuosity for sulfate and lactate were 11.5 and 3.4×10^{-6} cm^2 sec^{-1}.

Values for the leaching rate parameters were derived from Harrison and Mann (1975a, b) who used two size classes of eelgrass particles. For 0.1 cm particles, 15% of the dry weight (54% of the DOM) was leached within 20 days with no apparent additional release during the next 82 days. Particles which were 2-4 cm long released only 5% of the dry weight (18% of the DOM) in 20 days. These data suggest that small particles lose DOM at an exponentially decreasing rate while larger particles release DOM at a more constant rate over time. Because the particles used by Jorgensen and Fenchel (1974) were intermediate to these two size classes, both types of release were tested in the model (equations 7 and 8). In equation 7, release rate parameter k1 equaled 0.1% h^{-1}. In equation 8, parameter k2 equaled 18.3 nmol C cm^{-3} sediment h^{-1}.

DOM that is released has several possible fates. It can be used by the sulfate reducers directly, it can be processed by other microbes into forms which are both usable and unusable by the sulfate reducers, it can be complexed by humics, or it can be absorbed by clays. For these reasons, the amount of DOM available to the sulfate reducers was assumed to equal 30% to 50% of the amount released.

These equations were written into Fortran language and processed on the digital PDP-11M RSX-11 Computer System at the Marine Biological Laboratory, Woods Hole, Mass. The computer program may be obtained from the author.

Results

In general, the simulation model predicted rates of sulfate reduction and sulfate concentrations which were similar to those observed by Jorgensen and Fenchel (1974) (Figs. 1 and 2). However, there is a discrepancy at the 1 cm depth which may have been due to formation of ice on days 19 and 41 in the microcosm. Below this depth, at 4 and 8 cm, predicted rates of sulfate reduction (Fig. 2) overlap observed rates. These results were based on a simulation in which the amount of DOM decreased exponentially with time (equation 7). As a result, sulfate reduction rate is high initially, then decreases to a low rate. In contrast, it appears from the observed data that a constant amount of DOM is released over time. When this observation was implemented in the model (equation 8), the predicted results were similar to the sulfate reduction rates as measured by Jorgensen and Fenchel (1974). The predicted results are equivalent to the rate shown for day 19 in Fig. 2. Because the release rate is constant over time, there is no temporal variability in the rate of sulfate reduction. The sulfate concentration profiles are similar to those shown in Fig. 1.

In order to determine sensitivity of results to changes in initial parameter values, output from models with such changes was compared to output from the 'best fit' model, which was described above. Results of the sensitivity analyses are:

1) Rates of sulfate reduction are directly proportional to the supply of DOM.
2) Type of substrate affected both the rate of sulfate reduction and microbial biomass. When the substrate was changed from lactate to a carbohydrate, a higher energy substrate, the rate of sulfate reduction decreased by 18% and microbial biomass increased by 66%. In contrast, switching to a low energy compound such as propionate caused the rate of sulfate reduction to increase by 27% and microbial biomass to decrease by 32%.
3) When cell decay or turnover time parameters were increased by a factor of 10, the rate of sulfate reduction increased 8%. When the parameters were decreased by a factor of 10, the sulfate reduction rate decreased by 1.4%.
4) When the maximum potential rate of substrate utilization, Vmax, was increased, the concentration of interstitial DOM decreased. The rate of sulfate reduction decreased only when the value of

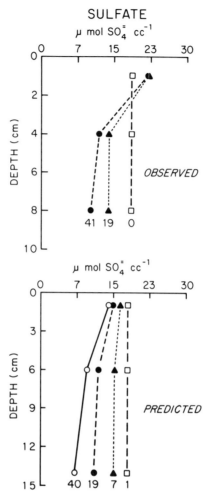

Figure 1. Observed (after Jorgensen and Fenchel 1974) and predicted
sulfate concentrations (μmol cm⁻³ of interstitial water in the sediment) at various times after the beginning of the experiment.
Numbers at the bottom of the lines refer to the day when
measurements were made.

Vmax was decreased to a level such that demands on the population by death and respiration exceeded the rate of supply of
energy. When this happened, microbial biomass eventually
became zero.

5) When microbial biomass was restricted to one sixth of the level in
the 'best fit' model, the rate of sulfate reduction was unchanged
and interstitial DOM concentration increased.

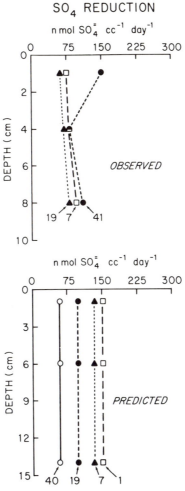

Figure 2. *Observed (after Jorgensen and Fenchel 1974) and predicted sulfate reduction rates (nmol cm⁻³ of sediment•day) in the sediment at various times after the beginning of the experiment. Numbers as in Fig. 1.*

6) When the half saturation coefficient was increased, the concentration of DOM in the interstitial water increased. When this coefficient was decreased, the DOM concentration decreased. For both cases, the sulfate reduction rate was unchanged.

Discussion and Conclusions

This simulation study was able to predict rates of sulfate reduction similar to those measured by Jorgensen and Fenchel (1974) by simply regulating the flow of DOM. This flow was a function of 1) rate of release of DOM from the eelgrass and 2) the fraction of DOM that was available to sulfate reducers. If the fraction available to the reducers is 100%, then sulfate reduction could be maintained at the observed average rate of 89 nmol SO_4 cm^{-3} day^{-1} for a period of 232 days. Because it is unlikely that all of the DOM is available to sulfate reducers, the actual rate will be maintained for a period less than 232 days. This is, in fact, what was found. Jorgensen and Fenchel (1974) observed this average rate on day 76, but when they next made a measurement (day 215) the rate had fallen off to 25 nmol SO_4 cm^{-3} day^{-1}. Sulfate reducers at this time are probably using DOM that is released by microbes which ferment the structural material of the eelgrass. These fermenters may be unimportant initially due to repression of cellulolytic enzyme synthesis by organic compounds in the leachate of eelgrass. A similar phenomenon is known to occur in fungi, where cellulolytic enzyme synthesis is repressed by glucose and cellobiose (Eriksonn 1978). Because these compounds may also be present in the leachate of eelgrass, enzymatic repression may also occur.

The process by which DOM becomes available, i.e. through leaching or microbial activity, cannot be determined with this model. The model only suggests that after an initial rapid leaching period the amount released over time is constant.

Type of substrate used by sulfate reducers affected the rate of reduction. This result was due to differences in energy yield for various substrates. A high energy substrate, such as a carbohydrate, gives a high yield (equation 3). As yield decreases, the proportion of substrate which is involved in the reduction of sulfate increases. For example, this proportion (equation 4) increases from 0.719 to 0.852 to 0.926 as the substrate is changed from a carbohydrate to lactate to propionate, respectively. These values are obtained when cell decay is 0.056% h^{-1} and turnover time is 180 h.

Changes in the half saturation coefficient and Vmax (equation 5) affected only DOM concentration. When the value of the half saturation coefficient was increased, concentration of interstitial DOM increased while microbial biomass and rate of sulfate reduction remained unchanged. When the half saturation coefficient was decreased, DOM concentration decreased. These changes in DOM maintain the quotient of S/(S + h) (equation 5) at a value such that the rate of uptake of DOM is unchanged. Since the microbes cannot take up DOM any faster than it is released, the rate of sulfate reduction is unchanged. Similarly, when Vmax was increased, microbial biomass and rate of sulfate reduction were unchanged but the concentration of DOM decreased. The explanation for this is similar to the previous explanation. Rate of uptake cannot increase since it is dependent on supply. In order for rate of uptake to stay the same, either microbial biomass must decrease or substrate concentration must decrease (equation 5). Since biomass does not change, DOM concentration must decrease.

Changes in microbial biomass had little effect on the rate of sulfate reduction. In the model, 10^8 microbes cm^{-3} sediment reduced sulfate at the observed rate. This value is 1 to 10% of the total bacterial numbers present in sandy sediments (Rublee and Dornseif 1978). When microbial biomass was restricted to a level one-sixth of the above value, the rate of sulfate reduction still remained unchanged, but DOM concentration increased. The reasons for this are analogous to the explanation given above for Vmax. Similar results have been found for continuous culture systems where predation by protozoa decreased bacterial biomass but rate of consumption of substrate remained unchanged (Straskrabova et al. 1977).

Tenfold changes in cell decay rate (0.056% h^{-1}) and turnover time (180 h) resulted in only an 8% increase (upward change) or a 1.4% decrease (downward change) in rate of sulfate reduction. These small changes are due to the effect that these parameters have on the fraction of the substrate (fs) that is used for cell synthesis (equation 3). For example, fs decreases when there are increases in either cell decay or turnover time. As fs decreases, fe approaches one (1.0) and most of the substrate, therefore, is used for energy, i.e. sulfate reduction. In the case of lactate, fe can only increase from 0.852 by 17% to equal 1.0. In contrast, as both cell decay and turnover time are decreased, fs approaches the yield, Y. For lactate, fe decreases by 1.6% from 0.852 to 0.838 as fs approaches yield.

As a result of the development of the model, the importance of the release of dissolved organic matter to sulfate reducers and to decomposition has become evident. Experiments are now in progress to determine the controls of solubilization by the cellulose fermenters in salt marsh sediments.

Summary

1. A model was developed which predicted rates of sulfate reduction for a sand-detritus system.

2. The model was based on energetics of biological reactions, substrate concentrations and microbial parameters, such as biomass, maximum rate of substrate uptake, half saturation coefficients, respiration and cell decay parameters.

3. It was found that rate of release of organic material was the primary factor which regulated rate of sulfate reduction.

4. Leaching of soluble material from detritus was sufficient to account for observed rates of sulfate reduction.

5. Parameters which are difficult to measure, such as respiration, death, numbers of active bacteria, or type of substrate are only of secondary importance to this model.

6. Interaction between cellulose fermenters and sulfate reducers is unclear and needs further investigation.

Acknowledgments

I would like to thank J. E. Hobbie, W. B. Bowden and B. J. Peterson for their suggestions. This research was supported by NSF grant DEB 7905127. Permission to use data from the paper by Jorgensen and Fenchel in Figures 1 and 2 was granted by Springer-Verlag.

References Cited

Berner, R. A. 1977. Stoichiometric models for nutrient regeneration in anoxic sediments. *Limnol. Oceanogr.* 22:781-786.

Bunnell, F. L., D. E. N. Tait, P. W. Flanagan and K. Van Cleve. 1977. Microbial respiration and substrate weight loss - I. A general model of the influences of abiotic variables. *Soil Biol. Biochem.* 9:33-40.

Christensen, D. R. and P. L. McCarty. 1975. Multi-process biological treatment model. *J. Water Poll. Control Fed.* 47:2652-2664.

Eriksonn, K. 1978. Enzyme mechanisms involved in cellulose hydrolysis by the rot fungus, *Sporotrichum pulverulentum. Biotech. Bioeng.* 20:317-332.

Fenchel, T. 1978. The ecology of micro- and meiobenthos. *Ann. Rev. Ecol. Syst.* 9:99-121.

Harrison, P. G. and K. H. Mann. 1975a. Chemical changes during the seasonal cycle of growth and decay in eelgrass (*Zostera marina*) on the Atlantic Coast of Canada. *J. Fish. Res. Board Can.* 32:615-621.

Harrison, P. G. and K. H. Mann. 1975b. Detritus formation from eelgrass (*Zostera marina* L.): The relative effects of fragmentation, leaching and decay. *Limnol. Oceanogr.* 20:924-934.

Hunt, H. W. 1977. A simulation model for decomposition in grasslands. *Ecology* 58:470-484.

Jorgensen, B. B. and T. Fenchel. 1974. The sulfur cycle of a marine sediment model system. *Mar. Biol.* 24:189-201.

McCarty, P. L. 1975. Stoichiometry of biological reactions. *Prog. Water Tech.* 7:157-172.

Rublee, P. and B. E. Dornseif. 1978. Direct counts of bacteria in the sediments of a North Carolina salt marsh. *Estuaries* 1:188-191.

Straskrabova, V., M. Legner, P. Puncochar, J. Benndorf and F. Cramer. 1977. Microbial self purification in an experimental continuous flow system. *Int. Rev. ges. Hydrobiol.* 62:573-590.

ARCTIC ESTUARIES

A STUDY OF PLANKTON ECOLOGY IN CHESTERFIELD INLET, NORTHWEST TERRITORIES: AN ARCTIC ESTUARY

John C. Roff, Robert J. Pett, Greg F. Rogers

Department of Zoology
College of Biological Science
University of Guelph
Guelph, Ontario

and

W. Paul Budgell

Ocean and Aquatic Sciences
Department of Fisheries and Oceans
Canada Centre for Inland Waters
Burlington, Ontario

Abstract: Chesterfield Inlet NWT, a long (200 km) deep estuary, was studied in summer 1978. Its drainage basin (2.9×10^5 km²) covers sparsely vegetated shield rocks of the barren grounds. Nutrient distributions generally paralleled salinity or temperature gradients. Soluble reactive phosphorus (maximum 0.6 μg-at l⁻¹) and total dissolved Kjeldahl nitrogen (maximum 13 μg-at l⁻¹) were positively correlated with salinity, whereas $NO_3 + NO_2$ and SiO_2 were not correlated to salinity. Minimum biomass (chlorophyll a = 0.3 μg l⁻¹) occurred close to the mouth of the estuary, and maximum biomass (chlorophyll a = 1.9 μg l⁻¹) was observed near the estuary head in an area of higher water residence time. The ratio of total chlorophyll a to phaeopigments was high in Baker Lake, Cross Bay and Hudson Bay, indicating that these were biomass sources, and low in the estuary, indicating senescent populations. Several parameters related to mixing processes, including a longitudinal dispersion coefficient, estuarine Richardson number, and the trap volume/channel volume ratio indicated the importance of physical processes in determining the distribution of biomass in the estuary.

Introduction

The few studies of arctic and sub-arctic estuaries are generally limited to semi-quantitative descriptions (Alexander et al. 1974; Schell 1974; Grainger et al. 1977; Legendre and Simard 1978), and have not attempted to relate biological or chemical features to physical processes. Previous studies of marine and freshwater arctic aquatic environments have suggested the importance of nutrients in limiting phytoplankton standing crops (Dunbar 1968; Kalff 1970; Hobbie 1973; etc.).

In contrast, in estuarine systems, numerous physico-chemical and biological factors may be responsible for observed phytoplankton distributions. Estuarine water circulation and mixing is most often cited as the controlling influence on distribution of particulates. The residual non-tidal circulation of a partially mixed estuary creates a "null-zone" (Hansen 1965) and associated turbidity maximum (Postma 1967; Festa and Hansen 1978). The development of a phytoplankton maximum in, or seaward of, the null-zone is believed to result from the local increase in water column residence time (Peterson et al. 1975a, b).

According to Okubo (1978) and Steele (1978), nutrients and grazing are the least likely factors controlling phytoplankton biomass in actively flushed systems. Irregular estuarine basin morphometry may significantly enhance longitudinal dispersion (Fischer 1969, 1976; Okubo 1973), and by a process of "trapping", populations of phytoplankton may develop in relatively calm embayments before being flushed into the main channel. Macrophyte beds or woody debris may also reduce flow of particulates between traps (embayments off the main channel) and the main channel or the ocean (Mann 1975; Naiman and Sibert 1978).

In the summer of 1978, a multi-disciplinary survey of Chesterfield Inlet was undertaken to investigate the physico-chemical and biological processes in this large arctic estuary. In particular, a number of parameters related to mixing processes were calculated including an estuarine Richardson number (Ri_E) (Fischer 1972), longitudinal dispersion coefficient (E_L) (Stommel 1953), and trap volume/channel volume ratio (r) (Okubo 1973), where trap volume represents the integrated volume of embayments taken between deep water headlands of the main channel. The relationships between biomass distributions and physical processes in Chesterfield Inlet led to the development of a simple conceptual model presented here.

Study Area

Chesterfield Inlet is a partially mixed estuary situated on the northwest coast of Hudson Bay, in the central part of the Keewatin District, Northwest Territories (Fig. 1). The inlet extends 200 km northwest from Hudson Bay to Baker Lake, a large freshwater lake. Depths range from approximately 110 m at the mouth to 6 m at Chesterfield Narrows, the major outflow from Baker Lake.

Tidal forcing is quite strong everywhere in the Inlet. The tidal range varies from a maximum of 5 m at Big Island to a minimum of 1.5 m at Baker Lake. Tidal action in Baker Lake itself is negligible. The currents in the inlet are tidally driven and current speeds in excess of 2.0 m sec^{-1} are observed throughout most of the estuary (Budgell 1976).

The morphometry of Chesterfield Inlet is complex, characterized by numerous sharp channel bends, shoals, embayments, islands and acute changes in channel width and depth. Its huge drainage basin (2.9×10^5 km^2) is predominantly sparsely vegetated shield rocks of the barren grounds (Wright 1967; Budgell 1976). The barren grounds is an area of climatic extremes with low temperatures, high winds and little precipitation (Environment Canada 1974a).

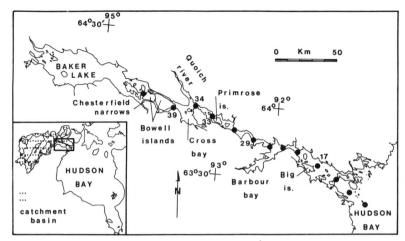

Figure 1. Chesterfield Inlet showing major sampling stations. Those referred to in text are numbered. Insert shows location and catchment basin.

Methods and Materials

During the neap tides of September 1978, 13 anchor stations in Baker Lake, Chesterfield Inlet and nearshore Hudson Bay were sampled within a twelve day period (Fig. 1). Most stations were occupied at least twice. Water samples were taken with 6 litre PVC Van Dorn bottles at several depths during high water slack tide to minimize strong current effects. Additional surface samples were obtained from the major bays and rivers.

The following variables were measured at each station: conductivity, temperature, current velocity, soluble reactive phosphorus (SRP), reactive nitrate + nitrite (NO_3 + NO_2), reactive silicate (SiO_2), total dissolved Kjeldahl nitrogen (TDKN), chlorophyll *a* (corrected for phaeopigments), phaeopigments, particulate organic carbon (POC), and phytoplankton. *In situ* measurements of conductivity, temperature and current velocity were made with a combined Guildline Mk IV CTD probe and Endeco model 110 current meter. Data were transmitted to an on-line Hewlett Packard 9825 A calculator for calculation of salinity. Measurements were made at 14 depths in each vertical profile, every half-hour over a thirteen hour period.

Nutrients, chlorophyll *a*, phaeopigments and POC were determined by methods detailed in Pett (1980), following procedures from Strickland and Parsons (1972) or Environment Canada (1974b). All filtrations used Whatman GF/C glass fibre filters. Nutrient and biomass values presented here were taken from the 10 m depth, and are representative of the upper water column (Pett 1980).

Phytoplankton samples were enumerated using Utermöhl's inverted microscope technique (Utermöhl 1958), according to methods outlined in Lund et al. (1958). Phytoplankton community indices were calculated as described in Kwiatkowski and Roff (1976). The percentage similarity of community (PSc) was calculated from the formula:

$$PSc = 100 - 0.5 \Sigma |a' - b'|$$

where a' and b' are, for each species, the respective percentages of the total organisms in Samples A and B. The coefficient of community (CC) was calculated from the formula:

$$CC = c/(a + b - c) \times 100$$

where a is the number of species in the first sample, b is the number of species in the second, and c is the number of species common to both.

Correlation (Pearson's r) and regression analyses were carried out on biomass and physio-chemical data using verified programs. All calculations were made in APL at the Institute of Computer Science, University of Guelph. Calculations on hydrographic data were performed on computing facilities at the Canada Centre for Inland Waters.

Results

Salinity and temperature gradients (Fig. 2) were generally paralleled by nutrient distributions (Fig. 3). Maximum concentrations of SRP (0.6 µg-at l^{-1}) and TDKN (13 µg-at l^{-1}) were observed in the cold saline waters of the lower estuary. Highest levels of $NO_3 + NO_2$ (0.7 µg-at l^{-1}) were observed in the warmer brackish waters of the upper inlet, while SiO_2 remained virtually constant and low (1-3 µg-at l^{-1}) over the entire length of the inlet.

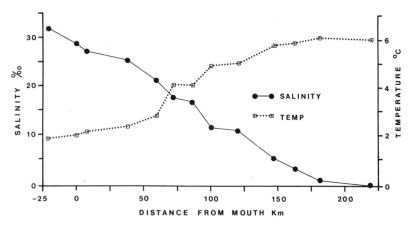

Figure 2. Temperature (C) and salinity ($^0/_{00}$) distributions at 10 m depth, September 1978 high water slack tides.

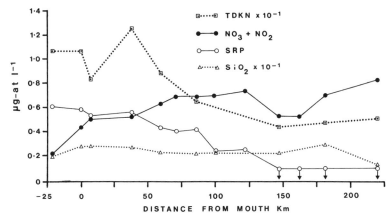

Figure 3. Nutrient distributions at 10 m depth, September 1978 high water slack tides.

Significant correlations (P < 0.05) between SRP, TDKN and the conservative tracers of water source, salinity and temperature (Table 1), indicate the importance of physical processes to chemical distributions. Neither SiO_2 nor $NO_3 + NO_2$ was significantly correlated with salinity, temperature, other nutrients or biomass. However, the $NO_3 + NO_2$ depression at 150 km (Fig. 3) was associated with the biomass peak (Fig. 4) at Stns 33 and 34.

Biomass levels as indicated by chlorophyll *a* and POC were highest in the warmer brackish waters of the upper inlet (chlorophyll *a* maximum = 1.9 μg l^{-1}), and at a minimum (chlorophyll *a* = 0.3 μg l^{-1}) close to the mouth of the estuary (Fig. 4). Chlorophyll *a* and POC were strongly

Table 1. Correlations (Pearson's r) between physico-chemical variables and dispersion coefficient.

SRP	TDKN	Chlor a	POC	E_L	Temp	
0.99	0.89	−0.91	−0.89	0.72	−0.99	Sal
	0.91	−0.93	−0.89	*	−0.98	SRP
		−0.84	−0.75	*	−0.91	TDKN
			0.95	−0.76	0.91	Chlor a
				*	0.87	POC
					*	E_L

P < 0.05 for all values; * not recorded; SRP = soluble reactive phosphorus; TDKN = total dissolved Kjeldahl nitrogen; Chlor *a* = corrected chlorophyll *a*; POC = particulate organic carbon; E_L = longitudinal dispersion coefficient; Temp = temperature C; Sal = salinity °/$_{oo}$

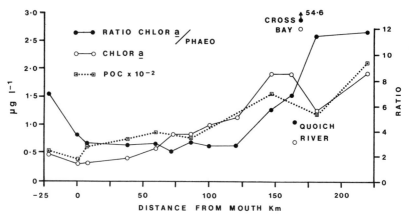

Figure 4. *Biomass distributions at 10 m depth, September 1978 high water slack tides, showing particulate organic carbon (POC), corrected chlorophyll a (chlor a) and ratio of total chlorophyll a to phaeopigments (chlor a/phaeo).*

negatively correlated to SRP, TDKN and salinity, and strongly positively correlated to temperature, also indicating the importance of physical processes to the distribution of particulate biomass (Table 1). Cross Bay and Hudson Bay were characterized by higher values of the total chlorophyll *a*/phaeopigment ratio (Fig. 4); Cross Bay in particular had a very high value. Phaeopigments were quantitatively more important in the middle and lower portions of Chesterfield Inlet.

Most of the variation in chlorophyll *a* (81.5%) was explained by simple regression on salinity. The next most important variable was the estuarine Richardson number (Ri_E), defined by Fischer (1972) as:

$$Ri_E = \frac{g\overline{\Delta\varrho}Q_f}{\overline{\varrho}b|\overline{U}|^3}$$

where:

g = acceleration due to gravity
$\overline{\Delta\varrho}$ = tidally averaged difference in density from top to bottom of water column
Q_f = freshwater discharge
$\overline{\varrho}$ = tidally averaged density
b = top width of channel
$|\overline{U}|^3$ = cube of the tidal average of the absolute value of the cross-sectionally averaged velocity

Residuals of the regression of biomass on salinity were significantly negatively correlated with the computed Ri_E and increased the multiple R^2

to 89.9%; however there was no significant correlation between Ri_E and biomass alone. The highest Ri_E was calculated for Stn 39 (km 180), in an area of high discharge from Baker Lake, and at Stn 17 (km 40), towards the mouth of the estuary; the lowest Ri_E was calculated for Stn 34 (km 165), the biomass peak (Figs. 4 and 5).

Biomass levels were significantly negatively correlated to the longitudinal dispersion coefficient (E_L), ($P < 0.05$, Table 1), defined by Stommel (1953) as:

$$E_L = \frac{Q_f \bar{s}}{A(d\bar{s}/dx)}$$

where:

$\bar{s}$ = tidally averaged salinity
A = channel cross section area
x = longitudinal distance

Highest biomass levels at Stns 33 (km 145) and 34 (km 165) corresponded to the lowest values of E_L (Figs. 4 and 5).

The biomass peak in the estuary was located immediately downstream of Cross Bay, which was characterized by the highest measured chlorophyll *a* levels (Fig. 4), a large trap volume, and a high trap volume/channel volume ratio (r), (Fig. 6). Retention times (k) for Cross Bay were on the order of 10 or more days and were derived from Okubo's (1973) formula:

$$E_L = \frac{E_L{}^1}{1+r} + \frac{U_o^2}{s(1+r)^2(1+r+o/k)}$$

where:

E_L = longitudinal dispersion coefficient including trap effects
$E_L{}^1$ = longitudinal dispersion coefficient of channel itself
U_o = amplitude of the cross-sectionally averaged tidal velocity
r = trap volume/channel volume ratio
s = salinity
o = tidal frequency (2Π/T) where T = time
k = reciprocal of trap retention time, i.e. exchange time

Other large embayments (the Big Island area, Quoich River estuary, and Barbour Bay) were flushed by river discharge or tides and had retention times on the order of 1-3 days. The Primrose Island area, despite its high r, had a very low trap volume and low retention time (Fig. 6). None of these areas showed elevated chlorophyll *a* levels or levels higher than nearby main channel values; indeed Quoich River levels were considerably lower (Fig. 4).

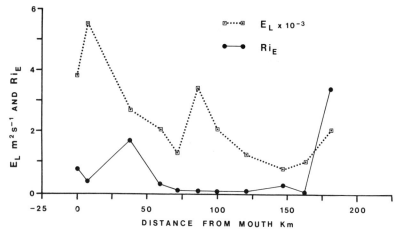

Figure 5. *Variation in estuarine Richardson number (Ri$_E$) and longitudinal dispersion coefficient (E$_L$) computed from tidally averaged data.*

An analysis of phytoplankton community structure (Table 2) showed that Stn 34 (the biomass peak at km 165) was abundant in *Chlamydomonas* and *Rhodomonas* species, and was most closely related to the communities of Baker Lake and Cross Bay. Station 34 was less like the Quoich River which contained significant proportions of *Dinobryon tabellariae* and *Tabellaria flocculosa,* or the downstream stations where marine *Skeletonema* and *Chaetoceros* species became important. The highest similarity between communities was achieved with Baker Lake and Cross Bay combined as sources when compared to Stn 34.

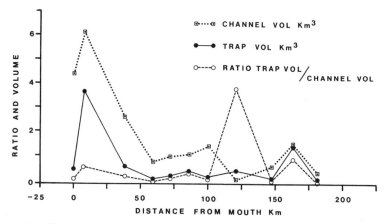

Figure 6. *Channel volumes, trap volumes, and ratios of trap volume/ channel volume in Chesterfield Inlet.*

Table 2. Coefficient of Community (CC) and Percentage Similarity of Community (PSc) between phytoplankton communities at Stn 34 and five other locations.

	Baker Lake	Cross Bay	Stn 29	Quoich River	Stn 2
CC	33.3	42.4	36.4	25.4	12.9
PSc	63.4	58.2	44.1	33.5	41.4

The null zone in Chesterfield Inlet must be located about 30-40 km above the estuarine biomass maximum. The exact location of the null zone could not be determined from examination of the residual non-tidal current data as it was located upstream of the last estuarine station sampled, Stn 39 (Fig. 7), where two layered flow still persisted.

Discussion

A complete explanation of biomass distributions in Chesterfield Inlet would necessitate information on primary production, nutrient turnover rates, and grazing, which is lacking here. A numerical model of this estuarine system encompassing biological and chemical data has not been attempted and must await further analysis of the physical data. However, a conceptual model can be proposed which may be used as a basis for

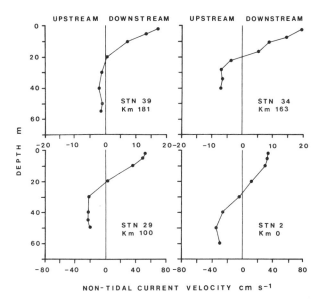

Figure 7. Non-tidal current velocities at four stations in Chesterfield Inlet.

testing various hypotheses, both specific to Chesterfield Inlet and with broader applications (Fig. 8).

The high negative correlation of chlorophyll a with salinity, and the high levels of chlorophyll a in Baker Lake and Cross Bay, together with the information from phytoplankton community analysis, indicate that these two areas are the dominant sources in the upper estuary. The depression in $NO_3 + NO_2$ at the chlorophyll a peak in the upper estuary is consistent with higher local production.

Both Baker Lake and Cross Bay have relatively high retention times (>10 days), and exhibit relatively high phytoplankton biomass with low phaeopigment content; other embayments in the estuary have much shorter retention times (< 3 days). Loss of phytoplankton populations from the source areas to estuarine regions of higher salinity coincides with an increased proportion of cell phaeopigment. Jensen and Sakshaug (1973) noted that a rise in degradation products (i.e. lower chlorophyll a/phaeopigment ratio) may indicate algal senescence.

In the highly dynamic Chesterfield Inlet, where tidal and discharge flushing occurs, the development and maintenance of a "healthy" estuarine phytoplankton population may be limited by the physical action of water mixing (Ketchum 1954; Hobbie 1976; Duxbury 1979). The chlorophyll a/phaeopigment ratio increases again in Hudson Bay, reflecting a marine phytoplankton source, as water column stability increases and flushing becomes negligible. The high covariance of chlorophyll a and salinity is explained if biomass is produced in "trap" areas (Baker Lake, Cross Bay),

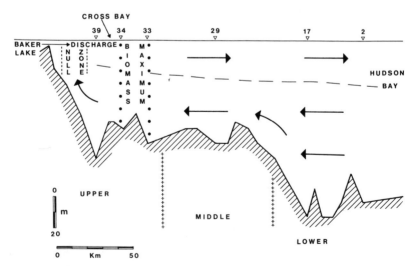

Figure 8. Diagrammatic model of water movements and biomass levels in upper, middle, and lower Chesterfield Inlet.

flushed into the estuary and diluted by Hudson Bay water of lower chlorophyll *a* content. Primary production in the middle and lower estuary does not appear to contribute significantly to biomass.

Okubo (1978) noted that ". . . physical effects become important in the distribution of organisms when the time scale of the process is comparable to (or smaller than) that of biological factors, such as the time scale of organisms' reproduction." Physical processes do indeed dominate the lower estuary; higher Ri_E values in this area imply more distinct two-layer flow. Peterson et al. (1975a) have shown that a more distinct two-layer flow results in a lower retention time in both layers. Thus, we expect that biomass should decrease with increasing strength of two-layer flow and with increasing Ri_E. The conceptual model is further supported by negative correlation between chlorophyll *a* and E_L. A lower value of E_L indicates a volume of water which may retain its identity for a longer period of time, i.e. permitting residence or accumulation of biomass.

Extensive glaciation of the Chesterfield Inlet area was responsible for the complex basin morphometry of this arctic estuary, and basin morphometry appears to influence particulate biomass distribution strongly here. In estuaries characterized by relatively rectangular basins, phytoplankton (or turbidity) maxima are generally found closely associated with the null zone (Peterson et al. 1975a; Festa and Hansen 1978). In Chesterfield Inlet, however, the phytoplankton maximum is located approximately 30 km downstream of the null zone. Any function of the null zone in this estuary is unclear, but it is subsidiary to the trap contributions of Cross Bay and Baker Lake which have longer retention times.

Acknowledgments

Many people contributed to this study. In particular we thank D. Brooks and the officers, crew and scientists of the M. V. Petrel IV, N. Watson, P. Philbert and K. Burnison of the Canada Centre for Inland Waters, and J. Gerrath of the University of Guelph. Financial assistance included a Natural Sciences and Engineering Research Council Canada grant and Fisheries and Environment Canada grant No. FMS 547-G1 to J. C. Roff. A fuller interpretation of the physical processes described in this paper will form part of a Ph.D. thesis by P. Budgell at the University of Waterloo, Ontario.

References Cited

Alexander, V., C. Coulon and J. Chang. 1974. Studies of primary productivity and phytoplankton organisms in the Colville River system, pp. 229-426. *In:* E. W. Shallock (ed.), Environmental studies of an Arctic estuarine system - Final report. Ecol. Res. Ser., U.S. Environmental Protection Agency, Corvallis, OR.

Budgell, W. P. 1976. Tidal propagation in Chesterfield Inlet, N.W.T. Man. Rep. Ser. No. 3. Ocean and Aquatic Sciences. Central Region, Canada Centre for Inland Waters, Environment Canada, Burlington, Ontario. 99 pp.

Dunbar, M. J. 1968. *Ecological Development in Polar Regions.* Prentice-Hall, Englewood Cliffs, NJ. 119 pp.

Duxbury, A. C. 1979. Upwelling and estuary flushing. *Limnol. Oceanogr.* 24:627-633.

Environment Canada. 1974a. *Sailing Directions: Labrador and Hudson Bay.* Fisheries and Marine Service, Ottawa. 335 pp.

Environment Canada. 1974b. *Analytical Methods Manual.* Inland Waters Directorate, Water Quality Branch, Environment Canada, Ottawa.

Festa, J. F. and D. V. Hansen. 1978. Turbidity maxima in partially mixed estuaries: A two dimensional numerical model. *Est. Coastal Mar. Sci.* 7:347-359.

Fischer, H. B. 1969. The effect of bends on dispersion in streams. *Water Resources Res.* 5:496-506.

Fischer, H. B. 1972. Mass transport mechanisms in partially stratified estuaries. *J. Fluid Mech.* 53:671-687.

Fischer, H. B. 1976. Mixing and dispersion in estuaries. *Ann. Rev. Fluid Mech.* 8:107-133.

Grainger, E. H., J. E. Lovrity and M. S. Evans. 1977. Biological oceanographic observations in the Eskimo Lakes, Arctic Canada. Physical, nutrient and primary production data, 1961-1975. Tech. Rep. No. 685, Fisheries and Marine Service, Environment Canada. Ste. Anne de Bellevue, Quebec. 108 pp.

Hansen, D. V. 1965. Currents and mixing in the Columbia River estuary. Ocean Sci. and Ocean Eng., Trans. Joint Conf. Marine Tech. Soc. Am. Soc. Limnol. and Oceanogr. pp. 943-955.

Hobbie, J. E. 1973. Arctic limnology: A review, pp. 127-168. *In:* M. E. Britton (ed.), *Alaskan Arctic Tundra.* Tech. Pap. No. 25. Arctic Institute of North America.

Hobbie, J. E. 1976. Nutrients in estuaries. *Oceanus* 19:41-47.

Jensen, A. and E. Sakshaug. 1973. Studies on the phytoplankton ecology of the Trondheimsfjord. II. Chloroplast pigments in relation to abundance and physiological state of the phytoplankton. *J. Exp. Mar. Biol. Ecol.* 11:137-155.

Kalff, J. 1970. Arctic lake ecosystems, pp. 651-663. *In:* M. W. Holdgate (ed.), *Antarctic Ecology Vol. 2.* Academic Press, New York.

Ketchum, B. H. 1954. Relation between circulation and planktonic populations in estuaries. *Ecology* 35:191-200.

Kwiatkowski, R. E. and J. C. Roff. 1976. Effects of acidity on the phytoplankton and primary productivity of selected northern Ontario lakes. *Can. J. Bot.* 54:2546-2561.

Legendre, L. and Y. Simard. 1978. Dynamique estivale du phytoplancton dans l'estuaire de la Baie de Rupert (Baie de James). *Naturaliste Can.* 105:243-258.

Lund, J. W. G., C. Kipling and E. D. LeCren. 1958. The inverted microscope method of estimating algal numbers and the statistical basis of estimations by counting. *Hydrobiologia* 11:143-170.

Mann, K. H. 1975. Relationship between morphometry and biological functioning in three coastal inlets of Nova Scotia, pp. 634-644. *In:* L. E. Cronin (ed.), *Estuarine Research, Vol. 1.* Academic Press, New York.

Naiman, R. J. and J. R. Sibert. 1978. Transport of nutrients and carbon from the Nanaimo River to its estuary. *Limnol. Oceanogr.* 23:1183-1193.

Okubo, A. 1973. Effect of shoreline irregularities on streamwise dispersion in estuaries and other embayments. *Neth. J. Sea Res.* 6:213-224.

Okubo, A. 1978. Horizontal dispersion and critical scales for phytoplankton patches, pp. 21-42. *In:* J. H. Steele (ed.), *Spatial Pattern in Plankton Communities.* Plenum Press, New York.

Peterson, D. H., T. J. Conomos, W. W. Broenkow and P. C. Doherty. 1975a. Location of the nontidal current null zone in northern San Francisco Bay. *Est. Coastal Mar. Sci.* 3:1-11.

Peterson, D. H., T. J. Conomos, W. K. Broenkow and E. P. Scrivani. 1975b. Processes controlling the dissolved silica distribution in San Francisco Bay, pp. 153-187. *In:* L. E. Cronin (ed.), *Estuarine Research, Vol. 1.* Academic Press, New York.

Pett, R. J. 1980. Chesterfield Inlet oceanographic data report 1978, Volume 1 - Nutrient and Seston Data. Ocean and Aquatic Sciences. Central Region, Canada Centre for Inland Waters, Environment Canada. Burlington, Ontario. (in press).

Postma, H. 1967. Sediment transport and sedimentation in the estuarine environment, pp. 158-179. *In:* G. H. Lauff (ed.), *Estuaries,* Publ. No. 83, Amer. Assoc. Adv. Sci., Washington, D.C.

Schell, D. M. 1974. Seasonal variation in the nutrient chemistry and conservative constituents in coastal Alaskan Beaufort Sea waters, pp. 233-298. *In:* E. W. Shallock (ed.), Environmental studies of an Arctic estuarine system - Final report. Ecol. Res. Ser., U.S. Environmental Protection Agency, Corvallis, OR.

Steele, J. H. 1978. Some comments on plankton patches, pp. 1-20. *In:* J. H. Steele (ed.), *Spatial Pattern in Plankton Communities.* Plenum Press, New York.

Stommel, H. 1953. Computation of pollution in a vertically mixed estuary. *Sewage Ind. Wastes* 25:1065-1071.

Strickland, J. D. H. and T. R. Parsons. 1972. *A Practical Handbook of Seawater Analysis.* Bull. 167. Fisheries Research Board of Canada, Ottawa. 310 pp.

Utermöhl, H. 1958. Zur Vervolkommung der Quantitativen Phytoplankton. *Methodik. Mitt. Int. Ver. Limnol.* 9:1-38.

Wright, G. M. 1967. Geology of the southeastern Barren Grounds, parts of the districts of Mackenzie and Keewatin (Operations Keewatin, Baker, Thelon). Memoir 350. Geol. Surv. Canada Dept. of Energy, Mines, and Resources, Ottawa. 91 pp.

ZOOPLANKTON IN A CANADIAN ARCTIC ESTUARY

M. S. Evans

Great Lakes Research Division, University of Michigan
Ann Arbor, Michigan

and

E. H. Grainger

Arctic Biological Station, Ste. Anne De Bellevue, Quebec

Abstract: Zooplankton was collected as part of a biological investigation of the Eskimo Lakes and Liverpool Bay, the first study of its kind carried out on one of the major estuaries in the Canadian arctic. Of the 43 zooplankton species found in the system, copepods, polychaete and cirripede larvae, and rotifers were the most abundant. Principal component analysis showed five geographical groups of stations in the estuary, which was dominated by the character of the inflowing waters at both extremities, and in which salinity and water source were the strongest determinants of zooplankton composition. Essentially freshwater species originating in waters which feed the head of the system dominated at the innermost stations. Marine species from the Beaufort Sea prevailed at the outermost stations. The most abundant copepods in the system, *Acartia clausi, Eurytemora herdmani, Pseudocalanus* sp., *Limnocalanus macrurus* and *Drepanopus bungei,* occurred widely but reached maximum numbers in different parts of the estuary. The annual cycle of standing stock is one of extreme seasonality, with biomass peaks occurring in mid-summer. A mean annual biomass of 80 mg (wet weight) per m³ represents an especially low standing stock. This and the additional factors of low primary productivity and phytoplankton standing stocks suggest extreme oligotrophy within the estuary.

Introduction

One of the major estuaries in the Canadian arctic, formed by Liverpool Bay and the Eskimo Lakes, extends southward about 200 km from the Beaufort Sea (Fig. 1). Arctic climatic conditions prevail. Yearly mean air temperature is about −11C and ice covers the system for about 8 months of the year and achieves a thickness of more than 2 m. Waters are cold, less than −1C over the long winter and rising only occasionally to as high as 12C in summer. Salinity ranges generally from less than 1⁰/₀₀ at the head of the system to about 20⁰/₀₀ in the northernmost part of Liverpool Bay. In Liverpool Bay salinity is influenced by the adjacent Beaufort Sea in which low salinity surface water is formed by the plume of the nearby Mackenzie River. Low nutrient concentrations and chronically low levels of subsurface light severely limit primary production which was not found to exceed 80 mg C per m² per day in summer. Small standing stocks of phytoplankton (represented by chlorophyll *a* levels consistently less than 2 mg per m³), low water temperatures and salinities, and short ice-free sum-

199

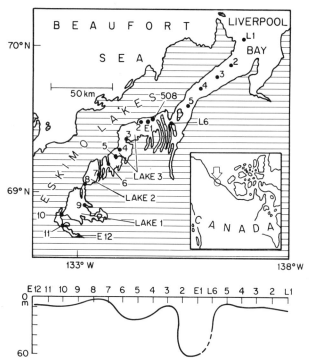

Figure 1. The Eskimo Lakes and Liverpool Bay, showing Lakes 1 to 3, positions of stations occupied and a depth profile from innermost station E12 to the outermost station L1.

mers characterize this arctic estuarine environment (Grainger, Lovrity and Evans 1977; Evans and Grainger 1979).

Methods and Materials

Zooplankton was collected on 38 dates during 1971, 1972, and 1973. The most intensive series of samples was collected at station 508 (Fig. 1) where zooplankton was collected at approximately weekly intervals in summer and at longer intervals (2-3 months) in winter. During the ice-free season, zooplankton was taken over a wider area of the estuary. Of particular interest to this paper is a collecting program conducted over a 2-week period in August 1973. At that time, a total of 18 stations were sampled (Fig. 1) along the entire length of the estuary.

Zooplankton was collected with a 30-cm diameter net of 0.073 mm mesh. Most samples were hauled vertically from the estuary bottom (3-60 m) to the surface. On occasion (for example, the August 1973 18-station series) sequential vertical net hauls were made from 5, 10, 15, and 20 m to the surface. Flowmeters were not used and plankton densities were

calculated assuming 100% filtration efficiency. Calculated densities consequently were underestimated but probably by less than a factor of two.
Literature values for filtration efficiencies for #20-mesh nets range from
37% to 86% (Rawson 1956; Schindler 1969; Hall, Cooper and Werner
1970). Since there was no evidence of net clogging in any of the collection
series, the true filtration efficiency for this study was probably closer to
86%. All samples were preserved with buffered formalin.

Standing stock of zooplankton, expressed as wet weight per m³, was
calculated on the basis of wet weight estimates of taxa and volumetric concentrations. Wet weights were generally determined by directly weighing
preserved specimens. Some were obtained from values in the literature or
from published length-weight regressions (Bogorov 1959; Pertsova 1967;
Fulton 1968). In order to subdivide the estuary into regions with similar
zooplankton assemblages, principal component analysis (Sprules 1977) was
performed on the data collected from 18 stations in August 1973.
Although station depths varied from 3 to 60 m along the estuary, only
data collected in the upper 10 m of the water column were used. Data
from station 508 were not used since the entire water column (20 m) was
sampled. The analysis was performed using the variance-covariance matrix
of the log transformed (numbers per m³ + 1) taxa data. Correlations between the principal components and the log-transformed original variates
were performed to assist with the interpretation of the principal components. Correlations between the principal components and physical and
chemical variates were also calculated.

Results

Species Composition and Distribution

More than 40 species of zooplankton have been recorded so far
from the Eskimo Lakes and Liverpool Bay (authors' unpublished data).
Twenty-three of these were collected during the August 1973 sampling
period and are listed in Table 1. The five dominant copepod species were
Acartia clausi, Eurytemora herdmani, Drepanopus bungei, Pseudocalanus
sp. and *Limnocalanus macrurus.* Rotifers were not routinely identified to
species but at least three species *(Keratella cruciformis, K. longispina* and
Asplanchna sp.*)* were present. Mollusc larvae were rare whereas
polychaete and cirripede larvae were frequently important components of
the zooplankton during the ice-free season. Of cumaceans and amphipods,
three species *(Diastylis* sp.*, Boeckosimus affinis* and *Dulichia porrecta)* and
the mysid *Mysis relicta* were occasionally collected in the plankton,
although not in the August 1978 18-station series. Tintinnids were considered only as a group, and they were exceedingly numerous in Liverpool
Bay and in Lakes 2 and 3 in August 1973. Hydromedusae and
chaetognaths were most common in Liverpool Bay, although the medusa
Halitholus cirratus was found as far inland as Lake 3.

Table 1. Zooplankton (numbers per m³) and some physical and chemical quantities in the upper 10 m in early August 1973 in the Eskimo Lakes (E) and Liverpool Bay (L). Station locations are shown on Figure 1.

	E12	E11	E10	E9	E8	E7	E6	E5	E4	E3	E2	E1	L6	L5	L4	L3	L2	L1
Tintinnids	39	–	–	4	–	–	1052	294	3293	2769	815	3960	3042	229473	85569	46229	12642	12877
Halitholus cirratus	–	–	–	–	–	–	–	–	–	–	–	–	–	–	–	–	–	–
Obelia sp.	–	–	–	–	–	–	–	–	–	–	–	1	–	2	6	6	7	9
Aglantha digitale	–	–	–	–	–	–	–	–	–	–	–	–	–	–	–	–	–	1
Rotifers	175	92	2581	32661	23661	4339	23	15	80	72	17	23	–	–	–	–	–	–
Polychaete larvae	–	–	–	–	566	66	394	23	192	193	79	158	2784	579	–	–	–	132
Bosmina sp.	118	2	–	–	–	–	–	–	–	–	–	–	–	–	–	–	–	–
Daphnia longiremis	99	4	–	–	–	–	–	–	–	–	–	–	–	–	–	–	–	–
D. middendorffiana	3	6	–	–	–	–	–	–	–	–	–	–	–	–	–	–	–	–
Podon leuckarti	–	–	30	189	75	9	34	56	11	8	28	–	45	322	2318	483	136	132
Evadne nordmanni	–	–	–	–	–	–	23	–	23	40	6	–	204	129	258	–	–	–
Calanus glacialis	–	–	–	–	–	–	–	–	–	–	–	–	–	–	–	–	–	5
Pseudocalanus p.	–	–	–	2	–	–	11	11	45	169	45	23	2736	9915	24145	4893	5477	5528
Drepanopus bungei	–	–	–	2	–	–	79	79	–	–	–	–	191	–	–	–	–	–
Derjuginia tolli	–	–	–	–	–	–	–	–	–	–	–	–	–	4	–	–	–	12
Eurytemora herdmani	11	2	30	90	133	18	79	79	46	781	537	1652	2354	13328	27364	14680	2761	1189
E. arctica	9	2	4	46	–	–	–	–	–	–	–	–	–	–	–	–	–	–
Diaptomus sicilis	397	366	32	46	–	–	–	–	–	–	–	–	–	–	–	–	–	–
Epischura lacustris	28	28	62	–	38	–	–	–	–	–	–	–	–	–	–	–	–	–
Limnocalanus macrurus	110	178	6	70	9472	3236	5760	2931	3100	3300	1550	10139	–	–	–	–	–	–
Acartia clausi	3	4	–	50	–	–	–	–	–	–	–	–	6945	9529	7211	1287	1629	283
A. longiremis	–	–	–	–	–	–	–	–	–	–	–	–	–	2	2	–	23	32
Cypris larvae	–	–	–	–	–	–	–	–	–	–	–	–	136	–	–	–	16	16
Cyclopoid	115	91	229	304	264	38	34	–	11	16	23	23	–	–	–	–	–	–
Harpacticoid	–	–	–	–	57	–	–	–	–	16	45	45	23	–	–	–	–	–
Crustacean nauplii	8428	3340	6984	14181	23717	4613	2399	1324	962	1915	1997	5522	15749	80483	83122	181505	20074	12868
Eggs	530	354	2491	5038	1132	198	747	860	690	1038	413	2376	407	3982	3670	2704	1743	1038
Temperature (C)	9.5	9.1	9.1	9.6	8.9	10.0	10.9	11.0	11.5	10.6	9.0	8.8	12.0	12.8	12.6	12.1	12.1	9.5
Salinity (‰)	0.7	2.5	3.2	3.2	6.7	7.2	12.9	12.7	11.4	12.4	13.6	13.7	14.0	18.0	17.5	16.5	16.6	18.8
Phosphate (µg-at P/L)	0.0	0.0	0.0	0.0	0.0	0.0	0.0	0.0	0.0	0.1	0.1	0.1	0.1	0.3	0.4	0.4	0.5	0.2
Nitrite (µg-at N/L)	0.0	0.0	0.0	0.0	0.0	0.0	0.0	0.0	0.0	0.1	0.1	0.1	0.1	0.1	0.0	0.0	0.5	0.1
Nitrate (µg-at N/L)	0.2	0.7	0.1	0.0	0.0	0.0	0.0	0.0	0.0	0.1	0.6	0.6	0.6	0.0	0.0	0.0	0.1	0.7
Silicate (µg-at Si/L)	9.9	10.6	9.8	9.6	6.9	6.5	2.3	2.5	2.5	3.0	3.5	3.5	5.8	18.5	17.9	20.6	19.1	22.0
Chlorophyll a (mg/m³)	0.9	0.5	0.5	0.7	0.6	0.3	0.5	0.5	0.4	0.5	0.4	0.6	1.2	1.2	1.7	1.5	1.2	1.4

STATION

Each zooplankton taxon had a unique distribution pattern along the 200 km length of the estuary. Some, such as *Aglantha digitale, Obelia* sp. and *Acartia longiremis*, were restricted to Liverpool Bay whereas others such as *Diaptomus sicilis, Epischura lacustris* and *Daphnia middendorffiana*, were confined to the low salinity waters of Lake 1. *Limnocalanus macrurus* was found in all lakes and in Liverpool Bay (its apparent absence from Lakes 2 and 3, according to Table 1, is a consequence of its occurrence there only below 10 m). *Acartia clausi, Eurytemora herdmani* and *Pseudocalanus* sp. were taken along most of the estuary. *Drepanopus bungei* occurred primarily in the cold, deep water of Lake 3, avoiding the more extreme salinities elsewhere.

Regional Subdivisions of the Estuary

The first principal component (PCI) accounted for 54.3% of the variance while the second principal component (PC2) accounted for an additional 19.3% of the variance. The third principal component accounted for only an additional 8% of the variance and it was not considered further.

Plotting of the 18 stations by their PC1 and PC2 values revealed five groups of stations (Fig. 2). Group I, with low PC1 and high PC2 values, consisted of the four stations in Lake 1. Group II had low PC2 and slightly higher PC1 values; the two stations in this group were located in Lake 2. Groups III and IV had higher PC1 values and were located respectively in Lake 3 and lower Liverpool Bay. Group V had high PC1 and PC2 values and consisted of the four stations in the outer half of Liverpool Bay.

As expected, PC1 was positively correlated (r = +0.96) with salinity (Table 2) which apparently was the major environmental factor determining zooplankton assemblages along the estuary. Correlations with phosphate (0.77) and chlorophyll *a* (0.72) were also high because concentrations of those variables were greatest in the high salinity waters of the bay. Temperature was highly correlated with PC1 (0.76) but less so than salinity. Taxa correlations with PC1 were as expected. Rotifers (−0.88), *Diaptomus sicilis* (−0.70) and such cyclopoid copepods as *Cyclops scutifer* and *C. vernalis* which were most numerous in the low salinity of Lake 1 had negative correlations with PC1. Taxa such as tintinnids (0.95), *Obelia* sp. (0.55), *Acartia clausi* (0.65) and *Pseudocalanus* sp. (0.87) which attained maximum numbers in the saline waters of Liverpool Bay had positive correlations with PC1. Silicate was positively correlated with PC2 (0.81), and occurs in highest concentrations at both extremities of the estuary where it enters from adjacent waters.

Seasonal Biology at Station 508

The most detailed information on the seasonal biology of zooplankton comes from station 508 located near the geographical centre

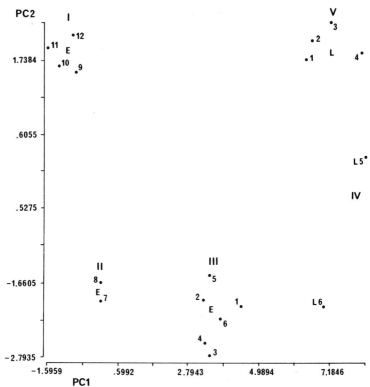

Figure 2. *Coordinate plot of 18 stations sampled along the full length of Liverpool Bay (L1 to L6) and the Eskimo Lakes (E1 to E12) in early August 1973. The stations are separated into five groups (I to V), representing five regions of the estuary.*

of the estuary. Additional seasonal data from Lake 3 confirm the major features of seasonal patterns interpreted at station 508.

The most abundant copepod was *Acartia clausi*, a species widely distributed in subarctic and temperate waters of low salinity (Brodskii 1967). Examination of data for the six copepodite stages (Fig. 3) revealed evidence of two consecutive generations within a single summer. The first, originating from overwintering adults, appeared as nauplii (not shown in Fig. 3), then as early copepodites which entered the plankton in June. Adults attained maximum numbers in late July. This was followed by another pulse of young copepodites in August. Some adults survived the winter to give rise to the next generation the following spring.

Eurytemora herdmani is distributed mainly in subarctic and temperate waters, less widely than *A. clausi* (Brodskii 1967). Overwintering *E. herdmani* (Fig. 3) was dominated by immature (primarily late stage)

Table 2. Correlation (r) between numbers of taxa along with physical and chemical variables used in the principal component analyses of the 18-station data from August 1973 and the first (PC1) and second (PC2) components.

	PC1	PC2
Tintinnids	0.95	−0.01
Halitholus cirratus	0.08	−0.23
Obelia sp.	0.55	0.51
Aglantha digitale	0.22	0.24
Rotifers	−0.88	−0.16
Polychaete larvae	0.28	−0.81
Bosmina sp.	−0.39	0.34
Daphnia longiremis	−0.42	0.36
D. middendorffiana	−0.49	0.39
Podon leuckarti	0.58	0.37
Evadne nordmanni	0.54	−0.55
Calanus glacialis	0.22	0.24
Pseudocalanus sp.	0.88	0.39
Drepanopus bungei	0.16	−0.78
Derjuginia tolli	0.39	0.24
Eurytemora herdmani	0.91	0.10
E. arctica	−0.62	0.47
Diaptomus sicilis	−0.70	0.54
Epischura lacustris	−0.12	0.41
Limnocalanus macrurus	−0.66	0.57
Acartia clausi	0.65	−0.63
A. longiremis	0.33	0.38
Cyclopoids	−0.87	−0.02
Harpacticoids	−0.11	−0.54
Cypris larvae	0.43	−0.06
Crustacean nauplii	0.47	0.59
Eggs	0.34	0.42
Temperature (C)	0.76	0.07
Salinity ($^0/_{00}$)	0.96	−0.12
Phosphate (μg-at P/L)	0.77	0.51
Nitrite (μg-at N/L)	0.44	−0.34
Nitrate (μg-at N/L)	−0.03	0.07
Silicate (μg-at Si/L)	0.45	0.81
Chlorophyll a (mg/m^3)	0.72	0.59

copepodites. This overwintering population matured into adults and reproduced in July. The summer generation developed through July and August although development was not completed until the following year.

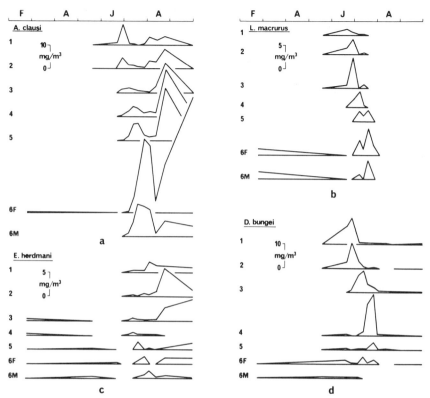

Figure 3. Seasonal occurrence as wet weight (mg per m³) of six copepodite stages of four copepod species at station 508 between February and September 1972. a, Acartia clausi; b, Limnocalanus macrurus; c, Eurytemora herdmani; d, Drepanopus bungei.

Drepanopus bungei, an estuarine species which seemed to avoid the highest and the lowest salinities in the estuary, is restricted generally to fairly low salinity arctic waters. Unlike the two species discussed above, D. bungei is unable to survive in warmer temperate seas. In the estuary it was confined to the deeper waters of Lake 3 when surface waters were warmest. The species is known only from the Kara to the East Siberian Sea along the Siberian coast, the Chukchi Sea, the south Beaufort Sea and the Ellesmere Island coast (Bowman and Long 1968; Grainger 1975). At station 508, D. bungei (Fig. 3) produced a single generation each year with overwintering adults producing young copepodites in June.

Limnocalanus macrurus is a circumpolar species found in waters of low salinity around the periphery of the Arctic Ocean, in certain far northern fresh waters, and in temperate fresh waters such as the Great Lakes

(Bowman and Long 1968; Watson and Carpenter 1974). In the estuary it was found in Liverpool Bay and as far inland as Lake 1. A single generation was produced each year (Fig. 3), with overwintering adults giving rise to the youngest copepodites in June, and these maturing to adulthood in July.

Pseudocalanus sp. was not as tolerant of low salinity water as *Acartia clausi, Eurytemora herdmani* and *Limnocalanus macrurus,* and appeared to be more successful than they were in Liverpool Bay. In Lake 3, it was most abundant in the deeper waters. A single generation was produced each year, with an extended reproductive period in spring followed by summer-long emergence of young copepodites.

The copepods as a group showed a typical far northern annual cycle, as exemplified by the herbivorous copepods shown in Fig. 4. Abundance rose from close to zero in February to a maximum of nauplii in May, followed by several peaks representing successive copepodite development in different species. Polychaete larvae were numerous during the ice-free season with high population levels in late June followed by a decline and then a second peak in late July. Cirripede nauplii were most abundant in late June and diminished during July and August. In contrast to the herbivores, the few obligatory carnivores (medusae) showed no clear seasonal pattern.

The calculated mean annual biomass over the two years at station 508 was 80 mg wet weight per m³. This consisted mainly of herbivorous copepods, polychaete larvae and cirripede nauplii in that order of importance. Certain late copepodites, especially of *Acartia clausi* and *Limnocalanus macrurus,* along with *Mysis relicta* and *Boeckosimus affinis,* are probably mixed feeders. The principal invertebrate carnivore is *Halitholus cirratus.* This pelagic invertebrate food structure contributes to the support of 4 main pelagic vertebrates, the carnivorous fishes *Clupea harengus, Coregonus autumnalis* and *C. sardinella* (unpublished data), and the primarily fish- and plankton-feeding ringed seal, *Phoca hispida* (McLaren 1958).

Discussion

Distribution of zooplankton in the Eskimo Lakes and Liverpool Bay shows a typical estuarine pattern. Principal component analysis demonstrated five groups of stations arranged along the longitudinal axis of the estuary. Salinity was the environmental factor on which the distribution of the various zooplankton species of the estuary was most directly dependent. Other factors also showed fairly strong correlations with certain zooplankton but most probably had little direct influence on the animal plankton. Silicate in itself probably does not affect zooplankton distribution within the estuary, but the correlations are indicative of two sources of zooplankton populations in the estuary. Fresh water flowing into the head of the system contained high amounts of silicate and zooplankton typical of fresh waters. At the other extreme, the outer reaches of Liverpool Bay

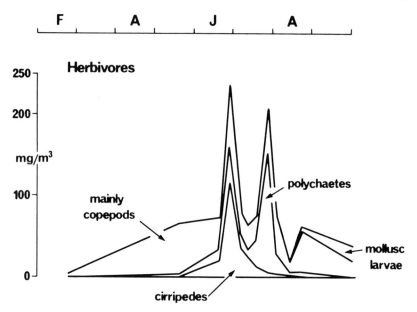

Figure 4. Herbivorous zooplankton wet weight (mg per m³) at station 508 between February and September 1972.

through direct connection with the Beaufort Sea were high in silicate and supported Beaufort Sea zooplankton which was never found within the Eskimo Lakes.

The estuary was thus dominated by the character of the inflowing waters, with salinity and water source both affecting the zooplankton composition. In the Eskimo Lakes, the peninsulas separating Lakes 1 and 2 and Lakes 2 and 3 apparently form strong barriers to water mixing, as shown by differences in salinity and zooplankton populations on opposite sides of them. In Liverpool Bay, variations in the inflow of Beaufort Sea subsurface water affects the species composition within the bay. The division of the bay into groups IV and V therefore is probably not a consistent feature of this part of the estuary.

It was shown that *Acartia clausi* may pass through as many as two generations in a year in the Eskimo Lakes. This is contrary to expectations for arctic copepods both in the sea (Ussing 1938; Grainger 1959, 1965) and in fresh waters (Comita 1956; McLaren 1961), where production of a single generation in a year is the rule. However, although abundant in the Eskimo Lakes, *A. clausi* is not an arctic species; it is characteristically more southerly in distribution.

Temperature has been shown to be a major factor in affecting growth and development rates in copepods (Corkett and McLaren 1978).

It may be of consequence then that the water temperature range found in the Eskimo Lakes during the moulting period of *Acartia clausi* was from about 5 to 12C, and therefore notably higher than the near-zero levels in which most of the arctic development rate studies have been carried out. The Eskimo Lakes in fact were not much colder than waters where *A. clausi* has been shown to pass through from two annual generations on Martha's Vineyard (Deevey 1948) to as many as five or six in the English Channel (Digby 1950). Generation length in the Eskimo Lakes (40-50 days) relates quite reasonably to estuarine water temperatures at the time of development (mean around 8.5C) when compared, for example, to the 42-day generation time at 8.5C derived by McLaren (1978) from Marshall's Loch Striven data.

The calculated average annual biomass level of 80 mg per m³ (wet wt) at station 508 is low by world standards. It is similar to levels reported from seas bordering the Arctic Ocean elsewhere as well as from certain open oceanic waters of lower latitudes (FAO 1972). It is notably low however when compared, for example, with the 2-year mean of 62 mg dry weight (convertible to around 380 mg wet weight) measured in lower Narragansett Bay, a U.S. east coast estuary (Hulsizer 1976). Zooplankton standing stock levels shown here along with earlier demonstrations of low primary productivity and phytoplankton stocks (Grainger et al. 1977; Evans and Grainger 1979) indicate severe oligotrophic conditions in the Eskimo Lakes-Liverpool Bay estuary.

Acknowledgments

We thank our colleagues at the Arctic Biological Station, especially J. E. Lovrity and B. Canning for their assistance in the field, and A. A. Mohammed, K. Robinson, B. Emmett and M. Shum for their work in the laboratory.

Contribution No. 257, Great Lakes Research Division, University of Michigan.

References Cited

Bogorov, B. G. 1959. On the standardization of marine plankton investigations. *Int. Rev. ges. Hydrobiol.* 44:621-642.

Brodskii, K. A. 1967. *Calanoida of the Far Eastern Seas and Polar Basin of the USSR.* Keys to the fauna of the USSR, no. 35. Zoological Institute, Academy of Sciences of the USSR. Jerusalem, 440 p. (Translation by the Israel Program for Scientific Translations).

Bowman, T. E. and A. Long. 1968. Relict populations of *Drepanopus bungei* and *Limnocalanus macrurus grimaldi* (Copepoda: Calanoida) from Ellesmere Island, N.W.T. *Arctic* 21:172-180.

Comita, G. W. 1956. A study of a calanoid copepod population in an arctic lake. *Ecology* 37:576-591.

Corkett, C. J. and I. A. McLaren. 1978. The biology of *Pseudocalanus. Adv. Mar. Biol.* 15:1-231.

Deevey, G. B. 1948. The zooplankton of Tisbury Great Pond. *Bull. Bingham Oceanogr. Coll.* 12:1-44.

Digby, P. S. B. 1950. The biology of the small planktonic copepods off Plymouth. *J. Mar. Biol. Ass. U.K.* 29:393-438.

Evans, M. S. and E. H. Grainger. 1979. Seasonal variations in nutrients and chlorophyll *a* in a Canadian arctic estuary. Abstr. 42nd annual meeting Amer. Soc. Limnol. Oceanogr., Stony Brook, New York.

FAO. 1972. *Atlas of living resources of the seas.* FAO, Dept. of Fisheries. Rome. 16 pp. 120 plates.

Fulton, J. 1968. A laboratory manual for the identification of British Columbia marine zooplankton. Fish. Res. Board Can., Tech. Rep. No. 55: 141 p.

Grainger, E. H. 1959. The annual oceanographic cycle at Igloolik in the Canadian arctic. I. The zooplankton and physical and chemical observations. *J. Fish. Res. Board Can.* 16:453-501.

Grainger, E. H. 1965. Zooplankton from the Arctic Ocean and adjacent Canadian waters. *J. Fish. Res. Board Can.* 22:543-564.

Grainger, E. H. 1975. Biological productivity of the southern Beaufort Sea. The physical-chemical environment and the zooplankton. Beaufort Sea Project, Tech. Rep. No. 12a, Victoria, B.C. 82 p.

Grainger, E. H., J. E. Lovrity and M. S. Evans. 1977. Biological oceanographic observations in the Eskimo Lakes, arctic Canada. Physical, nutrient, and primary production data, 1961-1975. Environment Canada, Fisheries and Marine Service, Tech. Rep. No. 685. 108 p.

Hall, D. J., W. E. Cooper and E. E. Werner. 1970. An experimental approach to the production dynamics and structure of freshwater communities. *Limnol. Oceanogr.* 15:839-928.

Hulsizer, E. E. 1976. Zooplankton of lower Narragansett Bay, 1972-1973. *Chesapeake Sci.* 17:260-270.

McLaren, I. A. 1958. The biology of the ringed seal (*Phoca hispida* Schreber) in the eastern Canadian arctic. Bull. Fish. Res. Board Canada No. 118. 97 p.

McLaren, I. A. 1961. A biennial copepod from Hazen Lake, Ellesmere Island. *Nature* 189: 774.

McLaren, I. A. 1978. Generation lengths of some temperate marine copepods: estimation, prediction, and implications. *J. Fish. Res. Board Can.* 35:1330-1342.

Pertsova, N. M. 1967. Average weights and sizes of abundant species of zooplankton in the White Sea. *Oceanology* 7:240-243.

Rawson, D. S. 1956. The net plankton of Great Slave Lake. *J. Fish. Res. Board Can.* 13:53-127.

Schindler, D. W. 1969. Two useful devices for vertical plankton and water sampling. *J. Fish. Res. Board Can.* 26:1948-1955.

Sprules, W. G. 1977. Crustacean zooplankton communities as indicators of limnological conditions: an approach using principal component analysis. *J. Fish. Res. Board Can.* 34:962-975.

Ussing, H. 1938. The biology of some important plankton animals in the fjords of East Greenland. *Medd. om Gronland* 100:1-108.

Watson, N. H. F. and G. F. Carpenter. 1974. Seasonal abundance of crustacean zooplankton and net plankton biomass in Lakes Huron, Erie, and Ontario. *J. Fish. Res. Board Can.* 31:309-317.

PRIMARY PRODUCTION
AND PHOTOSYNTHESIS

EFFECTS OF HARVESTING ON THE ANNUAL NET ABOVEGROUND PRIMARY PRODUCTIVITY OF SELECTED GULF COAST MARSH PLANTS

Judy P. Stout

University of South Alabama
Dauphin Island Sea Lab
Dauphin Island, Alabama

Armando A. de la Cruz

Department of Zoology
Mississippi State University
Mississippi State, Mississippi

and

Courtney T. Hackney

Department of Biology
University of Southwestern Louisiana
Lafayette, Louisiana

Abstract: A winter harvest was applied to three common Gulf Coast tidal salt marsh plants (*Juncus roemerianus* and *Spartina alterniflora* in Alabama; *S. cynosuroides* and *J. roemerianus* in Mississippi). Aboveground living and dead plant material was manually clipped and removed ("harvested") from study plots in winter 1977, and from the same plots (reharvest) and additional plots in winter 1978. Annual net aboveground primary productivity was determined by applying a predictive periodic model, corrected for loss of dead material, to monthly biomass samples. A single harvest resulted in an increase in annual aboveground net productivity of as much as 50% for the Mississippi *J. roemerianus* and 140% for *S. cynosuroides*. When the species were reharvested, net aboveground productivity increased up to 100% for Mississippi *J. roemerianus*, 45% for *S. cynosuroides* and 250% for *S. alterniflora*. Alabama *J. roemerianus* showed consistently lowered levels of productivity in harvested plots.

Introduction

No harvest of marsh plants of significant economic impact is employed in the U.S. However, if any of the recent work on chemical derivatives (Miles and de la Cruz 1976) and on the pulping potential of marsh plants (de la Cruz and Lightsey Unpubl. ms.) proves to be of economic value, the prospect exists of regularly harvesting certain marsh plants for chemicals, cellulose, and other by-products and employing agriculture techniques to farm marsh grass species.

Various marsh plants are grazed by cattle along the eastern and Gulf Coasts (Chabreck 1976). This repeated harvest of biomass and the trampling effect of cattle has been shown to diminish yield of grazed plants (Williams 1955; Reimold et al. 1975). Reimold et al. (1974) found an increase in mean dry weight biomass following simulated grazing in a Georgia *Spartina alterniflora* marsh. Clipping experiments in a Mississippi mixed brackish marsh indicated that neither repeated clippings nor single clippings altered the growth rate of the harvested plants (Gabriel and de la Cruz 1974).

Increased human occupancy of coastal areas has historically and increasingly jeopardized the existence of tidal marsh resources. Dwindling marsh acreage results in a proportionately greater importance to the estuarine ecosystem of those marshlands remaining. Activities conducted within marshes must, therefore, be compatible with maintenance of the natural system function and not diminish the production role of the plants. In addition, the potential harvest of marshlands for products of direct value to mankind implies a cropping system that will produce maximum biomass yield.

The objective of our study was to determine the effect of a single harvest and repeated annual harvests on the annual aboveground primary productivity of *Juncus roemerianus*, *Spartina alterniflora* and *S. cynosuroides*. Effects upon *J. roemerianus* were examined in two environmental settings.

Materials and Methods

Marshes dominated by each of the study plant species were selected in Alabama and Mississippi because of accessibility, species composition and availability of a historical data base. Two study sites, one in each state, were chosen for *J. roemerianus* to reflect both a saline environment supporting an essentially monospecific community and a brackish, more heterogeneous plant community. The Mississippi study area, for *J. roemerianus* and *S. cynosuroides*, was located on a marsh island on the western side of St. Louis Bay, Hancock County, an area previously described by de la Cruz (1973) and Gabriel and de la Cruz (1974). In Alabama, two marshes on the leeward shore of Dauphin Island, Mobile County, were utilized for the study of *J. roemerianus* and *S. alterniflora*. Species composition and environmental setting of the two Dauphin Island study sites have been described by Stout (1978). The substrates of the two Alabama marshes are of higher salinity, greater sand content, and lower organic matter concentration than the Mississippi marsh.

Both *J. roemerianus* communities were of short-medium plants (average 1.0 m in height), although the one in Mississippi is more of a high marsh community (terminology of Subrahmanyam, Kruczynski and Drake 1976) than the one in Alabama. The *S. cynosuroides* plants were about

2-3 m in height while the S. *alterniflora* community was composed of short plants (0.5 m).

Study plots 10 m × 10 m (100 m²) were staked and marked, three in 1977 and four in 1978, within each of the four study areas. Corridors with a minimum width of 5 m were maintained between plots to minimize sampling impact. One study plot in each area was used as a control. A second plot was cleared of aboveground material and ground litter by burning in February 1977 and 1978. This plot was monitored to determine production and loss of dead material over the study period. One of the study plots was manually clipped, using grass shears, in mid-February, 1977. All clipped material and litter on the substrate was removed from the plot. In 1978, an additional plot was similarly harvested and the 1977 experimental plot was reharvested. Study plots for each study species are summarized below:

1977 Plots	*1978 Plots*
Control	Control
Burned	Burned
Single Harvest	Single Harvest
	Reharvest

Six 0.25 m² (0.5 m × 0.5 m) samples were taken monthly from each study plot from April through December. Samples were taken at random along non-overlapping transects across each plot, a new transect used each month. Samples were sorted into living and dead plant material, dried at 100C to constant weight, and weighed.

A predictive periodic model (PPM) was applied to monthly live biomass data from control and harvested plots (Hackney and Hackney 1978). The model was also utilized on monthly dead biomass data from each burned plot. After the data were reviewed, a two-harmonic model was utilized for Mississippi J. *roemerianus* data from 1978, in order to accommodate two growth peaks. The two-harmonic model fits a periodic regression curve of:

$$y = c_o + c_1 \sin (ct_i) + c_2 \cos (ct_i) + c_3 \sin (2 ct_i) + c_4 \cos (2ct_i)$$

where

y = dependent variable
c_o = overall mean
c_1, c_2 = coefficients of the first harmonic function
c_3, c_4 = coefficients of the second harmonic function
$c = 2 \pi/n$
t_i = i th independent variable.

A single harmonic model was utilized for all other study plots and consisted of

$$y = c_o + c_1 \sin (ct_i) + c_2 \cos (ct_i)$$

having the same component definitions as in the two-harmonic model. Table 1 summarizes the predictive periodic model for each study plot.

The periodic models for all study plots were compared by analysis of variance (ANOVA) as provided by Hackney and Hackney (1977) to test whether the productivities were different (p = 0.05). This is an added advantage of the PPM max-min technique because, when using the standard max-min procedure, one has only two numbers to compare and no way to make a statement about any statistically significant differences between two communities.

Annual productivity values for live biomass in each plot and production of dead material for that species were obtained by the max-min procedure using maximum and minimum values extracted from the respective PPM. Net annual primary productivity (NAPP) of each study plot was estimated by summing the annual live biomass and loss of dead biomass values derived for each. The periodic max-min technique applied to monthly live biomass changes and losses due to death of new plant growth provides an estimate based on every sample collected during the study and an estimate accounting for live biomass lost between samples, hence a more accurate and reliable estimate.

Results and Discussion

Table 2 summarizes the estimated annual net primary productivity of both control and harvested plots. Clipping resulted in a significant increase in primary production of 26-106% for the Mississippi *J. roemerianus* community for both years of the study, 46-141% for *S. cynosuroides* during 1978, and for *S. alterniflora* an increase of 248% when reharvested only. The Alabama *J. roemerianus* community exhibited consistently lowered primary productivity in all clipped plots.

The net aboveground primary productivity of 460-658 g m^{-2} yr^{-1} for the control *J. roemerianus* plot is comparable to values reported for high marsh *J. roemerianus* (595 g m^{-2} yr^{-1}) in North Florida (Kruczynski et al. 1978). The 571 g m^{-2} yr^{-1} NAPP for control *S. alterniflora* in 1978 is comparable to values (581 g m^{-2} yr^{-1}) for a similar marsh in Barataria Bay, Louisiana (Kirby and Gosselink 1976), although the 1977 value of 240 g m^{-2} yr^{-1} is significantly lower. Previous investigations by de la Cruz (1975) of similar marshes yielded values of NAPP (2190 g m^{-2} yr^{-1}) comparable to the values of the control *S. cynosuroides* (1430-1740 g m^{-2} yr^{-1}) in our study.

Growth patterns of the two *J. roemerianus* communities differed as did the response of each to the clip treatment (Figs. 1 and 2). The Mississippi community yielded peaks in live biomass in June and again in October. The Alabama *J. roemerianus* marsh did not show a biomodal pattern, reaching a single peak in biomass in early fall. The NAPP of both communities was not significantly different between the control plots. However, when clipped, the Alabama marsh responded by a lowering in

Table 1. Periodic models for four natural marsh communities and harvested study plots in each. c_o = annual mean; c_1, c_2 = coefficients of first harmonic function; c_3, c_4 = coefficients of second harmonic function; r^2 = coefficient of determination; MS = Mississippi; AL = Alabama.

Community Treatment	c_o	c_1	c_2			r^2
1977						
J. roemerianus control (MS)	687.0	−143.2	−182.6			0.3936
J. roemerianus clip (MS)	546.3	−217.8	−19.7			0.6056
J. roemerianus control (AL)	646.0	144.0	−177.4			0.2924
J. roemerianus clip (AL)	377.2	−85.0	105.6			0.3622
S. alterniflora control	227.7	16.3	−126.5			0.3521
S. alterniflora clip	194.0	19.4	19.5			0.0246
S. cynosuroides control	606.4	−344.5	−675.8			0.6958
S. cynosuroides clip	804.1	−427.7	−691.4			0.7403
1978	c_o	c_1	c_2	c_3	c_4	r^2
J. roemerianus control (MS)	858.3	−209.0	−54.7	−106.6	33.8	0.3730
J. roemerianus clip (MS)	567.9	−221.0	49.0	71.7	−2.6	0.5200
J. roemerianus reclip (MS)	635.5	−265.6	137.2	−11.9	147.0	0.7200
J. roemerianus control (AL)	129.8	−17.6	26.4	−	−	0.2078
J. roemerianus clip (AL)	66.4	−41.4	−5.9	−	−	0.7208
J. roemerianus reclip (AL)	52.8	−32.1	10.6	−	−	0.6032
S. alterniflora control	107.2	−52.6	19.7	−	−	0.7200
S. alterniflora clip	103.1	−75.6	9.0	−	−	0.7130
S. alterniflora reclip	94.7	−82.6	29.9	−	−	0.8340
S. cynosuroides control	847.1	−764.6	−441.5	−	−	0.7500
S. cynosuroides clip	1040.4	−694.5	−801.8	−	−	0.6300
S. cynosuroides reclip	843.7	−508.6	−374.0	−	−	0.4000

Table 2. Summary of estimated annual net aboveground primary productivity (NAPP) of 1977 and 1978 control and harvested plots in four marsh communities. Total NAPP was estimated from max-min values of periodic models, corrected for loss of dead material.

Marsh Community	NAPP ($g\ m^{-2}\ yr^{-1}$)	
	Control	Harvested
1977: Single clip application		
Juncus roemerianus (MS)	580	877
J. roemerianus (AL)	460	270
Spartina alterniflora	240	33
S. cynosuroides	1740	1755
1978: Single clip application		
Juncus roemerianus (MS)	540	681
J. roemerianus (AL)	658	414
Spartina alterniflora	571	305
S. cynosuroides	1430	3450
1978: Reclip of 1977 experimental plot		
Juncus roemerianus (MS)	540	1113
J. roemerianus (AL)	658	257
Spartina alterniflora	571	1987
S. cynosuroides	1430	2089

NAPP while the Mississippi marsh showed increased productivity, as great as four times that found in Alabama for the same treatment. Extensive peat deposits beneath the Mississippi marsh, absent in Alabama, may indicate a greater age for the marsh. Removal of accumulated standing material, both living and dead, from the older marsh may have provided greater space and sunlight for new leaf production, relative to preharvest condition, than in the younger, less dense Alabama marsh. Within the study period, live standing biomass reached control levels only in the Alabama 1977 clip plot and the Mississippi 1978 plot.

S. *cynosuroides* exhibits a very distinct annual growth pattern, with winter death of aboveground culms. February harvest removed mostly dead material. Growth patterns in all plots were similar (Fig. 3). Standing live biomass peaked in mid- to late-summer with levels in harvested plots exceeding control plots from the onset of annual regrowth. Reduction of shading and crowding by the previous year's dead culms may have been a significant factor in the increased productivity from clipping.

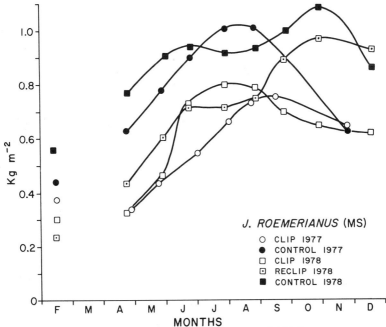

Figure 1. *Periodic max-min curves for Mississippi (MS) Juncus roemerianus based on monthly standing crop of live plants. February data points represent preharvested biomass levels within each study plot.*

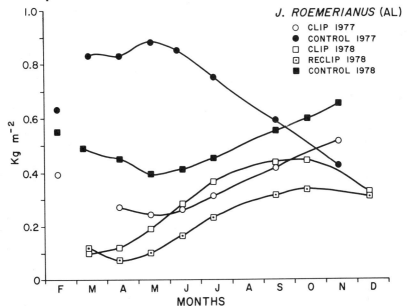

Figure 2. *Periodic max-min curves for Alabama (AL) Juncus roemerianus as in Fig. 1.*

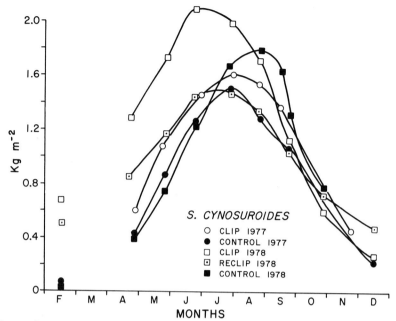

Figure 3. Periodic max-min curves for Spartina cynosuroides *as in Fig. 1.*

The *S. alterniflora* community was the most frequently flooded of the four areas studied. Water often covered the marsh to depths of several inches, even at low tide. Trampling and compaction of the substrate due to harvesting activities was very severe in this marsh. Many plants showed no regrowth and rotted areas were common within the plots as the study progressed. Lowered productivity in both single clip plots was the result of fewer plants due to plant death and not lowered individual plant biomass production.

Growth patterns were markedly different between the study years but similar between plots within each year (Fig. 4). Peak 1977 biomass was achieved in June, but occurred much later in September to October, 1978. Both 1978 harvested plots exceeded control in biomass production during the latter portion of the growing season. Reharvest of the 1977 plot in February, 1978 was easily and rapidly accomplished without significant damage because of the lowered plant density. Recovery and vegetation expansion in this plot resulted in the largest percentage increase (250%) over control of any community.

If harvesting is determined to be of economic value, techniques with a low disruption factor must be found for marshes such as the *S. alterniflora* marsh with soft, saturated, easily disturbed substrates.

Effects of harvesting plant materials vary between species of marsh plants. There appears to be little detrimental effect on productivity and

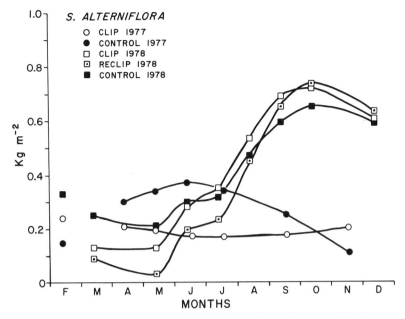

Figure 4. Periodic max-min curves for Spartina alterniflora *as in Fig. 1.*

there may even be an increase in biomass production as a result of harvesting. The effects of different harvesting methods and of repeated annual harvest beyond the second year need to be examined before the impact upon the plant communities can be adequately evaluated. In addition, the impact upon other biotic components and system interaction must be determined. Long range alteration of inputs to peat production, *in situ* nutrient regeneration and loss of detrital sources due to removal of plant biomass need to be considered.

Acknowledgments

For assistance with field sampling and processing we are grateful to Eileen Duobinis, Susan Ivester, Keith Parsons, Charles Harp and Julian Stewart. We are grateful to Olga Hackney for her assistance and counseling with statistical and modelling work. This research was supported by a grant from the NOAA Office of Sea Grant, Department of Commerce under grant no. 04-8-MOI-92, and the Mississippi-Alabama Sea Grant Consortium grant no. M-ASGL-78-039. The U.S. Government is authorized to produce and distribute reprints for governmental purposes notwithstanding any copyright notation that may appear hereon.

Dauphin Island Sea Lab Contribution Number 033.

References Cited

Chabreck, R. H. 1976. Management of wetlands for wildlife habitat improvement, pp. 226-233. In: M. L. Wiley (ed.), *Estuarine Processes, Vol. I.* Academic Press, New York.

de la Cruz, A. A. 1973. The role of tidal marshes in the productivity of coastal waters. *Assoc. of S. E. Biol. Bull.* 20:147-156.

de la Cruz, A. A. 1975. Primary productivity of coastal marshes in Mississippi. *Gulf Res. Repts.* 4:351-356.

Gabriel, B. C. and A. A. de la Cruz. 1974. Species composition, standing stock and net primary production of a salt marsh community in Mississippi. *Chesapeake Sci.* 15:72-77.

Hackney, C. T. and O. P. Hackney. 1978. An improved, conceptually simple technique for estimating the productivity of marsh vascular flora. *Gulf Res. Repts.* 6:125-129.

Hackney, O. P. and C. T. Hackney. 1977. Periodic regression analysis of ecological data. *J. Miss. Acad. Sci.* 22:25-33.

Kirby, C. J. and J. C. Gosselink. 1976. Primary production in a Louisiana Gulf Coast *Spartina alterniflora* marsh. *Ecology* 57:1052-1059.

Kruczynski, W. L., C. B. Subrahmanyam and S. H. Drake. 1978. Studies on the plant community of a north Florida salt marsh. Part I. Primary production. *Bull. Mar. Sci.* 28:316-334.

Miles, D. H. and A. A. de la Cruz. 1976. Pharmacological potential of marsh plants, pp. 267-276. In: M. L. Wiley (ed.), *Estuarine Processes, Vol. I.* Academic Press, New York.

Reimold, R. J., R. A. Linthurst and T. L. Wolf. 1975. Effects of grazing on a salt marsh. *Biol. Conserv.* 8:105-125.

Stout, J. P. 1978. An analysis of annual growth and productivity of *Juncus roemerianus* Scheele and *Spartina alterniflora* Loisel in coastal Alabama. Ph.D. Dissertation, U. Alabama, Tuscaloosa, AL. 95 p.

Subrahmanyam, C. B., W. L. Kruczynski and S. H. Drake. 1976. Studies on the animal communities in two north Florida salt marshes. Part II. Macroinvertebrate communities. *Bull. Mar. Sci.* 26:172-195.

Williams, R. E. 1955. Development and improvement of coastal marsh ranges. U.S. Department of Agriculture, Yearbook of Agriculture 1955. p. 444-450.

A COMPARISON OF *SPARTINA ALTERNIFLORA* PRIMARY PRODUCTION ESTIMATED BY DESTRUCTIVE AND NONDESTRUCTIVE TECHNIQUES

Michael A. Hardisky*

Coastal Resources Division
Georgia Department of Natural Resources
1200 Glynn Avenue
Brunswick, Georgia

Abstract: Individual live and dead culms of highmarsh *Spartina alterniflora* Loisel. within permanent 0.1 m² quadrats were tagged. Growth data reveal leaf production, senescence and abscission are continuous processes throughout the year in this Georgia salt marsh. Leaf production of culms > 10 cm in height peaked in June (0.023 leaves culm⁻¹ day⁻¹) and of culms ≤ 10 cm in height in December (0.031 leaves culm⁻¹ day⁻¹). Leaf senescence peaked in August (0.047 leaves culm⁻¹ day⁻¹) as did leaf abscission (0.023 leaves culm⁻¹ day⁻¹). Production of new leaves from shoots ≤ 10 cm exceeded leaf production from older culms during fall and early winter. Recruitment of new shoots occurred throughout the year with a maximum rate of 71 culms m⁻² 56 days⁻¹ in April and a total annual recruitment of 277 culms m⁻² yr⁻¹. Total annual mortality was estimated at 248 culms m⁻² yr⁻¹. Live standing biomass in grams dry weight (gdw) was estimated using the model, $\hat{y} = 0.63 - 0.05h + 0.002h^2$ where h = culm height. Field testing of the model suggested agreement between destructively and nondestructively estimated biomass within 95% confidence limits throughout the year, except during spring and early summer. Nondestructive estimation of net aerial primary production (NAPP) consisted of summing biomass losses to the dead component resulting from culm mortality and abscission of leaves from live culms. Leaf abscission from live culms accounted for 31% of the annual NAPP. Destructively estimated NAPP was 931 gdw m⁻² yr⁻¹ whereas the nondestructively estimated NAPP was 635 gdw m⁻² yr⁻¹. The difference in NAPP estimates appears to be related to sampling variability and leaf loss during the year, both of which are addressed by the nondestructive method.

Introduction

Net aerial primary productivity (NAPP) in salt marshes has traditionally been estimated by clear-cut harvesting of plant tissues (Smalley 1958; Milner and Hughes 1968; Valiela et al. 1975; Reimold and Linthurst 1977). Reviews of primary productivity estimates for salt marsh plants by Keefe (1972) and Turner (1976) reveal the variability of NAPP estimates. Differences in NAPP methodologies have been investigated by Kirby and Gosselink (1976), Linthurst and Reimold (1978) and Shew et al. (1980),

---

*Present Address: College of Marine Studies, University of Delaware, 103A Robinson Hall, Newark, Delaware 19711.

and the results confirm the variability of NAPP for a marsh estimated by different harvest techniques.

Harvest methodologies rely upon changes in live and dead standing crop over defined intervals to record tissue production. Destructive sampling requires a homogeneous stand of sufficient size to allow repeated sampling. Large numbers of replicate harvests are often required to attain statistically comparable estimates between sampling intervals. Thus, local variation within a marsh plant community represents a significant consideration in determining NAPP by harvest techniques (Nixon and Oviatt 1973).

If a large sampling area is not available and if variations in plant density or height within a stand are suspected, a nondestructive technique for estimating primary productivity might be appropriate. Tagging individual stalks of marsh plants to determine changes in live and dead tissues has been performed in the past (Williams and Murdoch 1972; Bernard 1975; Hatcher and Mann 1975; Hardisky and Reimold 1977). Tagging allows the investigator to monitor production and abscission of leaves from individual plants which approximates the rate of tissue transition from the live to dead component for the entire marsh. Using these rates, observed changes obtained by harvest techniques can be validated.

The work described herein had three objectives:

1. to compare estimates of net aerial primary production for short *Spartina alterniflora* via destructive and nondestructive techniques;
2. to estimate recruitment, growth and mortality of *S. alterniflora* culms within stationary quadrats using tagging procedures; and
3. to assess the feasibility of estimating *S. alterniflora* biomass using regression models.

Methods and Materials

The study site was located north of the F. J. Torras Causeway just west of the Back River bridge near Brunswick, Georgia, U.S.A. (Fig. 1). The area supports an extensive short *S. alterniflora* Loisel. highmarsh with a mean culm height of about 35 cm. A visual survey of differences in plant canopy assisted in delineation of an appropriate area for study. Although every effort was made to select a homogeneous area, slight local variations in plant community structure were detected.

Ten replicate 0.1 m² Monel wire circular quadrats were permanently secured along a transect perpendicular to the ebb and flood of the tide onto the marsh. Within each quadrat all live and dead culms > 10 cm were tagged with numbered plastic cable ties (Ty-rap No. Ty-553M) at or near the soil surface. Beginning in November 1977, each quadrat was visited at approximately 56-day intervals and each live and dead culm was measured for height and the numbers of live and dead leaves counted through October 1978. Height was determined to the nearest centimeter

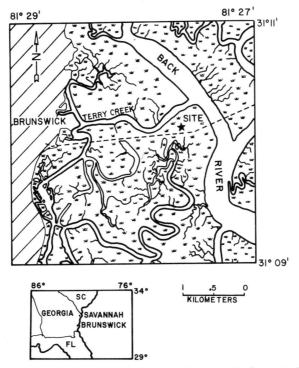

Figure 1. Location of tagging study site near Brunswick, Georgia, U.S.A.

by gently collecting all leaves at the top of the culm and measuring the distance from soil level to the tip of the tallest leaf or infloresence. Leaves lacking green coloration were assumed to be dead and those possessing green coloration were recorded as live. Differences in height and number of live and dead leaves for individual tagged plants between successive intervals were analyzed using the method of Hardisky and Reimold (1977) to determine leaf production, senescence and abscission. Recruitment was assessed by counting and measuring all new sprouts too short to be tagged. Culms which died during a sampling period were used as an estimate of mortality.

One hundred live *S. alterniflora* culms were collected from an adjacent marsh three times during the 1976 growing season from a clearcut harvest area and measured for height as described above. Selecting plants from a clearcut area provided height class frequencies proportional to the existing community. Each live culm and any attached dead leaves were individually packaged for return to the laboratory, dried to a constant weight at 100C and weighed on a Mettler H31 balance to the nearest 0.1 mg. Dead culms were collected from the marsh containing the quadrats six times during the 1977 growing season yielding a total of 600 culms. Dead culms were dried and weighed in a manner identical to live culms.

Regression analysis was performed equating culm height and biomass. Linear, power, exponential and polynomial models were tested using the general linear model procedure outlined by Barr et al. (1976). Data were first linearized for power and exponential models. The resultant models were then screened by selecting the two models with the largest coefficients of determination (r^2).

Data for selecting the best regression model were collected during the summer of 1977. All plants harvested for validation of the regression models were collected from the marsh containing the tagged plant quadrats. Five 0.1 m² areas were harvested, all live and dead culms measured and the biomass determined after drying to the nearest 0.1 g. The biomass predicted from the two models for each of the live and dead culms was compared with the destructively estimated biomass. The closest agreement between destructively and non-destructively determined bio-mass, as determined by the t-statistic and without regard to r^2 value, was the model selected.

Further field testing of the models was initiated. Three times between June and October 1978, ten replicate 0.1 m² areas of S. alterniflora were harvested, all culms measured and the dry weight biomass determined by weighing. Live and dead standing crop biomass was compared with predicted biomass from the models. The live culm biomass was further tested for seasonal performance by measuring all live culms from destructive productivity harvests from April 1977 to March 1978.

Net aerial primary productivity was destructively estimated by clear-cut harvesting of five replicate 0.1 m² quadrats randomly placed in the vicinity of the tagged areas. This procedure was conducted concurrent with the tagged plant monitoring (every 2 months). Samples were separated in-to live and dead components, weighed to the nearest 0.1 g, and the dry weight determined from a sub-sample dried at 100C to a constant weight. Primary productivity was estimated using the procedure presented by Smalley (1958).

Nondestructive estimation of net aerial primary productivity was ac-complished by calculating the abscission rate of leaves from live culms and by determining culm mortality from tagged culms which died during the study interval. Biomass of the dead culms was predicted by the live culm regression model using the maximum height (either live or dead) attained by the culm. Total number of leaves lost from live culms was multiplied by the average dry weight of a dead leaf as estimated from 200 dead leaves collected four times during the growing season in 1977. The primary pro-ductivity was then determined by adding the biomass contributions by abscission of dead leaves plus the mortality of live culms as these tissues transposed to the dead component. The total biomass was summed over the sampling intervals comprising an annual cycle. This yielded annual grams dry weight production.

Results

Models

The predictive ability of the best two regression models was tested using 5 replicated 0.1 m² samples for live and 10 replicated 0.1 m² samples for dead marsh plants. A polynomial regression (1) provided the best estimate for live culms with a mean biomass of 793 gdw m⁻² predicted and a mean biomass of 785 gdw m⁻² estimated by weighing. The best dead culm model was also a polynomial regression (2) with a predicted biomass mean of 130 gdw m⁻² and a weight-estimated biomass mean of 151 gdw m⁻². In both cases, the predicted versus estimated means were not significantly different at the 0.05 level.

(1) Live Culms $\quad \hat{Y} = 0.63 - 0.05h + 0.002h^2$

$\quad\quad r^2 = .89 \quad\quad\quad\quad n = 298$

(2) Dead Culms $\quad \hat{Y} = -0.34 + 0.025h + 0.00026h^2$

$\quad\quad r^2 = .65 \quad\quad\quad\quad n = 603$

where h = culm height in cm and n = number of plants.

An analysis of the seasonal results using equation (1) is illustrated in Fig. 2. The predicted live biomass means fell within the harvest confidence intervals during the major portion of the year. However, variable predictions were observed during that portion of the year (Nov.- April) where data were not collected for derivation of the regression model. Peak standing crop of live material occurred in November.

A second validation of the regression models indicated that the live culm model (1) yielded destructively and nondestructively determined biomass means which were not significantly different. The dead culm model (2) produced biomass means significantly different at the 0.05 level when estimated by both techniques (Table 1). The inconsistency of the predictive ability of the dead culm model (2) directed my attention to the live component which could be more accurately estimated.

Growth and Mortality

Annual rate curves for production of live leaves, senescence of live leaves and abscission of dead leaves were computed in average numbers of leaves for all tagged culms and are depicted in Fig. 3. Leaf production continued all year long in this Georgia salt marsh. The peak leaf production rate for mature and young culms (Fig. 3) coincided with recruitment highs (Fig. 4). A comparison of leaf production rates for mature culms and the combination of mature and young shoots suggests that each is a mirror image of the other. This may indicate a paced seasonal allocation of food reserves for leaf production between mature and new shoots. Leaf

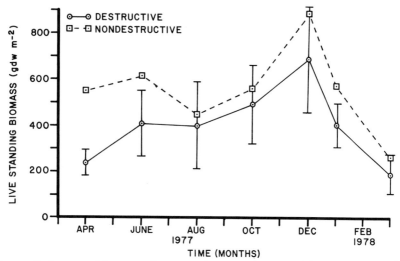

Figure 2. Seasonal live standing crop biomass estimated by destructive and nondestructive techniques. Point symbols represent means of 5 replicates and bisect the interval they represent. The bars on the destructive means represent the 95% confidence interval of the mean.

Table 1. Seasonal comparison of destructively and non-destructively estimated live and dead standing crop biomass. Data represented are means ± 1 standard error. n = number of samples; gdw = grams dry weight; NS = not significant, * = 0.05, * = 0.001.**

		Destructively Estimated Biomass (gdw m⁻²)		Nondestructively Estimated Biomass (gdw m⁻²)		t-test Significance	
1978	n	Live	Dead	Live	Dead	Live	Dead
June	10	484±38	385±55	597±76	223±34	NS	*
August	10	444±42	206±17	464±55	101±17	NS	***
October	10	717±69	170±22	675±79	112±16	NS	*

senescence rates were highest during summer and early winter months. Leaf abscission rate peaked during summer.

Recruitment rate in culms m⁻² based on tagged plants indicated that recruitment occurred all year with a peak in early spring; however, cumulatively more new shoots were produced during fall (Fig. 4). Live

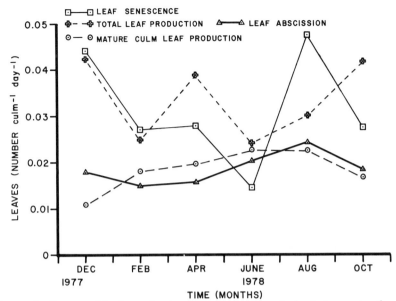

Figure 3. *Seasonal leaf production, senescence and abscission rates for S. alterniflora* culms. *Point symbols represent means of all tagged plants and bisect the interval they represent.*

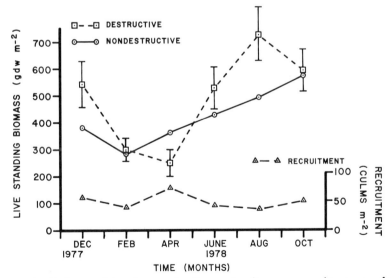

Figure 4. *(Left scale) Seasonal live standing crop biomass from destructively sampled areas and from tagged plants within permanent quadrats. Point symbols represent means and bisect the interval they represent. Bars are standard error of the mean.*

(Right scale) Seasonal recruitment of new shoots. Point symbols as above.

standing crop biomass peaked at 576 gdw m^{-2} in October for tagged areas and at 725 gdw m^{-2} in August for harvested areas. The live standing crop biomass of the destructively harvested areas experienced a greater magnitude of seasonal fluctuations than did the tagged areas.

Primary Production

Net aerial primary production for destructively harvested plants was assessed by comparing differences of live and dead standing crop with time (Smalley 1958). Based on the biomass comparison, a net annual primary production rate of 931 gdw m^{-2} yr^{-1} was obtained. Only the March through July sampling period showed positive production values.

Net aerial primary productivity nondestructively estimated from the tagged quadrats was composed of two components; 1) the mortality of live culms, and 2) the abscission of dead leaves from live culms. In essence, production was estimated as living tissue passed into the dead component. Although dead leaves were present on live culms, they were accounted for by estimates of live standing biomass. Therefore, until the dead leaf detached, it was not considered in the dead component for the purposes of this method. Mean dead leaf biomass was found to be 0.18 ± 0.006 gdw based on samples described earlier. Summation of biomass of culm mortality estimated by equation (1) and any abscissed leaves multiplied by 0.18 gdw yielded an annual net production of 635 gdw m^{-2} yr^{-1}. Leaf abscission accounted for 31% of the annual net production.

Annual recruitment estimated from the tagged areas was 277 culms m^{-2} yr^{-1}. Culm mortality for the year estimated from the tagged areas was 248 culms m^{-2} yr^{-1} which represents a 10% discrepancy between the two processes. Average annual live culm density divided by the average recruitment indicated a 1.4 annual turnover rate for number of live culms. The turnover rates for biomass in the destructively harvested areas was 1.9 whereas the rate for nondestructively sampled areas was slightly less at 1.5. Similarity of the numerical and biomass turnover rates for the tagged areas suggests the 1.9 turnover rate for destructive sampling may represent an overestimate.

Discussion

Salt marsh plant height has been equated to average standing crop biomass by a number of investigators (Williams and Murdoch 1969; Nixon and Oviatt 1973; Turner and Gosselink 1975). I have applied this relationship to individual culms with good success. The deviation of the predicted biomass from destructively estimated biomass during winter and spring months suggests that seasonal regression models might best predict biomass. Seasonal variation of the proportion of live and dead leaves on individual culms caused the observed differences. Failure of the regression model to predict dead culm biomass results from large variability in the weight of dead culms. Large portions of photosynthate are potentially translocated from the aerial culms to the roots during culm senescence (Turner and Gosselink 1975). This translocation of materials plus leaching of minerals from the standing dead matter (Gallagher et al. 1976; Gallagher and Pfeiffer 1977) contribute to substantial weight loss with time. Precise measurement of culm height for dead culms was difficult as the brittle dead leaves were easily broken. These difficulties suggest monitoring of tagged plants for biomass or productivity measurements is best suited to the live component of the community.

The tagged plants have allowed documentation of year-round production of leaves for *S. alterniflora* in this Georgia marsh. Recruitment of new shoots is continuous throughout the year unlike more northerly marshes (Williams and Murdoch 1969). Continual production and addition of dead and decaying plant tissue to the estuary tends to complicate primary productivity assessment.

Williams and Murdoch (1972), Odum and Fanning (1973), Hatcher and Mann (1975), Bradbury and Hofstra (1976) and Williamson (1976) all compensated for leaf loss during the growing season in estimating primary productivity. Odum and Fanning (1973) estimated that 8% of the total primary production accumulated by September in highmarsh *S. alterniflora* in Georgia (GA) was exported in the form of dead leaves. Hatcher and Mann (1975) calculated that 35% of the end-of-season biomass for medium *S. alterniflora* in Nova Scotia (NS) resulted from leaf loss. My estimate of 31% of the total NAPP originating from leaf loss approximates that of Hatcher and Mann. The NAPP [710g (NS) versus 635g (GA)] and average end-of-season biomass [559g (NS) versus 502g (GA)] for the two areas were similar.

The destructive harvest estimate of NAPP of 931 gdw m^{-2} yr^{-1} can be contrasted with 973 gdw m^{-2} yr^{-1} (Smalley 1958) and 1300 gdw m^{-2} yr^{-1} (Gallagher et al. 1980) reported for other Georgia short *S. alterniflora* marshes. Gallagher et al. (1980) utilized a paired plot method outlined by Wiegert and Evans (1964) which estimates disappearance of dead material during sampling intervals. Without addition of the disappearance component, their NAPP estimate was 700 gdw m^{-2} yr^{-1}. These previous estimates are similar to the destructive NAPP estimate I obtained.

The nondestructive technique for estimating NAPP resembles, in theory, the method set forth by Lomnicki et al. (1968) and a modification of the Lomnicki et al. method described by Shew et al. (1980). The Lomnicki et al. (1968) method uses live plots to estimate mortality from living plants and then adds to this any change in live biomass for an interval. The method presented here estimates leaf mortality and estimates live culm production in the form of culm mortality as it passes to the dead component. Lomnicki et al. (1968) also estimated leaf mortality and live culm production directly. Tagged culms in stationary quadrats allow long term monitoring of mortality specifically for NAPP estimation.

Biomass turnover rate of 1.5 calculated from nondestructively sampled areas was less than the 1.9 rate estimated by destructive means. Shew et al. (1980) found the harvest methods of Lomnicki et al. (1968) and Wiegert and Evans (1964) to produce higher turnover rates for highmarsh *S. alterniflora* than the potential turnover rate estimated at 1.5 by mean longevity of one crop determined from tagged plants. Unique tidal flushing problems during the study by Shew et al. (1980) caused a lower turnover rate of 0.9 calculated from the Smalley (1958) method. The variability of destructively estimated turnover rates from those estimated by tagging exemplify the diversity of environmental and community parameters influencing harvest techniques.

The Smalley (1958) technique for estimation of NAPP is generally considered an underestimate, since decomposition losses occurring between sampling intervals are not considered. The NAPP estimated by the tagging procedure was smaller than that estimated by the Smalley method. This relationship appears to exemplify errors associated with destructive harvest techniques rather than a further underestimate of NAPP by the tagging procedure. The destructive harvests yielded positive production values from March through July. This is contrary to the tagged plant data which indicates recruitment and leaf production throughout the year. Biomass variability among sample replicates associated with destructive harvests within a small area was also evident. The close agreement between numerical and biomass turnover rates from the tagged plants suggests the higher turnover rates calculated from destructive harvests were in error. Thus, reduction of potentially large sampling errors of destructive harvest techniques through repeated sampling of tagged plants produced an apparent underestimate of NAPP in this study. Based on the discrepancies between productivity estimated by harvests and the biomass production and mortality actually occurring at a given point in time as indicated by tagged plants, the tagged plants in permanent quadrats should provide a more realistic estimate of NAPP.

The differences in NAPP between the destructive and nondestructive methods for this study must be further weighed based on the errors inherent in each method. The destructive method assumes homogeneity of the plant community, neglects leaf loss and relies on gross differences in live and dead standing crop biomass between intervals for estimation of NAPP. The nondestructive technique is limited by accuracy of the predictive model, probably entails more human error in measurement, and could induce undetected leaf loss or other plant damage resulting from frequent manipulation of the plants. The regression model failed to provide statistically acceptable estimates during a portion of the year; however, 82% of the NAPP was estimated when this was not the case. The destructive sampling produced much larger standard errors about means indicating that site specific community variations may be the source. Since the same group of plants for nondestructive analysis was monitored continually over the year, there was no sampling population variability. Thus, tagging procedures provide a repeatable measure of culm numbers and tissue biomass.

There are several advantages to the nondestructive technique: 1) The same population of plants is continually monitored, eliminating sampling errors which large numbers of replicates attempt to diminish in destructive harvest. Site specific variability in plant stands can increase error. This phenomenon is illustrated by Fig. 4 where nondestructively estimated standing crop biomass provides a rather smooth curve as compared to the destructively estimated standing crop biomass; 2) Recruitment can effectively be monitored within stationary quadrats with tagged plants; 3) Size of a homogeneous stand of plants necessary for a nondestructive assessment is much less than would be required by a destructive procedure. This is particularly important when multispecies plant communities or confined greenhouse communities are under investigation.

By tagging a small cohort of plants in a particular locale, NAPP can be determined nondestructively by deriving a simple model equating live culm height and dry weight biomass, and estimating leaf abscission, culm mortality and a mean weight for dead leaves. In areas not available to destructive harvest or when general surveys of marsh health and production are required, this technique can provide reasonable estimates of NAPP which normally fall within the error limits of more extensive destructive efforts.

Acknowledgments

The many hours spent measuring and counting the plants through wind, rain and snow flurries were cheerfully endured by Kathy Smith, Brenda Sailors, Nancy Bevan, Milledge Smith, Jim Vernon, Barb Harrington and Jim Kowalchuk. A special acknowledgment to the field assistance of the late Patrick C. Adams. Thanks also to Mr. Rick Linthurst who provided statistical assistance and reviewed the primary productivity calculations. The continual encouragement, advice and logistical support from Dr. Robert J. Reimold was greatly appreciated. Dr. Vic Klemas provided valuable assistance during manuscript preparation. Secretarial expertise was provided by Earline Thomas. This project was supported by the U.S. Army Corps of Engineers, Waterways Experiment Station, Vicksburg, MS under contract no. DACW20-75-C-0074.

References Cited

Barr, A. J., J. H. Goodnight, J. P. Sall and J. T. Helwig. 1976. *A Users Guide to SAS 76*. Sparks Press, Raleigh. 329 pp.

Bernard, J. M. 1975. The life history of shoots of *Carex lacustris. Can. J. Bot.* 53:256-260.

Bradbury, I. K. and G. Hofstra. 1976. Vegetation death and its importance in primary production measurements. *Ecology* 57:209-211.

Gallagher, J. L. and W. J. Pfeiffer. 1977. Aquatic metabolism of the communities associated with attached dead shoots of salt marsh plants. *Limnol. Oceanogr.* 22:562-565.

Gallagher, J. L., W. J. Pfeiffer and L. R. Pomeroy. 1976. Leaching and microbial utilization of dissolved organic carbon from leaves of *Spartina alterniflora*. *Est. Coastal Mar. Sci.* 4:467-471.

Gallagher, J. L., R. J. Reimold, R. A. Linthurst and W. J. Pfeiffer. 1980. Aerial production, mortality and mineral accumulation—export dynamics in *Spartina alterniflora* and *Juncus roemerianus* plant stands in a Georgia salt marsh. *Ecology* (in press).

Hardisky, M. A. and R. J. Reimold. 1977. Salt-marsh plant geratology. *Science* 198:612-614.

Hatcher, B. G. and K. H. Mann. 1975. Above-ground production of marsh cordgrass *(Spartina alterniflora)* near the northern end of its range. *J. Fish Res. Board Canada* 32:83-87.

Keefe, C. W. 1972. Marsh production: a summary of the literature. *Contrib. Mar. Sci.* 16:163-181.

Kirby, C. J. and J. G. Gosselink. 1976. Primary production in a Louisiana Gulf Coast *Spartina alterniflora* marsh. *Ecology* 57:1052-1059.

Linthurst, R. A. and R. J. Reimold. 1978. An evaluation of methods for estimating the net aerial primary productivity of estuarine angiosperms. *J. Appl. Ecol.* 15:919-931.

Lomnicki, A., E. Bandola and K. Jankowska. 1968. Modification of the Weigert-Evans method for estimation of net primary production. *Ecology* 49:147-149.

Milner, C. and R. E. Hughes. 1968. *Methods for the Measurement of the Primary Production of Grasslands.* IBP Handbook No. 6, Blackwell Scientific Publ., Oxford. 70 pp.

Nixon, S. W. and C. A. Oviatt. 1973. Analysis of local variation in the standing crop of *Spartina alterniflora. Bot. Mar.* 16:103-109.

Odum, E. P. and M. A. Fanning. 1973. Comparison of the productivity of *Spartina alterniflora* and *Spartina cynosuroides* in Georgia coastal marshes. *Bull. Ga. Acad. Sci.* 31:1-12.

Reimold, R. J. and R. A. Linthurst. 1977. Primary production of minor marsh plants in Delaware, Georgia and Maine. Report No. D-77-36, U.S. Army Corps of Engineers, Vicksburg, MS. 138 pp.

Shew, D. M., R. A. Linthurst and E. D. Seneca. 1980. Comparison of production methods in a southeastern North Carolina *Spartina alterniflora* salt marsh. *Estuaries* (in press).

Smalley, A. E. 1958. The role of two invertebrate populations, *Littorina irrorata* and *Orchelimum fidicinum* in the energy flow of a salt marsh ecosystem. Doctoral Dissertation, University of Georgia, Athens. 126 pp.

Turner, R. E. 1976. Geographic variations in salt marsh macrophyte production: a review. *Contrib. Mar. Sci.* 20:47-68.

Turner, R. E. and J. G. Gosselink. 1975. A note on standing crops of *Spartina alterniflora* in Texas and Florida. *Contrib. Mar. Sci.* 19:113-118.

Valiela, I., J. M. Teal and W. J. Sass. 1975. Production and dynamics of salt marsh vegetation and the effects of experimental treatment with sewage sludge. *J. Appl. Ecol.* 12:973-981.

Wiegert, R. G. and F. C. Evans. 1964. Primary production and the disappearance of detritus on three South Carolina old fields. *Ecology* 56:129-140.

Williams, R. B. and M. B. Murdoch. 1969. The potential importance of *Spartina alterniflora* in conveying zinc, manganese and iron into estuarine food chains, pp. 431-439. *In:* D. J. Nelson and F. C. Evans (eds.), *Proceedings of the Second National Symposium on Radioecology.* United States Atomic Energy Commission. CONF-670503.

Williams, R. B. and M. B. Murdoch. 1972. Compartmental analysis of the production of *Juncus roemerianus* in a North Carolina salt marsh. *Chesapeake Sci.* 13:69-79.

Williamson, P. 1976. Above-ground primary production of chalk grassland allowing for leaf death. *J. Ecol.* 64:1059-1075.

AN EVALUATION OF AERATION, NITROGEN, pH AND SALINITY AS FACTORS AFFECTING *SPARTINA ALTERNIFLORA* GROWTH: A SUMMARY

Rick A. Linthurst

North Carolina State University
Department of Botany
Raleigh, North Carolina

Abstract: In a greenhouse experiment simulating a variety of salt marsh substrate aeration regimes, growth of *Spartina alterniflora* was positively correlated with increasing redox potentials. A field experiment supported these findings except where high redox potentials were associated with drying of natural marsh substrate with a subsequent decrease in pH to below 5. Interactions between pH and salinity were observed in another greenhouse experiment making main effect conclusions difficult; generally, however, biomass was significantly greater in pH 6 vs pH 4 and 8 treatments. The average effect of aerated in comparison to unaerated substrate was a 2.49 times increase in *S. alterniflora* biomass. In the same experiment, there was a 2.01 times increase in *S. alterniflora* biomass with an addition of 168 kg/ha equivalent $(NH_4)_2SO_4$-N to the natural marsh substrate. The combined effect of N and aeration increased *S. alterniflora* biomass 4.53 times. An increase in salinity from $15^o/_{oo}$ to $45^o/_{oo}$ decreased biomass by 66%. Low salinity ($15^o/_{oo}$), nitrogen enriched, aerated treatments produced 11 times more biomass than high salinity ($45^o/_{oo}$), nitrogen poor, unaerated treatments. These experiments suggest that nitrogen and aeration (drainage) are both able to reduce detrimental effects of high salinity. It is also suggested that, in addition to nitrogen and salinity, pH and aeration may regulate *S. alterniflora* growth in the field.

Introduction

In 1975, several sites characterized by dead *Spartina alterniflora* stubble and ponded water were found in the lower Cape Fear estuarine marshes in North Carolina. Similar dieback sites were described by Goodman et al. (1959); Goodman (1960); and Goodman and Williams (1961) in Great Britain for *Spartina townsendii.* The dieback sites varied in size with the largest site being 40 ha in area (Linthurst 1979a).

Dieback of *S. alterniflora* potentially represents extensive losses to the input of organic materials necessary for the maintenance of the detritus based food web of the ecosystem (Odum 1961; Teal 1962; Odum and de la Cruz 1967; Cooper 1969; Gosselink et al. 1974). With this in mind, it seemed appropriate that the phenomenon be studied in order to obtain enough information to provide possible suggestions for management of these areas.

Revegetation by man-initiated plantings did not appear to speed recovery of the affected sites appreciably (Linthurst 1979a). Natural recolonization occurred readily after the sites were able to drain regularly as decomposing root mats of the dead *S. alterniflora* were eroded and small drainage channels formed. The successional patterns prevalent at these dieback sites provide little toward the understanding of the phenomenon (Linthurst 1979a).

Smith (1970) suggests that dieback of *S. alterniflora* in Louisiana was probably caused by (1) excess salinity, (2) trash wrack coverage by dead plant material, (3) waterlogging due to poor drainage, (4) lack of available iron, (5) hydrogen sulfide toxicity, (6) oxygen deficiency in roots, (7) change in tidal regime, and/or (8) pollution. Many of these factors are interrelated and determination of a specific causal agent is difficult. However, based on Smith's observations, dominant factors may have been waterlogging and anaerobiosis. The objective of this work was to examine the response of *S. alterniflora* to anaerobiosis in both greenhouse and field experiments. Known *S. alterniflora* response factors, i.e., salinity and nitrogen, and changes in pH as a result of ponding and drying of the salt marsh substrate were also investigated. In this paper the results of these experiments are summarized and compared to related information in the literature.

Methods and Materials

Detail on the study methods involved in each experiment can be obtained as referenced. For clarity, each study will be summarized briefly below. Field culms from a creekbank and silt loam creek delta substrate, with an average NH_4-N concentration of 38 kg/ha, were collected from Oak Island, North Carolina and used in all greenhouse studies. This substrate did *not* acidify upon oxidation as would some high marsh substrates.

(1) Aeration Effects (Linthurst 1979a, b)

Four simulated marsh systems using creek delta substrate were established in a greenhouse to evaluate the effects of various waterlogging and aeration regimes on the growth of *S. alterniflora*. System 1 was filled to 10 cm above the marsh surface and the water was allowed to stand stagnant and unaerated (SU). System 2 was regularly flooded (RF) diurnally from 10 cm above the marsh surface to 15 cm below and the water was continuously aerated. System 3 (AS) was filled and the water was allowed to stand throughout the study as in System 1; however, substrate and water were both aerated continuously. System 4 (SA) was also filled, as were Systems 1 and 3, but only the water was aerated, not the substrate directly. Aeration was accomplished using aeration stones and an air compressor. Water salinity was maintained at $15^o/_{oo}$.

(2) Aeration, Nitrogen and Salinity (Linthurst and Seneca 1980a)

Based on the previous experiment, aerated substrate and stagnant unaerated simulated marsh systems had maximum growth differences. Therefore, only these two aeration treatments were selected for study.

Water salinities of 15, 30, and 45⁰/₀₀ and two nitrogen levels (a natural substrate N level and natural substrate plus 168 kg/ha $(NH_4)_2SO_4$-N equivalent) were also chosen for investigation. The experimental design was 3 salinities × 2 nitrogen × 2 aeration treatments. All treatments were established using natural creek delta substrate with 15 cm of standing water on the surface of each simulated system.

(3) Aeration, pH and Salinity (Linthurst and Blum 1980)

This experiment differed from the previous experiments in that it was conducted with washed river sand rather than natural substrate. A factorial arrangement of treatments was used: 3 pH (4, 6, and 8) × 4 salinities (15, 25, 35, and 45⁰/₀₀) × 2 aeration (aerated and unaerated) treatments. Again, standing water was associated with all treatments as in the previous experiments (1 and 2).

(4) Standing Water—Drainage Potential (Linthurst and Seneca 1980b)

In lined and unlined pots (undrained and drained) plant and substrate plugs of S. alterniflora from a short S. alterniflora zone on Oak Island were placed intact in the Oak Island marsh at five elevation levels. Pots were placed with their surfaces 30, 20, and 10 cm below, even with, and 10 cm above the natural marsh surface. Those pots below the surface held water throughout the study similar to the ponding of the high marsh dieback sites.

Results

(1) Aeration Effects (Linthurst 1979b)

Substrate pH was affected by aeration (Fig. 1a). Reducing potential of the soil decreased with increased aeration and redox potentials adjusted to pH 7 were positive in the aerated substrate system (Fig. 1b). There was no significant difference between redox potentials in the regularly flooded and stagnant aerated system. This suggests that the ionic forms of the elements, as affected by redox potentials, might be similar in these two systems. Molar sulfide concentrations of the substrate were highest in the stagnant unaerated (SU) and stagnant aerated (SA) systems (Fig. 1c).

Examination of the biological responses indicated that average culm height of the tallest plants per treatment was clearly representative of a height gradient (Fig. 1d). Stem densities were not significantly different in the aerated systems but they were lower in the unaerated greenhouse marsh (Fig. 1e). Both aerial and root biomass were enhanced by the aerated substrate system (Figs. 1f and 1g). Root/shoot ratios were highest in the unaerated system (Fig. 1h).

(2) Aeration, Nitrogen and Salinity (Linthurst and Seneca 1980a)

All biological variables were significantly affected ($a \leq 0.01$) by salinity, nitrogen, and aeration treatments except for root biomass (MOM) which

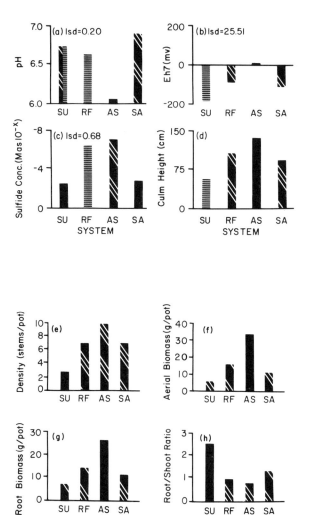

Figure 1. Measurements of the physical and biological parameters associated with various aeration treatments. SU - stagnant unaerated system; RF - regularly flooded system; AS - aerated substrate system; and SA - stagnant aerated water system. Bars of the same design are not significantly different. Eh7 = 241 + Eh + (7-pH)(-60) (Allam et al. 1972). Sulfide concentrations (molarity). (From Linthurst 1979b).

was not significantly different between nitrogen treatments. There were no significant interactions between any of the factors; however, for the reader's benefit, I chose to present the interaction means here rather than just main effect means (Table 1).

Salinity increases of 15°/oo decreased biomass, density, mean height, and MOM an average of 42, 32, 22, and 37%, respectively. Reductions in every biological variable measurement were greater when salinity was raised from 30 to 45°/oo than when it was raised from 15 to 30°/oo. Increase of salinity from 15 to 45°/oo decreased biomass, density, mean height and MOM 66, 53, 38, and 61%, respectively.

Increasing nitrogen from a natural level to 168 kg/ha higher N increased biomass, density and mean height 2.02, 1.46, and 1.26 times, respectively. An increase of O_2 concentration from essentially zero to saturation stimulated biomass, density, mean height and MOM, 2.31, 1.89, 1.49, and 1.89 times, respectively.

The high level of N and aeration together in comparison to the low level of these two factors, averaged over salinity, produced increases of 4.53, 2.71, 1.88, and 2.24 times in biomass, density, mean height and MOM, respectively. Eleven times more biomass was observed in low salinity (15°/oo), nitrogen enriched, aerated treatments than in high salinity (45°/oo), nitrogen poor, unaerated treatments.

Table 1. Interaction treatment means for biological response variables as affected by salinity, nitrogen and aeration (From Linthurst and Seneca 1980a). MOM = macroorganic matter (> 2 mm); w = with; o = without.

Treatment				Biological variable means			
Salinity °/oo	Nitrogen (kg/ha)	Aeration	n	Biomass (g/pot)	Density (culms/ pot)	Mean height (cm)	MOM (g/pot)
15	0	o	4	11.7	13.5	73.7	15.9
15	0	w	4	23.5	22.8	113.1	19.3
15	168	o	4	28.1	20.5	114.7	15.0
15	168	w	4	52.8	34.8	129.9	38.6
30	0	o	4	7.3	8.5	62.1	8.8
30	0	w	4	21.0	19.8	98.8	21.2
30	168	o	4	10.5	10.5	84.4	10.5
30	168	w	4	35.0	25.0	118.4	21.7
45	0	o	4	4.8	7.3	53.3	7.4
45	0	w	4	7.5	8.8	67.6	8.9
45	168	o	4	6.7	7.0	36.9	6.6
45	168	w	4	20.1	19.5	107.1	11.7
Standard error (n=4)				2.9	1.8	5.1	4.6

(3) Aeration, pH, and Salinity (Linthurst and Blum 1980)

No aeration effect was observed. However, since sand was used instead of natural substrate, redox potential differences are not maximized as they are when marsh substrate is utilized.

The pH and salinity treatments affected all biological variables. Interactions were also significant. Factor analysis (Barr et al. 1976) was used to weight the biological variables such that a linear combination of the variables produced a single plant performance score. These scores are plotted in Fig. 2.

Plant performance score increase is nearly linear with decreasing salinity at pH 6. Because this intermediate pH level is also the high point on the response surface (Fig. 2), the variable response is quadratic. Degree of curvature must increase rapidly as salinity and pH changes, accounting for the high number of interactions observed for plant performance, salinity, and pH.

The plant performance was best near field pH (6.0) and low salinity. Plants grown at pH 8 had higher performance scores than plants grown in

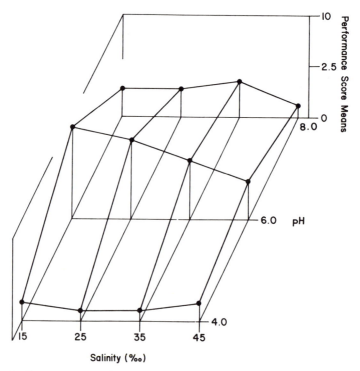

Figure 2. Response surface diagram of performance score means representing the response of S. alterniflora to pH and salinity (From Linthurst and Blum 1980).

pH 4 nutrient culture. Salinity effect is maximized at pH 6 with little effect at pH 4 and 8. At the pH extremes, the salinity effect is apparently minimized by the general pH effect.

(4) Standing Water—Drainage Potential (Linthurst and Seneca 1980b)

Mean *in situ* pH, reducing potential, free sulfides and salinity significantly decreased with increasing pot surface elevation. Soluble salts, Na, P, K, Ca, and Mg concentrations also decreased with increasing elevation. Organic matter, NH_4-N, sulfate, Mn, Zn and Cu were not significantly different between elevation treatments. Buffer acidity (pH 6.6) was low in all treatments except the control.

Lined and unlined treatments produced fewer detectable differences than were observed in the elevation treatments. In general, all of the soil variable concentration means were highest in lined vs unlined treatments. Factor analysis (Barr et al. 1976) was used to convert the soil variables into weighted means descriptive of each treatment. These values are plotted on Fig. 3.

Living aerial biomass is also plotted on Fig. 3. Living dry weight biomass and density had a similar pattern. Those treatments at 30 and 20 cm below the marsh surface had no growth. As these two systems were influenced by pore waters (as opposed to flood waters), gas exchange inhibition of the plants and the surrounding water rather than air, and decreased light intensity through standing water, it is not possible to attribute their

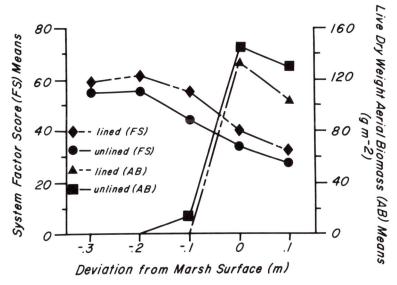

Figure 3. Factor score means and live dry weight aerial biomass plotted against deviation from marsh surface for lined and unlined treatments (From Linthurst and Seneca 1980b).

death to anaerobiosis or the substrate selective properties (Linthurst and Seneca 1980b). The lined, 10 cm treatment also permitted no growth. Peak *S. alterniflora* biomass was observed in the control treatment and decrease in biomass in the 10 cm above the marsh surface pots was associated with a decrease in pH from 6.0 to 4.7. Such a pH decrease is expected in a drying catclay type substrate like that used in this study (Bloomfield and Coulter 1973).

Discussion

Subsections are numbered to represent the four experiments just described.

Salinity (2 and 3)

The effect of salinity on the growth of *S. alterniflora* is one of the most widely studied of the factors affecting distribution and zonation of this species (Mooring et al. 1971; Woodhouse et al. 1974; Haines and Dunn 1976; Smart and Barko 1978). Mooring et al. (1971) found that *S. alterniflora* grows best at $10^0/_{oo}$ salinity. Similar results were observed by Haines and Dunn (1976). The salinity response as investigated here was found to be strongly pH dependent (Linthurst and Blum 1980). At higher salinities (30 and $45^0/_{oo}$) *S. alterniflora* growth is still stimulated by aeration and nitrogen additions.

At high salinities, osmotic stress or cell membrane damage is likely to be the primary growth regulating factor. Osmotic stress could result in reduced water uptake (Levitt 1972). Membrane permeability changes could reduce the influx of necessary nutrients and/or cause leakage of nutrients from the roots to the surrounding substrate. Increased permeability may also decrease the effectiveness of any selective ion uptake mechanisms in addition to increasing the potential for losses of needed oxygen from the roots. Because of nitrogen, aeration, and pH effects observed in these experiments, the influence of salinity on *S. alterniflora* may be far more complex than previously considered.

Nitrogen (2)

Numerous authors have investigated the effects of nitrogen on *S. alterniflora* growth (Sullivan and Daiber 1974; Valiela and Teal 1974; Broome et al. 1975; Gallagher 1975; Haines and Dunn 1976; Mendelssohn 1979a). Mendelssohn (1979b) found higher levels of ammonium present in the short height form *S. alterniflora* and suggested that something may be interfering with nitrogen uptake and/or metabolism.

In this experiment (2), it is hypothesized that NH_4-N and Na may be competing for uptake sites. At a given aeration-salinity level, biomass of *S. alterniflora* increased as NH_4-N was added to the system. This is the type of effect which might be expected if N and Na were competing for the same sites (Levitt 1972). At $45^0/_{oo}$ salinity, considerably more nitrogen than was added here may be necessary to compensate totally for this competition effect. However, since a combination of aeration and nitrogen was more effective in increasing biomass at all salinities, it is proposed that more than competition between these two ions is operative.

Our data suggest that growth is enhanced by additions of NH_4-N possibly by: (1) eliminating a portion of the NH_4-N vs Na uptake competition in the substrate; (2) creating a larger N gradient, subsequently increasing the rate of diffusion to the root zone; and/or (3) increasing concentrations of a naturally nitrogen deficient system.

Aeration enhanced the nitrogen response, particularly at the high salinity. I hypothesize that continued aerobiosis is beneficial to the plant and significantly enhances the potential for nutrient uptake and for the general nitrogen response. Aeration may eliminate the need for energetically expensive metabolic adjustments by the plant at low oxygen concentrations and/or minimize damage to root systems caused by extreme anaerobiosis. As a result, more energy is available for growth, and root integrity is maintained for normal uptake processes.

Aeration (1, 2, 3, and 4)

The effects of aeration on *S. alterniflora* have not been studied extensively. Our results show growth to be positively correlated with redox potentials. Mendelssohn and Seneca (1980) reported that the redox potentials in the substrate of the tall form *S. alterniflora* were greater than those in the substrate of the short form, again suggesting a positive correlation. However, at high redox potentials (associated with drying substrate and decreased pH) the growth of *S. alterniflora* is inhibited. Therefore, pH changes can be more influential in determining the growth response than is aeration under certain conditions.

Crawford (1978) suggests that organisms withstand anoxia in four general ways: (1) control of metabolic rate, (2) diversification of the end products of glycolysis, (3) provision of adequate carbohydrate supplies, and (4) coupling of metabolic pathways to facilitate proton disposal and provide additional ATP. Crawford also suggests that (1) some flood-tolerant plants exhibit no acceleration of glycolysis; (2) in marsh plants, malate, as opposed to ethanol, seems to be a common end product of anaerobic respiration; (3) well developed rhizomes store carbohydrates; and (4) couplings of carbohydrate and amino acid metabolism can increase ATP yield (e.g. by increasing nitrate reductase activity). The first two mechanisms have not been examined for *S. alterniflora*. The third possible tolerance control (3) can be supported. During the winter months, *S. alterniflora* must respire anaerobically in an oxygen deficient (or absent) atmosphere since no (or reduced numbers of) living aerial culms are present for oxygen transport. Assumedly, all three may play a part in species survival.

There is little doubt that aerenchyma in *S. alterniflora* are a primary pathway for oxygen maintenance of roots through diffusion of oxygen to the root zone from the aerial portion of the plants. Teal and Kanwisher (1966) suggest that this diffusion prevents internal anoxia. Vartapetian

(1973) demonstrated, however, that the roots of rice are highly sensitive to anoxia which caused irreparable cellular ultrastructure damage and subsequently organelle damage and breakdown. Therefore, if even for short periods of time the root is not able to maintain adequate internal oxygen concentrations, cellular damage may be initiated. Knott (1977) found that lysogenous space size decreases in *S. alterniflora* as salinity increases which in turn must physically decrease the potential supply of oxygen to the roots.

It has also been demonstrated that oxygen produced in photosynthesis improves the oxygen status of the submerged root (Armstrong 1978). This suggests that during periods of sunless weather, oxygen concentrations may significantly decrease in the roots (Vámos and Köves 1972; Armstrong 1978). As a result, the normal oxidizing activity of the roots could be impaired. In fact, M. L. Gleason (U. Virginia, Charlottesville pers. comm.) has demonstrated that the base of *S. alterniflora* culms become anoxic under some conditions which necessitates anoxia of the roots at least during these periods. It is likely then that, under prolonged anaerobiosis, oxygen deficiencies would occur, particularly in a greenhouse experiment where sunlight may not be at optimal levels and where standing water covers the substrate surface. Mendelssohn and Seneca (1980) and Linthurst and Seneca (1980a, b) found that *S. alterniflora* growth decreased when drainage was impaired in the field. Finally, mixing associated with aeration could renew nutrient concentrations around the roots, decreasing diffusion resistance of nutrients from soil water to the roots, and thus could have contributed to the growth enhancement. The effect of mixing has not been investigated. Therefore, it is not possible to quantify conclusively what contribution the mixing effect of the greenhouse aeration system makes. However, since a single site of aeration within the pots normally developed as the substrate consolidated, only the water was being mixed, not the substrate. Therefore, the mixing effect was assumed to be minor in relation to the aeration effect.

pH (3 and 4)

The effect of pH was studied in only one greenhouse experiment. Clearly growth was optimal at pH 6 in comparison to that at pH 4 and 8 treatments. A pH drop in the catclay type marsh substrate, as observed in the field study (Linthurst and Seneca 1980b), could be responsible for the poor growth response observed in the high marsh. Short *S. alterniflora* with its shallow root system (Gallagher et al. 1978) may be subjected to changing pH when periods of high temperatures and low tides and rainfall prevail. Creekbank *S. alterniflora* substrate in the North Carolina marshes studied were not catclays and therefore are not subject to changing pH.

S. alterniflora growth was also inhibited at pH 8, a pH observed in dieback substrate when the areas were ponded. This pH increase could again, in part, contribute to the dieback phenomenon. As with the other factors, pH influences the substrate complex and resultant substrate changes are difficult to describe. In addition, pH has a direct effect on the root membrane which directly influences plant performance. Both aspects of pH effects are important to *S. alterniflora* growth and there appears to be enough evidence here to warrant further work on this variable as an *S. alterniflora* growth factor.

Conclusions

It is apparent from these studies that aeration, nitrogen, salinity, and pH are determinants of *S. alterniflora* growth. High salinities, low redox potentials and generally low nitrogen concentrations are all observable in the field. High or low pH is only observable under isolated circumstances. Based on these studies, these four factors are adequate to explain differences between the height forms of *S. alterniflora* and the dieback phenomenon observed in the North Carolina marshes. Unfortunately, each of these factors contributes to changes in the *S. alterniflora* substrate complex. Therefore, much more work on the salt marsh substrate and the physiological response of *S. alterniflora* to this changing complex is required.

Acknowledgments

This work is a portion of a Ph.D. dissertation under the direction of Dr. E. D. Seneca, Department of Botany, North Carolina State University. The work was supported by Carolina Power and Light Company under contract No. 99227. L. L. Hobbs is gratefully acknowledged for his technical assistance and logistics support. Drs. U. Blum, E. J. Kamprath, and J. O. Rawlings are acknowledged for their assistance in the project manuscript preparations. Special thanks are extended to A. S. Linthurst and C. S. Harp for their secretarial expertise.

The following are reprinted with permission: Table 1 (Estuarine Research Federation); Figure 1 (American Journal of Botany); Figure 2 is reprinted by permission of the Journal of Experimental Marine Biology and Ecology. Copyright by Elsevier/North Holland Biomedical Press; Figure 3 with permission from Estuarine and Coastal Marine Science. Copyright by Academic Press Inc. (London) Ltd.

Paper No. 6286 of the Journal of the North Carolina Agricultural Research Service, Raleigh, North Carolina 27650.

References Cited

Allam, A. I., G. Pitts and J. P. Hollis. 1972. Sulfide determination in submerged soils with an ion-selective electrode. *Soil Sci.* 114:456-467.

Armstrong, W. 1978. Root aeration in the wetland condition, pp. 269-297. *In:* D. D. Hook and R. M. M. Crawford (eds.), *Plant Life in Anaerobic Environments.* Ann Arbor Science, Ann Arbor, MI. 564 pp.

Barr, A. J., J. H. Goodnight, J. P. Sall and J. T. Helwig. 1976. *A User's Guide to SAS 76.* SAS Institute, Inc., P. O. Box 10066, Raleigh, North Carolina.

Bloomfield, C. and J. K. Coulter. 1973. Genesis and management of acid sulfate soils, pp. 265-326. *In:* Brady, N. C. (ed.), *Advances in Agronomy.* Amer. Soc. of Agronomy. Vol. 25.

Broome, S. W., W. W. Woodhouse, Jr. and E. D. Seneca. 1975. The relationship of mineral nutrients to growth of *Spartina alterniflora* in North Carolina: II. The effects of N, P, and Fe fertilizers. *Soil Sci. Soc. Am. Proc.* 39:301-307.

Cooper, A. W. 1969. Salt marshes, pp. 563-611. *In:* H. T. Odum, B. J. Copeland and E. A. McMahon (eds.), *Coastal Ecological Systems of the U.S. Vol. 1.* Report to the Water Pollution Control Administration, Washington, D.C.

Crawford, R. M. M. 1978. Metabolic adaptation to anoxia, pp. 119-137. *In:* D. D. Hook and R. M. M. Crawford (ed.), *Plant Life in Anaerobic Environments.* Ann Arbor Science, Ann Arbor, MI. 564 pp.

Gallagher, J. L. 1975. Effect of an ammonium nitrate pulse on the growth and elemental composition of natural stands of *Spartina alterniflora* and *Juncus roemerianus. Am. J. Bot.* 62:644-648.

Gallagher, J. L., F. G. Plumley and P. L. Wolf. 1978. Underground biomass dynamics and substrate selective properties of Atlantic coastal salt marsh plants. U.S. Army Engineer Waterways Expt. Sta. Vicksburg, Miss. Report: TR-D-77-28:131 pp.

Goodman, P. J. 1960. Investigations into 'die-back' in *Spartina townsendii* Agg. II. The morphological structure and composition of the Lymington Sward. *J. Ecol.* 48:711-724.

Goodman, P. J., E. M. Braybrooks and J. M. Lambert. 1959. Investigations into 'die-back' in *Spartina townsendii* Agg. I. The present status of *Spartina townsendii* in Britain. *J. Ecol.* 47:651-677.

Goodman, P. J. and W. T. Williams. 1961. Investigations into 'die-back' in *Spartina townsendii* Agg. III. Physiological correlates of 'die-back'. *J. Ecol.* 49:391-398.

Gosselink, J. G., E. P. Odum and R. M. Pope. 1974. The value of the tidal marsh. LSU-SG-74-03, Center for Wetland Resources, Louisiana State University, Baton Rouge, LA. 30 pp.

Haines, B. L. and E. L. Dunn. 1976. Growth and resource allocation responses of *Spartina alterniflora* Loisel. to three levels of NH_4N, Fe, and NaCl in solution culture. *Bot. Gaz.* 137:224-230.

Knott, W. M. 1977. The response of *Spartina alterniflora* Loisel. to various salinities under simulated marsh conditions in the greenhouse. Ph.D. Thesis. North Carolina State University, Raleigh, NC. 93 pp.

Levitt, J. 1972. *Responses of Plants to Environmental Stresses.* Academic Press, New York. 697 pp.

Linthurst, R. A. 1979a. Aeration, nitrogen, pH and salinity as factors affecting *Spartina alterniflora* growth and dieback. Ph.D. Thesis. North Carolina State University, Raleigh, NC. 159 pp.

Linthurst R. A. 1979b. The effects of aeration on the growth of *Spartina alterniflora* Loisel. *Am. J. Bot.* 66:685-691.

Linthurst, R. A. and U. Blum. 1980. Growth modifications of *Spartina alterniflora* Loisel. by the interactions of pH and salinity under controlled conditions. *J. Expt. Mar. Biol. Ecol.* (in press).

Linthurst, R. A. and E. D. Seneca. 1980a. Aeration, nitrogen and salinity as determinants of *Spartina alterniflora* Loisel. growth response. *Estuaries* (in press).

Linthurst, R. A. and E. D. Seneca. 1980b. The effects of standing water and drainage potential on the *Spartina alterniflora*—substrate complex in a North Carolina salt marsh. *Est. Coastal Mar. Sci.* (in press).

Mendelssohn, I. A. 1979a. The influence of nitrogen level, form and application method on the growth response of *Spartina alterniflora* in North Carolina. *Estuaries* 2: 106-118.

Mendelssohn, I. A. 1979b. Nitrogen metabolism in the height forms of *Spartina alterniflora* in North Carolina. *Ecology* 60:574-584.

Mendelssohn, I. A. and E. D. Seneca. 1980. The influence of soil drainage on the growth of salt marsh cordgrass, *Spartina alterniflora. Est. Coastal Mar. Sci.* (in press).

Mooring, M. T., A. W. Cooper and E. D. Seneca. 1971. Seed germination response and evidence for height ecophenes in *Spartina alterniflora* from North Carolina. *Am. J. Bot.* 58:48-55.

Odum, E. P. 1961. The role of tidal marshes in estuarine production. *New York State Conservationist* 15:12-15.

Odum, E. P. and A. A. de la Cruz. 1967. Particulate organic detritus in a Georgia salt marsh-estuarine ecosystem, pp. 383-388. *In:* G. H. Lauff (ed.), *Estuaries.* Amer. Assoc. Advance Sci., Publ. No. 83, Washington, D.C.

Smart, R. M. and J. W. Barko. 1978. Influence of sediment, salinity and nutrients on the physiological ecology of selected salt marsh plants. *Est. Coastal Mar. Sci.* 7:487-495.

Smith, W. G. 1970. *Spartina* 'die-back' in Louisiana marshes. Coastal Studies Bull. No. 5. Louisiana State University, Baton Rouge, pp. 89-95.

Sullivan, M. L. and F. C. Daiber. 1974. Response in production of cord grass, *Spartina alterniflora* to inorganic nitrogen and phosphorus fertilizer. *Chesapeake Sci.* 15:121-123.

Teal, J. M. 1962. Energy flow in a salt marsh ecosystem of Georgia. *Ecology* 43:614-624.

Teal, J. M. and J. W. Kanwisher. 1966. Gas transport in the marsh grass, *Spartina alterniflora. J. Expt. Bot.* 17:13-37.

Valiela, I. and J. M. Teal. 1974. Nutrient limitation in salt marsh vegetation, pp. 547-564. *In:* R. J. Reimold and W. H. Queen (ed.), *Ecology of Halophytes.* Academic Press, New York.

Vámos, R. and E. Köves. 1972. Role of the light in prevention of poisoning action of hydrogen sulphide in the rice plant. *J. Appl. Ecol.* 9:519-525.

Vartapetian, B. B. 1973. Aeration of roots in relation to molecular oxygen transport in plants. Proc. Uppsala Symp. (Ecology and Conservation 5) New York: UNESCO, 1973.

Woodhouse, W. W., Jr., E. D. Seneca and S. W. Broome. 1974. Propagation of *Spartina alterniflora* for substrate stabilization and salt marsh development. U.S. Army, Corps of Engineers, CERC, Ft. Belvoir, Va. Report: TM-46: 155 pp.

INFLUENCES OF ESTUARINE CIRCULATION ON THE DISTRIBUTION AND BIOMASS OF PHYTOPLANKTON SIZE FRACTIONS

Thomas C. Malone, Patrick J. Neale and David Boardman

Marine Biology
Lamont-Doherty Geological Observatory
Palisades, New York

Abstract: Biomass of netplankton diatoms in the lower Hudson estuary is highest when surface temperature is below 15C and depends on the development of diatom blooms in adjacent coastal waters. Peaks in netplankton biomass occur between pulses in freshwater flow due to entrainment of diatoms in high salinity bottom water advected into the estuary. As the seasonal thermocline develops and surface temperature approaches 20C, coastal diatom blooms become less frequent and diatoms become rare in the estuary. Nanoplankton blooms occur during the summer in association with pulses of freshwater or when freshwater flow is low. Peaks in nanoplankton biomass are due to growth within the estuary and are dissipated by tidal mixing, nontidal seaward flow of the surface layer, and grazing. These seasonal patterns were reflected in chlorophyll budgets of the lower Hudson estuary. The estuary was a chlorophyll sink during winter-spring when netplankton dominated and a source during summer when nanoplankton dominated.

Introduction

Nutrient cycling and dynamics of phytoplankton productivity in estuaries are receiving considerable attention because of the importance of estuaries as natural resources and the threat of "cultural" eutrophication (Ketchum 1969; Loftus et al. 1972; McCarthy et al. 1975; Taft and Taylor 1976; Taft et al. 1978). Partially mixed estuaries, such as the lower Hudson estuary (Abood 1974), are typically well fertilized systems as a consequence of freshwater runoff, waste disposal, and a two-layered circulation pattern that transports nutrient-rich coastal water into the estuary and increases the residence time of particulate organic matter decomposing within the estuary (Ketchum 1967). Effects of these nutrient inputs on phytoplankton biomass depend on how rapidly phytoplankton populations are growing and on a variety of environmental processes that influence distribution and fate of biomass once produced. Among the latter, circulation is a major factor and the focus of this contribution. Estuarine circulation not only influences the distribution and residence time of biomass produced within an estuary, but may also introduce new populations of phytoplankton from contiguous water bodies.

249

Within the region of salt intrusion, circulation in the lower Hudson estuary consists of tidal and non-tidal components (Dyer 1973; Abood 1974). Seaward flow of the surface layer is driven by the freshwater flow of the Hudson River (0.5 to 25 × 10^7 m^3 d^{-1}), and upstream flow of the bottom layer is driven by the density difference between freshwater and seawater inputs. Tidal velocities are greater than net velocities and provide most of the energy for turbulent mixing between seawater and freshwater inputs. Distribution of salt reflects the two-layered flow pattern and is similar to that of the James River, Virginia, a partially mixed estuary in which non-tidal advection and turbulent mixing between surface and bottom layers are the important terms required to formulate a salt balance (Pritchard 1954).

Concentrations of dissolved inorganic nutrients are high and usually distribute conservatively as a consequence of the large amount of sewage-derived nutrients discharged into the lower estuary (1.6 × 10^5 kg N d^{-1} and 1.3 × 10^4 kg P d^{-1}), and of low phytoplankton productivity (Garside et al. 1976; Malone 1977a; Simpson et al. 1977). Most sewage-derived nutrients are transported into adjacent coastal waters where they support high levels of phytoplankton productivity in the coastal plume (Malone and Chervin 1979). Some proportion of these nutrients is returned to the estuary as phytoplankton biomass.

Preliminary observations suggest that the lower estuary is a sink for phytoplankton biomass during the winter and early spring when chain-forming diatoms are most abundant, and a source of phytoplankton biomass during the summer when non-diatomaceous nanoplankton are most abundant (Malone 1977a). This paper evaluates when and under what conditions the lower Hudson estuary is a source or a sink of phytoplankton biomass relative to adjacent coastal waters.

Methods and Materials

Samples were collected at approximately weekly intervals from February to June and from July to September during 1977 and 1978. Surface chlorophyll (*in vivo* fluorescence) and salinity (conductivity and temperature) were monitored continuously with and against the tide along a transect between the lower Bay and the Tappan Zee (Fig. 1). Vertical profiles of temperature, salinity and chlorophyll *a* were obtained at 7 stations (Fig. 1) with a conductivity-temperature-depth sensor, submersible pump and bottle casts. Vertical profiles were obtained every 1-3 h over two tidal cycles on 9 occasions at MP-7. Bottles were used to collect samples for extracted chlorophyll *a* and primary productivity measurements as described by Malone (1977a). Netplankton and nanoplankton refer to phytoplankton populations that were retained and passed by a 20 μm mesh screen, respectively.

Freshwater flow of the Hudson River at Green Island (250 km north of Upper Bay) was provided by the Water Resources Division of the

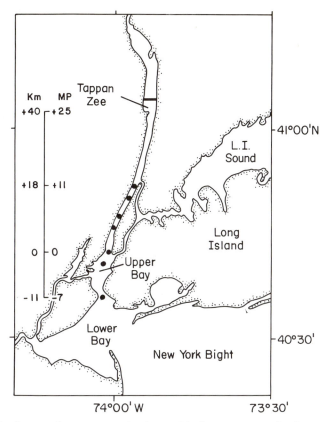

Figure 1. *Station locations in the lower Hudson estuary. Surface properties were monitored from the Verrazano Narrows (MP −7) to the Tappan Zee (MP = mile point, miles north (+) or south (−) of the Battery).*

Geological Survey, U.S. Department of the Interior. Total freshwater flow in the lower estuary was calculated as described by Hammond (1975).

Results and Discussion

Freshwater flow (Q_f) at Green Island varied seasonally with rates above 5×10^7 m³ d⁻¹ during spring and below 2×10^7 m³ d⁻¹ during summer (Fig. 2). Spring runoff began earlier and was greater in magnitude in 1977 than in 1978. Variations in Q_f were reflected in salinity distributions both seasonally and interannually (Figs. 3 and 4). Peaks in Q_f were expressed as downstream movements of surface isohaline with lag times of 4 to 15 d and a mean of 7 d. Similar lags have been reported for the Hudson (Stewart 1958; Hammond 1975). Given a 7 d lag, surface salinity (S)

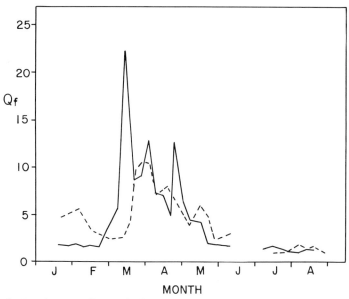

Figure 2. Freshwater flow of the Hudson River at Green Island during 1977 (———) and 1978 (----); $Q_f = 10^7 \, m^3 \, d^{-1}$.

averaged over a 13 km segment of the estuary between MP 0 and MP 10 was related to Q_f by an exponential equation having the same form as one-dimensional, advection-diffusion models used to describe salinity distributions in estuaries. Under low flow conditions ($Q_f < 5 \times 10^7 \, m^3 \, d^{-1}$),

$$S = 40.1 \, e^{-0.50 \times 10^7 \, Q_f}$$

$$(r = 0.99, < 0.01)$$

compared to

$$S = 13.7 \, e^{-0.14 \times 10^7 \, Q_f}$$

$$(r = 0.75, < 0.01)$$

under high flow conditions. Coefficients of dispersion calculated from the exponential constants were $2.1 \times 10^2 \, m^2 \, s^{-1}$ and $7.5 \times 10^2 \, m^2 \, s^{-1}$, respectively. Advection-diffusion models for the lower Hudson suggest coefficients of $1.6 \times 10^2 \, m^2 \, s^{-1}$ (Pritchard et al. 1962; O'Connor 1970; Abood 1974; Hammond 1975).

Distributions of chlorophyll and the partitioning of chlorophyll among nanoplankton and netplankton size fractions varied in response to changes in Q_f as indicated by the distribution of salt. During February-May, increases in surface chlorophyll were due to chain-forming diatoms retained in the netplankton fraction (Fig. 3). Chlorophyll concentrations were highest between pulses in freshwater flow and increases were propagated upstream from the mouth of the estuary. Vertical salinity and chlorophyll

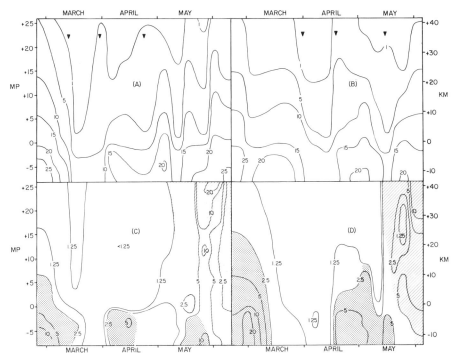

Figure 3. Distributions of surface salinity ($^o/_{oo}$) and chlorophyll a (mg m^{-3}) during spring between mile point -7 (-11 km) and mile point 27 (43 km); A - salinity 1977, B - salinity 1978; C - chlorophyll a 1977; D - chlorophyll a 1978 (shaded area netplankton > 80%, cross-hatched area nanoplankton > 80%). Peaks in freshwater flow are indicated (▼).

sections (Fig. 5) suggest that increases in chlorophyll during this period were caused by upstream transport of netplankton populations in the bottom layer.

Since phytoplankton growth rates and grazing by macrozooplankton appear to be low during winter and early spring (Malone 1977a), variations in chlorophyll should be positively correlated with salinity if advective transport from adjacent coastal waters is the primary mechanism by which phytoplankton biomass increased during this period. This was found to be the case (Fig. 6). Salinity accounted for 60% or more of the variance in chlorophyll distributions when netplankton dominated. Under these conditions, the quantity of chlorophyll in the lower estuary was determined by the chlorophyll content of coastal water transported into the estuary through Verrazano Narrows at MP -7 (Fig. 7). Thus the distribution of chlorophyll within the estuary was determined by patterns of estuarine circulation whereas chlorophyll concentration was related to the magnitude of

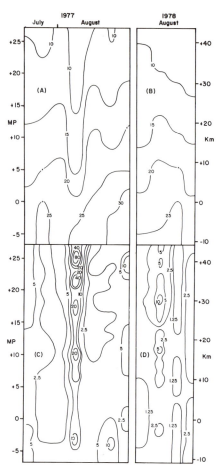

Figure 4. Distributions of surface salinity (°/₀₀) and chlorophyll a (mg m⁻³) during summer when nanoplankton accounted for more than 80% of total chlorophyll; A - salinity 1977; B - salinity 1978; C chlorophyll a 1977; D - chlorophyll a 1978.

coastal blooms. That such blooms are dominated by netplankton diatoms which achieve high concentrations during winter-spring is well documented (Malone 1977b; Malone and Chervin 1979).

As the season progressed, the incidence of netplankton populations in the estuary declined as the seasonal thermocline developed in coastal waters (Malone and Chervin 1979). By late May, nanoplankton populations accounted for most of the chlorophyll in the estuary. The transition from netplankton to nanoplankton dominance appeared to be related to temperature with netplankton dominating at temperatures below 15C and nanoplankton dominating at temperatures above 15C (Fig. 8). A similar transition temperature has been found for adjacent coastal waters (Malone 1977b).

Increases in chlorophyll during the period of nanoplankton dominance were generally associated with pulses in freshwater flow and were propagated downstream from within the estuary (Fig. 4). Vertical salinity and chlorophyll sections (Fig. 9) suggest that increases in chlorophyll were due to growth and downstream transport of nanoplankton populations in the surface layer. Although chlorophyll concentration was poorly correlated with salinity (Fig. 6), high chlorophyll concentrations occurred when chlorophyll was inversely related to salinity, i.e. peaks in chlorophyll coincided with increases in Q_f. Since biological activity is greatest during this period, deviations from conservative behavior were probably the result of high phytoplankton growth rates in the surface layer, high grazing rates by zooplankton populations, and low Q_f (Malone 1977a).

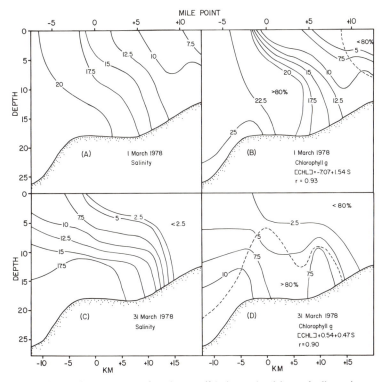

Figure 5. Vertical sections of salinity (⁰/₀₀) and chlorophyll a (mg m⁻³) between mile point −7 (−11 km) and mile point 11 (18 km) when netplankton accounted for most of the chlorophyll in the estuary (---- 80% netplankton chlorophyll).

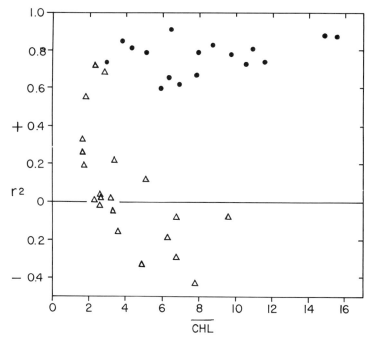

Figure 6. Coefficients of variation (r^2) for regressions of chlorophyll a on salinity in relation to the mean concentration of chlorophyll a (mg m^{-3}) in the lower estuary. $\overline{CHL}$ = mg m^{-3}; • - netplankton accounted for > 60% of total chlorophyll; △ - nanoplankton accounted for > 60% of total chlorophyll.

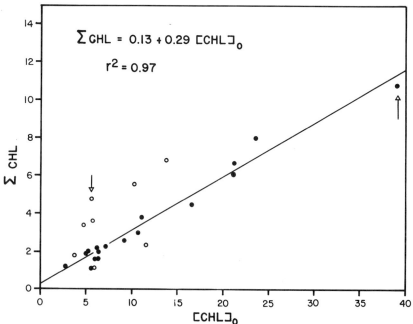

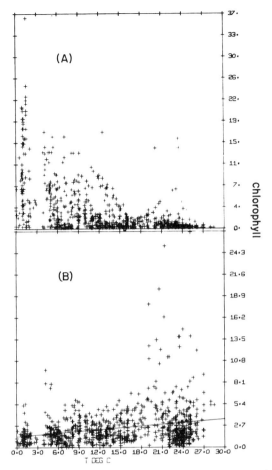

Figure 8. Variations in netplankton (A) and nanoplankton (B) chlorophyll a concentrations (mg m⁻³) in relation to temperature (C) in the lower estuary.

Figure 7. (opposite). Chlorophyll a content (10³ kg) of the lower estuary (MP + 11 to MP −7) as a function of chlorophyll a concentration in the bottom layer at MP −7. [CHL]₀ = mg m⁻²; • - netplankton dominated February-May (data used for regression analysis); ○ - nanoplankton dominated May-August; arrows indicate examples used to illustrate details of salinity and chlorophyll distributions and fluxes (Fig. 10).

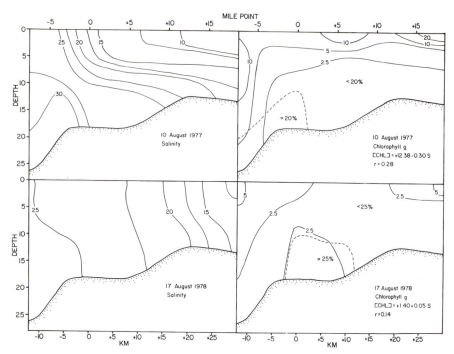

Figure 9. Vertical sections of salinity ($^{0}/_{00}$) and chlorophyll a (mg m^{-3}) between mile point −7 (−11 km) and mile point 18 (29 km) when nanoplankton accounted for most of the chlorophyll in the estuary (----20% and 25% netplankton).

These observations lead to the hypothesis that the lower Hudson estuary is a "sink" for phytoplankton during winter-spring and a "source" during summer. Chlorophyll fluxes were calculated using a two-dimensional box model similar to that described by Pritchard (1969). Given Q_f, end member concentrations of chlorophyll, and the tidally averaged salinity field, the model calculates a conservative distribution of chlorophyll which, when compared to the observed distribution, yields estimates of the magnitude and direction of fluxes between boxes and layers. Taft et al. (1978) used a similar approach to calculate nitrate and ammonia fluxes in Chesapeake Bay.

Tidally averaged salinities were not available except from Verrazano Narrows. Comparison of volume transports calculated from tidally averaged salinity with those calculated from salinity at a given point in the tidal cycle gave coefficients of variation of 15% to 50%. Deviations from the mean were less than 30% under most conditions (66 of 86 cases) and tended to be negative, i.e. volume transports were underestimated more often than they were overestimated.

Table 1. Sources (+) and sinks (−) of chlorophyll in the upper and lower layers of the lower Hudson estuary between mile point 11 and mile point −7. Values are in Kg Chl Day⁻¹.

Month	1977			1978		
	Upper	*Lower*	*Total*	*Upper*	*Lower*	*Total*
Feb.	− 867	+ 232	− 635	− 273	+ 167	− 106
				− 652	+ 430	− 222
Mar.	− 440	− 139	− 579	−7970	−2990	−10960
				− 242	− 663	− 905
Apr.	− 876	+ 177	− 699	+ 42	− 28	+ 14
	−1080	− 86	−1166	−5703	+ 813	− 4890
	−1416	−1025	−2441			
May	−1079	− 172	−1251	− 60	− 561	− 621
	+ 66	− 201	− 135	+ 384	− 199	+ 185
	+ 304	+ 512	+ 816			
	−2146	−1117	−3263			
July	+2060	−1124	+ 936			
Aug	+ 165	− 242	− 77	+ 971	− 777	+ 194
	+1235	− 688	+ 547	+ 126	− 546	− 420
	+ 269	− 943	− 674			
	+3522	−2857	+ 665			

Table 2. Sources (+) and sinks (−) of phytoplankton-carbon (calculated from chlorophyll *a* using C:Chl = 50) in the upper and lower layers and mean phytoplankton productivity (PP) of the upper layer in the lower Hudson estuary.

Season	10^4 Kg C Day⁻¹			
	Upper	*Lower*	*Total*	*PP*
Winter	− 2.99	+ 1.38	+ 1.61	0.86
Spring	− 7.22	− 2.03	− 9.25	3.45
Summer	+ 5.96	− 5.13	+ 0.83	13.18

Results of the model are summarized in Table 1. Although the magnitude of Q_f was substantially different between 1977 and 1978, the seasonal development of sources and sinks was similar during both years. The estuary was a sink for phytoplankton biomass produced in adjacent coastal waters during the winter and spring and a source during summer (Table 2). The upper and lower layers exhibited opposite trends in that the upper layer became an increasingly important source as the bottom layer became an increasingly important sink as the seasons progressed. The former trend reflects the seasonal cycle of nanoplankton productivity in the surface layer (Malone 1977a) while the latter reflects (1) resuspension of netplankton diatoms during the winter when the lower layer is a source of chlorophyll, and (2) an increase in decomposition and grazing rates as the biomass and metabolism of heterotrophs increase during spring and summer.

Note that the estuary was a sink during spring even though Q_f was high. This occurred because of the development of large concentrations of netplankton diatoms in adjacent coastal waters when the water column was poorly stratified (Malone and Chervin 1979). Consequently, chlorophyll concentrations in bottom water entering the estuary were sufficiently higher than surface concentrations to cause a net flux into the estuary (Fig. 10A). Increases in Q_f were associated with decreases in chlorophyll, but the effect was minimized by the relatively low chlorophyll content of the surface layer. In addition, intrusions of salt water occur rapidly following pulses in freshwater flow (Simpson et al. 1974) so that increases in netplankton chlorophyll developed between peaks in freshwater flow (Fig. 3).

These trends were reversed during the summer and the estuary became a source of chlorophyll for adjacent coastal waters although Q_f was typically low (Fig. 10B). This resulted from the development of nanoplankton blooms in the surface layer (Fig. 4) combined with low concentrations of chlorophyll in bottom water entering the estuary.

Implications

The effects of freshwater flow on material fluxes in partially mixed estuaries are complicated by two-layered flow patterns and feedback interactions between estuarine and coastal waters. Although the model calculations presented must be considered rough approximations, it is clear that the allochthonous flux of phytoplankton biomass into the estuary can be large relative to other inputs of organic carbon. Sewage disposal accounts for about 24.2×10^4 kg C d^{-1} (Garside and Malone 1978) compared to an annual mean of 3.8×10^4 kg C d^{-1} (0.66 g C m^{-2} d^{-1}) produced by phytoplankton within the lower estuary. Net fluxes of phytoplankton-carbon into the estuary from adjacent coastal waters can be significant relative to these rates, especially during spring (Table 2). This is a direct consequence of phytoplankton production that depends almost entirely on nutrients exported from the estuary (Malone 1980). Assuming N:C = 0.14 (w/w), the average spring flux (Table 2) is equivalent to 1.3×10^4 kg N d^{-1}. Since most of the 1.6×10^5 kg N d^{-1} of sewage input is exported (Garside et al 1976; Malone 1980), it appears that a maximum of 8% of sewage-derived nitrogen is returned to the estuary in the form of phytoplankton biomass.

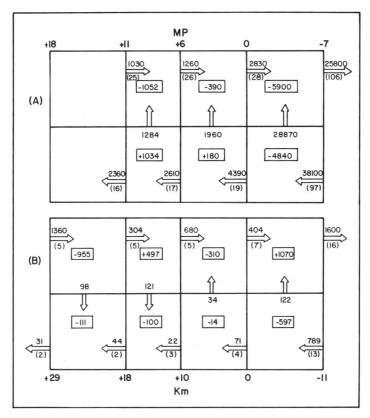

Figure 10. Chlorophyll fluxes (kg d⁻¹) when netplankton dominated on 1 March 1978 (A) and when nanoplankton dominated on 10 August 1977 (B); numbers in parentheses are advective transports (10^7 m^3 d^{-1}) and those in rectangles are net transports into (−) or out of (+) each box.

This confirms the occurrence of a process which Redfield (1955) termed "biochemical" circulation. It is an important process in terms of particulate organic carbon fluxes to the lower estuary, but is relatively insignificant in terms of the return of nutrients. The lower Hudson estuary is an effective conduit of sewage-derived nutrients, and, since exported nutrients are rapidly assimilated into metazoan food chains (Malone and Chervin 1979; Malone 1980), the New York Bight functions as an efficient tertiary treatment system under most conditions.

Acknowledgments

This research was supported by NSG Grant OCE76-80883. Lamont-Doherty Geological Observatory Contribution No. 2902.

References Cited

Abood, K. A. 1974. Circulation in the Hudson Estuary, pp. 35-111. *In:* O. A. Roels (ed.), *Hudson River Colloquium,* Ann. N. Y. Acad. Sci. Vol. 250.

Dyer, K. R. 1973. *Estuaries: A Physical Introduction.* John Wiley & Sons, New York. 140 pp.

Garside, C. and T. C. Malone. 1978. Monthly oxygen and carbon budgets of the New York Bight Apex. *Est. Coastal Mar. Sci.* 6:93-104.

Garside, C., T. C. Malone, O. A. Roels and B. A. Sharfstein. 1976. An evaluation of sewage-derived nutrients and their influence on the Hudson Estuary and New York Bight. *Est. Coastal Mar. Sci.* 4:281-289.

Hammond, D. E. 1975. Dissolved gases and kinetic processes in the Hudson River Estuary, Ph.D. Thesis, Columbia University, New York. 161 pp.

Ketchum, G. H. 1967. Phytoplankton nutrients in estuaries, pp. 329-335. *In:* G. H. Lauff (ed.), *Estuaries.* Amer. Assoc. Adv. Sci. Publ. No. 83, Washington, D.C.

Ketchum, G. H. 1969. Eutrophication of estuaries, pp. 197-209. *In: Eutrophication Causes, Consequences, Correctives.* National Academy of Sciences, Washington, D.C.

Loftus, M. E., D. V. Subba Rao and H. H. Seliger. 1972. Growth and dissipation of phytoplankton in Chesapeake Bay. I. Response to a large pulse of rainfall. *Chesapeake Sci.* 13:282-299.

McCarthy, J. J., W. R. Taylor and J. L. Taft. 1975. The dynamics of nitrogen and phosphorus cycling in the open waters of the Chesapeake Bay, pp. 664-681. *In:* T. M. Church (ed.), *Marine Chemistry in the Coastal Environment.* Amer. Chem. Soc. Symposium Series, No. 18, New York.

Malone, T. C. 1977a. Environmental regulation of phytoplankton productivity in the lower Hudson Estuary. *Est. Coastal Mar. Sci.* 5:157-171.

Malone, T. C. 1977b. Light-saturated photosynthesis by phytoplankton size fractions in the New York Bight, USA. *Mar. Biol.* 42:281-292.

Malone, T. C. 1980. Factors influencing the fate of sewage derived nutrients in the lower Hudson Estuary and New York Bight. *In:* G. F. Mayer (ed.), *Ecological Stress and New York Bight: Science and Management,* Estuarine Research Federation, Columbia, SC (in press).

Malone, T. C and M. Chervin. 1979. The production and fate of phytoplankton size fractions in the plume of the Hudson River, New York Bight. *Limnol Oceanogr.* 24:683-696.

O'Connor, D. J. 1970. Water quality analysis for the New York Harbor complex, pp. 121-144. *In:* A. A. Johnson (ed.), *Water Pollution in the Greater New York Area.* Gordon and Breach, New York.

Pritchard, D. W. 1954. A study of the salt balance in a coastal plain estuary. *J. Mar. Res.* 13:133-144.

Pritchard, D. W. 1969. Dispersion of flushing of pollutants in estuaries. *J. Hydraulics Div.,* Proc. Am. Soc. Civil Engineers, 95:115-124.

Pritchard, D. W., A. Okubo and E. Mehr. 1962. A study of the movement and diffusion of an introduced contaminant in New York Harbor waters. Technical Report 31, Chesapeake Bay Institute, Johns Hopkins U., Baltimore, MD. 89 pp.

Redfield, A. C. 1955. The hydrography of the Gulf of Venezuela. *Deep-Sea Res.* 2(Suppl.):115-133.

Simpson, H. J., R. Bopp and D. Thurber. 1977. Salt movement patterns in the lower Hudson. Lamont-Doherty Geological Observatory Contribution No. 2086, 34 pp.

Stewart, H. B. 1958. Upstream bottom currents in New York Harbor. *Science* 127:1113-1115.

Taft, J. L. and W. R. Taylor. 1976. Phosphorus dynamics in some coastal plain estuaries, pp. 79-89. *In:* M. L. Wiley (ed.), *Estuarine Processes. Vol. 1.* Academic Press, New York.

Taft, J. L., A. J. Elliott and W. R. Taylor. 1978. Box model analysis of Chesapeake Bay ammonium and nitrate fluxes, pp. 115-130. *In:* M. L. Wiley (ed.), *Estuarine Interactions.* Academic Press, New York.

CHARACTERIZATION OF LIGHT EXTINCTION AND ATTENUATION IN CHESAPEAKE BAY, AUGUST, 1977

Michael A. Champ, George A. Gould III, William E. Bozzo

Department of Biology
The American University
Washington, D.C.

and

Steven G. Ackleson and Kenneth C. Vierra

College of Marine Studies
The University of Delaware
Lewes, Delaware

Abstract: When solar radiation penetrates the sea, it is both scattered and absorbed at rates which depend upon the concentration of dissolved organic substances and suspended particulate matter present in the water column. Both scattering and absorption act in unison to produce a rate of attenuation. It is well known that as a transformation from clear oceanic waters to more turbid coastal waters occurs, short wavelengths of light are attenuated more quickly while yellow light penetrates the deepest. This phenomenon is termed "Yellow Shift." Light penetration measurements at selected wavelengths (430, 470, 514.5, 540, and 600 nm ± 10 nm) taken in August, 1977 in the upper 200 km of Chesapeake Bay from the mouth of the Susquehanna River, suggest that there may be an estuarine "Orange Shift." It was found for 50% of all light penetration measurements that orange light (600 nm) exhibited the smallest average attenuation coefficient (i.e., orange light penetrated the deepest). Beam transmission volume attenuation coefficients (α) calculated for stations from the mouth of the Susquehanna River to just offshore (6.3 km) from the mouth of Chesapeake Bay indicate that Chesapeake Bay can be divided into three turbidity zones: Estuarine Turbidity Maximum (α values of 3.50 to 4.77 ln/m); Estuarine Turbidity Normal (2.79 to 2.91); Estuarine Turbidity Minimum (2.73 to 2.76). The Turbidity Maximum of northern Chesapeake Bay is examined by comparing the relationship of contours of attenuation coefficient, turbidity, suspended sediment, salinity and chlorophyll a.

Introduction

Light penetration in a water mass is regulated by two physical processes: (1) absorption and (2) scattering (Williams 1970). Absorption is the conversion of light into heat energy, while scattering is defined as the change in direction of light energy by particulate matter. Studies of the extinction rates of solar radiation in the open ocean and in coastal waters have found that, in the open ocean, blue light penetrates most deeply while red light is attenuated completely in the first few meters. Yellow light penetrates most deeply in coastal waters and the penetration of blue light is greatly decreased. Red light is again completely attenuated in the upper few meters of the water column; however, the depth of penetration of red light in coastal waters is very close to its depth of penetration in the open

263

ocean. This phenomenon is termed "Yellow Shift" and is simply a trend of longer wavelengths penetrating to the greater depths, relative to the entire spectrum, as turbidity and dissolved organic compounds increase in concentration, selectively attenuating shorter wavelength energy.

Historical Chesapeake Bay Data

The photic zone of Chesapeake Bay has not been studied adequately. Most data have been collected using Secchi Discs. Seliger and Loftus (1974) have described the estimation of absorption by the Secchi Disc as being, at best, poor. Because the method is visual it measures relative absorption of a range of wavelengths of the human eye which has a peak sensitivity at 555 nm. This peak lies between the absorption peaks of chlorophyll and most of the blue-absorbing accessory pigments. The Secchi Disc measures some function of both α (attenuation) and k (extinction), and there is no way of separating them in the measurement. If the irradiance is to serve as illumination, then relative irradiance (measured by the extinction coefficient k) is paramount, whereas under conditions of image formation, beam transmittance (measured by the attenuation coefficient α) is important. Although beam transmittance is a measure of both absorption and scattering, in turbid estuaries where scattering predominates it is a good measure of the scattering alone (Williams 1973).

Hulburt (1945) of the Naval Research Laboratory was the first to report the measuring of attenuation and scattering in Chesapeake Bay water samples. Burt (1953) collected over 100 water samples that were analyzed for the presence of "dissolved coloring matter in order to determine whether coloring matter in solution is of importance in the extinction of light in Chesapeake Bay Waters." He found significant absorbency below 550 nm and attributed it to the presence of suspensoids not retained on fine filters. These suspensoids, when combined with dissolved coloring matter, act as a yellow substance by absorbing and scattering light from the shorter wavelength end of the visible spectrum (violet and blue). This has also been reported by Jerlov and Kullenberg (1953). The color curves for samples from the Wicomico and Nanticoke Rivers of Chesapeake Bay, both of which drain extensive swamp areas, indicate the presence of greater amounts of organic material than are found in samples from the lower Bay and the James River.

Burt (1955a) reported that some 25,000 extinction readings were made aboard ship on freshly drawn samples as part of regular survey work in the Chesapeake Bay and its tributaries. Observations covered the entire range of depths and seasons over the whole of the Bay system. Burt (1955b) also investigated the distribution of suspended materials in Chesapeake Bay. He found large local time changes (seasonal and tidal) as well as large horizontal gradients of suspended inorganic materials caused by wind mixing, tidal scouring and river discharge. He found that concentrations of suspended materials ranged from less than 1 to approximately 60 ppm by volume.

Seliger and Loftus (1974), using a pressurized underwater spectrometer, reported on the spectral distribution for the Rhode River (July

27, 1972), a subestuary of Chesapeake Bay. In comparing incident surface intensities to 0.8 m depth, they found a significant distortion of the original sunlight spectrum with a marked reduction of light in the photosynthetically significant region (400 nm to 500 nm). Jerlov (1968, 1976) has also reported a significant shift in transmission peak from clear oceanic to turbid coastal waters in the 400-500 nm wavelengths. Seliger and Loftus (1974) also concluded that during the months of April through October, light appears to be a limiting factor for phytoplankton growth in the Rhode River and the West River, both subestuaries of the Chesapeake Bay. Our studies in the Potomac Estuary for the past several years have found the compensation point (1%) to vary from 9 cm to 480 cm, with over 90% of the measurements being less than 200 cm.

We report the results of a study of the spectral distribution of irradiance extinction coefficients (k) and beam transmittance attenuation coefficients (a) for white light in Chesapeake Bay. Optically related water quality parameters (turbidity, suspended sediment, salinity and chlorophyll a) are compared to attenuation coefficients (a) for the Turbidity Maximum in upper Chesapeake Bay.

Methods and Materials

Station Locations

Irradiance measurements were taken at all stations occupied during daylight hours, whereas transmissometer measurements were taken day and night at all hydrocast stations. Irradiance measurements were made at all stations (except 1, 5, 16, 91, 93, 96, 99, 102 and 105) from the mouth of the Susquehanna River to 6.3 km offshore from the mouth of Chesapeake Bay during August 17-22, 1977. The position of each station (Fig. 1) has been plotted on NOS Nautical Charts (scale 1:40,000) and located by the R/V Annandale using the Decca TM 626 True Motion Radar, for shoreline features in the upper Bay in conjunction with aids to navigation, and Simrad Internav 101-Loran C with plotter in the lower Bay. Nautical distances for each station from the designated entrance to Chesapeake Bay (37°00'00"N., 76°00'00"W.) have been measured in nautical miles on an latitude basis only (due to the vertical orientation of the Bay) and converted into km.

Light Extinction

Solar irradiance (light penetration in the water column) was measured with a Montedoro/Whitney Solar Illuminance Meter, Model LMD-8A. This instrument originates from early work of Whitney (1938, 1941) in Northern Wisconsin Lakes. The spectral response of this instrument is presented in Fig. 2. To measure the distribution of specific wavelengths, light penetration measurements were made at selected wavelengths (430, 470, 514.5, 540 and 600 ± 10 nm) with Ditric Optical Filters. These filters are hydrophilic with moisture voiding the wavelength

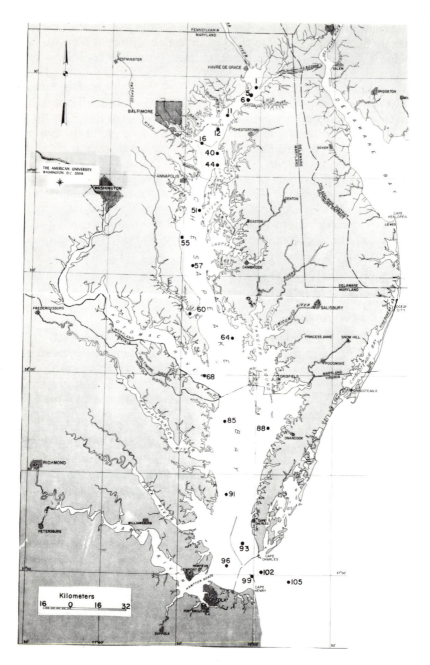

Figure 1.   Station locations in Chesapeake Bay.

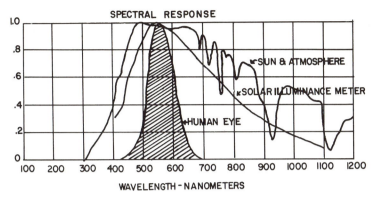

Figure 2. *Spectral response of Montedoro/Whitney Solar Illuminance Meter, Model LMD-8A.*

specificity; we recommend plexiglass housings for their underwater protection.

Extinction coefficients (k) were calculated by:

$$I_z = I_o e^{-kz} \tag{1}$$

(McLellan 1965; Newman and Pierson 1966) where:

I_z = Intensity at depth z

I_o = Intensity at the surface (for precise measurements use wetted photocell surface suspended above the water's surface)

k = Extinction coefficient (ln/m)

The extinction coefficient for a column of water was calculated by:

$$k = - \frac{1}{n} \ln \left(\Sigma \frac{I_n}{I_{z\text{-}n}} \right) - \ln x \tag{2}$$

where:

k = Extinction coefficient (ln/m)

n = Constant vertical distance between two successive light intensity measurements

I_n = Intensity at depth z

$I_{z\text{-}n}$ = Intensity at depth z-n

x = The number of successive pairs averaged together

Light Attenuation

Beam transmittance was measured with an *in situ* transmissometer designed and built by Hydroproducts (Model 612-S). The optical system consisted of a cylindrically limited beam of light 10 cm long with a diameter of 1.5 cm and the mean acceptance angle of 5.0° ± 1.0°. The optical system was calibrated so that air transmittance was 92.0%. The transmittance data were transformed into values for volume attenuation coefficients (α) according to Bouger's Law (Duntley 1963; Kiefer and Austin 1974):

$$\alpha = \ln \frac{p}{T} \qquad (3)$$

where:

   α = Attenuation coefficient (ln/m)
   T = Transmissivity (transmittance per m)
   p = Path length in m

Attenuation lengths were calculated by:

$$\text{Attenuation length} = \frac{1}{\alpha} \quad \text{(m/ln)} \qquad (4)$$

Optically Related Water Quality Parameters

Water samples were collected at specific depths selected and coordinated with light penetration and transmission. The following selected physical, chemical and biological optically related water quality parameters were determined by standard methods:

   Suspended Sediment - (American Public Health Association 1975)
   Turbidity - HACH Laboratory Turbidimeter (Nephelometer)
   Chlorophyll a - (Strickland and Parsons 1972)
   Temperature/Salinity - Beckman Laboratory Salinometer

We are in agreement with Pickering's (1976) views on 'Turbidity' as a non-quantitative term; however, we did make turbidity measurements in FTU's and report them for relative information.

Results and Discussion

Light Extinction

Station (water column) extinction coefficients (k) are presented in Table 1. Stations 6, 11 and 12 in the Turbidity Maximum had the highest extinction coefficients. At 50% of the stations for irradiance measurements, orange light (600 nm) penetrated the deepest of all wavelengths measured. At the rest of the stations, either yellow-green light (540 nm) or green light (514.5 nm) penetrated the deepest. These findings (far from conclusive, due to the limited number of stations and wavelengths measured) suggest that there is a continuation in the trend of deeper penetration of the longer wavelengths from oceanic to coastal waters ("Yellow Shift") to an "Orange Shift" from coastal to estuarine waters. Further evidence of this shifting trend may be seen in Fig. 3 which compares extinction coefficients reported by Sverdrup et al. (1942) for Utterback's data (1936) for the open ocean and coastal waters to our August, 1977 Chesapeake Bay data.

The trend observed by Schenck and Davis (1973) of a continual decrease in k values for white light as one traverses from North to South in

Table 1. Water column extinction coefficients (k) for specific wavelengths (± 10 nm band widths) at indicated stations. Means were calculated using data from stations 11 (except 600 nm), 12, 40, 44, 64 and 85. ND = no data.

Station No.	W/O Filter	Wavelength				
		430	470	514.5	540	600
006	ND	13.80	13.80	3.63	3.69	3.77
011	2.48	4.37	3.90	2.20	1.52**	2.09
012	2.49	4.29	2.35	1.86	1.63**	1.64
040	1.60	3.47	2.80	1.98	1.49	1.17**
044	1.48	2.30	1.88	1.41	1.35**	1.61
060*	1.25	4.73	1.97	2.76	2.18	1.18**
064	0.88	2.28	1.70	0.98	0.91	0.81**
068*	ND	2.40	2.90	1.97	1.65	1.35**
085	0.60	1.61	1.99	0.96	0.97	0.84**
088*	1.15	1.78	1.87	1.14**	1.25	1.24
Means	1.59	3.05	2.44	1.56	1.31	1.21

* Station located at river mouth
** Station where specific wavelength penetrated the deepest

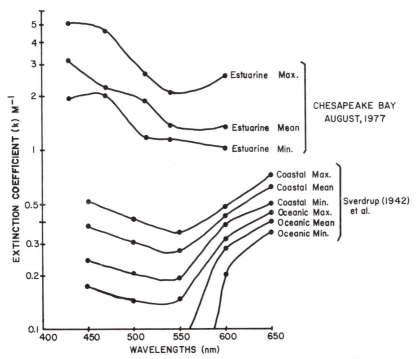

Figure 3. Comparison of attenuation coefficients reported by Sverdrup et al. (1942) for oceanic and coastal waters with those for Chesapeake Bay, August, 1977.

Narragansett Bay also occurs in Chesapeake Bay. It should be noted that the k values calculated by Schenck and Davis (1973) for the high turbidity West Passage of Narragansett Bay are approximately 60% lower than the ones calculated for the Turbidity Maximum (0.955 to 2.49) in Upper Chesapeake Bay. However, for the middle region of Chesapeake Bay, the k values are similar. The difference in the k values is due to the higher inorganic and organic loadings into the upper Chesapeake Bay from the Susquehanna River. The suspended sediment loadings are generally one order of magnitude higher, even in low flow drought conditions, as August of 1977 was, (comparing our suspended sediment data and data from Schubel 1972 and Gross et al. 1978 to Schenck and Davis 1973 for Narragansett Bay).

The 1.0% light penetration depths (in percent) for white light and the specific wavelengths measured are plotted in Fig. 4.

Exponential Extinction

According to the expression of the attenuation of light as it propagates through a column of water (eq. 1), light intensities will diminish exponentially with depth. To test the data collected with the solar illuminance meter for the estuary against this theory, light intensities at each station were graphed with respect to the depth at which each measurement was made in semilog form. The theoretical exponential curve

$$I_z \, / \, I_o = e^{-n}$$

was then drawn in on the same graph. If in fact the light intensity measurements collected were the results of an exponential extinction, the graph would demonstrate a strong correlation between the plots of the station measurements and the exponential curve. The results are presented in Fig. 5, a composite of all the data. We conclude from this plot that light in the more turbid estuarine waters of Chesapeake Bay is being attenuated according to theory.

Light Attenuation

Beam transmittance attenuation coefficients (α) are presented in Table 2. The data can be grouped into three estuarine turbidity zones if the stations located at the mouth of rivers (60 and 88) or very close to shore (55 and 91) and station 96 (below the Bay Bridge Tunnel) are excluded: (1) Estuarine Turbidity Maximum with α's that ranged from 3.50 to 4.77 with the smallest attenuation length of 0.20 m/ln; (2) Estuarine Turbidity Normal with α's that ranged from 2.79 to 2.91 with an attenuation length of approximately 0.35 m/ln; and (3) Estuarine Turbidity Minimum with α's that ranged from 2.73 to 2.76 with the best light attenuation length of 0.40

Figure 4. (opposite) Irradiance depths (1.0%) for white light (w/o filter) and specific wavelengths, Chesapeake Bay, August, 1977.

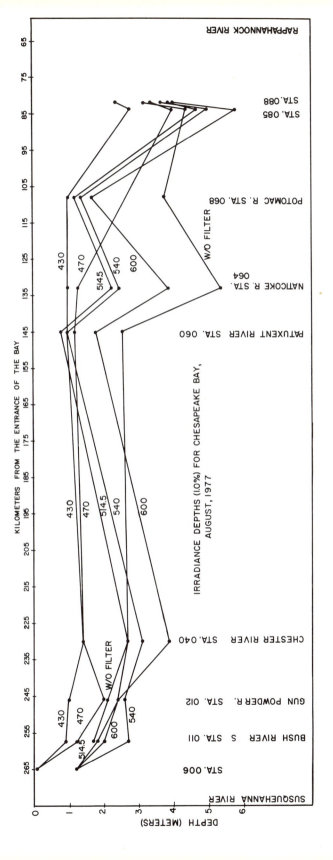

IRRADIANCE DEPTHS (1.0%) FOR CHESAPEAKE BAY, AUGUST, 1977

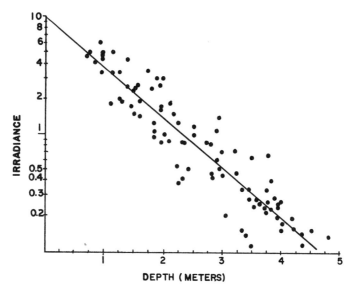

DEPTH (METERS)

Figure 5. *Correlation of measured light extinctions for all stations and the theoretical exponential curve.*

m/ln. Station 57 with an α of 2.71 had the smallest attenuation coefficient determined for the Bay proper, excluding coastal water near the mouth with an α of 2.49. Station 6 with an α of 4.77 and a light attenuation length of 0.20 m/ln was the highest calculated for the entire study. The highest α value reported by Schenck and Davis (1973) for Narragansett Bay was 3.31 which is approximately 40% lower than the Turbidity Maximum for Chesapeake Bay. The average α for their study was 2.18 with a light attenuation length of 0.46 m/ln, indicating that Narragansett Bay has approximately 30% (on the average) less light attenuation than Chesapeake Bay.

Turbidity Maximum

We have located the Turbidity Maximum described by Schubel (1968) using optical methods and optically related water quality parameters. The Turbidity Maximum was located in the upper reaches of the Bay between stations 1 and 12. Figure 6 is a vertical profile of salinity for this region of the Bay starting at station 1 and extending 25.9 km to station 12. Station 6 is at the front of the tidal salinity interface (0.5 ‰). Figure 7 presents the corresponding contours for attenuation coefficients (α) for the same stations in the Turbidity Maximum. The highest attenuation occurred between stations 5 and 6. Station 6 also exhibited the highest extinction coefficient for the selected wavelengths of irradiance measured. Contours of suspended sediment concentrations (gm/l) are presented in Fig. 8 with Turbidity (FTU's) in Fig. 9 and Chlorophyll *a* in Fig. 10. Com-

Table 2. Mean water column attenuation coefficients (α) and attenuation lengths ($1/\alpha$) for indicated stations. Data from stations marked with asterisks were not used to calculate zone means.

	Station No.	Km from Bay entrance	α (ln/m)	Zone Mean	$1/\alpha$ (m/ln)	Zone Mean
Estuarine	001	272.3	3.69		0.27	
turbidity	005	267.5	4.25		0.24	
maximum	006	264.6	4.77	3.95	0.20	0.26
	011	257.2	3.92		0.26	
	012	246.4	3.57		0.28	
	016	239.0	3.50		0.28	
Transition	040	230.7	3.03	3.08	0.33	0.32
zone	044	230.5	3.14		0.32	
Estuarine	051	203.5	2.85		0.35	
turbidity	055**	187.8	3.07		0.33	
normal	057	172.8	2.71		0.37	
	060*	144.7	3.01	2.81	0.33	0.36
	064	133.4	2.91		0.34	
	068*	108.2	2.79		0.36	
	085	83.8	2.80		0.36	
Transition	088*	82.0	3.11		0.36	
zone	091**	43.8	3.13		0.32	
Estuarine	093	19.2	2.76	2.74	0.36	0.36
turbidity	096***	5.9	3.02		0.33	
minimum	099	4.3	2.73		0.37	
Bay Mouth	102	2.4	2.49	2.50	0.40	0.40
Offshore	105	6.3	2.50		0.40	

   * Station located at river mouth
  ** Station located close to shore
 *** Station located close to Bay Bridge Tunnel

parison of Fig. 6 through 10 presents a synoptic overview of the interrelationship of a physical process to its biological expression. Jerlov (1976) has also "concluded that the distribution of particles and yellow substance, and ultimately of irradiance attenuance, is controlled by those dynamical processes in the sea which have an important bearing on productivity."

Optical measurements (qualitatively and quantitatively) have tremendous potential as a water resource monitoring management tool. They are relatively quick and easy to take and should provide very useful data for

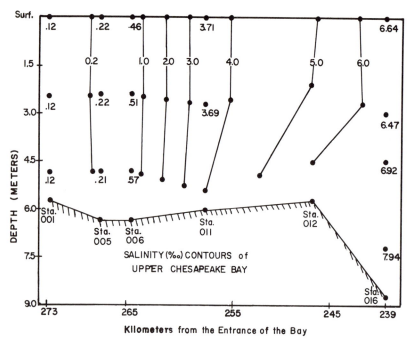

Figure 6. Salinity contours (°/₀₀) for indicated stations in upper Chesapeake Bay.

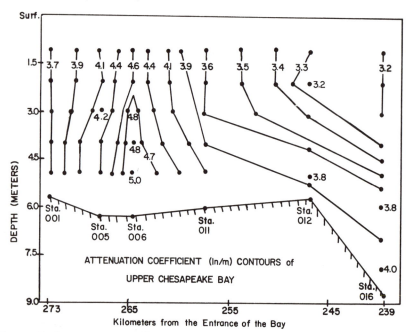

Figure 7. Attenuation coefficient (ln/m) contours for indicated stations in upper Chesapeake Bay.

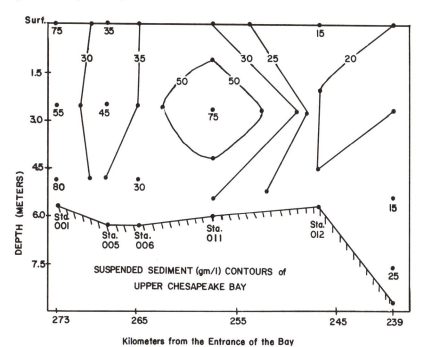

Figure 8. *Suspended sediment (gm/l) contours for indicated stations in upper Chesapeake Bay.*

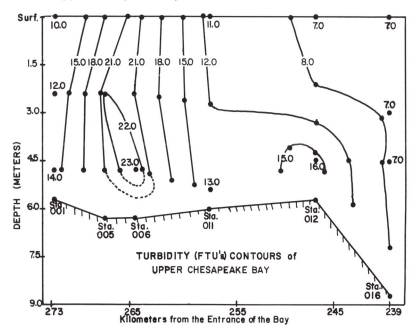

Figure 9. *Turbidity (FTU's) contours for indicated stations in upper Chesapeake Bay.*

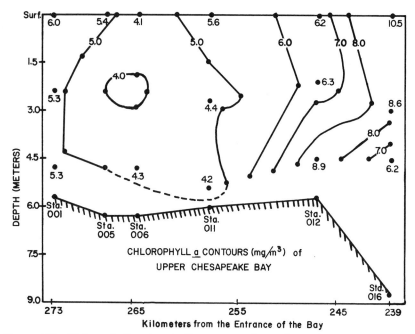

Figure 10. Chlorophyll a (mg/m³) contours for indicated stations in upper Chesapeake Bay.

interpretation of productivity and optically related water quality parameters as they become a regular parameter in field surveys.

Acknowledgments

We thank The Marine Science Consortium for shiptime on the R/V Annandale; Dr. Frank Hoge and Robert N. Swift (NASA, Wallops) for the loan of their transmissometer; the following for their assistance in data collection: Captain Gary Van Tassel, Hilton H. Smith, Samuel R. Heller III, Mary Gunn, Rhoda Twombly, William Renz, Michael D. Conrad, Mark Byrnes, Larry E. Parson, Roger Murray; Tommie R. Champ for illustrations and editing; and Jerome Williams and David F. Bleil for their comments on an earlier draft of the manuscript.

References Cited

American Public Health Association. 1975. Standard Methods for the Examination of Water and Wastewater. 14th Ed., APHA, Washington, D.C. 1193 p.

Burt, W. V. 1953. Extinction of light by filter passing matter in Chesapeake Bay waters. Science 118:386-387.

Burt, W. V. 1955a. Interpretation of spectrophotometer readings on Chesapeake Bay waters. J. Mar. Res. 14:33-46.

Burt, W. V. 1955b. Distribution of suspended materials in Chesapeake Bay. J. Mar. Res. 14:47-62.

Duntley, Seibert O. 1963. Light in the sea. *J. Opt. Soc. Amer.* 53:214-233.

Gross, Grant M., M. Karweit, William B. Cronin and J. R. Schubel. 1978. Suspended sediment discharge of the Susquehanna River to Northern Chesapeake Bay, 1966 to 1976. *Estuaries* 1:106-110.

Hulburt, E. O. 1945. Optics of distilled and natural water. *J. Opt. Soc. Amer.* 35:698-705.

Jerlov, N. G. 1968. *Optical Oceanography.* Elsevier Publishing Co., New York. 194 p.

Jerlov, N. G. 1976. *Marine Optics.* Elsevier Publishing Co., New York. 231 p.

Jerlov, N. G. and B. Kullenberg. 1953. The Tyndall effect of uniform menerogenic suspensions. *Tellus* 5:306-307.

Kiefer, D. A. and R. W. Austin. 1974. The effect of varying phytoplankton concentration on submarine light transmission in the Gulf of California. *Limnol. Oceanogr.* 19:55-64.

McLellan, H. G. 1965. *Elements of Physical Oceanography.* Pergamon Press, New York. 150 p.

Newman, G. and W. J. Pierson. 1966. *Principles of Physical Oceanography.* Prentice-Hall, Englewood Cliffs, NJ.

Pickering, R. J. 1976. Measurement of "Turbidity" and related characteristics of natural waters. U.S. Geol. Surv. Open-File Rept. 76-153. 7 p.

Schenck, H., Jr. and A. Davis. 1973. A turbidity survey of Narragansett Bay. *Ocean. Engng.* 2:169-178.

Schubel, J. R. 1968. Turbidity maximum of the Northern Chesapeake Bay. *Science* 161:1013-1015.

Schubel, J. R. 1972. The physical and chemical conditions of Chesapeake Bay; an evaluation. The John Hopkins University. Chesapeake Bay Institute Spec. Rept. 21, Ref. 72-1, 73 p.

Seliger, H. H. and M. E. Loftus. 1974. Growth and dissipation of phytoplankton in Chesapeake Bay. II. A statistical analysis of phytoplankton standing crops in the Rhode and West Rivers and an adjacent section of the Chesapeake Bay. *Chesapeake Sci.* 15:185-204.

Strickland, J. D. H. and T. R. Parsons. 1972. *A Practical Handbook of Seawater Analysis.* Fish. Res. Board Canada, Bull. No. 167. Ottawa. 310 p.

Sverdrup, H. U., M. W. Johnson and R. H. Fleming. 1942. *Oceans, Their Physics, Chemistry and General Biology.* Prentice-Hall, Inc., Englewood Cliffs, NJ. 1087 p.

Utterback, C. L. 1936. Spectral bands of submarine solar radiation in the North Pacific and adjacent inshore waters. *Conseil Perm. Intern. l'Explor. Mer, Rapp. et Proc.-Verb.,* v. 101 pt. 2, no. 4, 15 p.

Whitney, L. V. 1938. Transmission of solar energy and the scattering produced by suspensoids in lake waters. *Trans. Wisc. Acad. Sci., Art Letters.* 31:201-213.

Whitney, L. V. 1941. The angular distribution of characteristic diffuse light in natural waters. *J. Mar. Res.* 4:122-131.

Williams, J. 1970. *Optical Properties of the Sea.* U.S. Naval Academy, Annapolis, MD. 123 p.

Williams, J. 1973. Optical properties of the ocean. Reports on Progress in Physics. Institute of Physics, London. 36:1567-1608.

CONTEMPORARY TECHNIQUES IN
ESTUARINE RESEARCH

MEASURING EFFECTS OF PREDATION ON BENTHIC COMMUNITIES IN SOFT SEDIMENTS

Robert W. Virnstein

Harbor Branch Foundation, Inc.
RR 1, Box 196
Fort Pierce, Florida

Abstract: Most studies on effects of predation on benthic communities in soft sediments have relied either on correlation and pattern analysis or, more recently, on the use of mesh cages to exclude potential or suspected predators from areas of the bottom. Because such cages usually produce other physical effects besides excluding predators, results from such studies are often difficult to interpret. I propose that a combination of techniques is mandatory. The approaches can be broadly classified as either non-manipulative or manipulative. The non-manipulative approaches include 1) observation; 2) correlation and pattern analysis, such as abundance and distribution of predators and prey; 3) gut content analysis; and 4) analysis of age class structure. None of these methods can demonstrate that predation causes certain effects, but they can show that predation does occur and they should be used to formulate hypotheses which can then be tested with manipulative techniques. Manipulative approaches have the advantage that both controls and replication are possible; they include 1) laboratory studies, 2) microcosm studies, and 3) field studies. Where they are possible and can be done correctly with controls and replication, field studies are preferred because their results relate directly to the natural system. The use of more than one technique is necessary since each single technique tells only a part of the story. No single approach is both controlled enough to be interpreted without question and natural enough to be extrapolated to the field.

Introduction

A central question in ecology has long been, what factors control the distribution and abundances of organisms? For communities in soft sediments, experimental evaluation of the role of predation has been investigated only recently, primarily using cages which can result in many problems (Virnstein 1978). Most prior studies relied primarily on correlation or pattern analysis. In this paper, I argue that a number of additional approaches, used in combination, are a much more powerful tool for measuring the effect of predation than is any single method.

General Approaches

In broad terms, the approach to determining the effects of predation can be either non-manipulative or manipulative. Non-manipulative approaches (inferences based on systematic sampling or observation) include strictly observational studies, gut analysis studies, size class analysis, and correlation, regression, and pattern analysis studies of predator-prey abundances. Such studies may show 1) that a predator does feed on a given prey or 2) that prey abundance decreases when predators are abundant. However, 1) this feeding may have an insignificant effect on the prey, and 2) the predator and prey may both be responding to the same external factor(s), rather than one being causally related to the other.

Manipulative approaches generally consist of altering predator densities in some manner and then testing for a response in the prey populations. Such studies may be easier to do in aquaria in the laboratory, but conditions are not natural. Field studies afford the most natural conditions but little control. This lack of control is both an advantage (Connell 1974, 1975) and disadvantage (Virnstein 1978). With respect to degree of control and naturalness, microcosm studies offer an approach intermediate to field and aquarium studies.

Specific Approaches

I. Non-manipulative Techniques

The first three approaches considered are non-manipulative, and are based on observations or systematic sampling. All should lead to the formulation of testable hypotheses.

Observation

Careful observation could possibly tell how, where, when, and perhaps on what a species is feeding. As such, it can be a powerful tool. Fiber optics and X rays may provide views within the sediments. A dissecting microscope mounted above an aquarium or dish of sediment can show activity and behavior patterns of small predators. Remote cameras in the field may provide important evidence of predator activity. However, for infauna of soft sediments in turbid estuaries, observational capabilities in the field are usually severely limited. Such information is not sufficient to determine the effect on the prey assemblage but could distinguish actual predation from physical disturbance, which other methods cannot do.

Gut Content Analysis

Analysis of the gut contents of suspected predators can confirm that a certain species of prey is actually eaten. However, interpretive caution must be exercised here, because many animals, if ingested whole, may pass live through the gut of a predator. If prey are damaged beyond recognition, serological techniques may be necessary to determine which species of prey have been eaten (Feller et al. 1979) and can provide quantitative data on amounts present. The use of $^{13}C/^{12}C$ ratios (Haines and Montague 1979), usually employed in relating trophic structure to primary producers, permits an evaluation of the relative quantitative importance of each prey type in the predator's diet.

The number or biomass of prey eaten per unit area per day can be determined from gut analyses if the following are known: the density of predators, percentage of foraging time in the habitat, and the residence time of food in the gut (see Peterson and Bradley 1978). By comparing this figure with prey densities, the percentage of the prey population eaten per day can be calculated. The inverse of this value would be equal to either 1) the time required to replace the prey population (turnover time), assuming adequate recruitment and constant population size, or 2) the length of time to eliminate the prey, assuming no recruitment. For example, if 2% of the prey are eaten per day, then the turnover time must be not more than 50 days in order to maintain the prey population.

Obviously, if a large percentage of the prey is removed in a short time (relative to turnover time), the effect will be significant. Such work has been done for a mesohaline Chesapeake Bay macrobenthic community fed on by spot, *Leiostomus xanthurus* (A. F. Holland, Martin Marietta Labs, Baltimore, MD pers. comm.), and for a continental slope community fed on by the red crab, *Geryon quinquedens,* and other predators (Farlow 1980). Because the benthos of estuaries is so variable, both spatially and temporally (Boesch et al. 1976; Coull and Fleeger 1977), I would stress that predators collected for gut analyses should ideally be collected from the same area (habitat) and during the same time period as the benthic prey population is sampled.

Analysis of Age Class Structure

Assuming that age can be estimated (e.g., from size or from growth rings or other clues) and that growth rates are known, the total mortality rate for a size class within a time interval can be calculated as the difference between the actual numbers at the end of the time interval and the potential numbers in the next size class if there had been 100% survival. If high mortality is found in size classes well below the maximum size and cannot be attributed to other sources, perhaps a large portion of this mortality can be attributed to predation. However, such a method depends on a number of assumptions and recognition of cohorts. By itself, this method has little value, and would require some supporting information on actual rates of predation.

Correlation Analysis

Correlation analyses (and regression, pattern analysis, etc.) are powerful tools that can show statistically significant relationships between predator and prey abundances (Arntz 1980; Livingston 1980). They may imply (but do not and cannot demonstrate) cause for the relationship, since predator and prey may both simply be responding to the same external factor, such as temperature. It is critically important that sampling be on an appropriate time scale, i.e., shorter than the life cycle of the prey; otherwise, results of predation may be confused with effects of competition or with recruitment events. Lag times may have to be considered.

The functions of non-manipulative and manipulative approaches are complimentary and non-overlapping. Associations and relationships derived from non-manipulative techniques should be used to formulate hypotheses which can then be tested by manipulative techniques. Since the manipulative techniques are dependent on the prior formulation of testable hypotheses, one should not, for example, simply set out a cage without any prior knowledge or hypotheses as to what is happening.

II. Manipulative Techniques

The following three approaches use manipulative as opposed to observational or sampling techniques and represent a progression from most controlled and least natural to least controlled but most natural situations. In all cases, the density of predators is somehow artificially manipulated and then the response of the prey is determined (Table 1). Additionally, the investigator can act as a predator, removing selected animals. Any one of the manipulative approaches can be used in combination with any of the response measurements (see below). A major advantage of manipulative approaches is that a control and replication are possible and are always necessary.

Laboratory Aquarium Studies

Studies in aquaria in the laboratory provide for the tightest control of experimental conditions. In the simplest approach, prey are added to an aquarium and then a predator is added and allowed to feed for some period of time. From such a study, one could determine whether and at

Table 1. A summary of approaches for measuring the effect of predation.

Approaches	Factors Responding
Non-manipulative	
Observation	
Gut analysis	
Age-class structure	— —
Correlation analysis	
Manipulative	
Laboratory studies	Density
Microcosm studies	Distribution
Field studies	Reproduction
	Growth

what maximum rate a given predator could consume prey. However, such a one prey/one predator system is artificial and unnatural, especially for soft-sediment infaunal communities. The results obtained probably cannot be extrapolated to the field. Such studies can be useful for determining predator preferences for certain species or for sex or size of given species. Additionally, observation becomes an available tool.

Microcosm Studies

A microcosm, by retaining the whole assemblage of animals along with associated sediment and other physical structures (e.g., seagrass, animal tubes and burrows), is a much more natural system, and results can be extrapolated more validly to the field. The manipulation of predator densities is still quite controllable, but prey densities are more difficult, although not impossible, to manipulate.

Such multispecies assemblages may be set up by placing undisturbed cores of sediment from the field into tanks, or by putting sediment into tanks with running unfiltered seawater and then adding prey and/or allowing recruitment to occur naturally. Microcosm studies may represent the next-best alternative when field studies are too complicated to interpret (Bell and Coull 1978; Virnstein 1978).

Field Studies

Field experimental studies (primarily caging, but other methods may work, depending on the character of the predator) have the advantage of retaining the system in an essentially natural state and letting all factors but the one under control vary naturally in both the control and experimental treatments. Assuming adequate controls are provided, problems of caging studies (see Virnstein 1978) are primarily related to changes in current structure above the sediment (Hulberg and Oliver 1980) and the presence of a physical structure, and their effects on recruitment and attraction of animals. In seagrass beds, for example, exclusion of decapod crustacean predators for any but a very short term seems nearly impossible, due both to recruitment and subsequent fast growth, and to attraction to the cage structure (unpubl. data; R. J. Orth, Virginia Institute of Marine Science; K. L. Heck, Benedict Estuarine Lab, MD; W. G. Nelson, University of Bergen, Norway; and S. A. Woodin, Johns Hopkins University, MD, all pers. comm.). Addition of predators is easier than exclusion, but densities above normal may not reflect normal activities or food selection of predators. This method cannot show that predators are normally important.

If regular seasonal fluctuations of prey occur and one wishes to test whether these are due to predators, the best time to set out a cage would be just before the decline in prey abundance. In this way, if predators are producing the decline, then prey abundance should not decline inside the cage, but would outside the cage. To set out a cage at other times and expect an increase in prey density inside the cage requires not only that predators be excluded and are important, but also that recruitment of prey occur during the time that the cage is in place.

To separate effects of differential recruitment from predation, reopening (removing) a cage may provide an answer. The abundance of predators at this time is important in determining the results.

What to Measure

Once predation intensity is manipulated, whether in the laboratory, microcosm, or field, the responses of the soft-sediment community can be measured (Table 1). Most commonly, changes in adult population density are measured; this is the most immediate effect of the mortality. If size selection is important, the size-class structure of a prey population can be altered. Changes in the spatial distribution of prey, both horizontally (Vargas 1979; N. K. Mountford, Chesapeake Biological Lab, MD pers. comm.) and vertically, could occur in response to predation. Predators could cause prey to shift to more cryptic or other microhabitats where predation is less effective, such as at the base of rooted plants.

Although I am considering predation here to be the eating of whole animals, non-mortality effects due to the loss of body parts (e.g., siphons, palps, lophophores, etc.), whether by natural or artificial (by the investigator) predation, may also be important. These effects could be measured as changes (presumably decreases) in reproductive output and growth rate (see Peterson 1980) due to the increased energy expenditure necessary to replace lost tissues.

Other indirect effects might be changes in recruitment success due either to decreased reproductive output, or to changes in adult population density. However, such changes are probably not measurable for species with planktonic larvae (this does not include most meiofauna). Although only adult populations are usually censused, effects of predator activity on larval or juvenile recruitment may be even more important. Harder to detect would be changes in prey populations that occur as changes in the timing of events, e.g. seasonal abundance peaks and time of spawning.

Some Examples and Problems of Applicability

Because field methods lack control and laboratory methods lack naturalness, some combination of methods is necessary for a proper balance of control versus naturalness. I doubt that any one method is both controlled enough to be interpreted without question and natural enough to be applicable to the field situation. For example, Nelson (1979; pers. comm.) presents a convincing story that predation by pinfish (*Lagodon rhomboides*) on amphipods in seagrass beds in North Carolina and Florida has a significant effect, but only as a result of a combination of types of evidence. Amphipod abundance decreases soon after pinfish become abundant and increases again after pinfish leave the area. Pinfish do eat amphipods presented as food items in aquaria and will significantly reduce amphipod numbers in microcosm experiments. Additionally, field-collected pinfish had large numbers of amphipods in their stomachs. In field caging experiments, the results were muddled because pinfish also fed on decapod crustaceans which in turn feed on amphipods.

As another example of the importance of predators on macrofauna and vice versa, Holland et al. (1980) and Virnstein (1977) found that infaunal densities of macrobenthos in Chesapeake Bay decrease and stay depressed while spot *(Leiostomus xanthurus)* are abundant and that spot do feed on the infauna in the areas studied. Based on daily food rations (calculated from growth rates and efficiency estimates) and density of spot (Holland pers. comm.), the amount of infauna consumed by spot accounts for 95% of the seasonal decline in infaunal biomass. I have shown that infaunal densities increase inside cages where spot are excluded and decrease in cages where spot are included (Virnstein 1979). The most adundant species in spot stomachs were the same species that showed large density increases inside cages. Taken together, the above evidence strongly suggests that predation by spot does have a significant effect on infaunal abundance. I must stress, however, that any single line of evidence by itself is weak.

The presumption in much of the above discussion has been that the predators on the infauna are animals that are much larger than most infauna and that are motile and epibenthic, such as fishes, portunid crabs, gastropods, and starfishes. Recent work by Reise (1977) in the Wadden Sea and by Nelson and Virnstein (unpub. data) in seagrass beds in the Indian River, Florida, has demonstrated the importance of predation by small decapod crustaceans, including penaid and palaemonid shrimps. Holland (pers. comm.) and J. A. Commito (University of Maine, Machias, ME pers. comm.) both found that species of the polychaete *Nereis* could significantly reduce densities of some infauna. Holland worked with *N. succinea* in microcosms; Commito worked with *N. virens* in containers of sediment in the field. No doubt there are also numerous other infaunal animals that are themselves predators on other infauna, e.g. glycerid polychaetes, nemerteans, and turbellarians. Polychaetes from numerous families have a rather formidable jaw apparatus, and although many of these species are found to have mainly sediment or detritus in their guts, there seems little reason to doubt that they could supplement their diet with an occasional animal. High rates of biological activity at the sediment-water interface (Rhoads 1974; Black 1980) must make it difficult (Woodin 1976), or even unlikely, that a newly settled larva can pass through this barrier without being eaten or killed by burial, etc. Indeed, this initial settlement and establishment in the sediment may be the most critical stage of recruitment.

Although all the above statements apply to and are based on studies of macrofauna, I see no reason that any of the statements and procedures could not also be applied to the meiofauna—the latter are just smaller and their predators are smaller. However, meiofauna are not easily manipulated, sediment parameters are critical, and they have different life history patterns and reproductive rates (Coull and Bell 1979). Numerous species of fish, including spot and gobies (Buzas and Carle 1979) have been collected with meiofauna in their guts. Bell and Coull (1978, 1980) have shown that predation by palaemonid shrimp has a significant regulatory effect on the density and population dynamics of meiofauna in the high marsh where macroinfaunal density was extremely low (Bell 1980).

Many macrobenthic deposit feeders that feed by selectively or non-selectively eating sediments undoubtedly ingest large numbers of meiofauna. Where high densities of macrofaunal surface deposit feeders occur, the entire upper few centimeters of sediment might be reworked (i.e., passed through an animal's gut) every few months (Rhoads 1974; Black 1980). Since the life cycles of many meiofauna are on this order of time, the meiofauna could be heavily cropped, but the rapid turnover rates and the ability to pass through guts live could still maintain relatively high population densities. From the above recent evidence, it appears likely that the meiofauna act not just as remineralizers (Tenore et al. 1977) and do not represent a dead-end closed food web (Heip and Smol 1975), but rather are intimately linked with the macroinfauna and motile epibenthic predators (see Coull and Bell 1979). The separation of the meiofauna from macrofaunal studies may be not only artificial but also unrealistic and invalid. The evaluation of all such predator-prey interactions is of utmost importance for describing and understanding an ecosystem.

Acknowledgments

I thank S. S. Bell, H. Lee, and C. H. Peterson for thoughtful and thorough reviews of the manuscript. J. A. Commito, K. L. Heck, H. Lee, N. K. Mountford, R. J. Orth, and S. A. Woodin offered comments and/or unpublished data.

Contribution No. 185 of the Harbor Branch Foundation.

References Cited

Arntz, W. E. 1980. Predation by demersal fish and its impact on the dynamics of macrobenthos. *In:* K. R. Tenore and B. C. Coull (eds.), *Marine Benthic Dynamics.* U. South Carolina Press, Columbia, SC (in press).

Bell, S. S. 1980. Meiofauna-macrofauna interactions in a high salt marsh habitat. *Ecol. Monogr.* (in press).

Bell, S. S. and B. C. Coull. 1978. Field evidence that shrimp predation regulates meiofauna. *Oecologia* 35:141-148.

Bell, S. S. and B. C. Coull. 1980. Experimental evidence and a model for juvenile macrofauna-meiofauna interactions. *In:* K. R. Tenore and B. C. Coull (eds.), *Marine Benthic Dynamics.* U. South Carolina Press, Columbia, SC (in press).

Black, L. F. 1980. The biodeposition cycle of a surface deposit-feeding bivalve, *Macoma balthica* (L.), pp. 389-402. *In:* V. S. Kennedy (ed.), *Estuarine Perspectives*. Academic Press, New York.

Boesch, D. F., M. L. Wass and R. W. Virnstein. 1976. The dynamics of estuarine benthic communities, pp. 177-196. *In:* M. L. Wiley (ed.), *Estuarine Processes, Vol. 1*. Academic Press, New York.

Buzas, M. A. and K. J. Carle. 1979. Predators of foraminifera in the Indian River, Florida. *J. Foram. Res.* 9:336-340.

Connell, J. H. 1974. Ecology: Field experiments in marine ecology, pp. 21-54. *In:* R. N. Mariscal (ed.), *Experimental Marine Biology*. Academic Press, New York.

Connell, J. H. 1975. Some mechanisms producing structure in natural communities: a model and evidence from field experiments, pp. 460-490. *In:* M. L. Cody and J. M. Diamond (eds.), *Ecology and Evolution of Communities*. Harvard U. Press, Cambridge, MA.

Coull, B. C. and S. S. Bell. 1979. Perspectives on marine meiofaunal ecology, pp. 189-216. *In:* R. J. Livingston (ed.), *Ecological Processes in Coastal and Marine Systems*. Plenum Publishing Corp., New York.

Coull, B. C. and J. W. Fleeger. 1977. Long-term temporal variation and community dynamics of meoibenthic copepods. *Ecology* 58:1136-1143.

Farlow, J. O. 1980. Natural history and ecology of a megabenthic invertebrate-demersal fish assemblage from the upper continental slope off southern New England. Ph.D. dissertation, Yale Univ., New Haven, CT.

Feller, R. J., G. L. Taghon, E. D. Gallagher, G. E. Kenny and P. A. Jumars. 1979. Immunological methods for food web analysis in a soft-bottom benthic community. *Mar. Biol.* 54:61-74.

Haines, E. B. and C. L. Montague. 1979. Food sources of estuarine invertebrates analyzed using $^{13}C/^{12}C$ ratios. *Ecology* 60:48-56.

Heip, C. and N. Smol. 1975. On the importance of *Protohydra leuckarti* as a predator of meiobenthic populations. *Tenth European Symp. Mar. Biol., Ostend* 2:285-296.

Holland, A. F., N. K. Mountfort, M. H. Hiegel, K. R. Kaumeyer and J. A. Mihursky. 1980. The influence of predation on infaunal abundance in upper Chesapeake Bay. *Mar. Biol.* (in press).

Hulberg, L. W. and J. S. Oliver. 1980. Caging manipulation in marine soft-bottom communities: the importance of animal interactions or sedimentary habitat modifications. *Can. J. Fish. Aquatic Sci.* 37:1130-1139.

Livingston, R. J. 1980. Ontogenetic trophic relationships and stress in a coastal seagrass system in Florida, pp. 423-435. *In:* V. S. Kennedy (ed.), *Estuarine Perspectives*. Academic Press, New York.

Nelson, W. G. 1979. Experimental studies of selective predation on amphipods: Consequences for amphipod distribution and abundance. *J. Exp. Mar. Biol. Ecol.* 38:225-245.

Peterson, C. H. 1980. Approaches to the study of competition in benthic communities in soft sediments, pp. 291-302. *In:* V. S. Kennedy (ed.), *Estuarine Perspectives*. Academic Press, New York.

Peterson, C. H. and B. P. Bradley. 1978. Estimating the diet of a sluggish predator from field observations. *J. Fish Res. Board Can.* 35:136-141.

Reise, K. 1977. Experiments on epibenthic predation in the Wadden Sea. *Helgoländer wiss. Meeresunters.* 31:55-101.

Rhoads, D. C. 1974. Organism-sediment relations on the muddy sea floor. *Oceanogr. Mar. Biol. Ann. Rev.* 12:263-300.

Tenore, K. R., J. H. Tietjen and J. J. Lee. 1977. Effect of meiofauna on incorporation of aged eelgrass, *Zostera marina*, detritus by the polychaete *Nephthys incisa*. *J. Fish. Board Can.* 34:563-567.

Vargas, J. A. 1979. Predation and community structure of soft-bottom benthos in Rehoboth Bay, Delaware. M.S. thesis, U. Delaware, Newark, DE.

Virnstein, R. W. 1977. The importance of predation by crabs and fishes on benthic infauna in Chesapeake Bay. *Ecology* 58:1199-1217.

Virnstein, R. W. 1978. Predator caging experiments in soft sediments: Caution advised, pp. 261-273. *In:* M. L. Wiley (ed.), *Estuarine Interactions.* Academic Press, New York.

Virnstein, R. W. 1979. Predation on estuarine infauna: Response patterns of component species. *Estuaries* 2:69-86.

Woodin, S. A. 1976. Adult-larval interactions in dense infaunal assemblages: Patterns of abundance. *J. Mar. Res.* 34:25-41.

APPROACHES TO THE STUDY OF COMPETITION IN BENTHIC COMMUNITIES IN SOFT SEDIMENTS

Charles H. Peterson

University of North Carolina at Chapel Hill
Institute of Marine Sciences
Morehead City, North Carolina

Abstract: A review of methodologies employed by benthic ecologists to investigate the importance of interspecific competition in soft-sediment communities reveals use of "natural experiments" (correlative approaches), manipulative experimentation in the field, and manipulative experimentation in laboratory microcosms. All three approaches have been applied in studying each of the major competitive processes potentially operating in infaunal communities: 1) direct interference competition for space, 2) exploitative competition for food, 3) adult interference with larval settlement, and 4) indirect interference through alteration of the physical environment ("trophic group amensalism"). Soft-bottom systems are usually difficult to manipulate in the field or laboratory because infauna cannot be counted or removed without excavation. Such excavation destroys sedimentary structures and selectively eliminates many delicate taxa, thus limiting any subsequent experiment to an impoverished biota in an unnatural environment. An alternative to excavation is addition of food, potential competitors, or surrogate organisms to undisturbed bottom plots. However, this approach is frequently inapplicable because of high spatial heterogeneity in soft-bottom communities, which makes difficult the initial estimation of infaunal densities in experimental plots and necessitates extreme treatments and/or excessive replication to achieve statistically significant results. Nonetheless, because infaunal animals do not require firm attachment to a substratum, they can be manipulated successfully in a variety of ways. When viewed in the broader context of all past work on community-level effects of competition ("diffuse competition") in terrestrial and aquatic systems, results of experimental manipulation on suites of benthic infauna suggest that competitive effects may be additive only where a single type of competitive mechanism prevails.

Introduction

I plan to review briefly the methods and approaches currently being used by ecologists to assess the significance of interspecific competition in the organization of estuarine and marine benthic communities in soft sediments. To provide structure to this review of methodologies, I treat separately each major type of competitive interaction thought to play a role in some soft-sediment system. Because two very recent reviews (Peterson 1979; Dayton and Oliver 1980) provide critical interpretations of results from studies of the significance of biological interactions in organizing soft-sediment benthic communities, I consider only the methodologies of such studies without attempting any evaluation of their results and conclusions.

The basic approach to the study of interspecific competition is the same for any ecological system. One looks to vary either the density of a presumed competitor or the abundance of a potentially limited resource. Only rarely are both types of factors manipulated simultaneously, although such experiments promise substantial reward for the added effort and complexity. To evaluate the importance of competition, one examines the effects of the manipulation on 1) growth, 2) mortality, 3) recruitment, 4) migration, or 5) reproductive effort. Each of these parameters relates loosely or more directly to competitive success (fitness under a variety of competitive regimes).

Most studies of interspecific competition involve only two or perhaps three competing species. Yet knowledge of the impact of all competitors is clearly necessary for accurately predicting whether a given species can invade a community and successfully co-exist there. Diffuse competition is the term used (MacArthur 1972) to indicate these cumulative effects of all potentially competing species in a given system. Because of the potential and realized complexity of multi-species interactions, diffuse competition can most easily be assessed wherever the effects of interacting species are additive (wherever second- and higher-order interaction terms are insignificant). Additivity implies that the impact of diffuse competition can be evaluated by a reasonably small number of treatments. Because benthic ecologists rarely mention diffuse competition despite its possible significance, I conclude this review by relating past work on competition in soft sediments to the broader body of ecological approaches to multi-species competition. Specifically, I use results from terrestrial and aquatic literature to help address the question of when one might expect the effects of interacting species to be additive in soft-sediment systems.

Methods of Assessing the Importance of Interspecific Competition among Infauna

(1) Natural Experiments

In a "natural experiment", one makes opportunistic use of the variation present in nature to provide a test of an hypothesis. For example, Sanders (1968, 1969) examined the species diversity of infaunal polychaetes and bivalves at different ocean depths to test whether increased environmental stability, such as characterizes deeper sea bottoms, produces predictable changes in the community structure of benthic infauna. Ordinarily, this usage of nature's variability to provide "built-in" experiments must be accompanied by the application of a conceptual or mathematical model. The model provides a focus by identifying (1) the factors which must be measured to ensure that an experiment has taken place, and (2) the parameters needed to test its impact.

Natural experiments have been used widely in combination with conceptual models to evaluate the role of interspecific competition in estuarine and marine soft sediments. Each major type of competitive interaction has been studied on several occasions by means of natural experiments (Table 1). For the purposes of this review, I recognize four types of competitive (in a broad sense) interactions in soft-sediment systems: 1) direct interference competition for space, 2) exploitative competition for food, 3) adult interference with larval settlement, and 4) indirect interference through alteration of the physical environment ("trophic group amensalism"). Direct interference competition for space has been invoked to explain complementary and largely non-overlapping distributions of various pairs of infaunal macroinvertebrates: *Mya arenaria* and *Gemma gemma* (Bradley and Cooke 1959; Sanders et al. 1962); *Macoma nasuta* and *Macoma secta* (Vassallo 1971). Exploitative competition for food is probably the best explanation for the observed character displacement in the sizes of the mud snails *Hydrobia ulvae* and *Hydrobia ventrosa* wherever they are found sympatrically (Fenchel 1975a, b). Woodin (1976) has interpreted the prevalence of monotypic functional groupings in areas of high densities of adult infauna as evidence for adult interference with subsequent larval settlement. Natural experiments also have provided the basis for the pioneering work on the importance of indirect interference through alteration of sedimentary characteristics. Rhoads and Young (1970) noticed that high abundances of deposit feeders in Long Island Sound were associated both with fine sediments of high organic content and also with relatively low densities of suspension feeders. They suggested that the complementary distributions of suspension feeders and deposit feeders were a consequence of the high mobility and sediment reworking activity of the deposit feeders, which increase sediment instability and ultimately help to exclude suspension feeders by clogging their filtering apparatus.

It seems apparent from this brief review and from the lists in Table 1 that natural experiments have provided most of the data upon which are based our current notions of the importance of interspecific competition in soft-sediment environments. The major drawback to the use of natural experiments is the difficulty in finding proper controls for the natural treatments (Connell 1974). Many factors are often correlated so that the effect of varying the treatment variable is frequently confounded by contributions from other variables which are not held constant. In other words, nature does not do all possible manipulations so that, in order to utilize a natural experiment, one must ordinarily be content with a partial test of an hypothesis. Natural experiments, therefore, range widely in quality as a function of the strength of the test provided by nature's variability and of the cleverness of the scientist who is looking to exploit that variability.

A second problem in using natural experiments to test hypotheses is closely related to this lack of complete controls. The results of a natural experiment do not normally speak unambiguously to the question of mechanism. This difficulty is best circumvented by the use of supplementary observations and experiments to confirm the presumed nature of the

Table 1. Studies of interspecific competition in estuarine and marine soft-sediment environments, grouped simultaneously by experimental approach and by the mechanism of competitive interaction. A question mark indicates that the interaction mechanism is not clearly and unambiguously established.

Mechanism	Natural Experiments	Laboratory Experiments	Field Experiments
Direct interference competition for space	Bradley and Cooke 1959 (?) Sanders et al. 1962 (?) Vassallo 1971 Levinton and Bambach 1975 Peterson 1977 Levinton 1977 Reise 1978 (?)	Woodin 1974 Rees 1975 (?) Levinton 1977	Woodin 1974 (?) Ronan 1975 Peterson and Andre 1980
Exploitative competition for food	Sanders 1968 (?) Sanders 1969 (?) Fenchel 1975a Fenchel 1975b Whitlatch 1976	Fenchel and Kofoed 1976 Lopez and Levinton 1978	Young and Young 1978 various caging studies reviewed by Virnstein 1978 and Peterson 1979
Adult interference with larval settlement	Segerstråle 1962 Woodin 1976 Reise 1978 (?)	Segerstråle 1962	Peterson 1977 (?) Oliver 1979 Bell and Coull 1980 (?)
Indirect interference through alteration of the physical environment	Rhoads and Young 1970 Rhoads and Young 1971 Aller and Dodge 1974 Levinton and Bambach 1975 Levinton 1977 Myers 1977 (?)	Brenchley 1978	Rhoads and Young 1970

interactions thought responsible for the observed natural pattern. Of course, such confirmation is impossible in cases where the experiment has already been completed at some time in the past and the driving force is no longer active.

(2) Laboratory Experiments

Species manipulations under laboratory conditions have contributed much less than natural experiments to the literature on interspecific competition in marine infaunal communities (Table 1). While laboratory experiments possess the obviously advantageous characteristic that they can be carefully controlled, high spatial variability, a common feature of many marine infaunal systems, generally inhibits widespread use of laboratory manipulation. Such high local variability makes it extremely difficult to excavate sufficiently similar sections of the bottom with which to stock the necessary experimental and control aquaria. An alternative to this procedure of establishing relatively undisturbed, replicate microcosms to serve as experimental substrata is to purge the sediments of all benthic fauna (by freezing, drying, etc.). Animals can then be added in equal numbers to each replicate aquarium. This process, however, completely disrupts the natural (including biogenic) sedimentary structures and necessarily eliminates from the experiment all the microbiota, meiofauna, and delicate macrofauna that cannot be handled individually. Thus, experiments done in such reconstituted systems occur under unnatural conditions with an impoverished biota.

Although such problems have been responsible for the general paucity of laboratory studies on the role of interspecific competition in marine infaunal systems, the laboratory is the ideal site for study of interaction mechanisms and provides an acceptable means of testing various hypotheses that do not require the presence of a complete infaunal community. In the laboratory, animals can actually be observed by use of small glass aquaria or inverted microscopes so that interaction mechanisms can be tested visually. Basic natural history information of this sort is largely unavailable for infaunal organisms, which live buried and obscured from view in their natural environment. Consequently, laboratory experiments can perhaps best be used in combination with natural or field experimentation to provide confirmation of the presumed mechanisms of interaction. Because of the careful controls and increased viewing potential in the laboratory, this coordinated approach can provide powerful tests of hypotheses by combining the best attributes of both techniques.

Although laboratory studies of interspecific competition are sparse in the literature, all of the four major types of competitive interaction have been examined at least once in the laboratory (Table 1). Levinton (1977) observed direct interference competition among three Protobranch bivalves, *Nucula proxima, Solemya velum,* and *Yoldia limatula,* in laboratory aquaria. Fenchel and Kofoed (1976) carried out elegant laboratory experiments to test for exploitative competition for food between species of

Hydrobia. In the only published laboratory test of the effects of adult organisms on larval settlement in soft sediments, Segerstråle (1962) measured the survivorship of larvae of *Mytilus edulis* with and without adult amphipods of the genus *Pontoporeia.* Oliver (1979) tested for the effect of some predatory polychaetes on the survivorship of week-old post-larval infauna in the laboratory. Only one completed study (Brenchley 1978) has used the laboratory to examine the importance of indirect interference competition through sediment alteration.

(3) Field Experiments

The history of using field experimentation to study competition in estuarine and marine soft-sediment environments is a very short one dating only from about 1970. As Connell (1974) and Dayton and Oliver (1980) observe, field experiments differ from laboratory experiments in that, in the field, all factors vary uncontrollably except the one that is being manipulated, whereas in the laboratory all factors save one are usually held constant. Despite this basic difference, field experiments in soft-sediment environments share their most serious limitations with laboratory experiments. If one wishes to initiate field experiments in soft sediments, one must either 1) excavate an area and use freezing, drying, sieving, or shocking to remove all animals, returning only those which are wanted and are large and robust enough to be handled, or else 2) manipulate an undisturbed plot with unknown initial faunal constitution. As in the laboratory, excavation and "defaunation" in the field destroy important sedimentary structures and eliminate the microbiota and delicate macrofauna which may be of critical importance. Consequently, manipulation of undisturbed plots is ordinarily preferable. Sampling in areas nearby can serve to estimate the initial faunal composition of the study site; however, when spatial variability is large, as it often is in infaunal communities, a large measure of uncertainty about the initial faunal composition is inevitable. Such uncertainty precludes the demonstration of subtle differences between treatments and controls and thus limits the usefulness of this sort of experimentation in the field as well as in the laboratory.

Because of the usual desire to maintain the experimental system in a relatively undisturbed state, only a limited number of experimental manipulations are possible in soft-sediment communities. In studying competitive interactions in the field (or in the laboratory) in undisturbed systems, one is limited to (1) adding individuals of various manipulatable infaunal species (ordinarily presumed competitors), (2) adding food resources, (3) adding surrogate animals (as space occupiers or structural elements), or (4) excluding various epibenthic predators. Except under extremely rare circumstances (e.g., Woodin 1974), removal of selected infaunal groups or species is impossible without causing substantial disruption of the natural system. The usual manipulative technique in a field study of competition on marine hard substrata is removal and continued exclusion of a predator or of a competitor (Connell 1974). Thus, to test for the importance of a competitor in soft substrata its density is increased, whereas on rocky shores its density is reduced. Because in either case one is testing the influence of a possible competitor by varying its abundance, one might expect the experimental conclusions to be independent of the choice of manipulative direction (increase or decrease). However, this is not invariably true for two reasons. First, increasing the abundance of a successful competitor will necessarily fail to demonstrate its impact upon the poorer competitors if, at natural "control" densities, the poorer competitors are virtually excluded from the community. Second, high spatial variability in natural infaunal community composition may require the application of a radical treatment in order to render the results statistically significant. If such several-fold increases in the density of a presumed competitor represent levels that are never achieved in nature, the experimental treatment may well elicit responses that are unrelated to the usual competitive mechanisms which the experiment is designed to explore. Such high-density artifacts are a real risk in soft-sediment benthic experimentation where many animals can so greatly modify their own micro-environment.

Although only a few field experiments have been done to explore the role of interspecific competition in soft-sediment systems, at least one study has been published on each major type of competitive interaction. For instance, Peterson and Andre (1980) demonstrate interference competition between suspension-feeding bivalves which share the same living depths in the sediments of a California lagoon. Exploitative competition for food has been examined through field additions of organic material by Young and Young (1978). Similar food additions are the inadvertent outcome of several caging experiments (reviewed by Virnstein 1978; Peterson 1979) designed to measure the importance of predators; however, these experiments are flawed by the confounding contributions of numerous artifacts including especially the increased organic deposition inside cages. Oliver (1979) has tested the importance of adult infauna on larval settlement in trays of sediments suspended near the bottom in Monterey Bay. The only published attempt to test in the field the importance of indirect interference competition through the mechanism of sedimentary alteration is Rhoads and Young's (1970) experimental measurement of *Mercenaria mercenaria* growth in trays at varying heights off a bottom where deposit feeders were abundant.

Evaluating Diffuse Competition

Interspecific competition is thought to be the dominant organizing feature of some communities. I refer, in particular, to communities of forest birds (MacArthur 1958), grassland birds (Cody 1974), desert lizards (Pianka 1975), and desert mammals (Rosenzweig and Sterner 1970; Brown 1975). Competitive pressures can determine how many species can co-exist in a given system, how similar co-existing species can be, and whether a given community can be successfully invaded by a particular potential colonizing species. For example, MacArthur and Wilson (1967) observed that although the bananaquit is the most abundant fruit- and berry-eating bird on virtually all the Caribbean islands, it is absent from Cuba. Suitable fruits and berries occur on Cuba, and it is likely that bananaquits have at times invaded the island. The most reasonable explanation for the lack of bananaquits on Cuba is that they are excluded competitively. However, since there is no single ecological analogue of the bananaquit on Cuba, competitive exclusion is probably a consequence of diffuse competition, arising from the joint competitive pressures of several fruit- and berry-eating birds (MacArthur and Wilson 1967).

Competitive pressures at the community level have often been evaluated by comparing natural systems with predictions arising from various mathematical or empirical models. MacArthur's (1957) "broken-stick" model, Hutchinson's (1959) "2.3 rule", and various niche width and overlap models (e.g., MacArthur and Levins 1967; MacArthur 1972; Horn and MacArthur 1972; May 1973; Cody 1974; Pulliam 1975) have all served to assess the importance of competition at a community level. These largely theoretical approaches to understanding diffuse competition assume either that species can be ordered along a single resource axis or that resource dimensions can be combined through an additive or multiplicative manipulation. These approaches also assume the effects of species to be additive, as in a classic Lotka-Volterra representation of the mathematics of competing species. Given that controlled, experimental tests have demonstrated that species interactions are not simply additive in various aquatic systems (Hairston et al. 1969; Wilbur 1972), the impressive success of the additive models of diffuse competition in explaining various vertebrate distribution patterns (MacArthur 1958; Pulliam 1975; Brown 1975) seems somewhat surprising. Perhaps an additive model of diffuse competition is appropriate only in systems where a single resource is limiting and where species do not employ a complex repertoire of interactions.

In marine infaunal communities, potential negative interactions are quite varied, representing at least four major types (Table 1). It thus seems reasonable to question seriously whether species interactions are likely to be additive in soft-sediment systems. I hypothesize that in systems where only one type of competitive interaction predominates, additivity of interactive effects will prevail. A partial test of this hypothesis in marine soft sediments comes from Brenchley's (1978) laboratory experiments. She demonstrated by adding sediments to aquaria which contained a relatively unaltered sample of the bottom community that sedimentation was a potential source of mortality for many infaunal species. She then tested how much each of three active burrowers *(Callianassa californiensis, Upogebia pugettensis,* and *Abarenicola pacifica)* contributed independently to the sedimentation (resuspension) rate by suspending sediment traps in the waters of aquaria containing monocultures of each species. The joint effect of all species tested together in a single aquarium was equal to the resuspension rate that would be predicted by the sum of each independent effect. In other experiments, sediment stabilizers (sedentary suspension feeders) as well as sediment destabilizers (mobile deposit feeders) seemed to have additive effects on a community's total resuspension rate. Whether infaunal mortality is likewise linearly related to resuspension rate is not clear.

In Brenchley's (1978) experiments, the only interaction examined was indirect interference through sediment alteration. That effects on resuspension rates were additive provides some support for my suggestion that additivity will characterize diffuse competition that arises from a single interaction mechanism. Because sediment suspension is a threshold phenomenon, it seems unlikely that Brenchley's conclusions would apply over all ranges of infaunal densities. The only other test of additivity involves a system of interacting suspension-feeding bivalves which also appear to be competing by a single mechanism, i.e. interference competition for space (Peterson and Andre 1980). As in Brenchley's (1978) study, the effects were additive, although Peterson and Andre's (1980) test of additivity was not as strong as Brenchley's (1978) because the additive joint effect was composed of one significant, non-zero component and one null component. The bivalve, *Protothaca staminea,* had no significant effect on growth in *Sanguinolaria nuttallii* and, when added to bottom plots which contained a species which did reduce *Sanguinolaria's* growth, *Protothaca* again did not change the amount of that reduction. No test of additivity has been attempted in a system where multiple types of competitive interactions were operating. Such an experiment in Levinton's (1977) deposit-feeding community would provide valuable data because direct interference, competition for food, and sediment reworking are all thought to be important in this system.

An important question remains to be addressed. If, as I suggest, additivity characterizes only those competitive interactions in which species employ but a single interactive mechanism, then one would like to know how common such simple systems are in marine soft sediments. Most studies of benthic infauna focus upon a single interactive mechanism (Table 1). However, such narrowness is not necessarily a feature of most estuarine and marine systems but instead may be a reflection of our need to choose and understand the simple, first-order phenomena before we tackle true complexity. Nonetheless, in any given marine infaunal community only one of the four common mechanisms often seems to predominate whenever competitive interactions appear to play an important role. Perhaps benthic soft-sediment communities generally share the characteristic simplicity of some terrestrial vertebrate systems. Because marine and estuarine infauna can be manipulated experimentally, this benthic system may provide even more fertile ground for future tests of the nature and importance of diffuse competition.

Acknowledgments

I appreciate comments on this manuscript from J. A. Commito, J. H. Hunt, V. S. Kennedy, M. C. Watzin, and two anonymous reviewers. G. R. Lopez provided useful suggestions and W. G. Ambrose, Jr. compelled me to think more clearly about problems of experimental design in soft sediments.

References Cited

Aller, R. C. and R. E. Dodge. 1974. Animal-sediment relations in a tropical lagoon Discovery Bay, Jamaica. *J. Mar. Res.* 32:209-232.

Bell, S. S. and B. C. Coull. 1980. Experimental evidence for a model of juvenile macrofauna — adult meiofauna interactions. *In:* K. R. Tenore and B. C. Coull (eds.), *Marine Benthic Dynamics.* Univ. of South Carolina Press, Columbia, S.C. (in press).

Bradley, W. H. and P. Cooke. 1959. Living and ancient populations of the clam *Gemma gemma* in a Maine coast tidal flat. *Fish. Bull.* 58:305-334.

Brenchley, G. A. 1978. On the regulation of marine infaunal organisms at the morphological level: the interactions between sediment stabilizers, destabilizers, and their sedimentary environment. Ph.D. Thesis, The Johns Hopkins Univ., Baltimore, Md. 265 pp.

Brown, J. H. 1975. Geographical ecology of desert rodents, pp. 315-341. *In:* M. L. Cody and J. M. Diamond (eds.), *Ecology and Evolution of Communities.* Belknap Press, Cambridge, Massachusetts.

Cody, M. L. 1974. *Competition and the Structure of Bird Communities.* Princeton Univ. Press, Princeton, New Jersey. 318 pp.

Connell, J. H. 1974. Field experiments in marine ecology, pp. 21-54. *In:* R. Mariscal (ed.), *Experimental Marine Biology.* Academic Press, New York.

Dayton, P. K. and J. S. Oliver. 1980. Problems in the experimental analyses of population and community patterns in benthic marine environments. *In:* K. R. Tenore and B. C. Coull (eds.), *Marine Benthic Dynamics.* Univ. of South Carolina Press, Columbia, S.C. (in press).

Fenchel, T. 1975a. Factors determining the distribution patterns of mud snails (Hydrobiidae). *Oecologia* 20:1-17.

Fenchel, T. 1975b. Character displacement and coexistence in mud snails (Hydrobiidae). *Oecologia* 20:19-32.

Fenchel, T. and L. H. Kofoed. 1976. Evidence for exploitative interspecific competition in mud snails (Hydrobiidae). *Oikos* 27:367-376.

Hairston, N. G., J. D. Allan, R. K. Colwell, D. J. Futuyma, J. Howell, M. D. Lubin, J. Mathias and J. H. Vandermeer. 1969. The relationship between species diversity and stability: an experimental approach with protozoa and bacteria. *Ecology* 49:1091-1101.

Horn, H. S. and R. H. MacArthur. 1972. Competition among fugitive species in a harlequin environment. *Ecology* 53:749-752.

Hutchinson, G. E. 1959. Homage to Santa Rosalia or why are there so many kinds of animals? *Amer. Nat.* 93:145-159.

Levinton, J. S. 1977. Ecology of shallow-water deposit-feeding communities in Quisset Harbor, Massachusetts, pp. 191-227. *In:* B. C. Coull (ed.), *Ecology of Marine Benthos.* Univ. of South Carolina Press, Columbia, S. C.

Levinton, J. S. and R. K. Bambach. 1975. A comparative study of Silurian and Recent deposit-feeding bivalve communities. *Paleobiology* 1:97-124.

Lopez, G. R. and J. S. Levinton. 1978. The availability of microorganisms attached to sediment particles as food for *Hydrobia ventrosa* Montagu (Gastropoda: Prosobranchia). *Oecologia* 32:263-275.

MacArthur, R. H. 1957. On the relative abundance of bird species. *Proc. Nat. Acad. Sci. U.S.A.* 43:293-295.

MacArthur, R. H. 1958. Population ecology of some warblers of northeastern coniferous forests. *Ecology* 39:599-619.

MacArthur, R. H. 1972. *Geographical Ecology*. Harper and Row, N.Y. 269 pp.

MacArthur, R. H. and R. Levins. 1967. The limiting similarity, convergence, and divergence of coexisting species. *Amer. Nat.* 101:377-385.

MacArthur, R. H. and E. O. Wilson. 1967. *The Theory of Island Biogeography*. Princeton Univ. Press, Princeton, New Jersey. 203 pp.

May, R. M. 1973. *Stability and Complexity in Model Ecosystems*. Princeton Univ. Press, Princeton, New Jersey. 235 pp.

Myers, A. C. 1977. Sediment processing in a marine subtidal sandy bottom community: II. Biological consequences. *J. Mar. Res.* 35:633-647.

Oliver, J. S. 1979. Physical and biological processes affecting the organization of marine soft-bottom communities in Monterey Bay, California and McMurdo Sound, Antarctica. Ph.D. Thesis, Univ. California, San Diego.

Peterson, C. H. 1977. Competitive organization of the soft-bottom macrobenthic communities of southern California lagoons. *Mar. Biol.* 43:343-359.

Peterson, C. H. 1979. Predation, competitive exclusion, and diversity in the soft-sediment benthic communities of estuaries and lagoons, pp. 233-264. *In:* R. J. Livingston (ed.), *Ecological Processes in Coastal and Marine Systems*. Plenum Press, New York.

Peterson, C. H. and S. V. Andre. 1980. An experimental analysis of interspecific competition among marine filter feeders in a soft-sediment environment. *Ecology* 61: (in press).

Pianka, E. R. 1975. Niche relations of desert lizards, pp. 292-314. *In:* M. L. Cody and J. M. Diamond (eds.), *Ecology and Evolution of Communities*. Belknap Press, Cambridge, Massachusetts.

Pulliam, H. R. 1975. Coexistence of sparrows: A test of community theory. *Science* 189:474-476.

Rees, C. P. 1975. Competitive interactions and substratum preferences of two intertidal isopods. *Mar. Biol.* 30:21-25.

Reise, K. 1978. Experiments on epibenthic predation in the Wadden Sea. *Helgoländer wiss. Meeresunters.* 31:51-101.

Rhoads, D. C. and D. K. Young. 1970. The influence of deposit-feeding organisms on sediment stability and community trophic structure. *J. Mar. Res.* 28:150-178.

Rhoads, D. C. and D. K. Young. 1971. Animal-sediment relationships in Cape Cod Bay, Massachusetts. II. Reworking by *Molpadia oolitica* (Holothuroidea). *Mar. Biol.* 11:255-261.

Ronan, T. E., Jr. 1975. Structural and paleoecological aspects of a modern marine soft-sediment community: An experimental field study. Ph.D. Thesis, U. California, Davis. 220 pp.

Rosenzweig, M. L. and P. Sterner. 1970. Population ecology of desert rodent communities: body size and seed husking as bases for heteromyid coexistence. *Ecology* 51:217-224.

Sanders, H. L. 1968. Marine benthic diversity: A comparative study. *Amer. Nat.* 102:243-282.

Sanders, H. L. 1969. Benthic marine diversity and the stability-time hypothesis. *Brookhaven Symp. Biol.* 22:71-81.

Sanders, H. L., E. M. Goudsmit, E. L. Mills and G. R. Hampson. 1962. A study of the intertidal fauna of Barnstable Harbor, Massachusetts. *Limnol. Oceanogr.* 7:63-79.

Segerstråle, S. G. 1962. Investigations on Baltic populations of the bivalve *Macoma baltica* (L.) Part II. What are the reasons for the periodic failure of recruitment and the scarcity of *Macoma* in the deeper waters of the inner Baltic? *Commentat. Biol.* 24:1-26.

Vassallo, M. T. 1971. The ecology of *Macoma inconspicua* (Broderip and Sowerby, 1829) in central San Francisco Bay. Part II. Stratification of the *Macoma* community within the substrate. *Veliger* 13:279-284.

Virnstein, R. W. 1978. Predator caging experiments in soft sediments: caution advised, pp. 261-273. *In:* M. L. Wiley (ed.), *Estuarine Interactions.* Academic Press, New York.

Whitlatch, R. B. 1976. Seasonality, species diversity and patterns of resource utilization in a marine deposit-feeding community. Ph.D. Thesis, Univ. of Chicago, Chicago, Ill. 127 pp.

Wilbur, H. M. 1972. Competition, predation, and the structure of the *Ambystoma-Rana sylvatica* community. *Ecology* 53:3-21.

Woodin, S. A. 1974. Polychaete abundance patterns in a marine soft-sediment environment: the importance of biological interactions. *Ecol. Monogr.* 44:171-187.

Woodin, S. A. 1976. Adult-larval interactions in dense infaunal assemblages: Patterns of abundance. *J. Mar. Res.* 34:25-41.

Young, D. K. and M. W. Young. 1978. Regulation of species densities of seagrass-associated macrobenthos: Evidence from field experiments in the Indian River estuary, Florida. *J. Mar. Res.* 36:569-593.

COMPUTER-PROCESSING OF ZOOPLANKTON SAMPLES

H. Perry Jeffries

Graduate School of Oceanography
University of Rhode Island
Kingston, Rhode Island

Kenneth Sherman, Ray Maurer

National Marine Fisheries Service
Northeast Fisheries Center
Narragansett, Rhode Island

and

Costantin Katsinis

Department of Electrical Engineering
University of Rhode Island
Kingston, Rhode Island

Abstract: Methods of identifying and counting zooplankton in preserved samples have changed little since Johannes Müller first towed a fine-mesh net through the ocean more than 100 years ago. Instrumentation basically consists of a microscope and a ruled counting chamber. Because time of analysis for a single sample ranges from hours to days, delay is inevitable. We are attempting to modernize the procedure, using improved image formation devices and computerized pattern recognition techniques. Total counts and size-frequency distributions can now be made in minutes with a simple, processor-controlled vidicon system, but calibration remains a problem awaiting sharpened contrast by optical edge enhancement and spatial filtering. Major species of North Atlantic zooplankton can be accurately classified to group (e.g. copepods, fish eggs, fish larvae, cladocerans, chaetognaths, euphausiids) by discriminant analysis of simple morphometric relations (length:width:perimeter:area). Identification of commonly occurring copepod species appears theoretically possible for samples taken with a 333 μm aperture net. Syntactic analysis, a complementary approach to pattern expression, may solve difficult problems of classifying developmental stages. Automated image analysis with improved photo-optical systems should meet present day needs.

Introduction

We are solely dependent on manual operations for sorting of preserved zooplankton samples, even though physical fractionation procedures exist (Price et al. 1977). A range of electronic approaches has been considered (Sutro 1974), e.g. particle discriminators, yet they work in limited ways (Beers 1976; Fawell 1976; Uhlmann et al. 1978). But the task is so arduous that few technicians can work continuously at counting, identifying and measuring organisms under a microscope. Personnel turnover is understandably high, and training periods further reduce sample output.

303

Consequently, effective average processing time per technician ranges up to four days per sample.

The MARMAP program alone has a capability for collecting 4,500 samples annually (Sherman 1980). Clearly, a technological breakthrough is required to support basic research and to meet requirements for improved fisheries forecasting. We are addressing the problem with tools that have been developed in widely disparate fields. The results are encouraging, but they also show the need for engineering advances in selected areas.

Among the several approaches applicable to analysis of samples collected with zooplankton nets, we believe that pattern recognition by combined optical and electronic means is the most promising. In the sense used here, pattern recognition consists of counting particles, extracting visual features from individual organisms (either alive or preserved), measuring these features automatically and then classifying the organisms into taxonomically meaningful or ecologically useful categories. Classification to species is the ideal, and our work on morphometrics indicates that it may be easier than one would imagine, but for present purposes we have identified the following groups as technologically amenable to feature extraction and classification:

1. copepods (copepodite stages I-VI)
2. nauplii (various)
3. zoeae (Brachyura)
4. zoeae (Macrura)
5. mysids, cumaceans, euphausids
6. amphipods
7. fish eggs
8. fish larvae and chaetognaths
9. lamellibranch larvae
10. polychaete larvae
11. medusae
12. particulate debris, organic aggregates

Useful descriptors of ecosystem state can be derived from total count and size-frequency distributions within these groups (Steele and Frost 1977).

Approaches to Pattern Recognition

The components of a pattern recognition system (Fig. 1) suitable for preserved zooplankton samples consist of : (1) a low power microscope or enlarging lens, with provisions for transmitted, incident and dark field illumination; (2) vidicon and input image monitor; (3) analog-to-digital converter; (4) computer-microprocessor; (5) output image monitor and line printer. The microprocessor is programmed to extract and measure specific features that are selected to characterize a large, but dimensional and finite, pattern space that has been reduced by the system's transducer (optical-vidicon system). The next step is a classification function that makes deci-

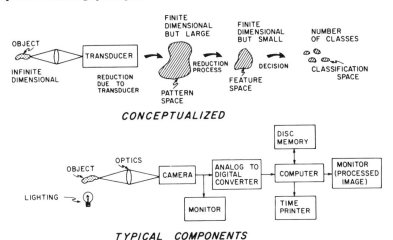

CONCEPTUALIZED

TYPICAL COMPONENTS

Figure 1. Theory and application of pattern recognition. In the lower figure, note the vidicon monitor from which tracings were made; the output monitor displays computer-generaged images, e.g. Figs. 2, 3.

sions for assignment of the input object to one of the predetermined classifications (Fu 1970).

The mathematics of pattern recognition vary from comparatively simple "template-matches" stored in the computer's memory to complex scene analyses consisting of so many features and patterns that the analysis must be described in hierarchical order from simpler sub-patterns. The decision-theoretic approach, which separates objects according to statistical models, has been most commonly used. In a statistical sense, ". . . each pattern is considered as a vector in n-dimensional space. The goal of the recognition system is to define partitions in this space such that each region can be identified with a class of pattern." (Viglione 1970). In the future, capabilities of the syntactic approach to description of structural pattern should become popular, especially for identifying biological objects (Fu 1978). Chromosome analysis, for example, has been automated by way of syntactic grammers. Our work to date has been limited to the discriminant-function approach, and although the classification ability for planktonic animals is surprisingly good, the two approaches will undoubtedly be combined in a semi-automatic instrument for classification, enumeration and sizing of planktonic organisms.

We used two instruments in this study. For developmental work and testing of computer programs, a sophisticated, custom-built system was adapted by Katsinis (1979) for application to zooplankton. It produced the results shown in Figs. 2, 3 and 5. The other instrument, a Bausch and Lomb QMS, was used for tests on counting capabilities, size-frequency determinations and staining techniques. This comparatively simple device relied on fixed circuits, as opposed to computer-directed flexibility, for system control. Both approaches have attributes, as were shown by the following results.

Image Formation by Electronic Means

In Fig. 2, the computer-generated outline of an adult female *Centropages typicus* viewed by vidicon is compared with a man-made drawing

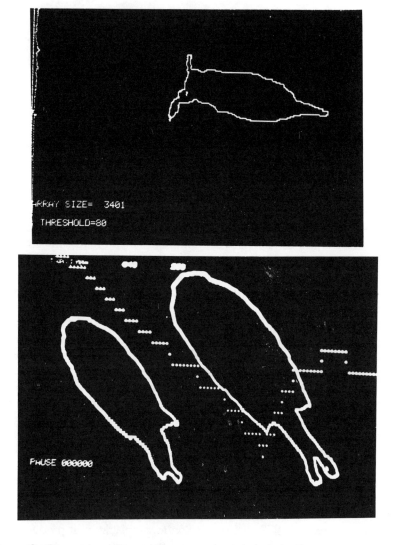

Figure 2. Computer-generated images of adult female Centropages typicus *on the output monitor. Top, input was an actual specimen; bottom, input was a silhouette manually traced from vidicon (input) monitor.*

traced from the input monitor and then resubmitted for computerized imaging. Note that edge definition of what man sees and interprets is far sharper than the computer generated pattern of an actual specimen. Herein lies the problem, much of it arising from the simple fact that low contrast makes planktonic animals difficult to discern electronically. This is certainly adaptive for organisms swimming in nature, but the problem it causes for imaging may require such refinement as a laser beam directed through individual specimens and spatial filtering of the coherent image at the frequency plane. Thus we are in the process of building a photoconductive-thermoplastic device in which the transparent electrode serves as a ground plane during charging and as a resistive heater during the erase cycle.

A second problem is the orientation of dead specimens. They assume various positions in the water-filled examination chamber. In Fig. 2 (lower) for example, the left specimen appeared to have an asymmetric urosome, but this was not so. Rather this was a visual aberration due to the copepod lying at about 45° on its major (length) axis, balanced tenuously on its pleopods, which in this case happened to be extended.

As shown in Fig. 3, computer-generated images can be manipulated in various ways. Length, perimeter, width and area of up to 10 organisms per field of view have been obtained. Because these measurements are made almost instantaneously, huge banks of data become available; thus we can at least minimize the problems of contrast and orientation by statistical techniques.

To complement the above electro-optical approach, we have also explored applications of silhouette photography (Ortner et al. 1979). Images shown in Fig. 4 (top) are probably better for input to currently available vidicons than are the organisms themselves. An advantage here is that this lens-less system of direct photography by electronic flash can be applied at sea with live organisms. Alive, they orient in rather repeatable ways from one individual to the next, but in a preserved sample random position is a problem. Alternatively, this latter problem might be solved with a two-camera system trained on organisms hydrodynamically oriented in a flowing, as opposed to static, observation medium.

When a pattern of an object formed by transmitted light is illuminated by coherent light, the light distribution at the focal plane of a double convex lens is its Fourier spectra, which is singularly characteristic for any particular shape. Typical optical spectra shown in Fig. 5 represent three of the categories that we have established for classification: copepod (left), Macrura zoea (middle) and fish larva (right). We are building an incoherent-to-coherent transducer to eliminate the intermediate step of exposing films.

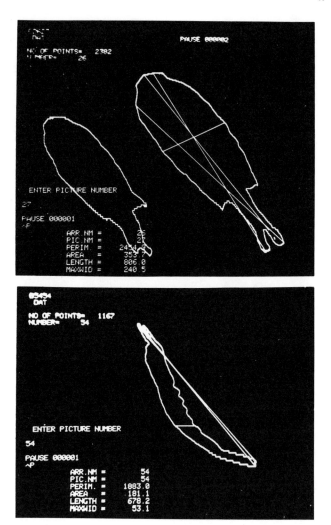

Figure 3. Several of the length-width-area options that are immediately available by automated programming.

Morphometric Relations of Common Species in the Western North Atlantic

A reference data base free of electronic imaging problems was needed to test the applicability of pattern recognition techniques to shapes represented by zooplankton. We photocopied published plates of 101 species, chiefly from Fiches d'Identification du Zooplankton of the Conseil

Figure 4. Silhouette photograph of zooplankton sample. Bottom, actual (natural) size; above, successively greater enlargements made from negative of bottom print.

Fourier Optical Spectra

 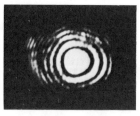 

Copepod	Macrura zoea	Fish larva
Tortanus discaudatus	*Crangon sp.*	*Limanda ferruginea*

Figure 5. Fourier transform spectra by coherent light.

International pour l'Exploration de la Mer. These species represented six of the 12 categories listed above.

The portion of each figure that would be perceived electronically if this were an actual specimen was estimated, and the figure was trimmed with scissors. This step usually involved elimination of antennae and feathery appendages. Measurements of body length, projected length, width and area were then made with the Bausch and Lomb QMS system. Projected length is the length along an organism's longest dimension plus the length of projections, such as extend from the posterior corners of the fifth thoracic segment of *Centropages typicus* (Fig. 2).

Six morphometric ratios were determined for each of the 101 species. The entire data matrix was then subjected to stepwise linear discriminant function analysis by the jack-knife procedure (Dixon and Brown 1977). Mean coefficients and f values for discriminatory power are shown in Table 1; the first three ratios listed were retained as significant descriptors. The ratio of width to projected length alone enabled 91.1% correct classification. Addition of a second variable (width to the square root of the area) increased discriminatory ability to 92.1%, and the third ratio (projected length to the square root of the area) brought improvement to 93.1%. Group means were plotted in Fig. 6 according to their positions on the first two canonical variates. These axes maximize distances between group means and show spatial relations among all six groups (Maxwell 1977). The main trend shown here is from left to right on the first variate. It results from increasing elongation, starting at the left with spherical fish eggs, then to somewhat elliptical lamellibranch larvae, progressing to elongate cladocerans and copepods. The relation turns downward on the axis of the second canonical variate, going from sausage-shaped copepods to even more elongate euphausids, finally to the most elongate chaetognaths and fish larvae. This then is a reference base—free of the major image formation problem in available instrumenta-

Table 1. Mean values of morphometric ratios for six shape categories. F-values for the initial discriminant function are also presented.

Morphometric parameter	Copepods	Euphausids	Fish eggs	Chaetognaths	Bivalve larvae	Cladocera	F
Width/ projected length	0.252	0.111	0.990	0.083	0.877	0.578	802.9
Width/ $\sqrt{\text{area}}$	0.645	0.403	1.125	0.313	1.065	0.984	235.3
Projected length/ $\sqrt{\text{area}}$	2.617	3.686	1.136	3.999	1.215	1.717	153.6
Length/ $\sqrt{\text{area}}$	2.379	3.472	1.136	3.999	1.215	1.643	141.0
Width/ length	0.278	0.119	0.990	0.083	0.877	0.606	589.4
Length/ projected length	0.912	0.940	1.000	1.000	1.000	0.956	10.0

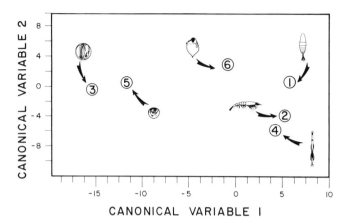

Figure 6. Results of discriminant analysis; positions of group means in canonical space. 1) copepod; 2) euphausids; 3) fish egg; 4) chaetognath; 5) lamellibranch larvae; 6) cladocerans.

tion—showing theoretical feasibility of a pattern recognition approach to zooplankton classification.

Taking a step toward operational reality, we next put together a collection of preserved specimens spanning 19 species and five of the predefined classification groups (12 copepods, 3 euphausids, 1 chaetognath, 1 pteropod, 2 salps) that collectively accounted for about 90% of all zooplankton (333 μm net fraction) in the Western North Atlantic. Species were selected for importance according to abundances reported by Bigelow (1926), Grice and Hart (1962), Sherman (1970) and Sherman et al. (1979). Each species was manually traced on an acetate sheet that was taped to a vidicon monitor; the vidicon was mounted on a stereo microscope set at 50-70×, and the specimens were viewed in seawater. These 326 tracings were chosen to duplicate the various positions assumed by each species in the observation chamber and thereby incorporated the error due to varying positions. The images were measured as before on the Bausch and Lomb QMS system and results were subjected to discriminant analysis using non-pooled covariance matrices (Blair et al. 1976). In this case, however, absolute values rather than ratios were used because specimen magnifications were known. The results, which are too extensive for presentation here, showed once again that the major planktonic organisms can be classified to group by simple morphometric relations. Table 2 lists error rates based upon single variables and their combinations. Width, along with either length or projected length, gave better than 93% correct classification to group. At the species level, 90% of the 326 images could be classified. This result was not the intention of the present analysis; rather, species identification by syntactic and metric statements of form is a task for the future.

Table 2. Discrimination to group of 326 silhouettes representing 19 major species in the zooplankton of the Western North Atlantic. Measurements were sequentially selected to give the most successful function at each step.

Step	Variables	Errors (out of 326)	Percent correct
1	Area	53	83.7
	Length	94	71.2
	Projected length	103	68.4
	Width	40	87.7
2	Width + Area	21	93.6
	Width + Length	19	94.2
	Width + Projected length	19	94.2
3	Width + Length + Area	10	96.9
	Width + Length + Projected length	13	96.0
4	Width + Length + Area + Projected length	7	97.9

Status of Practical Application: Commercial Instrumentation

Lengths of 300 preserved *Calanus finmarchicus* specimens, copepodite stages IV-VI, were measured once manually by examination under a stereo microscope and twice with the Bausch and Lomb QMS system. The machine consistently measured 0.1 mm lower than did the technician, but upon adjustment for systematic error the two size-frequency distributions were statistically identical according to the Kolmogorov-Smirnov test (Fig. 7). Furthermore, repeatability of machine measurements was high ($P < 0.01$).

A comparison of the time required for counting and length measurement by manual and by electronic means was made "pitting" an experienced technician against the QMS system. Five samples were prepared, each consisting of 25-100 stage V-VI *Calanus finmarchicus* copepodites, selected to yield a unimodal size frequency that would readily reveal minor deviations. The technician used a stereo microscope at 12× calibrated in the usual way with a stage micrometer. Copepods for electronic analysis were stained for about 16 h (overnight) in a dilute solution of methyl blue to enhance image contrast. A specially made template of ellipses in the general range of copepod lengths was used for calibrating the QMS system; detrital particles were rejected by the operator setting minimum size criteria; samples with up to 500 individuals required counting 10 successive fields; operator control was ensured by an enhanced outline automatically drawn around each object detected for measurement (Fig. 8).

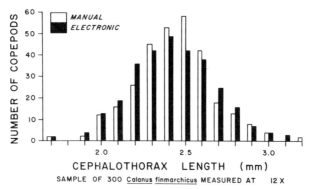

Figure 7. *Comparison of size frequency distributions in a prepared sample of* Calanus finmarchicus *determined by manual and by automated methods.*

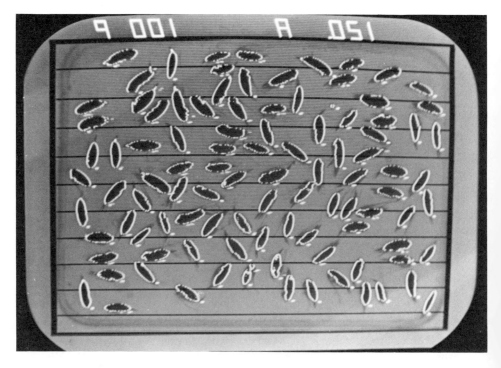

Figure 8. *Automated counting of 100* Calanus finmarchicus *with a commercially available instrument. The white line around the perimeter of each organism counted assists the operator in setting sensitivities and avoiding clumping.*

Electronic counting was nearly instantaneous. Manual counting required 0.7 min and measurement 40 min. Total time for all operations— from sample preparation through counting, data summation and plotting a size frequency distribution—was 54 min manually but only 12 min by the image analysis system, for a saving of 78% in time.

Manual processing of a 500-organism sample required 264 min vs. 72 min electronically for a 73% saving. Elimination of plotting the length distribution would reduce manual analysis by 36 min, but data summation is a necessary step and by hand it alone takes 21 min.

Even with a more elaborate image analyzer than the QMS system, time saving would probably not exceed 85% over manual. Sample preparation still requires up to 24 min for a 500-organism sample (examined as a series of subsamples in a multi-welled chamber).

With new devices for sample preconditioning and automatic scanning, we should be able to count, measure and group 500 organisms in 5-6 min. With allowances made for down time and systems upkeep, this translates to about 5,000 samples per year, which is near the requirements of large-scale projects. Perhaps more importantly, any ecological science draws its very substance from a continuing census of populations in nature. Although this seemed impossible for plankton biology a short while ago, a future direction is now clear.

Acknowledgments

This investigation was supported by NOAA/OOE Grant 04-8-MO1-108. We are indebted to Luther Bivins and Peter Ortner for advice on engineering approaches to a multidisciplinary problem. Vance McCollough and Lester Smith contributed broadly during consultation on preliminary feasibility estimates. Donald Tufts and Charles Polk advised on operations and made available the facilities of the Photoelectronics Laboratory in the Department of Electrical Engineering at the University of Rhode Island. William Johnson and Daniel O'Neill made measurements and statistical analyses. To all these contributors, we extend many thanks.

References Cited

Beers, J. R. 1976. Particle discriminators in studies of zooplankton biomass, p. 69-73. *In:* H. F. Steedman (ed.), *Zooplankton Fixation and Preservation.*UNESCO Press, Paris.

Bigelow, H. B. 1926. Plankton of the offshore waters of the Gulf of Maine. *Bull. U. S. Bur. Fish.* 40 (Part II):1-509.

Blair, A. J., J. H. Goodnight, J. P. Sall and J. T. Helwig. 1976. *A User's Guide to SAS 76.* SAS Inst., Raleigh, N. C. 329 p.

Dixon, W. J. and M. B. Brown. 1977. *BMDP-77, Biomedical Computer Programs P-Series.* U. California Press, Berkeley. 880 p.

Fawell, J. K. 1976. Electronic measuring devices in the sorting of marine zooplankton, p. 201-206. *In:* H. F. Steedman (ed.), *Zooplankton Fixation and Preservation.* UNESCO Press, Paris.

Fu, K. S. 1970. Statistical pattern recognition, p. 35-70. In: J. M. Mendel and K. S. Fu (eds.) *Adaptive Learning and Pattern Recognition Systems, Theory and Applications.* Academic Press, New York. 444 p.

Fu, K. S. 1978. Syntactic (linguistic) pattern recognition, p. 96-133. In: K. S. Fu (ed.), *Digital Pattern Recognition.* Springer-Verlag, New York. 206 p.

Grice, G. D. and A. D. Hart. 1962. The abundance, seasonal occurrence and distribution of the epizooplankton between New York and Bermuda. *Ecol. Monogr.* 32:287-308.

Katsinis, C. 1979. Digital image processing and identification of zooplankton. M.S. Thesis, U. Rhode Island Library, Kingston, Rhode Island. 108 p.

Maxwell, A. E. 1977. *Multivariate Analysis in Behavioral Research,* p. 94-105. John Wiley, New York. 164 p.

Ortner, P., S. R. Cummings, R. P. Aftring and H. E. Edgerton. 1979. Silhouette photography of oceanic zooplankton. *Nature* 277:50-51.

Price, C. A., J. M. St Onge-Burns, J. B. Colton and J. E. Joyce. 1977. Automatic sorting of zooplankton by isopycnic sedimentation in gradients of silica: performance of a rho spectrometer. *Mar. Biol.* 42:225-231.

Sherman, K. 1970. Seasonal and areal distribution of zooplankton in coastal waters of tne Gulf of Maine, 1967 and 1968. *U. S. Fish Wildlife Serv., Spec. Sci. Rep. Fish. No. 594.* Washington, D.C., 8 p.

Sherman, K. 1980. MARMAP, a fisheries ecosystem study in the NW Atlantic: Fluctuations in ichthyoplankton-zooplankton components and their potential for impact on the system. In: F. J. Vernberg (ed.), *Proceedings of the Workshop on Advanced Concepts in Ocean Management* I. Belle W. Baruch Institute for Marine Biology and Coastal Research, U. South Carolina, Columbia, South Carolina. Oct. 24-28, 1978 (In press).

Sherman, K., C. Jones and J. Kane. 1979. Zooplankton of continental shelf nursery and feeding grounds of pelagic and demersal fish in the Northwest Atlantic. Int. Council Explor. Sea, unpublished ms ICES C.M. 1979/L:27.

Steele, J. H. and B. W. Frost. 1977. The structure of plankton communities. *Phil. Trans. Royal Soc. London, Ser. B., Biol. Sci.* 280:485-534.

Sutro, L. L. 1974. Study of means of automatically classifying plankton. MIT Sea Grant Program, Rpt. No. MIT SG 74-11, Cambridge, MA.

Uhlmann, D., O. Schlimpert and W. Uhlmann. 1978. Automated phytoplankton analysis by a pattern recognition method. *Int. Revue ges. Hydrobiol.* 63:575-583.

Viglione, S. S. 1970. Applications of pattern recognition technology, p. 115-162. In: J. M. Mendel and K. S. Fu (eds.), *Adaptive Learning and Pattern Recognition Systems, Theory and Applications.* Academic Press, New York. 444 p.

COMPUTER SIMULATION OF AIR-ESTUARY THERMAL ENERGY FLUXES

Ned P. Smith

Harbor Branch Foundation, Inc.
RR 1, Box 196
Fort Pierce, Florida

Abstract: Meteorological data are used to develop and verify a numerical model of local heat exchanges in a shallow, bar-built estuary along South Florida's Atlantic coast. Examples are given to demonstrate the use of the model in estuarine research. Simulated temperatures follow observed day-to-day temperatures closely and cumulative errors of temperatures simulated over time intervals of up to two weeks have a standard deviation of 0.7C. Historical weather data are used to compute water temperatures of approximately 5C during mid January 1977 when a major fish kill was recorded in South Florida. Climatological data are used to calculate average midwinter temperatures of approximately 15.5C for a study site near Vero Beach, Florida. Model results suggest that conduction and evaporation are the primary heat exchange mechanisms during the winter months and that estuarine waters are in a continual state of readjustment to local meteorological conditions.

Introduction

Temperature variations in estuarine waters are characteristically large as a result of rapid response of shallow waters to air-estuarine heat exchange processes. Especially during winter months, estuaries along the southern tier of states may undergo dramatic temperature changes as frontal passages produce an alternation of maritime tropical and continental arctic air masses over time scales on the order of one to two weeks. Many field studies are based upon weekly or monthly sampling, and discrete measurements may miss temperature extremes which occur between sequential visits to study sites.

Smith (1977) documented estuarine temperature changes of as much as 10C in the first few days following severe cold air outbreaks over the central Texas coast. Hildebrand and King (1976) reported an extreme case in which the water temperature in upper Laguna Madre, Texas, fell 17.2C in only 18 h in late November 1974. Unpublished temperature data from a coastal lagoon along the Atlantic coast of Florida indicate temperature variations of 4-5C associated with frontal activity during the midwinter months of 1978. Even during the summer months, weekly temperature ranges may be as high as 2-3C (Hildebrand and King 1976).

The observed temperature changes are a direct result of local and advective heat exchange processes which are summarized by the heat budget equation:

$$Q_t = (Q_s + Q_v) - (Q_h + Q_b + Q_e)$$

where Q_t is net heat storage, Q_s is the insolation term, Q_v is advective heat flux, Q_h is the conduction term, and Q_b and Q_e are the effective back radiation and the evaporative cooling terms, respectively. Although none of these processes can be described mathematically with certainty, work carried out over the past 20 years has provided data which have been used in turn to formulate empirical equations to approximate closely each term in the heat budget equation. Heating by insolation can be estimated from time series of pyranometer data, or from empirical relationships based upon cloud cover (Reed 1976a, 1977), with a correction for surface reflectivity (Payne 1972). Outgoing longwave radiation can be estimated for water surfaces by the Stefan-Boltzmann equation, with corrections for clear-sky counter radiation (Swinbank 1963) and cloud cover (Reed 1976b). Both conductive and evaporative heat fluxes can be estimated from profile measurements; however, Priestley and Taylor (1972) have suggested that formulas of the bulk aerodynamic type are more satisfactory for over-water applications.

With a workable mathematical expression for each of the individual heat flux terms, one may not only simulate net heat storage but also investigate the magnitude and the relative importance of each term in producing the observed net effect. In addition, one can utilize a mathematical model to examine the response of an estuary to various types of hypothetical meteorological forcing.

In spite of a substantial body of literature on the subject of energy budgets in marine areas, relatively little work has been carried out in estuaries per se. Heath (1977) investigated the heat budget of a coastal inlet in New Zealand and found the shortwave and longwave radiation terms to be dominant and approximately in balance. Hsu (1978) reviewed both conductive and evaporative heat fluxes in estuarine settings and called attention to the dependence of these energy exchange processes upon the local wave statistics.

In many estuarine settings, advective heat fluxes may dominate, or at least seriously mask, local heat flux processes. Yet all estuarine temperature changes are strongly influenced by fluxes across the air-water interface, whether occurring locally or at some distance from the study site. Thus, it is important to focus upon local processes. Intracoastal bays and lagoons are well suited for such studies. These estuarine waters can be characterized by large surface area to volume ratios and relatively small tidal action due to restricted access to adjacent ocean waters. This tends to inhibit advective fluxes and thereby enhance local exchange processes. Most coastal bays and lagoons bordering the Gulf of Mexico, for example, have relatively little tidal flushing and many along the western edge of the

Gulf receive little fresh water run-off. In these areas, and at selected sites elsewhere, local air-estuary energy fluxes dominate.

The Indian River is an intracoastal lagoon which extends over 250 km along the east central Florida coast (Fig. 1). Water depths are on the order of 1-2 m; the width is generally between 2 and 3 km. Estuarine-shelf exchanges of water are through inlets spaced infrequently along the lagoon. The net internal circulation is primarily along the Atlantic In-tracoastal Waterway, but it is not well understood. The available data suggest that the circulation at points away from the Intracoastal Waterway is a highly variable wind drift. Thus, advection is small near the shore and appears to involve a relatively homogeneous water mass.

A study was conducted in the winter of 1977-78 at such a near shore site to provide hydrographic and meteorological data needed to develop and verify a numerical model of local energy exchanges. The study site (Fig. 1) was approximately 11 km north of the Ft. Pierce Inlet and well upstream of the 4.7 km tidal excursion computed from current measurements made in the waterway approximately 0.5 km away. The

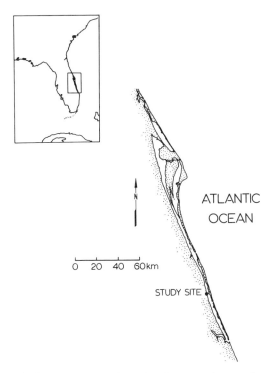

Figure 1. Site of the estuarine heat budget study in the Indian River lagoon, Florida, winter 1977-78.

study site was selected on the basis of its isolation from both tidal currents and the quasi-steady outflow along the Intracoastal Waterway. At present, the simulative capability of the model is such that the standard deviation of the cumulative error after a two-week period of time is 0.7C. It is within this precision that both the individual heat flux processes and the resultant temperature change can be described on a day-to-day basis. The distribution of this total error among the individual heat flux terms cannot be determined.

The purpose of this paper is to demonstrate how a numerical model of estuarine heat exchanges can be utilized to investigate the response of estuarine waters to local heat flux processes. As a research tool, a relatively simple model can simulate day-to-day temperature fluctuations in areas where local heat exchange processes dominate and thus can supplement field observations made over weekly or longer time intervals. In addition, one may use a modeling approach to test the sensitivity of estuarine waters to various changes in the overlying atmosphere. Several examples are offered following a brief description of the general form of the model.

The Model

The unique physical setting of each estuary necessarily makes a working model equally site-specific. Thus, one should not expect expressions developed and verified empirically to be universally applicable. However, because estuarine studies in general often tend to be site-specific, whether physical, biological, geological or chemical, this is not felt to be a serious drawback. In this section, some general remarks regarding the model are offered.

The insolation term is based upon measurements of sunlight made 0.5 km from the study site with an Eppley model 8-48 black and white pyranometer. Total daily insolation is corrected for reflectivity at the water surface (Payne 1972). Outgoing longwave radiation is estimated by the Stefan-Boltzmann equation, with a correction for counter radiation through clear skies (Swinbank 1963) and a further correction for cloud cover (Reed 1976b):

$$Q_b = 8.132 \times 10^{-11} \, \varepsilon \, T_w^4 \, (1-0.935 \times 10^{-5} \, T_w^2) \, (1-0.67C)$$

where ε is emissivity, T_w is the water temperature and C is cloud cover in tenths.

Both the conduction and evaporation terms are of the bulk aerodynamic type (Priestley and Taylor 1972). Meteorological measurements were made 0.5 km from the estuarine study site, except for dew point, which was recorded at the airport in Vero Beach, Florida, 10 km from the study site. Evaporation, or condensation, is determined by the vertical gradient of vapor pressure. The water temperature is used to estimate the saturation vapor pressure at the surface of the estuary (Murray 1967). The evaporative flux is then given by:

$$Q_e = 0.622 \, \varrho \, L \, A(V,\theta) \, (e_s - e)/p$$

where ϱ is the water density, L is the latent heat of evaporation, e_s and e are the saturation vapor pressure and the vapor pressure, respectively, and p is atmospheric pressure. The exchange coefficient, $A(V,\theta)$ is sensitive to both wind speed, V, and thermal stability, θ:

$$A(V,\theta) = 1.995 \, [1 + 0.03V^{0.75}] \, \{1 + [(T_w - T_a)/14.2]\}^3$$

In the above expression, T_w and T_a are the water and air temperature, respectively. The constants were determined empirically by successive temperature simulations until differences with observed temperatures were minimized. Any residual may include advection, but depending upon the study site selected the error will be comprised primarily of individual errors in the local heat flux terms.

Conduction is estimated from the air-water temperature difference, using the expression

$$Q_h = c_p \, A(V,\theta) \, (T_w - T_a)/Z$$

where c_p is the specific heat of air and Z is the level at which the air temperature is measured. Thus, to simulate water temperature changes, both air temperature and some measure of the moisture content of the air must be available for the computations.

Change in water temperature is obtained by monitoring the appropriate meteorological variables and evaluating each of the individual terms in the heat budget equation. These are then summed and the net effect is added to the previous day's water temperature. Heat flux terms are converted from units of cal cm^{-2} sec^{-1} to units of C/day to determine net heating or cooling effect of any or all of the local processes. The conversion factor will be directly proportional to water depth, which makes it site-specific; however, this is easily changed for other locations.

An important feature of the numerical model of estuarine heat storage is the exchange of heat between the water column and the underlying sediments. Conductive heat transfer through the water-sediment interface acts to reduce temperature extremes within the water column as heat is stored in, and later given up from, sediment layers. Unpublished data show temperature gradients as high as 2.2 C/10 cm through the upper 40 cm of sediments near the study site and during the winter months. A similar damping effect can be produced in the model by artificially increasing water depth until temperature extremes agree with those observed. This approach was taken to simulate what is undoubtedly a far more complex process. The water depth used in this study was 3.7 m, or about 2.5 m greater than the true depth.

Model Applications

Examples included in this section are intended to demonstrate how a heat exchange model might be incorporated into estuarine research in areas where local exchanges dominate. In general, applications fall into three categories. The first involves simulation of estuarine water temperatures. In its present form, the model works with 24-h averages of both meteorological data and water temperatures. Second, one may use a numerical model to hold constant terms that would normally vary in nature and thereby determine the time required to attain a thermodynamic equilibrium between the estuary and the overlying atmosphere. Finally, one can use a numerical model to test the sensitivity of estuarine waters to variations in individual meteorological variables. Results are useful for inferring the dominant driving forces in producing observed temperature changes.

Daily Mean Temperature Simulation

Many estuarine surveys involve weekly or biweekly visits to sampling sites. As mentioned above, temperature data obtained weekly may not include the extrema that can occur within the 5-7 day time scales normally associated with meteorological forcing. Hydrographic data obtained at weekly or longer time periods, however, can be used to verify simulated water temperatures based upon meteorological data recorded on a daily basis. Figure 2 is a composite of observed and computer-simulated daily mean temperatures for the Indian River at the study site. The time interval

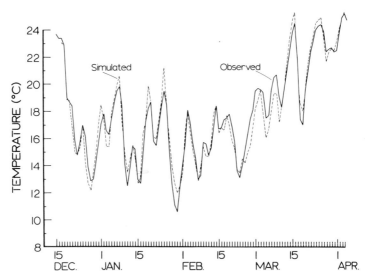

Figure 2. Observed and simulated estuarine water temperatures, December 15, 1977, to April 4, 1978, at study site in Indian River lagoon.

is a 110-day period in the winter of 1977-78. Although the match is poor at times, the fit is generally good, and it appears that the model can provide valuable information relating to high and low temperature extremes that might occur between regularly scheduled sampling trips.

A more specific example involves a 12-day period from 14-25 January 1977. During this time exceptionally low temperatures were produced by a cold front that moved down the Florida peninsula. This period of cooling was of particular interest due to associated hypothermal stress and mortality of 56 fish species in the Indian River (Gilmore et al. 1978).

Figure 3 is a composite of weather data from Vero Beach, Florida, approximately 10 km from the study site. In the bottom of Fig. 3 are the water temperatures computed from the available meteorological data. Simulated water temperatures appear to lag behind the daily average air and dew point temperatures by about three days. Lowest computed water temperatures are just below 5C. This is in good agreement with intake

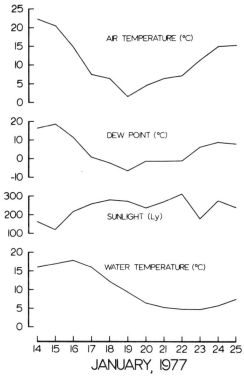

Figure 3. Composite time plots of daily average air temperature, dew point temperature, sunlight as computed from cloud cover measurements at the Vero Beach, Florida airport, and computed water temperature for the study site, 14-25 January 1977.

temperatures recorded at a power plant in Vero Beach. Note that dew point temperatures remained at or below 0C for five days. It is probable, then, that the low estuarine water temperatures and the length of time they remained low reflect in large part the effects of prolonged evaporative cooling.

Air-estuary Thermodynamic Equilibrium

In many coastal areas, temporal variability characteristic of time series of air temperature, humidity, and wind speed suggests that there may be a continual readjustment of estuarine waters to changes in local heat flux processes. To test this, the model can be used to find the daily average water temperature that is in thermodynamic equilibrium with specified atmospheric conditions. The time interval required to attain an equilibrium following a step-change in one or more atmospheric variables can then be found.

This exercise can be made more physically realistic by investigating the change in weather conditions that occurred with the passage of a cold front at the study site on 20 February 1978. Daily air temperature recorded 0.5 km from the study site decreased from 17.3C on 20 February to 6.7C on 21 February. The dew point dropped from 9.6 to −3.1C, while the average wind speed increased from 3.5 to 4.9 m/sec. Insolation decreased from 366 to 244 ly/day. The model suggests that a total of 20 days would have been required for the estuary to achieve an equilibrium state with these new atmospheric conditions, assuming it had been in equilibrium on 20 February. During this readjustment, water temperatures decrease exponentially from 15.8C to a final value of 2.8C.

The implication is that, at least during the winter, an estuary that is alternately exposed to markedly different air masses will rarely if ever be in thermodynamic equilibrium with the overlying atmosphere. The example given here is admittedly an extreme situation, but other calculations made with step changes representative of less intense frontal passages still suggested that complete equilibrium would require time intervals on the order of one to two weeks. The exponential nature of the water temperature change means that half of the eventual temperature change generally occurs within the first 2-3 days. Thus, it is apparent that the estuary, by virtue of its heat capacity, acts to filter temperature extremes in the overlying atmosphere. The filtering effect would vary, perhaps greatly, from one estuary to the next, but in any physical setting model results may be useful in quantifying this process.

Another application of the numerical model to studies of air-estuary thermodynamic equilibrium involves use of climatological data to calculate the seasonal temperature curve for the estuary in question. Just as weather conditions fluctuate about a very low-frequency, annual cycle, so do estuarine water temperatures slowly move through an annual cycle. This annual curve can be estimated for the winter months at least, because it is under such conditions that the model was verified. Figure 4 shows the an-

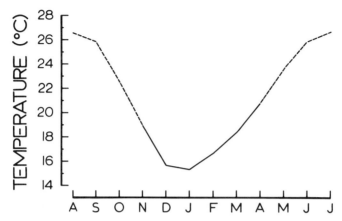

Figure 4. Simulated multi-annual mean monthly temperatures for the study site. Computations based upon climatological data from the Vero Beach, Florida airport. Summer temperatures (broken line) were not verified.

nual temperature curve computed for the Indian River at the study site. Calculations are based upon climatological data for the Federal Aviation Administration Office in Vero Beach, Florida, approximately 10 km from the study site. The midwinter low of approximately 15.5C in January compares favorably with the temperature data presented in Fig. 2. Although these calculations provide a good baseline against which a given winter may be judged "mild," "normal," or "severe," it is clear from Fig. 2 that attention should be focused upon periods of marked heating and cooling associated with frontal activity. Dominant time scales are on the order of several days, rather than months. Summer temperatures are traced with a broken line as the model was not verified under these conditions. The highest temperatures of just under 27C seem to be 2-3C too low for this area.

Estuarine Sensitivity to Individual Meteorological Variables

In nature, simultaneous and unrepeated variations of several variables mask the thermal response of the estuary to variations of any single variable. A model of thermal heat exchange processes allows one artificially to hold all but one variable constant to test the sensitivity of the estuary to that variable in particular. The results are useful for understanding the individual driving forces responsible for producing the observed net changes in specific estuaries.

Several series of computations have been made in which only one variable underwent a step change. In this section, thermal response of the estuary is compared quantitatively to these isolated changes to infer how the estuary responds to changes in temperature, dew point and insolation.

A number of generalizations can be made at the outset. First, test data suggest that the thermodynamic equilibrium temperatures vary only slightly with cloud cover. The direct effect of cloud cover is included in the radiative terms, and outgoing radiation appears to be largely insensitive to the insulating effect of clouds. Radiational cooling of estuarine waters varied only ±0.04 C/day during the 110-day study in the winter of 1977-78. The effect of clouds on heating by insolation is greater, but still relatively slight, during winter months. The standard deviation of the 110 computations in the 1977-78 study was 0.8C about a mean of 1.8 C/day. Extreme values ranged from 0.3 C/day to 3.5 C/day late in the study when the day length increased.

Second, the effect of variations in wind speed is, by itself, minimal. For example, at the study site an estuarine water temperature of 18.8C is in thermodynamic equilibrium with an air temperature of 20C, a dew point of 17C, 5/10ths cloud cover, 300 ly/day of insolation and a wind speed of 5 m/sec (representative winter conditions). Changing the wind speed to 0 or 10 m/sec resulted in new equilibrium water temperatures of 18.9C and 18.7C respectively. These results were basically unchanged through a wide range of weather conditions. It appears that the effect of wind is largely in determining how fast the new equilibrium temperature will be attained, rather than on temperature itself.

Greatest sensitivity was found for variations in air temperature and dew point. A crude generalization is possible. Sequential computations made around the conditions listed in the preceding paragraph show that there is an approximately 0.7C change in water temperature occurring with 1C changes in the dew point and a 0.2C water temperature change in response to a 1C change in air temperature. For comparison, water temperatures will vary approximately 0.3 C/day for each 50 langleys/day difference in sunlight. In winter, total daily insolation falls generally within the range of 150-300 ly.

Results thus suggest that latent and, to a lesser extent, sensible heat fluxes will dominate air-estuary local energy exchanges when air temperature and moisture content themselves vary significantly in the overlying air mass. During winter, when cold air outbreaks along the southern tier of states lead to alternating maritime tropical and continental arctic air masses, one may expect sensible and latent exchanges to be the dominant local heat flux terms.

Concluding Remarks

A few final comments are in order regarding the model and its applications. It should be apparent that use of the model for complementing estuarine research is determined largely by the user's needs. Where water temperatures alone are needed, and especially where long time intervals are not involved, it may indeed be simpler just to make *in situ* temperature measurements at the study site. Even for extended sampling periods, recording thermographs are available to provide temperature measurements closely spaced in time. Many field studies, however, involve several sampling stations, and in such cases the cost of recording equipment for each site may be prohibitive. It may then be more efficient to use regionally representative meteorological data in a numerical model in which coefficients can be adjusted for each location. For each site, there would be required only sufficient temperature data to verify the simulated temperatures to within some acceptable error.

The real advantage in a modeling approach to estuarine heat budget studies lies in the ability to simulate historical events, investigate hypothetical conditions, and decompose net heating or cooling into its individual components. Events such as fish kills can then be described using available meteorological data even if no hydrographic data exist for the time and place of interest. The investigation of the thermal response to contrived weather events may be more of an academic exercise, but it is of interest to describe temporal variations in temperature that would occur in response to sustained periods of clear skies, low temperatures, high humidities, etc. In this way, one can infer the probable extreme conditions at a study site. The ability to decompose the net heat flux into its individual components provides the only means to simulate temperatures, to understand the magnitudes and relative importance of the individual processes, and to improve the empirical expressions themselves. Although the modeling exercise is necessarily site specific, it can be used as a valuable research tool for heat budget studies in particular, and for estuarine investigations in general in areas where local fluxes are dominant.

Acknowledgments

The author would like to express his appreciation to Mr. George H. Kierspe for his help in writing the computer programs used in this study, and for processing the meteorological data. Harbor Branch Foundation, Inc. Contribution No. 183.

References Cited

Gilmore, R. G., L. H. Bullock and F. H. Berry. 1978. Hypothermal mortality in marine fishes of south-central Florida, January 1977. *Northeast Gulf Science* 2:77-97.

Heath, R. A. 1977. Heat balance in a small coastal inlet Pauatahanui Inlet, North Island, New Zealand. *Est. Coastal Mar. Sci.* 5:783-792.

Hildebrand, H. and D. King. 1976. A biological study of the Cayo del Oso and the Pita Island area of the Laguna Madre. Ann. Rept. 1975-76 to Central Power and Light Co., Corpus Christi, TX. 295 pp.

Hsu, S. A. 1978. Micrometeorological fluxes in estuaries. pp. 125-134. *In:* B. J. Kjerfve (ed.) Estuarine Transport Processes. U. South Carolina Press, Columbia, SC.

Murray, F. W. 1967. On the computation of saturation vapor pressure. *J. Appl. Meteor.* 6:203-204.

Payne, R. E. 1972. Albedo of the sea surface. *J. Atmos. Sci.* 29:959-970.

Priestley, C. H. B. and R. J. Taylor. 1972. On the assessment of surface heat flux and evaporation using large-scale parameters. *Monthly Weather Review* 100:81-92.

Reed, R. K. 1976a. An evaluation of cloud factors for estimating insolation over the ocean. NOAA Tech. Memoran. Pacific Marine and Environmental Lab. 20 pp.

Reed, R. K. 1976b. On estimation of net long-wave radiation from the oceans. *J. Geophys. Res.* 81:5793-5794.

Reed, R. K. 1977. On estimating insolation over the sea. *J. Phys. Oceanography* 7:482-485.

Smith, N. P. 1977. A note on winter temperature variations in a shallow seagrass flat. *Limnol. Oceanogr.* 22:1079-1082.

Swinbank, W. C. 1963. Long-wave radiation from clear skies. *Quart. J. Roy. Meteorol. Soc.* 89:339-349.

BOX MODEL APPLICATION TO A STUDY OF SUSPENDED SEDIMENT DISTRIBUTIONS AND FLUXES IN PARTIALLY MIXED ESTUARIES

Charles B. Officer

Earth Sciences Department
Dartmouth College
Hanover, New Hampshire

and

Maynard M. Nichols

Virginia Institute of Marine Science
College of William and Mary
Gloucester Point, Virginia

Abstract: Box model theory provides a simple method for the investigation of the behavior of nonconservative quantities in estuaries. It permits, within its limits, a quantitative examination of biological, chemical and geological distributions, transformations and other effects which depend, in part, on estuarine hydrodynamics for their explanation. The model inputs are salinity distribution, geometry of the estuary, and river flow, in addition to the distribution of the nonconservative quantity being investigated. The theory is applied here to a study of suspended sediment distributions and fluxes and to net deposition and erosion as a function of longitudinal distance along the estuary. Results are given for two partially mixed estuaries and for various river flow conditions within each estuary.

Introduction

Possible causes for observed suspended sediment distributions and their associated turbidity maxima have been discussed by Inglis and Allen (1957), Nichols and Poor (1967), Postma (1967), Schubel (1968, 1969), Dyer (1972), Krone (1972), Meade (1972), Peterson et al. (1975) and Krone and Ariathurai (1976) among others. Two principal controlling processes are cited. One is that the turbidity maximum is associated with gravitational circulation effects in which suspended particles moving seaward in the upper portion of the water column in the middle to lower reaches of an estuary sink and are carried back landward in the lower portion of the water column to provide a turbidity maximum in the upper to middle reaches of the estuary. The other process, acting independently or in combination with the first, is that the turbidity maximum is created by tidal resuspension of bottom sediments in the vicinity of the null zone of bottom sediment deposition and bottom water circulation. Characteristically, resuspension may become important elsewhere in reaches where the bottom shoals or where flow is constricted by narrow cross sections.

Festa and Hansen (1978) modeled the suspended sediment turbidity maximum related to gravitational circulation effects. The turbidity maximum modeling was based on the Festa and Hansen (1976) hydrodynamic numerical model for two dimensional, gravitational circulation. The model included longitudinal and vertical diffusion coefficients in addition to the longitudinal and vertical circulation velocities. It also assumed no net deposition or erosion of bottom sediments; in other words, it was a conservative model with respect to the estuary itself.

Officer (1980a) compared the numerical model results of Festa and Hansen with a two dimensional box model, or finite segment, approach. In the box model that was used, the assumption was made that longitudinal gravitational effects were dominant over longitudinal tidal exchange effects. In the classification system of Hansen and Rattray (1966) this is equivalent to assuming that their defining parameter, v, is nearly zero. The box model values predicted the same approximate location for the turbidity maximum; and the magnitude of the box model values, taken point by point, were similar to the numerical model values, particularly for small settling velocities.

In both cases, the turbidity maximum was associated with the upestuary portion of the salinity gradient region when the river sediment source was dominant or comparable with the ocean source. The suspended sediment maximum was associated with salinity values of around 0.10σ, where σ is the salinity of the ocean (or bay) reservoir into which the estuary empties. In the numerical model, the null zone for bottom water circulation is located at a salinity value of 0.01σ.

The null zone is sometimes associated with a depositional area for bottom boundary layer sediment transported upestuary from the ocean source with the upestuary, net circulation velocity and for bottom boundary layer sediment transported downestuary from the river source with advective river flow. There would appear to be ample evidence that local resuspension of bottom sediments by tidal currents in the vicinity of the null zone, depositional region, and elsewhere can contribute significantly to the suspended sediment distribution and the turbidity maximum, as discussed by Inglis and Allen (1957), Schubel (1968, 1969) and Krone and Ariathurai (1976).

Further, Allen et al. (1979) argue that in some estuaries there is an increase in the amplitude of the tidal current in an upestuary direction due to changes in the estuary geometry, followed further upestuary by a decrease in tidal current due to normal bottom frictional effects. They propose that in such an estuary the zone of maximum tidal current could provide a turbidity maximum even in the absence of a gravitational circulation. However, Nichols and Thompson (1973) state that if the turbidity maximum is seasonal, which is not uncommon, and related to the river input flux of suspended sediments, it cannot be associated solely with a local tidal resuspension.

Theoretically, it would, thus, seem appropriate to conclude that a suspended sediment turbidity maximum could be caused by a combination of gravitational circulation effects and local resuspension of bottom sediments by tidal currents or by either separately.

We propose here to investigate actual suspended sediment distributions using a box model approach. An advantage of a simple box model segmentation is that it can be easily applied to actual estuarine conditions and that it can be extended to include flux transfers with the bottom, i.e., net deposition of suspended sediments to the bottom or net erosion of bottom sediments to the suspended distribution.

Although the model permits the determination of suspended sediment fluxes and net deposition or erosion related thereto, it does not include longitudinal fluxes associated with bottom or bottom boundary layer sediment transport or transport related to fluid muds. Total bottom sediment balance including both suspended and bottom flux contributions cannot be made at this time.

Several investigators, including Odd and Owen (1972), Ariathurai and Krone (1976) and Kuo et al. (1978), have modeled the nonconservative behavior of suspended sediment distributions with varying assumptions as to the gravitational circulation flow and resuspension and settling effects. Our approach here has been not to assume what these nonconservative effects are in terms of bottom stresses and particle settling but rather to determine what the net depositional and erosional effects are in terms of the longitudinal fluxes.

Box models

The essentials of two dimensional box model formulation have been given by Pritchard (1969) and extended by Officer (1980b). The particular model used here assumes that horizontal nonadvective exchanges, due principally to tidal mixing effects, are small in comparison with horizontal net circulation exchanges, expressed here as advective quantities, i.e. $v = 0$. The model also assumes a steady state. The various longitudinal advective and vertical advective and nonadvective hydrodynamic exchange coefficients are determined directly from the observed salinity distribution.

For the definition of the suspended sediment distribution, a particle settling flux is added across the horizontal boundary between the upper and lower portion of each two-dimensional box. In addition, an exchange flux with the bottom representing the net flux of particle settling to the bottom and tidal resuspension is added.

Suspended sediment distributions, then, can be determined under various assumptions as to particle settling velocities and tidal resuspension fluxes. For our purposes here, net flux exchange with the bottom is given in terms of the difference of the calculated longitudinal fluxes across the vertical boundaries at each end of any given box. For details on this particular box model formulation the reader is referred to Officer (1980b).

Applications

The method has been applied to measurements of suspended sediment and salinity taken in the James and Rappahannock estuaries. Hansen and Rattray (1965) determined a value for the estuarine parameter, v, of 0.1 for the James estuary, which gives some justification for the application of the box model used here under the assumption that $v = 0$. Unfortunately, in dealing with estuarine phenomena one usually is constrained by (1) an imperfect or incomplete data set, related often to the tidal time scale within which one has to work, and (2) variable freshwater inflows and sediment influxes. In this case, measurements were taken along the estuary as a function of depth at the slack period before flood tide. Although they do not represent tidal averaged conditions, slack tide observations do provide a consistent picture within themselves.

James estuary. Six sets of observations of suspended sediment and salinity as a function of depth were taken in the James estuary during 1978. They divide into two groups corresponding to a period of moderate river flow, 12 and 25 April, and to a period of high river flow, 29 and 30 April and 2 and 7 May. The suspended sediment measurements represent total suspended material including lesser amounts of suspended organic constituents.

Contours of the suspended sediment distribution in mg/l and salinity in $^0/_{00}$ are shown in Fig. 1 for 19 April and in Fig. 2 for 29 April. For the moderate flow condition of 19 April, the turbidity maximum is centered at naut. mi. 34 (63 km), as measured from the estuary mouth, and for the high flow condition of 29 April at naut. mi. 23 (42 km). Correspondingly, the salinity values have moved downestuary from the moderate to high river flow condition; for 19 April the turbidity maximum locus resides near a salinity value of around 0.5 $^0/_{00}$ and for 29 April a value around 1 $^0/_{00}$.

Longitudinal flux relations from Officer (1980b) were used to estimate depositional and erosional fluxes. The estuary was divided into boxes of 5 naut. mi. (9.3 km) extent, centered over miles 0, 5, 10, etc. (0, 9.3, 18.5, —km). The results of these calculations are shown in Table 1. River flow values appropriate to the conditions at the time of measurement in the estuary were estimated from gauged river flows at Richmond, starting at a time two days prior to the measurement date and averaging over the preceding three days; these values were enhanced by a factor of 1.3 to account for additional drainage from Richmond to the estuary region in which the observations were made.

The individual depositional and erosional flux values show substantial variations from one box to the next and from one measurement period to the next. This expresses, in part, the variations in the individual slack tide,

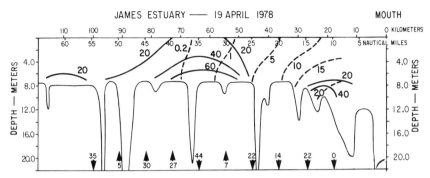

Figure 1. Slack tide suspended sediment contours in mg/l (solid lines) with superimposed salinity contours in °/oo (dashed lines) for the James estuary on 19 April 1978. Arrows indicate calculated slack tide deposition (pointing down) or erosion (pointing up) in gm/m²•day.

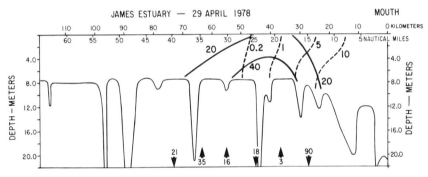

Figure 2. As Fig. 1, for James estuary on 29 April 1978.

rather than tidal averaged, concentration values. It is also related to the approximate calculation procedures used which assume steady state conditions with dominance of gravitational circulation over longitudinal tidal exchange effects. Still further, Pritchard (1978) and Elliott (1978) have demonstrated that the normal gravitational circulation, while dominant, may not be present during any given measurement period and may be replaced by other circulation patterns related to meteorological and other possible external forcing. It is clear from their observations and conclusions that measurements such as those discussed here should be taken over a considerably longer time period than one day to assimilate these transient effects. Three of our four data sets in the James and Rappahannock estuaries approach this criterion, and we consider that the averaged values

Table 1. James estuary suspended sediment fluxes with the bottom in units of tonnes/day for each box. Net deposition is negative and net erosion positive. Also included are the suspended sediment fluxes in from the river and in from Chesapeake Bay in tonnes/day, the estimated river flows (R) in m^3/sec, and the averaged net deposition or erosion at each location in $gm/m^2 \cdot day$. The number in parentheses is the total flux from both the River and Bay.

1978 data — moderate river flow

River mile	55	50	45	40	35	30	25	20	15	10	Flux in from River	Bay	R
19 April	−40	−20	50	230	−250	210	−390	−20	−50	−80	340	20	209
25 April	−250	40	440	580	−900	0	−310	−490	−780	80	750	840	205
average	−150	20	240	410	−570	100	−350	−260	−410	0	540	430	207
gm/m²·day	−35	5	30	27	−44	7	−22	−14	−22	0	(970)		

1978 data — high river flow

River mile	40	35	30	25	20	15	Flux in from River	Bay	R
29 April	0	590	330	−50	100	−530	360	−800	594
30 April	0	−990	520	480	−180	−2670	3560	−720	1350
2 May	−1040	750	220	370	−1260	−3130	5320	−1230	1734
7 May	−250	1430	−140	−1940	1600	−210	2640	−3130	651
average	−320	450	230	−290	70	−1640	2970	−1470	1082
gm/m²·day	−21	35	16	−18	3	−90	(1500)		

from each data set are valid indicators of net depositional, erosional and longitudinal flux effects.

The averaged values, as given in tonnes/day per box or gm/m²•day, show a more consistent pattern than the individual determinations. As illustrated in Fig. 1 and Table 1 for the moderate flow conditions of 19 and 25 April, there is net deposition downestuary from the turbidity maximum to the mouth. For the boxes centered at miles 25, 20, 15 and 10, the total deposition of 1020 tonnes/day is comparable with the total suspended sediment influx of 970 tonnes/day from the river and Bay combined. As illustrated in Fig. 2 and Table 1 for the high flow conditions of 29 and 30 April and 2 and 7 May, a similar longitudinal trend would appear to exist.

For the moderate flow condition there is a net flux of suspended sediment into the estuary from Chesapeake Bay of comparable magnitude to the river flux. For the high flow condition, in which the turbidity maximum is nearer to the estuary mouth, there is a net flux out of the estuary of about half the total suspended river flux into the estuary.

Rappahannock estuary. Four sets of observations were taken in the Rappahannock estuary during 1978, corresponding to a period of moderate to high river flow. Five other sets of measurements had been taken in 1972 following Tropical Storm Agnes.

Contours of suspended sediment and salinity are shown in Fig. 3 for 30 March 1978 and in Fig. 4 for 17 July 1972. In both cases the turbidity maximum is associated with a salinity value of around 0.5 °/oo.

Flux calculations are given in Table 2. The estuary river flow estimates were made in the same manner using the gauged flows at Fredericksburg. As shown in Table 2 and on Figs. 3 and 4, there is again net deposition downestuary of the turbidity maximum. For the 1978 measurements, the boxes centered at naut. mi. 25, 20, 15, 10 and 5 show a total deposition of 2520 tonnes/day as compared with a total influx of 2070 tonnes/day from the river and Bay combined; for the 1972 measurements the same boxes show a total deposition of 550 tonnes/day as compared with a total influx of 400 tonnes/day.

For the period of the 1978 measurements there is a net flux of suspended sediment into the estuary from Chesapeake Bay comparable in

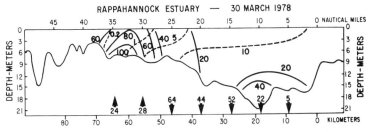

Figure 3. As Fig. 1, for Rappahannock estuary on 30 March 1978.

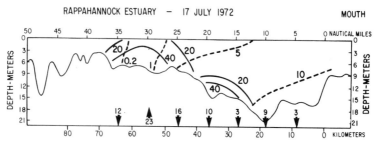

Figure 4. As Fig. 1, for Rappahannock estuary on 17 July 1972.

magnitude to the river flux. For the period of the 1972 measurements, following Tropical Storm Agnes, there is a slight flux out of the estuary.

Conclusions and Commentary

From these observations and calculations it would appear that the following conclusions can be made with regard to slack tide conditions in the James and Rappahannock estuaries. The turbidity maximum is an effect related to gravitational circulation, and it is associated with the upestuary portion of the salinity gradient region. In both estuaries there is net deposition downestuary from the turbidity maximum. For moderate flow conditions in both estuaries there is a comparable flux of suspended sediment into the estuary from both the river and the Bay. At the highest flow condition for the James estuary there is a net suspended sediment flux out of the estuary of about half the original river input flux. This conclusion with regard to a net flux into the estuaries from Chesapeake Bay itself is in agreement with that obtained by Schubel and Carter (1977) from an examination of the suspended sediment budget for Chesapeake Bay.

The slack tide conditions discussed above do not, however, give a complete picture of suspended sediment distribution and transport for at least three reasons:

First, resuspension of bottom sediments is a tidal dependent phenomenon, as illustrated by Nichols (1972) for the James estuary and by Schubel (1972) and Schubel et al. (1978) for Chesapeake Bay. From Nichols (1972, Figs. 9 and 10A), the slack tide values for the suspended sediment concentrations in the upper and lower portions of any given box underestimate the corresponding tidal averaged values by 30 and 50% in the vicinity of the turbidity maximum and by 0 and 20% downestuary from the maximum. Local resuspension is an important process in defining the tidal averaged suspended sediment distribution. After increasing the observed slack tide concentration values by the percentages given above, the calculations for Tables 1 and 2 were repeated. The influxes from the Bay are changed to 750, −1670, 1650 and −30 tonnes/day as compared

Table 2. Rappahannock estuary suspended sediment fluxes with the bottom in units of tonnes/day for each box. Net deposition is negative and net erosion positive. Also included are the suspended sediment fluxes in from the river and in from Chesapeake Bay in tonnes/day, the estimated river flows (R) in m³/sec, and the averaged net deposition or erosion at each location in gm/m²·day. The number in parentheses is the total flux from the river and Bay.

1978 – moderate to high river flow

River mile	35	30	25	20	15	10	5	Flux in from River	Flux in from Bay	R
28 March	220	-210	-340	-450	-60	-1030	-1530	650	2750	116
29 March	230	730	-1660	-440	-700	170	-150	1310	510	230
30 March	110	630	-1220	-360	-2900	240	1330	1370	800	289
10 April	90	20	-180	-920	630	-570	50	280	600	64
average	160	290	-850	-540	-750	-300	-80	900	1170	175
gm/m²·day	24	28	-64	-44	-52	-22	-5	(2070)		

1972 – adjustment period following Tropical Storm Agnes

River mile	35	30	25	20	15	10	5	Flux in from River	Flux in from Bay	R
29 June	-390	690	-590	-240	-440	-160	30	1180	-80	210
4 July	90	150	-370	-90	0	-280	-260	970	-210	224
14 July	-50	90	-30	-100	110	-90	0	110	-40	61
17 July	-220	370	-60	-150	70	-30	-30	60	-10	75
21 July	120	-80	-30	-10	10	-30	0	30	-10	66
average	-90	240	-210	-120	-50	-120	-50	470	-70	127
gm/m²·day	-12	23	-16	-10	-3	-9	-3	(400)		

with the corresponding values of 430, −1470, 1170 and −70 tonnes/day as given in Tables 1 and 2, but the deposition pattern within the estuary remains essentially the same.

Second, there can be important nonlinear tidal effects. Nichols (1972, Fig. 12) for the James estuary and Nichols and Thompson (1973) for the Rappahannock estuary illustrate that larger resuspension concentrations are often observed on flood than on ebb tide. This phenomenon could produce a substantial, and additional, suspended sediment flux in an upestuary direction for the middle to lower reaches of the estuaries.

Third, as stated previously, bottom boundary layer transport effects have not been included. Nichols (1972) and, more particularly, Schubel et al. (1978) show that concentrations within 1 m of the bottom can be several times the concentrations elsewhere in the estuarine water column. The net, nontidal, near bottom transport associated with these enhanced concentrations will be in an upestuary direction toward the null zone. Essentially this latter is the process modeled by Odd and Owen (1972) and by Krone and Ariathurai (1976). An additional depositional zone would be anticipated in the vicinity of the null zone, not indicated by the slack tide data and the box model formulation given here.

It is speculated that the additional fluxes, as represented by the combined effects of these three processes, will not substantially alter the depositional pattern indicated by the slack tide values for the lower reaches of the estuaries but that there will be an additional depositional region in the vicinity of the null zone, not indicated by the slack tide data. In addition, the turbidity maximum will be substantially enhanced in the vicinity of the null zone during periods of maximum flood and ebb by the local resuspension of bottom sediments.

These speculations as to two possible depositional regions, one in the vicinity of the null zone and the other downestuary from the turbidity maximum, are in accord with those reached by Nichols and Poor (1967, Fig. 9) for the Rappahannock estuary and by Nichols (1972, Fig. 14) for the James estuary from historical bathymetric data. Interestingly, the extensive investigations of the Thames estuary by Inglis and Allen (1957) also show two principal depositional regions; one, the Mud reaches, associated with the null zone and the other downestuary, the Gravesend shoal, associated with a change in estuarine geometry.

References Cited

Allen, G. P., J. C. Salomon, P. Bassoullet, Y. du Penhoat and G. de Grandpre. 1979. Effects of tides on hydrology and sedimentation in macrotidal estuaries, Proceedings of the International Association of Sedimentologists. (in press)

Ariathurai, R. and R. B. Krone. 1976. Finite element model for cohesive sediment transport. Amer. Soc. Civil Eng., *J. Hydraul. Div.* 102(HY3):323-338.

Dyer, K. R. 1972. Sedimentation in estuaries, pp. 10-32. *In:* F. S. K. Barnes and J. Green (eds.), *The Estuarine Environment.* Applied Science Publishers, London.

Elliott, A. J. 1978. Observations of the meteorologically induced circulation in the Potomac estuary. *Est. Coastal Mar. Sci.* 6:285-299.

Festa, J. F. and D. V. Hansen. 1976. A two dimensional numerical model of estuarine circulation: The effects of altering depth and river discharge. *Est. Coastal Mar. Sci.* 4:309-323.

Festa, J. F. and D. V. Hansen. 1978. Turbidity maxima in partially mixed estuaries: A two dimensional numerical model. *Est. Coastal Mar. Sci.* 7:347-359.

Hansen, D. V. and M. Rattray. 1965. Gravitational circulation in straits and estuaries. *J. Mar. Res.* 23:104-122.

Hansen, D. V. and M. Rattray. 1966. New dimensions in estuary classification. *Limnol. Oceanogr.* 11:319-326.

Inglis, C. C. and F. H. Allen. 1957. The regimen of the Thames estuary as affected by currents, salinities and river flows. *Proc. Inst. Civil Eng.* 7:827-868.

Krone, R. B. 1972. A field study of flocculation as a factor in estuarial shoaling processes. U.S. Army Corps of Engineers, Committee on Tidal Hydraulics. *Tech. Bull.* 19, 125 pp.

Krone, R. B. and R. Ariathurai. 1976. Applications of predictive sediment transport models. Proceedings of the Seventh World Dredging Conference, 259-272.

Kuo, A., M. M. Nichols and J. Lewis. 1978. Modeling sediment movement in the turbidity maximum of an estuary. Virginia Water Resources Center, Bull. 111, 76 pp.

Meade, R. H. 1972. Transport and deposition of sediments in estuaries, pp. 91-120. *In:* B. W. Nelson (ed.), *Environmental Framework of Coastal Plain Estuaries,* Memoir 133, Geol. Soc. Amer., Boulder, CO.

Nichols, M. M. 1972. Transport and deposition of sediments in estuaries, pp. 91-120. *In:* B. W. Nelson (ed.), *Environmental Framework of Coastal Plain Estuaries.,* Memoir 133, Geol. Soc. Amer., Boulder, CO.

Nichols, M. M. and G. Poor. 1967. Sediment transport in a coastal plain estuary. Amer. Soc. Civil Eng., *J. Waterways and Harbor Div.* 93(WW4):83-95.

Nichols, M. M. and G. Thompson. 1973. Development of the turbidity maximum in a coastal plain estuary. Virginia Institute of Marine Science, Final Report to U.S. Army Research Office Durham, NC. 47 pp.

Odd, N. V. M. and M. W. Owen. 1972. A two layer model of mud transport in the Thames estuary. *Proc. Inst. Civil Eng., Supplement IX,* Paper 75175, 175-205.

Officer, C. B. 1980a. Discussion of the turbidity maximum in partially mixed estuaries. *Est. Coastal Mar. Sci.* 10:239-246.

Officer, C. B. 1980b. Box models revisited. *In:* P. Hamilton (ed.), *Wetlands and Estuarine Processes and Water Quality Modeling,* Plenum Publishing Corp., New York. (in press)

Peterson, D. H., T. J. Conomos, W. W. Broenkow, and E. P. Scrivani. 1975. Processes controlling the dissolved silica distribution in San Francisco Bay, pp. 153-187. *In:* M. L. Wiley (ed.), *Estuarine Research, Vol. I.* Academic Press, New York.

Postma, H. 1967. Sediment transport and sedimentation in the estuarine environment, pp. 158-179. *In:* G. H. Lauff (ed.), *Estuaries,* Publ. No. 83, Amer. Assoc. Adv. Sci., Washington, D.C.

Pritchard, D. W. 1969. Dispersion and flushing of pollutants in estuaries. Amer. Soc. Civil Eng., *J. Hydraul. Div.* 95(HY1):115-124.

Pritchard, D. W. 1978. What have recent observations obtained for adjustment and verification of numerical models revealed about the dynamics and kinematics of estuaries?, pp. 1-9. *In:* B. J. Kjerfve (ed.), *Estuarine Transport Processes,* U. South Carolina Press, Columbia, SC.

Schubel, J. R. 1968. Turbidity maximum of the northern Chesapeake Bay. *Science* 161:1013-1015.

Schubel, J. R. 1969. Size distribution of the suspended particles of the Chesapeake Bay turbidity maximum. *Neth. J. Sea Res.* 5:252-266.

Schubel, J. R. 1972. Distribution and transport of suspended sediment in upper Chesapeake Bay, pp. 151-167. *In:* B. W. Nelson (ed.), *Environmental Framework of Coastal Plain Estuaries,* Memoir 133, Geol. Soc. Amer., Boulder, CO.

Schubel, J. R. and H. Carter. 1977. Suspended sediment budget for Chesapeake Bay, pp. 48-62. *In:* M. L. Wiley (ed.), *Estuarine Processes, Vol. II.* Academic Press, New York.

Schubel, J. R., R. E. Wilson and A. Okubo. 1978. Vertical transport of suspended sediment in upper Chesapeake Bay, pp. 161-175. *In:* B. J. Kjerfve (ed.), *Estuarine Transport Processes,* U. South Carolina Press, Columbia, SC.

TIME SERIES MEASUREMENTS OF ESTUARINE MATERIAL FLUXES

Björn Kjerfve

*Belle W. Baruch Institute for Marine Biology
and Coastal Research, Department of Geology,
and Marine Science Program, University
of South Carolina, Columbia, South Carolina*

and

H. N. McKellar, Jr.

*Belle W. Baruch Institute for Marine Biology
and Coastal Research, and Department of
Environmental Health Sciences, University
of South Carolina, Columbia, South Carolina*

Abstract: Annual material flux estimates were made in a major tidal creek cross-section of the North Inlet marsh-estuarine system, South Carolina. Fluxes were computed as a cross-correlation between instantaneous material concentration and instantaneous discharge. The concentrations were obtained from one daily water sample. A simultaneous estimate of the tidal discharge was simulated from a multiple regression model based on the tide record. A year-long time series of daily fluxes of organic carbon, total nitrogen, total phosphorus and chlorophyll-*a* resulted. By averaging these fluxes over 30 days the tidal variability was eliminated, yielding monthly mean values. Over the year, there was a highly significant export of organic carbon, total nitrogen, and a significant export of total phosphorus. Chlorophyll-*a* was simultaneously imported. The organic carbon flux plus the carbon accumulated in the sediment indicate a minimum annual net production of 503 g carbon fixed per m^2 within the North Inlet marsh-estuary system.

Introduction

Several recent studies have focused on the functional ecology of salt marshes and estuaries, particularly their role as material sinks or sources (e.g. Day et al. 1973; Heinle and Flemer 1976; Woodwell et al. 1977; Armstrong and Hinson 1978; Gardner and Kitchens 1978; Valiela et al. 1978). Whether or not marsh-estuarine systems supply carbon, nitrogen, and phosphorus to the coastal ocean remains a point of contention. Odum and de la Cruz (1967) estimated a significant export of detritus from a Georgia system, which led to the concept of *outwelling*. However, Haines (1975) and Haines and Dunstan (1975) suggested that most particulate organic material does not originate from estuarine outwelling of detritus, but rather originates *in situ*. There is little doubt that, over long time periods, U.S. east

341

coast estuaries are net material sinks (Redfield and Rubin 1962) because of the relative rise of mean sea level and the associated marsh accretion. Settlemeyer and Gardner (1977) found net estuarine export of organic suspended sediment but at the same time a net import of inorganic suspended sediment. Although this model seems to be an attractive one, detailed sediment measurements by Boon (1973) in a small Virginia salt marsh system indicate a net sediment export. Obviously, the question of the role of the marsh-estuarine systems has not yet been answered conclusively.

The difficulty in making assessments of magnitudes and directions of net material transport between marsh-estuaries and the coastal ocean rests with the large tidal variability. Most of the material in solution or suspension probably sloshes back and forth through the tidal inlet. On the average, only a small fraction of the maximum instantaneous flux is likely to be added to or lost from the system. Thus, it is difficult to determine net transport direction with any certainty. Also, a thorough knowledge of the hydrographic behaviour of the system is essential to material flux estimation. In much of the published work on material fluxes—in particular, assessments of carbon, nitrogen, and phosphorus transports—the lack of hydrographic measurements and insight is apparent. For example, if the net flux is computed as the difference between the ebb and flood mean discharge and concentration cross-product, the resulting flux may be in serious error (Boon 1980).

Here, we will outline a technique for estimating long-term, net estuarine material fluxes. It is based on (1) a daily water sample, subjected to chemical analyses, and (2) the corresponding tidal record from which the instantaneous discharge is determined. We will illustrate our technique by computing fluxes of organic carbon, total nitrogen, total phosphorus, and chlorophyll-a through an experimental cross-section in the North Inlet, South Carolina, marsh-estuarine system (Fig. 1). North Inlet is characterized by a 1.5 m mean semidiurnal tide, low fresh water input, salinities of 30-34 ppt, channel depths of 5 m, tidal currents as strong as 2.3 m/s, and well-mixed conditions at least with respect to temperature and salinity (Kjerfve 1978).

Sampling Design

As a part of a major interdisciplinary study of material fluxes through North Inlet, we collected daily water samples just below the surface in the middle of the experimental cross-section at 1000 (EDT) for 12 months, beginning 1 September 1978. The main purpose of this study is to examine the transport variability at a cross-section within and between tidal cycles and for different seasons. Some of the results have been reported by Kjerfve and Proehl (1979) and Chrzanowski et al. (1979). The daily water samples were initially only collected to monitor concentration levels of 1) total organic carbon, 2) total nitrogen, 3) total phosphorus, and 4) chloro-

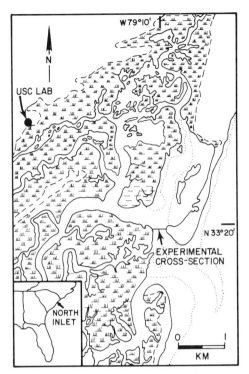

Figure 1. *North Inlet, S.C., area map showing location of experimental cross-section for which discharge and material fluxes were computed.*

phyll-*a*, in a time series fashion. These concentrations vary greatly but rather systematically over a tidal cycle with the concentrations largely depending on the stage of the tide.

As the average semidiurnal tidal cycle measures 12.42 h, it is obvious that we did not sample the same stage of the tide every day. However, we did systematically sample each phase of the tidal cycle (Fig. 2) as the sampling rate (=24.0 h) is slightly less than two tidal cycles (=24.84 h). In approximately 15 days the cycle was completed, i.e. if we started sampling at high tide, high tide was again sampled 15 days later. By averaging the concentrations over 15 days, or a multiple thereof, most of the bias due to tidal "noise" was eliminated. We chose to make concentration estimates at 30-day intervals to afford a more significant smoothing and thus greater reliability in each estimate than if we had used each 15-day estimate.

North Inlet may be classified as a well-mixed estuary based on the lack of salinity stratification, although net velocities show great cross-sectional variability (Kjerfve and Proehl 1979). Preliminary analyses of spatial nutrient concentration data indicate that, as a first order crude ap-

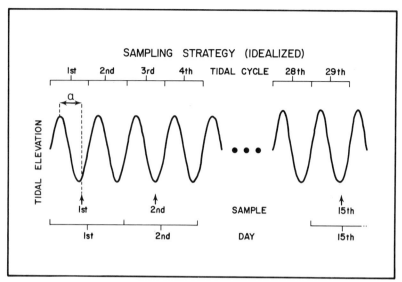

Figure 2. A schematic representation of the sampling strategy, indicating the relationship between tidal cycles, days, and sampling times. The phase lag, α, was determined empirically to be 1.0 hour.

proximation, it is reasonable to assume an approximately uniform concentration in the experimental cross-section at an instantaneous sampling time and for each parameter. Thus, we selected only one sample in the cross-section for this phase of the overall study.

Discharge

Stage curves are commonly used on rivers to relate water elevation to discharge. Because of the oscillatory nature of tidal currents this technique works poorly in estuaries. However a multiple regression scheme, regressing several tidal variables on discharge, yielded excellent results. It was selected as the most reasonable model after considerable trial and error computations with numerous other parameter choices. We found

$$Q(t) = 286 + 276\,h(t) + 1484\,b(t) + 986\,h(t)\,b(t) \tag{1}$$

with $R^2 = 0.96$, where $Q(t)$ is the instantaneous cross-sectional discharge (m^3/s) positive in the ebb direction and negative in the flood direction; t is time; $h(t)$ is instantaneous water elevation above (+) or below (−) annual mean sea level (cf. Kjerfve et al. 1978); and $b(t)$ is an instantaneous phase parameter which yields direction of the tidal current oscillation, where

$$b(t) = \sin\left[2\pi(t - \alpha)/T\right] \tag{2}$$

where T is the duration of the average tidal cycle (12.42 h) and α is a record-specific constant phase lag determined to be 1.0 h for our year-long time series. At 1000 (EDT) on 1 September 1978, t = 0 and α is the time lag of the first sample to the preceding high tide.

The regression model is obviously site specific and would have to be redeveloped for any other estuary or cross-section. It incorporates both fresh water input and tidal flow and is as good as the set of current meter data used in establishing the regression relationship. We used a 3-tidal cycle data set, consisting of 7,500 velocity measurements (Kjerfve and Proehl 1979), to derive the coefficients in eq. (1). A comparison between measured and simulated discharge is shown in Fig. 3. Spot checks between other measured and simulated data sets indicated that the model is a reasonably good one. The advantage of the model, once it has been derived, is that it allows simulation of the instantaneous cross-sectional discharge from a single reading of the tidal water elevation. Still, the model incorporates fresh-water discharge and tidal asymmetry effects.

Tidal elevation and computed discharge time series were constructed for the 1000 hours reading each day over the year. These series also suffer from tidal "noise" but by again averaging over 30-day periods the tidal signal was suppressed and monthly means resulted.

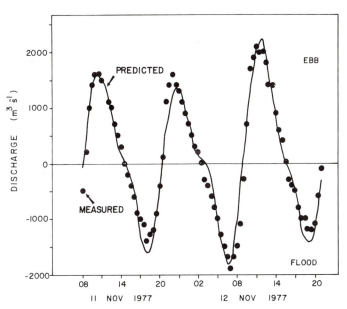

Figure 3. Comparison of measured discharge through the experimental cross-section and discharge predicted by the multiple regression equation. The multiple coefficient of determination, $R^2 = 0.96$.

Figure 4 shows the calculated variation of mean sea level and monthly net discharge. Two major points should be noted. First, the mean sea level varies annually, at least partially due to the heating cycle, from an early fall high level to a winter low level. This agrees with other North Inlet data (Kjerfve et al. 1978) and the typical picture for the U.S. east coast (Hicks and Crosby 1974). Second, during the fall, the net discharge from North Inlet is at a maximum and ebb-directed, coinciding with the falling mean sea level. From December through April the system imports water, some of which could be stored and some of which might be lost from the system to Winyah Bay in the south. Decrease in specific gravity of the North Inlet waters during the winter implies that there does not necessarily exist a relationship between net water import and the mean sea level.

Chemical Analyses

Each daily water sample was analyzed for total carbon, inorganic carbon, total nitrogen, total phosphorus and chlorophyll-*a*. Samples for carbon, nitrogen, and phosphorus determinations were stored in sterile glass containers in the dark at 4C for no longer than 1 week before analysis.

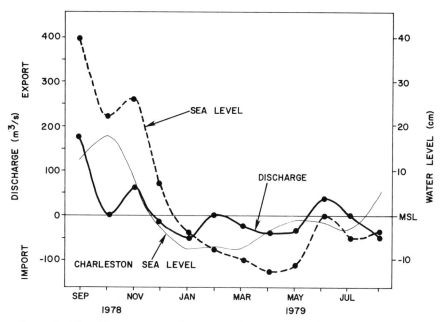

Figure 4. Monthly average of measured mean sea level at North Inlet (dashed line), monthly average of the discharge through the experimental cross-section (heavy line), and, for comparison, the 26-year measured mean sea level curve (light line) at Charleston, South Carolina.

Total organic carbon was calculated as the difference between total carbon and total inorganic carbon, which were determined through direct injection and infrared analysis with a Beckman 915-A TOC analyzer. The standard deviation of the TOC estimate for 11 replicates was 0.39 mg/l at concentrations around 4 mg/l.

Total phosphorus was determined by the persulfate oxidation method (Menzel and Corwin 1965) with a coefficient of variation of less than 3% in the 1-3 μg-at/l range. Total nitrogen was also determined by the persulfate reduction method (D'Elia et al. 1977) with a coefficient of variation of 6% in the 20-40 μg-at/l range. Both the total phosphorus and total nitrogen procedures were modified for Auto-Analyzer analysis (Loder 1978; Edwards 1979).

Pigment concentrations were determined with modifications of the standard fluorometric analysis, using a freeze-thaw acetone extraction procedure (Glover and Morris 1979). Sub-samples for pigment analysis were prepared within one hour of sampling by filtering 50 ml of water through a 25 nm GF/C filter. The filter was frozen under 1 ml of saturated $MgCO_3$ solution in a 30 ml plastic vial. Within one month the samples were thawed by adding 9 ml of acetone at room tempeature. The samples were placed in the dark at 4C for 48 h. The samples were then centrifuged and fluorometrically analyzed for chlorophyll-a. The average standard deviation for several triplicate analyses in the 3-7 mg/m^3 range was 0.38 mg/m^3 (R. G. Zingmark, Belle W. Baruch Institute, pers. comm.).

Material Fluxes

In general, instantaneous material fluxes, F(t), are computed from

$$F(t) = Q(t) \, C(t) \tag{3}$$

where C(t) is instantaneous material concentration (M/L^3) and Q(t) is instantaneous discharge (L^3/T) from eq. (1). Time series were formed for organic carbon, total nitrogen, total phosphorus, and chlorophyll-a fluxes, each series consisting of 360 instantaneous data points sampled at a rate of once a day (1000 EDT). These instantaneous material flux estimates were then subjected to the 30-day averaging to suppress the tidal signal, yielding monthly mean flux estimates for the 12-month sampling period.

The chosen averaging technique and sampling design work well whenever the instantaneous spatial concentration distribution is relatively uniform. The technique may, however, introduce a systematic bias (because of the daily sampling rate, always during daylight), if the material concentration changes diurnally in response to the day/night cycle. This is rather difficult to determine on data collected over a couple of tidal cycles, for example. Here, we have made no attempt to correct for this bias, if it exists.

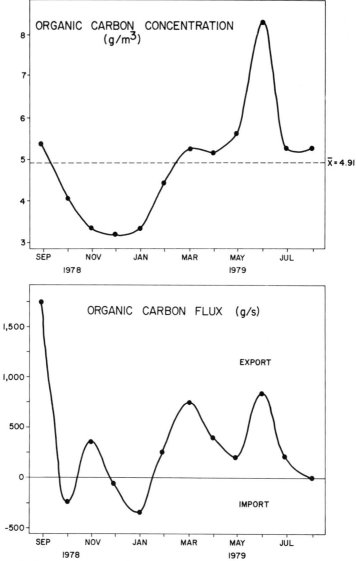

Figure 5. Plots of a) monthly mean concentration (g/m³) of tidal organic carbon in North Inlet, and b) the monthly mean total organic carbon flux (g/s) through the experimental cross-section.

a) Organic Carbon Flux

The concentration and flux estimates of organic carbon are shown in Fig. 5. Total organic carbon concentrations varied from a late fall minimum of 3.2 g/m³ to a late spring maximum of 8.2 g/m³. The seasonal trend represents changes in the sum of fine particulate organic matter (POM, as detritus and plankton) and the dissolved fraction of organic matter (DOM).

During late summer and early fall, the high sea level inundates the marsh surface to a greater extent than during other seasons (Kjerfve et al. 1978), allowing a greater exchange of organic matter between the marsh surface and the estuary. Thus, the large export of organic carbon in September occurs simultaneously with high mean sea level. Earlier data from North Inlet showed higher mean concentrations of POM in the estuary during summer and fall (Erkenbrecher 1976), indicating the importance of the higher sea levels in allowing transport of fine particulate organic matter from the marsh surface to the water. In contrast, during the winter months with lower organic carbon concentration levels, North Inlet imported organic carbon from the sea.

The spring and summer increases in TOC in the estuarine water were probably related to autochthonous organic production, higher rates of excretion, and particulate organic breakdown with increasing temperatures. Using a 50:1 carbon to chlorophyll ratio for phytoplankton, the sustained peak of 9 mg/m^3 of chlorophyll-*a* observed during the spring and summer accounts for less than 10% of the total organic carbon. Thus, the most significant fractions of organic carbon were probably finely divided particulate detritus and/or dissolved organic carbon. Erkenbrecher (1976) showed higher concentrations of dissolved organic carbon in the North Inlet during the spring and summer indicating the probable influence of accelerated excretion rates of estuarine biota and of first stage decomposition of particulate detritus.

High concentrations of total organic carbon from March through the summer corresponded to a sustained period of net organic export from the estuary to the coastal sea (Fig. 5). The timing of these trends of organic import and export have been simulated in an ecosystem model of carbon exchange in the North Inlet estuary (Summers and McKellar 1979; Summers et al. 1979). An 18-compartment, non-linear, deterministic model emphasized the interaction of sea level changes, tidal mixing, trophic transfers, metabolism and excretion in controlling seasonal trends of organic exchange in salt marsh ecosystems. The model predicted trends similar to those shown in Fig. 5 with net inwelling of organics (mainly POM) from the sea during winter and a net outwelling of organics (mainly DOM) during spring and summer.

b) Total Nitrogen Flux

The total nitrogen concentration varied around an annual mean of 378 mg/m^3 with autumn and winter low and spring and summer high concentration levels. Net export occurred from May through October (Fig. 6). Valiela et al. (1978) note that in late summer the *Spartina* growing season comes to an abrupt halt, with decreased nitrogen uptake by plants and an increased export of dissolved inorganic nitrogen. Although the *Spartina*

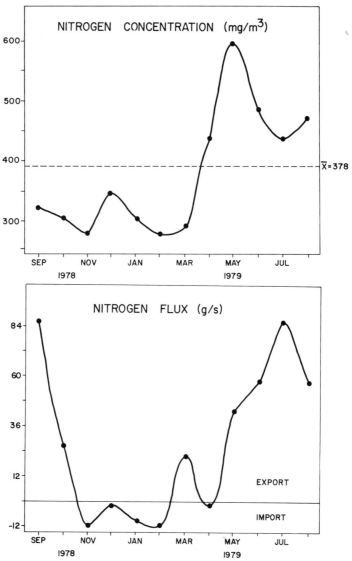

*Figure 6. Plots of a) monthly mean concentration (mg/m³) of total nitrogen
in North Inlet, and b) the monthly mean total nitrogen flux (g/s)
through the experimental cross-section.*

growing season does not end as abruptly in South Carolina, the decelera-
tion of *Spartina* growth in late summer may account for some of the
nitrogen export at that time. However, this does not explain the low concentrations and import of nitrogen during the late fall in particular.

   A significant fraction of total nitrogen in estuarine waters exists in dissolved forms; e.g. 60% as DON in the Great Sippewissett marsh (Valiela et al. 1978). Thus it is likely that the spring/summer peak in nitrogen concentration in North Inlet was due largely to an increase in dissolved organic nitrogen as a result of high rates of organic excretions from rapidly growing and metabilizing biota during the marsh production period.

c) Total Phosphorus Flux

The phosphorus concentration and flux estimates are shown in Fig. 7. Peaks were observed in September and October but a distinct seasonal trend in either concentration or flux is not apparent. Nitrogen is generally believed to be the most limiting nutrient for biological production in salt marsh environments and adjacent estuarine waters. Thus fluctuations of phosphorus are not easily interpreted in terms of seasonal trends in nutrient demand and release (Ryther and Dunstan 1971; Valiela and Teal 1974).

d) Chlorophyll-*a* Flux

The seasonal cycle of chlorophyll-*a* is an indicator of phytoplankton biomass. The concentration and flux curves indicate high export during times of high concentration (Fig. 8). These trends point to great rates of phytoplankton production coupled with microalgal production and outwash from the marsh during spring and summer. However, the chlorophyll-*a* picture is somewhat cloudy as phytoplankton blooms in the coastal ocean could easily influence the seasonal import/export pattern.

Overall Flux Means

Overall flux means were calculated for organic carbon, total nitrogen, total phosphorus, and chlorophyll-*a* and are shown in Table 1. To test the null hypothesis, $H_o:\mu=0$, i.e. there is no net flux over 12 months, we used a Student-t test, treating each monthly mean as a stratum (Mendenhall et al. 1971, pp. 53-92). The resulting t-values have been included in Table 1. The tabulated t-value in a two-tailed test at the 0.05 and 0.01 significance levels are 2.04 and 2.76, respectively, with 30 degrees of freedom.

The test indicates that organic carbon and total nitrogen yield a significant annual export at the 0.99 level, whereas the mean phosphorus export is significant at the 0.95 level. Similarly, the data indicate an annual export of water of 5 m^3/s; however, this is not significant. Chlorophyll-*a* displayed net import, but of questionable significance.

Net Ecosystem Production

The overall net production of organic matter by the marsh-estuarine ecosystem is the total gross photosynthetic production minus system respiration. This net production either accumulates within the system (as sediment accretion) or is exported by tidal action and by movement of large organisms, i.e. birds and nekton. Within our total project, the

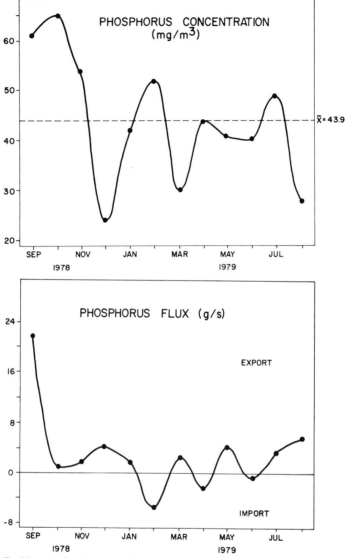

Figure 7. Plots of a) monthly mean concentration (mg/m³) of total phosphorus in North Inlet, and b) the monthly mean total phosphorus flux (g/s) through the experimental cross-section.

necessary data are being synthesized to allow magnitude estimation of mass transport by birds and nekton. These estimates are presently not available but assuming that the bird and nekton mass transport is relatively small, we can estimate the minimum net system production as the sum of tidal export and sediment accretion.

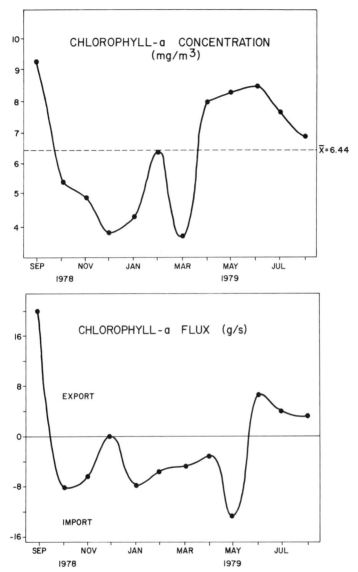

Figure 8. Plots of a) monthly mean concentration (mg/m³) of chlorophyll-a
in North Inlet, and b) monthly mean total chlorophyll-a flux (g/s)
through the experimental cross-section.

The mean annual flux of total organic carbon through the experimental cross-section is 330 g/s (Table 1). The North Inlet marsh-estuary covers
an area of 34 km² with approximately 70% of the system draining through
the experimental cross-section (Fig. 1). The annual export of organic car-

Table 1. Summary of annual mean fluxes through the experimental cross-section, standard error of the annual means, and corresponding t-statistics (*significant flux at the 95% and ** significant flux at the 99% level). Positive (+) flux is export, negative (−) flux represents import.

Parameter	Mean	S.E.	t-statistic
Discharge (m³/s)	+5	17	0.3
Organic carbon flux (g/s)	+330	90	3.7**
Nitrogen flux (g/s)	+28.6	7.7	3.7**
Phosphorus flux (g/s)	+3.0	1.1	2.7*
Chlorophyll-a flux (g/s)	−0.13	0.12	−1.1

bon then corresponds to a net annual carbon system production of 431 g/m².

To estimate the rate of carbon accumulation in the sediments, we assume that marsh sediments accumulate at a rate equal to the relative annual rise in mean sea level of 0.36 cm/yr (Hicks and Crosby 1974). The density of the marsh/peat sediment is 0.5 g/cm³ dry weight (Summers 1979). Settlemeyer and Gardner (1977) estimated the organic content of dry marsh sediment to be 10%. If organic matter is approximately 50% carbon, and 80% of the drainage area consists of marsh surface, the total annual organic carbon accumulation in the marsh sediment/peat is 90 g/m². This component corresponds to an average net annual systems carbon production of 72 g/m², as the system as a whole consists of both marsh and estuarine channels.

Thus the annual minimum net organic carbon production of the entire marsh-estuarine system is 503 g/m², of which 86% is transported out of the system by water movements. As the North Inlet system primarily consists of *Spartina* marsh, it is reasonable to compare our estimate to the range of *Spartina* production values for East and Gulf Coast marshes (Turner 1976). Although this comparison should be made carefully as our total organic carbon flux reflects metabolic activity in the marsh as well as in the water column, the agreement is very good.

Conclusions

The purpose of this paper was to outline a technique with which material fluxes through estuarine cross-sections can be estimated. To accomplish this, we:

1) simulated the instantaneous discharge from tidal parameters, based on a calibrated multiple regression curve;
2) made concentration measurements on a water sample collected at the same time once a day for 12 months;

3) cross-correlated instantaneous discharge and instantaneous material concentration to arrive at instantaneous fluxes; and

4) averaged the instantaneous flux estimates over 30-day periods to eliminate "noise" introduced by the semidiurnal tide.

A similar technique could be applied to other estuarine systems. However, a new discharge regression equation would have to be determined for each system. Also, it is important to verify that the variability of the cross-sectional concentrations is minimum, such that the one water sample can be used to represent the cross-sectional concentration. That will no doubt limit this technique to well-mixed estuaries.

It is important to realize that just because a concentration level is high, this does not necessarily imply a large net flux. The net flux is rather controlled by a large positive (export) or negative (import) cross-correlation of discharge and concentration over at least a number of tidal cycles.

Acknowledgments

This study was supported by National Science Foundation Grant No. DEP76-83010 and is Contribution No. 319 from the Belle W. Baruch Institute for Marine Biology and Coastal Research. Many individuals provided invaluable help. Marvin Marozas collected water samples and maintained the tide gage. R. T. Edwards, W. V. Johnson, H. G. Satcher III, and L. N. Henry carried out the chemical analyses, and Scott Dinnel did all flux computations. Part of the work on this manuscript was accomplished while the senior author was on leave with the Coastal Studies Unit, Department of Geography, The University of Sydney, Sydney, N.S.W. 2006, Australia.

References Cited

Armstrong, N. E. and M. O. Hinson, Jr. 1978. Influence of flooding and tides on nutrient exchange from a Texas marsh, pp. 365-379. *In*: M. L. Wiley (ed.), *Estuarine Interactions*. Academic Press, New York.

Boon, J. D., III. 1973. Sediment transport processes in a salt marsh drainage system. Ph.D. thesis. College of William and Mary, Williamsburg, Va. 226 pp.

Boon, J. D., III. 1980. Comment on paper by Valiela et al., entitled "Nutrient and particulate fluxes in a salt marsh ecosystem: Tidal exchanges and inputs by precipitation and groundwater." *Limnol. Oceanogr.* 25:182-183.

Chrzanowski, T. H., L. H. Stevenson and B. Kjerfve. 1979. Adenosine 5' triphosphate flux through the North Inlet marsh system. *Applied and Environ. Microbiol.* 37:841-848.

Day, J. W., Jr., W. G. Smith, P. R. Wagner and W. C. Stowe. 1973. Community structure and carbon budget of a salt marsh and shallow bay estuarine system in Louisiana. Publication No. LSU-SG-72-04. Center for Wetland Resources, Louisiana State U., Baton Rouge, La. 79 pp.

D'Elia, C. F., P. A. Steudler and N. Corwin. 1977. Determination of total nitrogen in aqueous samples using persulfate digestion. *Limnol. Oceanogr.* 22:760-764.

Edwards, R. T. 1979. A semiautomated technique for the determination of persulfate nitrogen and total persulfate phosphorus. Baruch Institute, U. South Carolina, Columbia, S.C. Unpublished manuscript.

Erkenbrecher, C. W., Jr. 1976. Influence of environmental factors on the distribution, composition, and transport of microbial biomass and suspended material in a salt marsh ecosystem. Ph.D. dissertation. U. South Carolina, Columbia, S.C. 215 pp.

Gardner, L. R. and W. M. Kitchens. 1978. Sediment and chemical exchanges between salt marshes and coastal waters, pp. 191-207. *In:* B. Kjerfve (ed.), *Estuarine Transport Processes.* U. South Carolina Press, Columbia, S.C.

Glover, H. E. and I. Morris. 1979. Photosynthesis carboxylating enzymes in marine phytoplankton. *Limnol. Oceanogr.* 24:495-509.

Haines, E. B. 1975. Nutrient inputs to the coastal zone: The Georgia and South Carolina shelf, pp. 303-324. *In:* L. E. Cronin (ed.), *Estuarine Research, Vol. 1.* Academic Press, New York.

Haines, E. B. and W. M. Dunstan. 1975. The distribution and relation of particulate organic material and primary productivity in the Georgia Bight, 1973-1974. *Est. Coastal Mar. Sci.* 3:431-441.

Heinle, D. R. and D. A. Flemer. 1976. Flows of materials between poorly flooded tidal marshes and an estuary. *Mar. Biol.* 35:359-373.

Hicks, S. D. and J. E. Crosby. 1974. Trends and variability of yearly mean sea level, 1893-1972. NOAA Technical Memorandum, NO3 13, COM-74-11012, Rockville, Md. 16 pp.

Kjerfve, B. 1978. Bathymetry as an indicator of net circulation in well mixed estuaries. *Limnol. Oceanogr.* 23:816-821.

Kjerfve, B., J. E. Greer and R. L. Crout. 1978. Low-frequency response of estuarine sea level to non-local forcing, pp. 497-513. *In:* M. L. Wiley (ed.), *Estuarine Interactions.* Academic Press, New York.

Kjerfve, B. and J. A. Proehl. 1979. Velocity variability in a cross-section of a well-mixed estuary. *J. Mar. Res.* 37:409-418.

Loder, T. C. 1978. A semiautomated total nitrogen and phosphorus method for low volume samples. Dept. of Earth Sciences, U. New Hampshire, Durham, N.H. Unpublished manuscript.

Mendenhall, W., L. Ott and R. Schaeffer. 1971. *Elementary Survey Sampling.* Duxbury Press, Belmont Ca. 247 pp.

Menzel, D. W. and N. Corwin. 1965. The measurement of total phosphorus in seawater based on the liberation of organically bound fractions by persulfate oxidation. *Limnol. Oceanogr.* 10:280-282.

Odum, E. P. and A. A. de la Cruz. 1967. Particulate organic detritus in a Georgia salt marsh-estuarine system, pp. 383-388. *In:* G. H. Lauff (ed.), *Estuaries.* Am. Assoc. Adv. Sci., Publ. No. 83, Washington, D.C.

Redfield, A. C. and M. Rubin. 1962. The age of salt marsh peat and its relationship to recent changes in sea level at Barnstable, Massachusetts. *Proc. Nat. Acad. Sci. U.S.* 48:1728-1735.

Ryther, J. H. and W. M. Dunstan. 1971. Nitrogen, phosphorus and eutrophication in the coastal marine environment. *Science* 171:1008-1013.

Settlemeyer, J. L. and L. R. Gardner. 1977. Suspended sediment flux through a salt marsh drainage basin. *Est. Coastal Mar. Sci.* 5:653-663.

Summers, J. K. 1979. A simulation model of estuarine subsystem coupling and carbon exchange with the sea. Ph.D. dissertation. U. South Carolina, Columbia, S.C.

Summers, J. K. and H. N. McKellar, Jr. 1979. A simulation model of estuarine subsystem coupling and carbon exchange with the sea. I. Model structure, pp. 323-366. *In:* S. E. Jorgensen (ed.), *State-of-the-art in Ecological Modelling.* Pergamon Press, Ltd., Oxford.

Summers, J. K., H. N. McKellar, Jr., R. F. Dame and W. M. Kitchens. 1979. A simulation model of estuarine subsystem coupling and carbon exchange with the sea. II. North Inlet model structure, output and validation. U. South Carolina, Columbia, S.C. Unpublished manuscript.

Turner, R. E. 1976. Geographic variations in salt marsh macrophyte production: a review. *Contrib. Mar. Sci.* 20:46-68.

Valiela, I. and J. M. Teal. 1974. Nutrient limitation in salt marsh vegetation, pp. 547-563. *In:* J. R. Reimold and W. H. Queen (eds.), *Ecology of Halophytes.* Academic Press, New York.

Valiela, I., J. M. Teal, S. Volkmann, D. Shaefer and E. J. Carpenter. 1978. Nutrient and particulate fluxes in a salt marsh ecosystem: Tidal exchanges and inputs by precipitation and groundwater. *Limnol. Oceanogr.* 23:798-812.

Woodwell, G. M., D. E. Whitney, C. A. S. Hall and R. A. Houghton. 1977. The Flax Pond ecosystem study: Exchanges of carbon in water between a salt marsh and Long Island Sound. *Limnol. Oceanogr.* 22:833-838.

REMOTE SENSING AS A TECHNIQUE FOR SYNOPTIC INVENTORIES OF FISHERIES RELATED RESOURCES

V. Klemas, D. S. Bartlett and W. D. Philpot

College of Marine Studies
University of Delaware
Newark, Delaware

Abstract: Remote sensors on aircraft and satellites are being used effectively to map fisheries related physical and biological surface properties synoptically over large coastal and estuarine areas. Tidal wetlands are known to be important fisheries habitat. With color and color infrared photography one can delineate tidal wetland boundaries and plant types. Multispectral analysis techniques have been developed for estimating the emergent biomass and species composition. Multispectral scanners and laser fluorosensors are being tested for monitoring primary productivity of coastal surface waters and the outflow of plankton and detritus from marshes and estuaries. Satellites can locate some high probability areas for fish availability, such as nutrient rich upwelling areas off the coasts of Peru, western United States and Africa, primarily due to strong thermal and spectral gradients caused by colder upwelling water and its spectrally different nutrient/chlorophyll content. There also appears to be some correlation between coastal water properties such as water color, turbidity, chlorophyll concentration and the presence of fish. A review of the state of the art indicates that remote sensing can be used effectively to map plant species, productivity and fishery-related quality of wetland habitat. However, techniques being developed for determining surface water productivity, detritus/nutrient flow and fish availability require additional field testing to establish their reliability.

Introduction

To manage production of a living marine resource such as fish, one must monitor and evaluate fish-habitat relationships and be aware of trends in extent and quality of available habitat including wetlands and associated water quality. Obtaining field measurements of such factors has proven to be costly in terms of time, manpower and funds. Also, the results are sample site specific and may not be representative of surrounding areas.

Remote sensing can provide valuable information at minimum cost and time when large coastal areas are to be surveyed. However, remote sensing, like any other technique, also has some serious limitations. The objective of this article is to give a fair appraisal of potentials and limitations of remote sensing as a technique for synoptic inventories of fishery-related resources such as wetlands and water quality and for direct estimation of fish populations.

Inventories of Fishery-Related Habitats

1. Wetlands

Coastal wetlands and estuaries are among the most productive ecosystems in the world. They provide food and shelter to finfish, shellfish, birds and mammals. Wetlands produce vitamins and growth regulators, as well as regenerating, recycling and storing nutrients. In addition, wetlands act as buffers to coastal erosion and, to some extent, as a control of water quality.

Both photographic and satellite data sources have advantages and limitations with respect to providing all data elements in an accurate cost-effective wetland mapping program. LANDSAT data processing is a least-cost method of producing coastal land cover maps and tabular data for large areas. Planning studies, however, often require more detailed coastal vegetation/habitat information at an accuracy level that is difficult to provide consistently over a range of categories through the LANDSAT data-extractive process. Manual interpretation of aerial photography is a more expensive and time-consuming process than digital multispectral processing, but it can yield the more detailed categorization of wetlands/habitats that many planning activities require.

Operational wetland mapping programs designed to meet rigorous cartographic standards typically employ photo-interpretation of low altitude color and color-infrared photographs supplemented by ground surveys. Most East Coast states in the USA have mapped their wetlands to define boundaries and to inventory major marsh plant species at scales ranging from 1:2,400 to 1:24,000 (Anderson and Wobber 1973; Bartlett et al. 1976; McEwen et al. 1976). The U.S. Fish and Wildlife Service is also conducting an inventory of the nation's wetlands. Aircraft have also been used in various ecological studies, including species composition of wetland vegetation, wetland productivity, wildlife habitat, diversity, and impact of man-made structures on wetlands and mosquito breeding habitats (Kolipinski et al. 1969; Egan and Hair 1973; Reimold et al. 1973).

More recently, the potential of Skylab imagery and LANDSAT multispectral scanner data for mapping and inventorying tidal wetlands has been demonstrated (Erb 1974; Anderson et al. 1975). The LANDSAT imagery is produced by the four channel multispectral scanner (MSS) having the following bands: band 4 = 0.5-0.6 μm; band 5 = 0.6-0.7 μm; band 6 = 0.7-0.8 μm; and band 7 = 0.8-1.1 μm. The satellite passes over each ground point every 18 days and, imaging from an altitude of 920 km, the MSS covers a swathwidth of 185 km. Visual analysis can be performed on MSS imagery or the data stored on magnetic tape can be analyzed directly, in digital form, using computers. Digital and visual analysis of LANDSAT data have been used in several studies of coastal wetland habitat (Carter and Schubert 1974; Anderson et al. 1975; Klemas et al. 1975). Results are generally good but are restricted by the limited spatial resolution of LAND-

SAT data (one picture element $\simeq$ 0.49 hectares = 1.1 acres) and by overlapping spectral signatures of some vegetation types at some times of the year. Klemas et al. (1975) found that classification accuracies derived by comparison of interpreted LANDSAT data with existing maps and photographs were above 80% for most categories.

In comparison, Skylab's S190A Multispectral Photographic Facility had a resolution of 0.09 hectare and its S190B Earth Terrain Camera about 0.02 hectare (Klemas 1976), providing valuable information for detailed wetland mapping. In a comparative study, Skylab photographs were used to identify five wetland plant species and drainage patterns were mapped in more detail than using LANDSAT (Anderson et al. 1975). Spatial resolution was still insufficient to satisfy most state inventory statutes which require geometric accuracy approaching National Map Accuracy Standards.

Data collected by orbital platforms seem more suited to periodic updating of inventories and to monitoring of large scale habitat dynamics. A significant advantage of the LANDSAT system in this regard is repetitive coverage which permits observations of physical and morphological characteristics of plants over their entire growing cycle. Although weather conditions and limitations in foreign data acquisition reduce the actual frequency of LANDSAT coverage to more than the nominal 18 day cycle, its repetitive nature is extremely valuable for inventory updates and when seasonal factors influence the spectral characteristics of the resource being monitored. An example of seasonal influences on spectral reflectance signatures has recently been noted during research on reflectance characteristics of wetland plant canopies in Delaware. Spectral discrimination of the two major wetland communities present, salt hay (a merged category containing *Distichlis spicata* and *Spartina patens*) and cordgrass *(Spartina alterniflora),* has been found to correspond to major differences in their canopy structures (Carter and Schubert 1974; Bartlett and Klemas 1977). However, seasonal changes in canopy height of cordgrass and the percentage of live and senescent vegetation in both communities can produce relative signature convergence or divergence during different portions of the year resulting in reduced or enhanced potential for discrimination in the LANDSAT spectral wave bands. Data developed recently suggests that optimal spectral discrimination of the salt hay from the cordgrass communities in Delaware will occur in November and December when a differential in rates of senescing vegetation in the two communities produces high reflectance contrast between them in all wave bands. Lowest contrast between these two communities occurred in May and June as a result of differing responses of canopy reflectance to increasing amounts of green vegetation during the spring. Traditional reliance on data collected during the growing season for analysis of wetlands vegetation at temperate latitudes is thus subject to reevaluation (Bartlett and Klemas 1979).

A further consequence of spectral response to canopy variables, particularly in cordgrass *(Spartina alterniflora)*, gives promise to estimation of emergent biomass by remote sensing. Reimold et al. (1973) found that general biomass classes could be established for this species using tonal variations in color-infrared aerial photography. Spectral measurements in the LANDSAT wave bands using a field radiometer have confirmed that various components of emergent biomass systematically affect canopy reflectance. Figure 1 shows the linear regression relationship between emergent green biomass, as measured by the harvest method, and the ratio of canopy reflectance in MSS bands 7 (0.8-1.1 μm) and 5 (0.6-0.7 μm). Band 5 (red) reflectance was shown to be inversely correlated with the percentage of green biomass in the canopy due to chlorophyll absorption of red radiation by green plant material. Band 7 (infrared) reflectance was positively correlated with canopy height. As plant leaves are generally highly transmissive and reflective in the near infrared, reflectance in this spectral region has been shown to be sensitive to the number of stacked

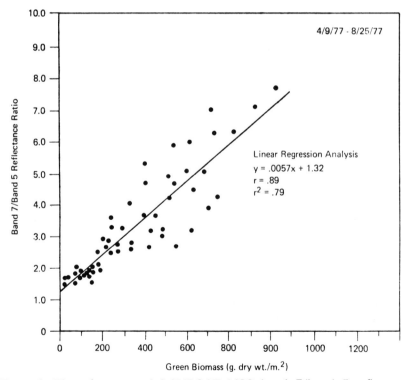

Figure 1. Plot of measured LANDSAT MSS band 7/band 5 reflectance ratio as function of green biomass in g dry wt m^{-2} of Spartina alterniflora.

leaf layers within a canopy and thus, in this case, to the thickness of the natural canopy. As the product of canopy thickness and percentage of green vegetation is highly correlated with green biomass, so is the product of infrared reflectance and the inverse of red reflectance as shown in Fig. 1. Thus, "remote sensing" would appear to hold promise for useful estimation of emergent green biomass of this wetland species, whether from satellites or in the field using a hand-held radiometer. In either case the speed of sampling inherent in the technique will allow more cost-effective and extensive sampling than is now possible by harvesting measured quadrants. Larger uncertainty in estimating the biomass of individual samples will be offset somewhat by more effective characterization of large areas than can be achieved using small numbers of harvested samples. In addition, as red reflectance is correlated with the percentage of green/total biomass ($r^2 = 0.72$), estimates of total emergent biomass may also be possible using multispectral data.

2. Surface Water Characteristics

There is a direct relationship between organic productivity of coastal areas and fish schooling. Aircraft and satellite imagery has been used with notable success in observing surface water features such as sediment load, and current patterns which are relevant to dynamics of phytoplankton distribution and productivity of surface waters. Remotely sensed data is limited to surface waters because of the limited penetration of light into water. Even with this limitation, there is a large amount of useful information available. Surface water characteristics may be classified by using single band imagery, spectral measurements or laser fluorosensing.

a. Single-band Imagery

One of the simplest and most immediately useful forms of remote sensing data is a single-band aircraft or satellite imagery. This category includes anything from black and white aerial photography to thermal or digital satellite imagery. In coastal and estuarine waters where sediment load in the surface waters is relatively high the sediment can be used as a tracer for surface currents. Patterns of sediment distribution are indicative of fronts, shoals, currents, eddies and upwelling regions. Klemas (1976) used LANDSAT imagery of Delaware Bay to observe variations in the current patterns over a tidal cycle. Klemas and Polis (1977) have also used LANDSAT imagery as well as aerial photography in a study of the location and dynamics of fronts in Delaware Bay. In the above work, remote sensing was an essential part of the study because the geographic regions were too large and/or the time scales too small to allow adequate coverage by any other method.

Color imagery adds another dimension to the observation of surface waters. The apparent water color is related to types of materials suspended or dissolved in the water column. Unfortunately, apparent water color is also affected by sun angle, atmospheric conditions and sea state. Nonetheless, properly treated ocean color analysis is potentially one of the most effective means of remote observation of water properties.

b. Classification of Coastal Water Characteristics Using Spectral Reflectance

The use of spectral reflectance characteristics to identify substances in the water has attracted considerable attention in the past several years (McCluney 1974; Gordon et al. 1975; Mueller 1976; Hovis 1977; Gordon 1978; Plass et al. 1978). Our approach is most similar to that suggested by Simmonds (1963) and Mueller (1976) — eigenvector (principal component) analysis. Perhaps the best and most complete presentation of this method can be found in Morrison (1976) who provides a detailed discussion of the technique.

Much of what is unique about our approach is an indirect result of the use of LANDSAT/MSS digital data. LANDSAT was not really intended for use in water observations. The gain is low, making the dynamic range of the sensor very limited. The four spectral channels were selected for land use applications and are hardly ideal for water observations. Yet there is a surprising amount of information in the LANDSAT imagery. LANDSAT data have been used to map sediment distribution patterns (Klemas et al. 1978), to observe occurrence of estuarine fronts (Klemas and Polis 1977) and to observe occurrence of internal waves (Apel et al. 1974), to cite only a few of the many papers in which LANDSAT imagery has been used in sensing of water. The history of utilization, coupled with the high-probability of the continuation of the LANDSAT program for many years to come, is sufficient incentive for trying to extract as much as possible from the data.

One major reason for using eigenvector analysis is that it allows the reduction of significant variates with minimal loss of information. With LANDSAT/MSS data there are only four spectral bands and therefore only four variables to begin with and the analysis will rarely reduce this number by more than one. However, eigenvectors can also provide an efficient representation of variations in water color which can be readily adapted to an automatic classification process. It is in this facile adaptation to automated classification that the eigenvector method carries the most promise for application to LANDSAT data.

It is likely that this approach can be extended with LANDSAT data to include several other substances, and that considerably better results could be achieved using spectral channels more appropriate for analysis of water, such as those on the Coastal Zone Coler Scanner (Hovis 1977).

c. Laser Fluorosensing

Laser fluorosensing systems have been used experimentally for several years for remote detection and characterization of oil slicks (Fantasia et al. 1971; Rayner and Szabo 1978) and phytoplankton (Mumola et al. 1973). We will be concerned here only with remote detection of phytoplankton.

Any fluorosensing material, when excited by light at one wavelength, will almost immediately reemit the energy as light at longer wavelength than the exciting wavelength. Generally, the emission is strongest over a rather narrow range of wavelengths and for a particular excitation wavelength. The success of a laser fluorosensing system depends on the fact that the excitation spectra and emission spectra are distinctly different for different materials. Thus, in phytoplankton, each pigment will have different fluorescence characteristics. Since each species of phytoplankton has its own characteristic balance of various pigments, it may be possible not only to detect the presence of phytoplankton, but also to identify the color group if not the specific species.

Mumola et al. (1973) have developed a multiwavelength laser system in order to make use of this species-dependent wavelength variability in the fluorescence. By using four excitation wavelengths and observing fluorescence at the four corresponding emission peaks, they have been able to detect the presence of chlorophyll and have also been able to make reasonable estimates of the chlorophyll concentration in the water.

3. Monitoring Coastal Upwelling

Upwelling is a process of vertical water motion in the sea whereby subsurface water moves toward the surface. Upwelled water can introduce large quantities of nutrients (phosphates, nitrates, etc.) to the euphotic or light zone; thus, upwelling is conducive to high organic production. Knowledge of the location and prevailing conditions of upwelling areas is important for fishing fleets. For example, extensive fishing areas and kelp beds are found in upwelling areas off the African and North and South American continents. In addition, considerable bird populations whose guano is of economic importance occur off Peru. Near the Antarctic Convergence, particularly in the Atlantic, abundant nutrients support an unusually large standing crop of diatoms and flagellates which ultimately support krill, the main food of whales, seals, and other species (Fairbridge 1966).

Upwelling may take place anywhere, but it is more common along western coasts of continents. Upwelling may be caused by wind displacing surface water away from the coast or by currents impinging on each other or on land masses. The most pronounced coastal upwellings are found off the western United States, Peru, Morocco, the Somali Republic and the west coast of Africa.

The ability to identify upwelling areas from aircraft and satellites depends on the fact that deep water which is brought up to, or near, the sea surface has different properties from surface water. The most distinguishing feature of the upwelled water is that it is colder and denser than the adjacent surface water and may contain chlorophyll and other nutrients at concentrations exceeding background levels.

When nutrients and cold water are brought into the sunlit layers, photosynthesis initiates a biological chain reaction that gives rise to accumulation of chlorophyll and other biochromes. In highly productive areas, the freshly upwelled water is initially cold and clear but gradually warms up and turns greenish with increased surface age. The fade-out of thermal contrast and the build-up of color contrast are supplementary processes (Clarke et al. 1969).

Remote sensing systems that simultaneously measure sea surface and temperature and chlorophyll coloration have provided valuable new information about distribution in space and time of the biological activity in upwelling areas (Huebner 1971). Particularly successful have been surface temperature observations from the NOAA series satellites (NOAA-2 to NOAA-5) which have been mapping temperatures of vast coastal regions with accuracies within several degrees Kelvin (Fig. 2). Under relatively cloud-free conditions, the Very High Resolution Radiometer (VHRR) aboard NOAA-5, which has a spatial resolution of 1 km at nadir, gathers data in both visible and infrared channels. The visible channel (0.6-0.7 μm) measures reflected solar radiation from the earth, while the infrared channel operates both day and night to measure radiation emitted from the earth's surface in the 10.5-12.5 μm wavelength region. The orbiting motion of the satellite (near-polar and sun-synchronous at an altitude of 1460 km), together with the day-night operation of the VHRR scanner, provides thorough coverage of North America and the adjacent ocean areas out to 100 km or more from shore twice daily at approximately 0900 and 2100 hours local time. The direct readout capability of this instrument (when the satellite is within range of the NOAA command and data acquisition ground stations at Wallops Island, VA; Fairbanks, AL; and San Francisco, CA) allows immediate used of the data. Information content of the gray-scale images is extracted to produce charts that display several significant ocean thermal features. Major water masses, fronts, upwelling areas, currents and eddies can be identified and located (LaViolette 1974; Maul et al. 1974). LANDSAT-III, which was launched in September 1977, will be particularly valuable for discriminating smaller coastal surface water features, since the multispectral scanner on this satellite has a thermal band (10.4 μm to 12.6 μm) capable of 100 m resolution on the ground. However, the thermal band is not functioning properly at this time.

Detection and monitoring of photosynthetic productivity in upwelling areas has been tried with aircraft, LANDSAT and Skylab, with some success in locating chlorophyll-rich upwelling areas (Clarke et al. 1970;

Figure 2. Portion of a NOAA-5 Satellite Very High Resolution Radiometer thermal image of Gulf stream, eddies, oceanographic fronts and colder continental shelf water along the U.S. east coast, with the coldest water shown over Georges Bank (13 April 1977).

Szekielda et al. 1975). Attempts to quantify chlorophyll concentrations in aquatic suspension have had mixed results. Inaccuracy is partly due to the fact that chlorophyll and inorganic sediment are indistinguishable. However, measurements of marine photosynthetic organisms show a "hinge-point" at approximately 0.52 μm, below which chlorophyll in suspension reflects strongly and above which absorption is dominant. On the other hand, sediment acts as a broadband backscatter. Thus, the use of two bands, separated at approximately 0.52 μm, may allow discrimination of chlorophyll from inorganic sediment. In summary, aircraft and satellite remote sensors can rapidly locate nutrient-rich upwelling areas for fishing fleets, but cannot reliably quantify chlorophyll concentration or photosynthetic productivity of the water.

Direct Estimation of Fish Populations

1. Observing Surface Currents and Fronts as Movers of Eggs and Larvae

Many fish and shellfish which spawn offshore depend on surface currents to transport their eggs and larvae into estuarine nursing grounds. This period of egg and larval drift represents the most critical survival period for certain marine fishes. When surface currents do not provide favorable transport or when fronts prevent egg and larval drift, the respective fishery may be severely affected. For example, a recent investigation by Nelson et al. (1977) demonstrated that surface transport was the most important oceanographic factor affecting menhaden recruitment along the Atlantic Coast.

Presently, surface transport for fishery applications is calculated by estimating geostrophic wind field from surface atmospheric pressure field maps prepared by groups such as the U.S. Fleet Numerical Weather Service. Sea surface stress is normally inserted into the appropriate Ekman formulation, including the Coriolis parameter, to obtain an estimate of the surface transport. However, this approach is limited by about 300 km in space and by about one month in time (Brucks and Leming 1977).

Satellite remote sensors can be used to monitor certain oceanographic features synoptically over wide areas and with frequent temporal coverage. Specifically it appears that a SEASAT type scatterometer could be used to map sea surface stress with the following advantages: a) synoptic coverage over a large area (2 × 500 km swathwidth); b) suitable resolution (50 km); c) a 36-hour repetition frequency; d) a direct measurement of sea surface stress which can be converted into surface transport for correlation with actual fish larvae movement (drift). SEASAT-A scatterometer algorithms for estimating wind-induced surface layer transport are being modified for coastal conditions and their accuracy evaluated. Presently we are proposing to track the actual movement of water masses containing fish eggs and larvae along the east coast of the U.S. in order to calibrate and verify in real time the circulation and transport model being developed by the National Marine Fisheries Service (Brucks and Leming 1977).

Another important aspect to be considered is the influence of coastal and estuarine fronts on the movement of fish eggs and larvae. Various scientists including those in our own group have found that coastal and estuarine fronts seriously influence drift and dispersion of oil slicks, phytoplankton and other suspended matter. In recent experiments in Delaware Bay, oil slicks and phytoplankton were found to line up along convergent fronts rather than follow the drift pattern predicated by a model using wind and current information (Klemas and Polis 1977). Convergent fronts have been observed regularly along the east coast on the shelf and in Delaware and Chesapeake Bays. Fronts and their movement can be monitored by NOAA-5, VHRR, LANDSAT, MSS and aircraft cameras. The aircraft/satellite/ship techniques proposed will help determine the influence of such fronts on fish egg and larvae drift as was done with oil slicks and phytoplankton in previous studies (Klemas and Polis 1976).

2. Location of Coastal Fish Schools

Spotter pilots flying light aircraft are regularly used to guide twin purse seine boats and other fishing vessels to large schools of fish such as menhaden *(Brevoortia tyrannus)*. Pilots direct boats to a particular fish school by radio and notify the captains when to encircle a school with a purse seine. An actual school of fish cannot be readily observed from satellite altitudes owing to problems of resolution, atmospheric transmission, and surface reflection. However, satellites and aircraft can detect secondary indicators or highly productive coastal waters, such as chlorophyll-*a*, sea-surface temperature, and turbidity (Maul et al. 1974). The assumption that fish distribution is governed by certain oceanographic parameters detectable from satellites has been substantiated (Savastano and Leming 1975; Kemmerer 1976). Oceanographic conditions reflected in ground truth measurements of surface temperature, chlorophyll-*a*, salinity, water color (Forel-Ule), and turbidity (Secchi disc transparency) from the two study areas were compared to determine which ones correlated with fishery data collected from fishing vessels at sites of menhaden capture. The assumption was that, if menhaden were caught in the same kind of water with respect to one or more of the parameters, the parameters showing consistency probably were affecting fish distribution.

Forel-Ule water color, turbidity, and chlorophyll-*a* concentrations were similar at locations of menhaden capture in both study areas; salinity and temperature were not. As the first three parameters can be identified in LANDSAT MSS imagery, a spectral pattern recognition technique was used to determine if water containing menhaden could be recognized from space. Locations of menhaden schools were translated into the LANDSAT coordinate reference system so that areas with and without menhaden could be identified. Radiance values from each of the four spectral channels were extracted from the data for these areas so that a computer algorithm could be developed. Digital LANDSAT MSS data were then classified into high and low probability fishing areas with the algorithm.

Menhaden school locations used to develop the classification algorithm were limited to those identified within ± 2 h of satellite coverage. Twenty-five of the 29 school locations satisfying this temporal limitation fell within or immediately adjacent to the high probability fishing areas, and 16 out of 19 other schools located outside the allocated time period fell within or next to these areas. A correlation analysis applied to menhaden and MSS spectral data provided correlation coefficients of 0.65, 0.75, 0.67 and 0.61 for bands 4, 5, 6 and 7, respectively, with all coefficients significant at the 99% confidence level (Savastano and Leming 1975; Kemmerer 1976).

In one mission, computer classification of LANDSAT MSS data into high probability fishing areas was completed and disseminated to the fishing fleet 21 h after satellite reception. The fleet reported that menhaden were concentrated in these areas and the test was successful. This report was verified by plotting locations of menhaden captures and observations on the prediction chart. Most locations fell into or adjacent to the high probability areas (Kemmerer 1976).

Summary and Conclusions

Remote sensing can provide considerable information for a minimal cost and time when large coastal areas are to be surveyed. Potential applications of remote sensing to determine available habitat, nutrient flow and fish biomass are shown in Table 1. Color and color infrared photography have been used to map tidal wetland boundaries, vegetation species and net primary production. Multispectral digital satellite techniques have been employed to map plant types, density and height of the standing crop, and other properties related to the quantity and quality of marsh biomass.

Both photographic and satellite data sources have advantages and limitations with respect to providing all data elements in an accurate cost-effective manner. LANDSAT data processing is a least-cost method of producing coastal land cover maps and tabular data for large areas. However, planning studies often require more detailed coastal vegetation/habitat information at an accuracy level that is difficult to provide consistently over a range of categories through the LANDSAT data extractive process. Manual interpretation of aerial photography is a more expensive and time-consuming process than digital multispectral processing, but it yields a more detailed categorization of wetlands/habitats that many planning activities require.

As shown in Table 1, multispectral scanners and laser fluorosensors are being tested for monitoring of primary productivity of coastal surface waters and the outflow of plankton and detritus from marshes and estuaries. Four selected laser excitation wavelengths have been used to in-

Table 1. Summary of remote sensing techniques used to determine wetland habitat, water productivity and fish availability.

AVAILABLE WETLAND HABITAT

Habitat size, location vegetation species, tidal conditions
- Aircraft color & color infrared photography
- Landsat MSS
- GT: Field checks (S)

Marsh productivity, plant vigor and water quality
- Aircraft color infrared photography
- Digital MSS from aircraft and satellites
- GT: Harvest & water sampling (L)

GT = Ground truth required
S = Small amount
M = Moderate amount
L = Large amount
MSS = Multispectral Scanner

SURFACE WATER PRODUCTIVITY

Gross flow of organic detrital turbid water into estuaries and bays
- Aircraft multiband photography
- Satellite MSS
- GT: Current measurements & water sampling (M)

Concentration of chlorophyll and phytoplankton in estuaries, and coastal waters
- Digital MSS from aircraft or satellite
- Laser fluorosensing low altitude aircraft
- GT: Water sampling (L)

SMS = Synchronous Meteorological Satellite
CZCS = Coastal Zone Color Scanner
HRIR = High Resolution Infrared Radiometer
NOAA = NOAA Series Satellites

FISH AVAILABILITY

Upwelling and other water masses having unique spectral/thermal signatures
- SMS, NOAA thermal infrared scanners (HRIR)
- Satellite MSS chlorophyll, turbidity and color (CZCS)
- GT: Boat fish catch (M)

Detection of fish schools and related properties
- Fish oil detection by aircraft spectrometers
- Fish-induced luminescense detected by sensitive TV cameras
- Fish egg and larvae drift into estuaries as a function of winds and currents measured by microwave sensors
- GT: Fishing boat reports, spotter planes (L)

Table 2. Remote sensing systems for fisheries applications.

	Temperature	Salinity	Chlorophyll	Color	Suspended Sediment	Sea State	Fronts	Patchiness	Oil
LANDSAT MSS	0	0	1	2	3	1	2	2	2
NIMBUS - G (CZCS)	0	0	2	3	3	1	2	1	1
NOAA - 5 HRIR	3	0	0	0	0	1	3	1	1
DMSP	3	0	0	0	0	1	3	1	1
HCMM	3	0	0	0	0	0	2	1	1
SEASAT	3	1	0	0	0	3	2	2	2
HAA MSS (OCS)	0	0	2	3	3	1	2	3	3
HAA PHOTOGRAPHIC	0	0	1	1	2	1	2	3	2
MAA PHOTOGRAPHIC	0	0	1	2	2	2	2	3	2
MAA MSS (M2S)	3	0	2+	3	3	2	3	3	3
MAA INFRARED (THERMAL)	3	0	0	0	0	1	3	2	2
MAA MICROWAVE	2	2	0	0	0	3	2	1	2
MAA RADAR	0	1	0	0	0	3	1	2	2
HELICOPTER FLUOROSENSOR	0	0	2+	1	1	1	2	2	3
SMALL AIRCRAFT PHOTOGRAPHIC	0	0	1	1	2	2	2	3	3

0 = Not Applicable
1 = Limited Value (Future Potential)
2 = Needs Additional Field Testing
3 = Reliable (Operational)

HAA = High Altitude Aircraft (U-2)
MAA = Medium Altitude Aircraft (C-130)
HCMM = Heat Capacity Mapping Mission Satellite
DMSP = Defense Meteorological Satellite
MSS = Multispectral Scanner
OCS = Ocean Color Scanner (10 bands)
M2S = Modular Multispectral Scanner (11 bands, including thermal infrared)
Photographic = Zeiss and Mitchell - Vinten Cameras

duce chlorophyll fluorescence that is indicative of both the concentration and diversity of phytoplankton. To map concentrations of organic and inorganic substances in coastal waters, multispectral scanner data is being analyzed by using the angular separation of eigenvectors representing each mapped substance in spectral signature space. Both techniques are still experimental and will require considerable field testing before their reliability is established. Satellite, aircraft and drogue techniques developed for monitoring the drift and dispersion of pollutants can be adapted to chart the flow and dispersion of detritus and other suspended matter.

Estimates of pelagic fish availability (and biomass in special cases) are normally obtained from biological studies, fishing fleet results or fish product company reports. Small spotter planes are used to guide certain fishing boats to schools of fish, such as menhaden. From satellite altitudes fish schools cannot be seen directly. However, satellites have been employed to locate areas of high probability for fish availability, such as the highly productive, nutrient rich, upwelling areas off the coasts of Peru, western United States and Africa. Upwelling areas can be mapped by multispectral scanners due primarily to their strong thermal and spectral gradients caused by the colder upwelling water and its spectrally different nutrient/chlorophyll content. There also appears to be a correlation between certain coastal water properties such as water color, turbidity, chlorophyll concentration and the presence of fish. Computer classification of LANDSAT MSS data into high probability fishing areas off the coasts of Louisiana and Mississippi concentrated in these areas. The applicability of similar techniques to locating other fish types such as thread herring and croakers is being investigated. A summary of remote sensor applicability to fisheries resources studies is presented in Table 2.

From these results it is reasonable to conclude that remote sensing techniques can be used effectively and reliably to map the location, size, and quality of wetland habitat; to chart fronts, slicks and upwelling regions; to observe the dispersion of certain pollutants; and to monitor sea state conditions. However, remote sensing techniques being developed for determining surface water productivity, detritus/nutrient flow and fish availability require additional field testing to establish their reliability.

References Cited

Anderson, R. R., L. Alsid and V. Carter. 1975. Comparative utility of Landsat-1 and Skylab data for coastal wetland mapping and ecological studies, pp. 469-478. *In:* O. Smistad (ed.), Proc. NASA Earth Resources Survey Symp., Vol. I-C. NASA Report TMX-58168. NASA Lyndon B. Johnson Space Center, Houston, TX.

Anderson, R. R. and F. J. Wobber. 1973. Wetlands mapping in New Jersey. *Photogram. Eng.* 39:353-358.

Apel, J. R., R. V. Charnell and R. J. Blackwell. 1974. Ocean internal waves off the North American and African coasts from ERTS-1, pp. 1345-1351. *In:* J. J. Cook (ed.), Proc. Ninth International Symp. on Remote Sensing of Environment, Vol. II. Envir. Res. Inst. Mich., Ann Arbor, MI.

Bartlett, D. S. and V. Klemas. 1977. Variability of wetland reflectance and its effect on automatic categorization of satellite imagery, pp. 1345-1352. *In:* Proc. of Amer. Soc. of Photogram. 43rd Annual Meeting. Amer. Soc. of Photogram. Washington, D.C.

Bartlett, D. S. and V. Klemas. 1979. Quantitative assessment of emergent biomass and species composition in tidal wetlands using remote sensing. Proc. of Workshop on Wetland and Estuarine Processes and Water Quality Modeling. U.S. Army Corps of Engineers, Vicksburg, MS.

Bartlett, D. S., V. Klemas, O. W. Crichton and G. R. Davis. 1976. Low-cost aerial photographic inventory of tidal wetlands. University of Delaware Marine Studies Report CRS-2-76, pp. 1-29.

Brucks, J. T. and T. D. Leming. 1977. Seasat-A wind stress measurements as an aid to fisheries assessment and management. National Marine Fisheries Service. MARMAP Contrib. No. 149:1-4.

Carter, V. and J. Schubert. 1974. Coastal wetlands analysis from ERTS MSS digital and field spectral measurements, pp. 1241-1260. *In:* J. J. Cook (ed.), Proc. Ninth International Symp. on Remote Sensing of Environment, Vol. II. Envir. Res. Inst. Mich., Ann Arbor, MI.

Clarke, G. L., G. C. Ewing and C. J. Lorenzen. 1969. Remote measurement of ocean color as an index of biological productivity, pp. 991-1002. *In:* J. J. Cook (ed.), Proc. Sixth International Symp. on Remote Sensing of Environment. Envir. Res. Inst. Mich., Ann Arbor, MI.

Clarke, G. L., G. C. Ewing and C. J. Lorenzen. 1970. Spectra of backscattered light from the sea obtained from aircraft as a measure of chlorophyll concentration. *Science* 167:1119-1121.

Egan, W. G. and M. E. Hair. 1973. Automated delineation of wetlands in photographic remote sensing, pp. 2231-2251. *In:* J. J. Cook (ed.), Proc. Seventh International Symp. on Remote Sensing of Environment. Envir. Res. Inst. Mich., Ann Arbor, MI.

Erb, R. B. 1974. The ERTS-1 investigation (ER-600): ERTS-1 coastal/estuarine analysis. NASA Report TMX-58118. Lyndon B. Johnson Space Center, Houston, TX. 258 pp.

Fairbridge, R. W. 1966. *The Encyclopedia of Oceanography.* Van Nostrand Reinhold Co., New York. 1021 pp.

Fantasia, J. F., T. M. Mard and H. G. Ingrao. 1971. An investigation of oil fluorescence as a technique for the remote sensing of oil spills. U.S. Coast Guard Report TSC-USCG-71-7.

Gordon, H. R. 1978. Remote sensing of optical properties in continuously stratified waters. *Applied Optics* 17:1893-1897.

Gordon, H. R., O. B. Brown and M. M. Jacobs. 1975. Computer relationships between the inherent and apparent optical properties of a flat homogeneous ocean. *Applied Optics* 14:417-427.

Hovis, W. A., Jr. 1977. Remote sensing of water pollution, pp. 351-362. *In:* J. J. Cook (ed.), Proc. Eleventh International Symp. on Remote Sensing of Environment. Envir. Res. Inst. Mich., Ann Arbor, MI.

Huebner, G. L. 1971. Proc. of the Symp. on Remote Sensing in Marine Biology and Fishery Resources. Texas A&M University Publication TAMU-SG-71-106. Remote Sensing Center, College Station, TX. 299 pp.

Kemmerer, A. J. 1976. An application of color remote sensing to fisheries, pp. 1-10. *In:* A. J. Kemmerer (ed.), Summary Report of the Ocean Color Workshop. NOAA Atlantic Meteorological and Oceanographic Lab. Miami, FL.

Klemas, V. 1976. Remote sensing of wetlands vegetation and estuarine water properties, pp. 381-403. *In:* L. E. Cronin (ed.), *Estuarine Processes, Vol. II.* Academic Press, New York.

Klemas, V., D. S. Bartlett and R. Rogers. 1975. Coastal zone classification from satellite imagery. *Photogram. Eng. and Remote Sensing* 41:499-512.

Klemas, V., D. S. Bartlett and W. Philpot. 1978. Remote sensing of coastal environment and resources, pp. 3-16. *In:* Proc. Coastal Mapping Symp., Amer. Soc. Photogrammetry, Falls Church, VA.

Klemas, V., G. Davis, J. Lackie, W. Whalen and G. Tornatore. 1977. Satellite aircraft and drogue studies of coastal currents and pollutants. *IEEE Transactions on Geoscience Electronics* GE-15:97-108.

Klemas, V. and D. F. Polis. 1977. Remote sensing of estuarine fronts and their effects on pollutants. *Photogram. Eng. and Remote Sensing* 43:599-612.

Kolipinski, M. C., A. L. Higer, N. S. Thomson and F. J. Thomson. 1969. Inventory of hydrobiological features using automatically processed multispectral data, pp. 79-96. *In:* J. J. Cook (ed.), Proc. Sixth International Symp. on Remote Sensing of Environment. Envir. Res. Inst. Mich., Ann Arbor, MI.

LaViolette, P. E. 1974. A satellite-aircraft thermal study of the upwelled waters off Spanish Sahara. *J. Phys. Oceanog.* 4:676-684.

Maul, G. A., D. R. Norris and W. R. Johnson. 1974. Satellite photography of eddies in the Gulf Loop current. *Geophysical Research Letters* 1:256-258.

McCluney, W. R. 1974. Ocean color spectrum calculations. *Applied Optics* 13:2422-2429.

McEwen, R. B., W. J. Kosco and V. Carter. 1976. Coastal wetland mapping. *Photogram. Eng. and Remote Sensing* 42:22-32.

Morrison, D. F. 1976. *Multivariate Statistical Methods.* McGraw-Hill, New York. 415 pp.

Mueller, J. L. 1976. Ocean color spectra measured off the Oregon coast: characteristic vectors. *Applied Optics* 15:395-402.

Mumola, P. B., O. Jarrett, Jr. and C. A. Brown, Jr. 1973. Multiwavelength laser introduced fluorescence of algae *in-vivo:* a new remote sensing technique, pp. 53-63. *In:* Second Joint Conference on the Sensing of Environmental Pollutants, Washington, D.C. Instrument Society of America, Pittsburgh, PA.

Nelson, W. R., M. C. Ingham and W. E. Schaaf. 1977. Larval transport and year-class strength of Atlantic menhaden: *Brevoortia tyrannus. Fish. Bull.* 75:23-41.

Plass, G. N., P. J. Humphreys and G. W. Katawar. 1978. Color of the ocean. *Applied Optics* 17:1432-1446.

Rayner, D. M. and A. G. Szabo. 1978. Time-resolved laser fluorosensors: a laboratory study of their potential in the remote characterization of oil. *Applied Optics* 17:1624-1630.

Reimold, R. J., J. L. Gallagher and D. E. Thomson. 1973. Remote sensing of tidal marsh. *Photogram. Eng.* 39:477-489.

Savastano, K. J. and T. D. Leming. 1975. The feasibility of utilizing remotely sensed data to assess and monitor oceanic gamefish, pp. 2023-2062. *In:* O. Smistad (ed.), Proc. NASA Earth Resources Survey Symp., Vol. I-C. NASA Report TMX-58168. NASA Lyndon B. Johnson Space Center, Houston, TX.

Simmonds, J. L. 1963. Application of characteristic vector analysis to photographic and optical response data. *J. Opt. Soc. Amer.* 55:968-974.

Szekielda, K-H., D. J. Suszkowski and P. S. Tabor. 1975. Skylab investigation of the upwelling off the northwest coast of Africa, pp. 2005-2022. *In:* O. Smistad (ed.), Proc. NASA Earth Resources Survey Symp., Vol. I-C. NASA Report TMX-58168. NASA Lyndon B. Johnson Space Center, Houston, TX.

ESTUARINE SEDIMENT: PHYSICAL AND BIOLOGICAL FACTORS

SEDIMENT INTRODUCTION AND DEPOSITION IN A COASTAL LAGOON, CAPE MAY, NEW JERSEY

Joseph T. Kelley

Department of Earth Sciences
University of New Orleans
New Orleans, Louisiana

Abstract: Mud rich sediment exists in northeastern Delaware Bay and crops out along the shoreface and on the inner continental shelf of southern New Jersey. Reworking of this sediment by tidal currents and storm waves generates most of the suspended sediment off Cape May Peninsula. The dispersed inorganic fraction of suspended sediment possesses a primary mode finer than $0.5\mu m$ and a secondary mode at $4\mu m$. The mineralogy of the clay fraction is dominated by illite, chlorite, and montmorillonite; the silt fraction contains an abundance of feldspar. A high degree of particle agglomeration is seen in Scanning Electron Microscope (SEM) photographs of suspended sediment collected on filters. Differential screening of seston *in situ* suggests that most composite particles present in the water column are $5\mu m$ and smaller. This offshore reservoir of very fine grained suspended sediment results from the deposition of all coarser components behind barrier islands on flood tides. As inner shelf suspended sediment is drawn through tidal inlets, waning tidal currents first permit deposition of discrete sand grains and large composite particles followed by progressively smaller composite particles moving in a landward direction.

Introduction

Shallow, tidally influenced coastal lagoons exist landward of barrier islands along the central and southern coast of New Jersey. The proportion of back barrier area occupied by lagoons decreases from north to south as the extent of salt marsh and tidal creeks increases. The small lagoons of southern New Jersey receive an insignificant input of freshwater from watersheds of Cape May Peninsula compared to larger, more estuarine bays to the north (Kran 1975). Tidal range on Cape May Peninsula is approximately 1.5 m. A low marsh dominated by *Spartina alterniflora* occupies all the marsh lands except for a fringe high marsh (*S. patens*) near the mainland. The region is thus microtidal and in a youthful stage of maturity according to the criteria of Davies (1964) and Frey and Basan (1978), respectively.

There have been no published reports on the geology of Cape May's lagoons since the pioneering investigations of Lucke (1934, 1935) and MacClintock (1943). Between 1973 and 1978 I conducted field and laboratory research focusing on the following problems:

1. What is the textural and mineralogical make up of sediment entering the back barrier regions of the Cape May Peninsula, and what is the source of this material?

379

2. What is the most important process which controls dispersal of sediment landward of the tidal inlets?

This paper is a brief summary of the most important results from that work.

Methods and Materials

Bottom samples were collected in Jenkins Sound, Great Sound, and adjacent tidal channels (Fig. 1) in the summers of 1973 and 1974 with a Dietz-Lafond grab sampler. Bottom sediment from each grab sample was texturally analyzed by sieve and pipette methods (Kelley 1975).

Suspended sediment was collected by pumping and filtering water (Kelley and Carson 1979) 1 m below the water surface and 1 m above the bottom offshore of Cape May Peninsula on August 5, 1977; February 16,

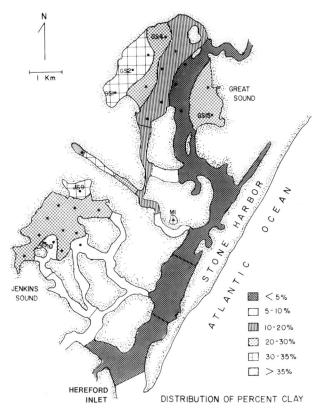

Figure 1. *Location of the core and grab samples and areal distribution of the percent clay in Great Sound and Jenkins Sound. Each dot represents a grab sample, or (where labeled) a core. Lines across tidal channels represent five equally spaced grab samples.*

1978; and April 18, 1978 (Fig. 2). Representative samples of suspensate were obtained by loading one or more 0.8μm membrane (Millipore) filters with suspended sediment at each near surface and near bottom station. In addition, size fractions of suspensate were obtained at the same stations and depths by pumping suspended sediment (separately) through various size nylon (Nitex) screens (5μm, 10μm, and 30μm). Because the screens were not loaded to capacity, the material collected on them was enriched in material that was coarser in suspension than the mesh opening of the screen (Sheldon and Sutcliffe 1969). Shortly after collection, select samples (still on screens and filters) were coated with carbon (Ladle and MacKay 1973) for examination under the SEM.

Bottom sediment was collected at each of the offshore suspended sediment stations (Fig. 2) in spring (1978) with a VanVeen grab sampler. Intertidal samples were also collected in the spring at 7 km intervals from

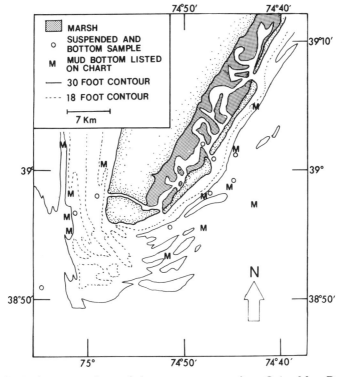

Figure 2. *Bathymetric chart of the water surrounding Cape May Peninsula from Coast and Geodetic Survey Charts 1217 and 1219 showing location of offshore samples. Samples of intertidal beach sediment were collected at 7 km intervals along the Atlantic and Delaware Bay shores in the spring.*

each of the beaches along the Atlantic and Delaware Bay margins of Cape May Peninsula (Fig. 2). Pleistocene age mudballs, cited as evidence of the onshore movement of continental shelf bottom sediment (Meza and Paola 1977), were collected from a washover fan at Stone Harbor.

Size distributions after oxidation of organic matter (NaOCl) and dispersal (NaPO$_4$) of each of the suspended, bottom, and beach samples were determined by Coulter Counter (62μm to 0.5μm, Behrens 1978) and pipette (percent of mud < 0.5μm) methods. The pipette estimate of the amount of material <11ϕ was verified by replicate determinations and extrapolated to 14ϕ on probability paper (Kelley 1980).

Five size fractions (32μm-8μm, 8μm-2μm, 2μm-0.5μm, 0.5μm-0.25μm, and <0.25μm) of each sample were separated by centrifugation (Jackson 1956) and evaluated for mineralogy by X-ray diffraction (Carroll 1970). Relative abundance of the minerals present was determined by a modification of Biscaye's method (1965).

Results

Figure 3 depicts the dispersed size distribution of suspensate collected on 0.8μm filters. More than 75% of the inorganic fraction of suspended sediment was finer than 0.5μm. No systematic seasonal differences in suspensate size distribution were noted, though regions of high turbidity (inlets, 60-200 mg/l) possessed slightly more silt than regions of lower suspended sediment concentration (offshore of inlets, <20 mg/l) (Kelley 1980).

An SEM examination of sediment on the filters revealed that most of the suspensate existed in the water column as composite particles (Kelley 1980). This high degree of particle agglomeration was also observed in material collected on screens (Kelley 1980). While the agglomerates were generally larger than mesh openings of the screens, individual grains comprising the composite particles were much smaller. After removal of organic matter and dispersal, the size distribution of suspensate collected on screens was similar, though significantly coarser grained than that collected on the membrane filters (Fig. 4). Each finer mesh screen collected material which on dispersal was finer grained. This supports the observation of Kranck (1975) that the dispersed size distribution of large composite particles is coarser than that of small composite particles.

In addition, the concentration of material collected by the various size screens and filters depended on the size of their mesh opening. At a station in the mouth of Delaware Bay (15 km southwest of Cape May, Fig. 2) where the total concentration of suspended sediment was 25.8 mg/l, the concentration of agglomerates larger than 5μm was 1.6 mg/l, with those larger than 30μm only 0.8 mg/l (Fig. 4). Thus, less than 10% of the suspended sediment in this area existed in the water column as particles larger than 5μm.

Grain size of back barrier sediment near Stone Harbor became progressively finer grained from the nearest tidal inlet (Hereford Inlet) toward

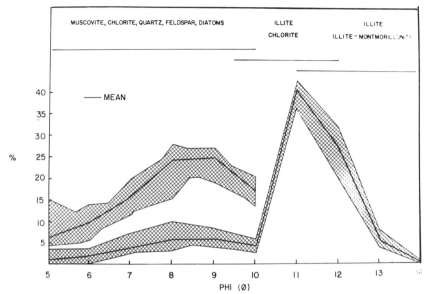

Figure 3. Dispersed size distribution of the inorganic fraction of material collected on 0.8μm membrane filters. The area within the patterned envelopes contains the size distributions of 15 samples collected in three seasons. The upper left distributions truncated at 10 φ were measured by Coulter Counter, whereas the lower distributions represent the same samples with the pipette estimate of the proportion of sample finer than 11 φ extrapolated to 14 φ.

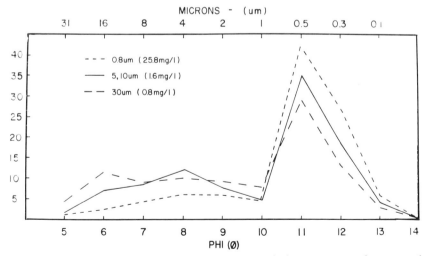

Figure 4. Average dispersed size distribution of the inorganic fraction of suspensate collected on 0.8μm filters (15 samples), a combination of 5 and 10μm screens (6 samples), and on a 30μm screen (6 samples).

the landward reaches of lagoons and tidal channels (Fig. 1). This was particularly evident in the long, straight tidal channel midway between Great Sound and Jenkins Sound (Fig. 1). As the mean grain size became finer than 4φ, sediment sorting became poorer (Fig. 5). This observation applies not only to grab samples of bottom sediment (Fig. 5) but to sediment accumulated in traps over only a few tidal cycles (Levy 1978). In addition, a gradient existed within Jenkins Sound and adjacent tidal channels from fine grained, less turbid near surface water to coarse grained, more turbid near bottom water (Kran 1975; Levy 1978). The finest sediment accumulating in back barrier regions, however, is coarser than the coarsest offshore suspended sediment collected on a 30μm screen (Fig. 5).

The mineralogy of offshore suspended, bottom, and beach sediment (mud fraction) was areally and seasonally uniform (Fig. 6). The finest size classes were dominated by illite and illite-montmorillonite mixed layer clays (Fig. 6). Coarser sizes contained increasing amounts of feldspar (and quartz), while the proportion of chlorite plus kaolinite remained nearly constant with respect to grain size. With the exception of the coarsest sizes, each size class was clearly distinguishable mineralogically.

Within size fractions, beach and offshore bottom sediment could clearly be distinguished from inlet and suspended sediment only in terms

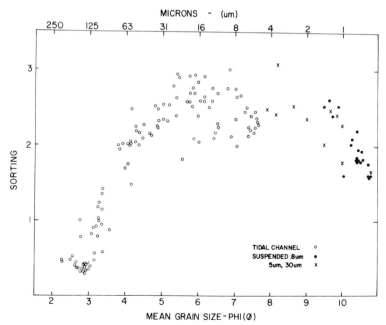

Figure 5. *Mean grain size versus sorting for tidal channel and lagoon samples (Fig. 1) and suspensate samples (Fig. 2).*

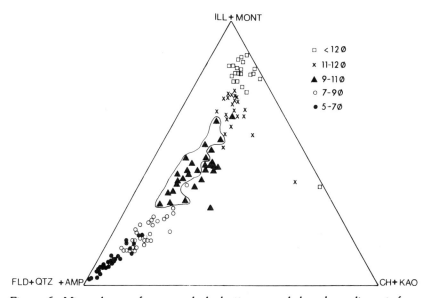

Figure 6. Mineralogy of suspended, bottom, and beach sediment from around the Cape May Peninsula. Triangle apices represent feldspar (fld) plus quartz (qtz) plus amphibole (amp); chlorite (ch) plus kaolinite (kao) and illite (ill) plus montmorillonite (mont). The encircled 9-11 φ values are all from suspended sediment.

of chlorite plus kaolinite in the 9-11 φ size class. While there appeared to be more chlorite plus kaolinite in the 9-11 φ beach and bottom fine fractions than in the suspended sediment, differences between environments was slight compared to overall variation in mineralogy within the 9-11 φ size class. These differences were insignificant when the precision of the semi-quantitative mineralogical determinations is considered (± 10%; Biscaye 1965).

Discussion

The mineralogical uniformity of nearshore suspended sediment and its similarity to such potential sources as Delaware Bay, continental shelf, and beach mud (Fig. 6) suggests that mud in the vicinity of Cape May Peninsula represents a single mineralogical population. When this population is size fractionated by sedimentary processes, the resulting sediment (bottom or suspended) will possess a mineralogy (within the trend of Fig. 6) which depends on its size distribution. Mineralogical variability encountered by Levy (1978) in Cape May lagoon and tidal channel regions, and by Meade et al. (1975) in suspension along the northeastern United States continental shelf probably resulted from variations in the size distribution of sediment analyzed by X-ray diffraction.

The most likely source of inner shelf suspended sediment (and ultimately Cape May marsh/lagoon mud) is resuspension of Delaware Bay (Oostdam 1971) and continental shelf (Meade et al. 1975; Meza and Paola 1977) bottom sediment. Oostdam (1971) has presented evidence for tidal current resuspension of northeastern Delaware Bay bottom sediment and transport around the southern tip of Cape May Peninsula. Meza and Paola (1977) have demonstrated that Pleistocene age continental shelf clay may be reworked during occasional storm events and transported landward. Though mud is not abundant on the continental shelf, several patches do crop out in topographic lows near Cape May Peninsula (Fig. 2) and elsewhere (Field et al. 1979).

Suspended material introduced to inner shelf water along the Atlantic side of Cape May Peninsula is drawn into tidal inlets on flood tides (DeAlteris and Keegan 1977). Only discrete sand size grains are deposited near the inlet and in channel thalwegs where water velocities are greatest (Fig. 1). Size analyses of this bottom sediment thus yield very well sorted distributions (Fig. 5). As suspended sediment moves landward in tidal channel and lagoonal regions, progressively finer sand grains and large composite particles settle out of suspension. This is indicated by an increase in the concentration and grain size of lagoonal and tidal channel near bottom suspended sediment over near surface values (Kran 1975; Levy 1978), as well as by progressively finer grained, more poorly sorted bottom sediment approaching the mainland. In the most landward reaches of lagoons and tidal channels, only small composite particles are deposited. On dispersal, these yield the finest, most poorly sorted sediment found in the back barrier regions (Fig. 5). These are coarser grained, however, (following dispersal, and by inference as settling, composite particles) than suspended matter offshore collected on $30\mu m$ screens. This suggests that inner shelf suspended sediment is purged of composite particles larger than $30\mu m$ (and their silt size components) by physical sedimentation landward of barrier islands. As a result, the inner shelf suspended sediment population represents a reservoir of very fine grained composite particles (finer than $5\mu m$) which settle too slowly to be deposited on flood tides behind barrier islands.

Conclusions

Reworking of Delaware Bay and continental shelf mud deposits provides fine sediment to inner shelf and back barrier regions of Cape May Peninsula. The mineralogy of this sediment is size dependent and ranges from feldspar-rich silt to illite-rich clay. Most of this material exists as composite particles while in suspension. The primary control on mud deposition landward of barrier islands is physical sedimentation of progressively smaller composite particles approaching the mainland. Removal of most silt size grains in this manner results in very fine grained suspended sediment along southern New Jersey's inner continental shelf.

Acknowledgments

This work was performed as part of a Ph.D. thesis at Lehigh University. I wish to thank Drs. Bobb Carson and James Parks of Lehigh University for technical assistance throughout this research. Financial support for this project from National Science Foundation Doctoral Dissertation Research Improvement Grant EAR77-15542, Geologic Society of American Research Grant 2212-77, as well as from the Center for Marine and Environmental Studies at Lehigh University is gratefully acknowledged. Thanks to Dr. A. Lee Meyerson for the Coulter Counter observations, and Alice A. Repsher for drafting assistance.

References Cited

Behrens, E. 1978. Further comparisons of grain size distributions determined by electronic particle counting and pipette techniques. *J. Sed. Pet.* 48:1213-1218.

Biscaye, R. 1965. Mineralogy and sedimentation of Recent deep sea clay in the Atlantic Ocean and adjacent seas and oceans. *Geol. Soc. Amer. Bull.* 76:803-832.

Carroll, D. 1970. Clay minerals—a guide to their identification. *Geol. Soc. Amer. Spec. Pap.* 126. 80 p.

Davies, J. 1964. A morphogenic approach to world shorelines. *Zeit. Geomorph.* 8:27-42.

DeAlteris, J. and R. Keegan. 1977. Advective transport processes related to the design of wastewater outfalls for the N.J. coast, pp. 63-89. *In:* R. J. Gibbs (ed.), *Transport Processes in Lakes and Oceans.* Plenum, New York.

Field, M., E. Meisburger, E. Stanley and S. Williams. 1979. Upper Quaternary peat deposits on the Atlantic inner shelf of the U.S. *Geol. Soc. Amer. Bull.* 90:618-628.

Frey, R. W. and P. B. Basan. 1978. Coastal salt marshes, pp. 101-169. *In:* R. A. Davies (ed.), *Coastal Sedimentary Environments.* Springer-Verlag, New York.

Jackson, M. L. 1956. *Soil Chemical Analysis—Advanced Course.* Published by author; Univ. Wisconsin, Madison, Wis.

Kelley, J. T. 1975. Sediment and heavy metals distribution in coastal lagoon complex, Stone Harbor, N.J. Lehigh Univ. Wetlands Reprint Ser. No. 3, Bethlehem, PA. 56 pp.

Kelley, J. T. 1980. Sources of tidal inlet suspended sediment, Hereford Inlet, N.J. Ph.D. Dissertation, Lehigh University, Bethlehem, PA. 150 pp.

Kelley, J. T. and B. Carson. 1979. A large volume suspended sediment sampler. *J. Sed. Pet.* 49:639-641.

Kran, N. 1975. Tidal controls on suspended sediment in a coastal lagoon, Stone Harbor, New Jersey. Lehigh Univ. Wetlands Reprint Ser. No. 2. Bethlehem, PA. 46 pp.

Kranck, K. 1975. Deposition from flocculated suspensions. *Sedimentology* 22:111-123.

Ladle, G. and D. McKay. 1973. The use of Millipore filters in the preparation of SEM mounts of particles less than 20 micrometers. *Amer. Miner.* 58:1082-1083.

Levy, J. 1978. Comparison of texture, mineralogy and organic content of suspended, accumulating, and bottom sediments within a coastal lagoon, Stone Harbor, N.J. M.S. Thesis, Lehigh University, Bethlehem, PA. 69 pp.

Lucke, J. 1934. A theory of evolution of lagoon deposits on shorelines of emergence. *J. Geol.* 42:561-584.

Lucke, J. 1935. Bottom conditions in a tidal lagoon. *J. Paleo.* 9:101-107.

MacClintock, P. 1943. Marine topography of the Cape May Formation. *J. Geol.* 51:458-472.

Meade, R., P. Sachs, F. Manheim, J. Hathaway and D. Spencer. 1975. Sources of suspended matter in waters of the Middle Atlantic Bight. *J. Sed. Pet.* 45:171-188.

Meza, M. and C. Paola. 1977. Evidence for onshore deposition of Pleistocene continental shelf clays. *Mar. Geol.* 23:M27-M35.

Oostdam, B. L. 1971. Suspended sediment transport in Delaware Bay. Ph.D. dissertation, University of Delaware, Newark, Delaware. 316 pp.

Sheldon, R. and W. Sutcliffe. 1969. Retention of marine particles by screens and filters. *Limnol. Oceanogr.* 14:441-444.

THE BIODEPOSITION CYCLE OF A SURFACE DEPOSIT-FEEDING BIVALVE, *MACOMA BALTHICA* (L.)

Luther F. Black

Zoology Department
Jackson Estuarine Laboratory
University of New Hampshire
Durham, New Hampshire

Abstract: The deposit-feeding behavior of *Macoma balthica* (L.) (Bivalvia: Tellinacea) is described relative to its role in sediment modification and the sedimentation cycle of Adams Cove, New Hampshire. Two size classes of *M. balthica* were selected for laboratory and field investigations of feeding behavior, sediment ingestion, sediment turnover, and fecal pellet decay. *M. balthica* feeds on surface deposits at low tide by sorting surface sediments in its mantle cavity. It ingests 12% (by weight) of the sediment and releases the remainder as a loose plume of pseudofeces. The ingested sediment is returned to the tidal flat surface within 3–6 h as compact fecal pellets. These pellets behave like fine sand grains. Laboratory and field experiments show that pellets remain intact for 1–2 weeks. Pellet decay is dependent on water agitation rather than abrasion or microbial decomposition. Estimated pellet deposition rates suggest that the *M. balthica* population of Adams Cove turns over the surface layer of the tidal flat 36 times per year. This rate of sediment processing indicates that *M. balthica* is grazing the tidal flat at a rate comparable to that of microbial recolonization, so the bivalve's deposit-feeding activity is an effective means of utilizing microbial productivity of the tidal flat.

Introduction

Deposit feeders live, burrow, and feed in sediments. Following ingestion, unassimilated sediment is usually deposited near or at the sediment surface as fecal pellets. Burrowing and pellet accumulation de-stabilize sediment so it is easily resuspended and transported by waves and tidal currents (Rhoads 1973). Haven and Morales-Alamo (1968) demonstrated that as much as 27% of the suspended material 1 m above the estuary bottom is fecal pellets.

Sediment ingested by deposit feeders is rich in microbiota (Newell 1965, 1970). Feeding activities of deposit feeders enhance microbial productivity when egestion releases fecal material enriched with nutrients (Hargrave 1977). Resuspension and settlement of fecal pellets facilitates microbial growth and subsequent increase in total nutritive value (Frankenberg et al. 1967). Microbial productivity in estuarine decomposer food chains depends on the turnover of sediment by macrofauna (Fenchel 1972; Hargrave 1977).

The substrate of the Adams Cove tidal flat (43°05'30"N; 70°51'55"W) on Great Bay Estuary, New Hampshire is fine-grained (Anderson

1973). The macrofauna is dominated by the capitellid polychaete *Heteromastus filiformis* (Clarapède 1864) and the tellinid bivalve *Macoma balthica* (L.) which produce prodigious quantities of fecal pellets at the tidal flat surface (Black and Anderson 1978). *H. filiformis* feeds on sediment at a depth of 5-20 cm while *M. balthica* feeds on surface sediments.

Although *M. balthica* has been reported to feed on seston (Brafield and Newell 1961; Tunnicliffe and Risk 1977), it utilizes benthic diatoms (Wernstedt 1942) and bacteria (Newell 1965; Fenchel 1972; Tunnicliffe and Risk 1977) from the surface sediments as a result of its deposit-feeding behavior. As this bivalve feeds, it sorts surface sediment in its mantle cavity and returns less cohesive material to the surface immediately as a loose plume of pseudofeces, or later as compact fecal pellets (Yonge 1949; Brafield and Newell 1961; Bubnova 1972; Gilbert 1977; Risk and Moffat 1977).

This study was undertaken to describe *M. balthica*'s feeding behavior and to understand the role of this bivalve in estuarine sedimentary processes.

Methods and Materials

From April 1976 to December 1977, seasonal benthic samples were taken from the upper half of Adams Cove tidal flat by using a metal frame (0.1 m² × 25 cm deep) and were screened on a 5-mm sieve. *M. balthica* retained on the sieve were measured along the antero-posterior axis, grouped into 5-mm size classes, and either preserved or maintained on a tidal schedule in running sea water at Jackson Estuarine Laboratory. Seasonal densities and size class structure were pooled and summarized (Black 1979). The modal size class, 20-25 mm, and the 10-15 mm size class were selected for further experimentation because, when combined, they represented 50.3% of the bivalve population during the 21 month sample period.

A tidal apparatus was constructed with a peristaltic pump and a 24 h timer to simulate the Adams Cove tidal schedule, 4 h of emersion and 8 h of immersion. A time lapse motion picture camera with flash attachment recorded surface feeding in plastic sediment containers (28 × 34 × 24 cm) of 25 large (20-25 mm) or 25-40 small (10-15 mm) clams. After the bivalves were acclimated to one of three temperatures (5, 10, 15C), the camera took an exposure every 15 s for a tidal cycle. The number of extended (feeding) siphons seen during each half hour period was tabulated as a percentage of the total number of siphons seen over the 12 h experiment. Data of all six experiments were normalized by the arcsine transformation and submitted to a three-way analysis of variance (ANOVA) with replication (2 bivalve sizes × 3 water temperatures × 3 tidal conditions; n = 8 half hour periods).

Adams Cove sediment which passed through a 177 μm sieve was dried and partially combusted at 400C for one hour. The oxidation of iron turned the sediment red-orange (Hylleberg and Gallucci 1975), a color which remained when oxidized sediment was enriched in a 1:3 ratio with unoxidized sediment. A five mm layer of red or natural dark sediment was introduced to and later removed from plastic containers (10 × 10 × 15 cm) of individual bivalves (Rhoads 1967) by means of a 9 × 9 cm piece of nylon window screen. The screen mesh (1.25 mm) was small enough to retain and support the cohesive sediment layer during transfer but large enough to permit the bivalve siphons to extend through to feed. Large and small bivalves were fed for one or two days on a red sediment surface layer, then given a dark layer. The times of first feeding and first dark fecal pellet appearance were noted to determine gut passage time.

Pseudofecal material and dark fecal pellets, easily distinguished from the surface layer, were collected by pipet every two hours over a 48 h period. Then the dark sediment layer was replaced with a red layer, and dark pellets were collected until red ones appeared. Both red and dark sediment layers were sieved (177 μm) to collect unnoticed pellets. Accumulated pseudofecal and fecal materials were dried on separate glass fiber filters (∼ 1 μm) and weighed to determine the ratio of rejected to ingested material.

Pellet Decay

Sets of 100 fecal pellets from laboratory-held bivalves of both sizes were introduced to 125 ml erlenmeyer flasks with 75 ml of sea water. Three flasks were placed on a reciprocating shaker (8.5 cm displacement, 2 cycles s^{-1}) and a fourth served as a stationary control (Risk and Moffat 1977). Pellets were counted and their condition noted at two or three day intervals. In the shaken flasks, the pellets fragmented and abraded, but were considered intact until they were less than half initial size. These experiments were run at 4, 9, and 20C for two or three weeks. A fourth experiment used pellets which had been preserved in 5% formalin to repeat the Risk and Moffat (1977) experiment.

Two series of pellet flasks fastened to a styrofoam float over the tidal flat experienced wave agitation on the rising and falling tide. A preliminary experiment used only untreated sea water in the flasks. The second field experiment introduced three treatments: 2 ml of fine (125-177 μm) sand, an antibiotic inoculum of penicillin (100 units ml^{-1} or 60 μg ml^{-1}) and streptomycin (10 μg ml^{-1}), and untreated sea water. The flasks were checked every two to three days for three weeks. Pellet data from the three experimental flasks were pooled to compute a mean breakdown time for 300 pellets released by the same bivalve during one tidal cycle.

Deposition

Fecal pellets of both bivalve size classes were collected from laboratory-held clams, sieved (177 μm) and introduced to a settling tube in five 7-ml replicates (Emery 1938). Cumulative frequency curves of particle settling size were plotted to describe settling characteristics of fecal pellets. Mean diameters of 300 pellets of both bivalve size classes were determined by measurement with a microscope.

Sets of six large and six small clams were acclimated at 5, 10, and 15C, fed red sediment for one day, then given a dark sediment layer. All dark pellets produced in the following three days were collected by pipet or sieve and counted. Two samples of 1000 pellets from each bivalve were dried and weighed to calculate a mean dry pellet weight. *M. balthica* individuals were also transplanted to the tidal flat in Nitex (100 μm)-covered containers. Pellets from each container were collected daily and after three days the surface layer was sieved. Any additional pellets were combined with those previously collected. Daily rates of sediment processed by individual bivalves were calculated by multiplying the mean number of pellets produced daily by the mean dry pellet weight (small, 10.9×10^{-6} g; large, 28.1×10^{-6} g). This ingestion rate was multiplied by the mean ratio of rejected to ingested material to estimate the amount of pseudofeces. The sum of these two calculations estimated total sediment processed daily. Because the estimated ingestion rates between temperatures were not significantly different, these rates were pooled to estimate daily and yearly rates of sediment ingestion and processing. A projected rate of sediment processing by all sizes of *M. balthica* was estimated conservatively by assuming the other bivalve size classes process sediment at half the rate of those measured.

In April and June 1978, a dry ice/acetone mixture was used to freeze the uppermost two mm of the surface sediment (Anderson and Black 1980). Frozen samples of 18.1 cm² area were thawed and sieved (177 μm) to retain fecal pellets and large detritus. This fraction was then decanted to separate pellets from detritus. All fractions were dried and weighed to compute the weight percent of fecal pellets in the surface layer.

Results

Feeding

Feeding by both large and small *Macoma balthica* was greater at low tide than at high tide (Fig. 1). Of the three main treatments of the ANOVA, the effects of tide ($F_{(2,126)}=106.6^{***}$) and water temperature ($F_{(2,126)}=3.79^{*}$) were significant. The first order interactions indicate that large and small clams responded differently to water temperature ($F_{(2,126)}=3.28^{*}$). Small *M. balthica* fed most often at 5C, but large individuals fed most often at 10C. Tidal condition dominated feeding behavior, and water temperature and clam size secondarily affected this behavior.

In all experiments over half the clams released dark pellets within 6 h of ingestion. Large *M. balthica* rejected 8.31 times as much material as they ingested (s=3.602, n=11); small clams rejected 7.16 times what they in-

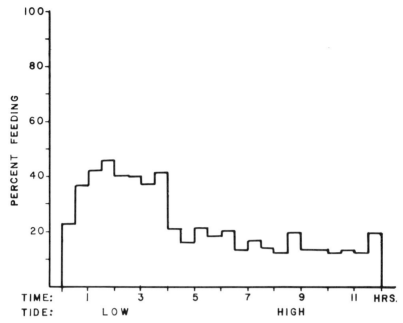

Figure 1. *Summary of feeding periodicity experiments. Mean percentage of* Macoma balthica *feeding is plotted from the pooled results of feeding experiments on large and small clams at all temperatures.*

gested (s=1.686, n=12). Because these ratios do not differ significantly ($t_{(21)}$=0.809; $t_{.05(21)}$=2.080), these data were pooled ($\bar{x}$=7.67, s=2.820, n=23) to indicate that *M. balthica* of any size will ingest only 11.5% by weight of the particles found in the surface sediments.

Pellet Decay

Time-to-breakdown for pellets is normally distributed so that the mean time of pellet breakdown occurs when the breakdown curve is in greatest decline (Fig. 2A). Mean time-to-breakdown may be estimated by the equation:

$$\bar{t} = \Sigma \frac{P_i}{N} t_i$$

where $\bar{t}$ is the mean breakdown time, P_i is the number of pellets that break down between t_{i-1} and t_i, N is the total number of pellets of the experiment, and t_i is the number of days the pellets (P_i) remain. Average decay times in the laboratory ranged from 8.0 to 14.4 days (Table 1). Small pellets consistently broke down faster than large pellets; at higher temperatures the time to decay of small pellets decreased, while for large pellets this decrease was not as evident. Pellets preserved in 5% formalin

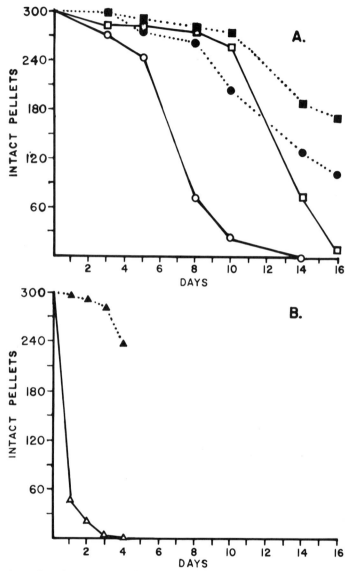

Figure 2. Pellet breakdown for preserved and untreated pellets from laboratory shaker experiments at 20C. A. Untreated pellets: (■) Stationary control for pellets from large clams; (●) Stationary control for pellets from small clams; (□) Untreated pellets from large clams; (O) Untreated pellets from small clams. B. Preserved pellets: (▲) Stationary control for preserved pellets from large and small clams; (△) Preserved pellets from large and small clams.

decayed within two days (Fig. 2B), which is similar to the report of Risk and Moffat (1977).

In field experiments, large pellets consistently broke down faster than small pellets (Table 1). These contrasting laboratory and field results are due to qualitative differences between agitation regimes of the shaker and the waves. There was a 50% decrease in decay time between first and second field experiments. Wind speeds recorded hourly at Pease Air Force Base, less than 2 km from Adams Cove, indicate that, during the second experiment, 75% of the days had mean wind speeds greater than 5 knots. During the first experiment only 50% of the days averaged greater than 5 knots.

Addition of sand to the flasks retarded decay of small pellets indicating that the relation of pellet size to grain size may affect decay rate. Addition of antibiotic mixture had no effect on decay of either pellet size.

Table 1. Mean times-to-breakdown for fecal pellets of two sizes of *Macoma balthica* under laboratory and field conditions. The estimate of sand-treated small pellets is based on a minimum value from one of three experimental flasks. n = number of pellets used in each experiment.

Laboratory		*4C*	*9C*	*20C*	*Preserved 20C*
Small	Mean Breakdown Time (days)	13.9	10.6	8.1	1.28
	n	300	300	300	300
Large	Mean Breakdown Time (days)	14.4	12.4	13.7	1.23
	n	300	300	200	300

		I *13-23C*	*II* *15-27C*		
Field		*Untreated*	*Untreated*	*Sand Treated*	*Antibiotic Treated*
Small	Mean Breakdown Time (days)	17.3	9.5	>12.0	9.4
	n	200	300	~100	300
Large	Mean Breakdown Time (days)	16.6	7.6	8.7	7.7
	n	200	300	300	300

Deposition

Measured diameters of *M. balthica*'s fecal pellets were 284 μm (1.81φ) and 393 μm (1.35φ) for small and large clams, respectively. Pellets settled in water at approximately the same rate as quartz grains of 2.0-2.5φ (fine sand) (Fig. 3).

Field estimates of daily and yearly sediment processing were, in all cases, greater than laboratory estimates (Table 2). There was no distinct relationship between pellet release and water temperature for either size of clam. There were no significant differences between large and small clams at the three temperatures because of great sample variability. The coefficient of variation ranged from 14 to 68%. Large clams could process up to three times more sediment than small clams despite the larger number of pellets released by small clams (Black 1979).

Each day, 5.8 g m^{-2} of pellets were released by the two size classes measured. By conservative projection, 8.7 g m^{-2} are released by the whole clam population (Table 3). Yearly estimates of 3.2 kg m^{-2} of sediment processed indicate the magnitude of sediment turnover by only one deposit-feeding species in Adams Cove. Freeze sampling of the surface sediment layer indicated that 50 g of the total 732 g dry wt m^{-2} was pelletal material; 6.8% of the uppermost 2 mm of sediment was maintained as compact pellets by all the fauna in Adams Cove.

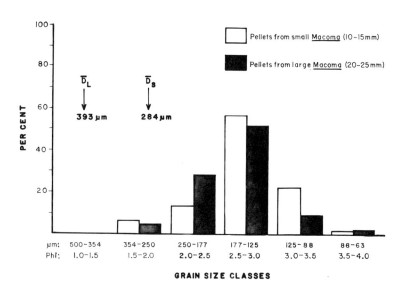

Figure 3. Size comparisons of fecal pellets from large and small clams measured by microscope and settling behavior. $\overline{D}_L$ = mean diameter of pellets from large clams; $\overline{D}_S$ = mean diameter of pellets from small clams.

Table 2. Laboratory and field mean estimates of total sediment processed by two sizes of *Macoma balthica* each day (10^{-3} g clam^{-1} day^{-1}). $\bar{x}$ = mean; s = standard deviation.

			Pellet		Pseudofeces		Total	
			Lab	Field	Lab	Field	Lab	Field
5C	Small	$\bar{x}$	10.1	18.2	77.5	139.6	87.6	157.8
		s	1.61	8.21				
	Large	$\bar{x}$	9.6	31.6	73.6	242.4	83.2	274.0
		s	6.54	19.38				
10C	Small	$\bar{x}$	8.9	12.7	68.3	97.4	77.2	110.1
		s	2.36	1.80				
	Large	$\bar{x}$	17.7	37.6	135.8	288.4	153.4	326.0
		s	7.27	3.68				
15C	Small	$\bar{x}$	14.3	26.0	109.7	199.4	124.0	225.4
		s	5.74	10.48				
	Large	$\bar{x}$	18.8	23.1	144.2	177.2	163.0	200.3
		s	9.54	11.81				

Table 3. Mean estimates of pellets and total sediment (= pellets + pseudofeces) processed by *Macoma balthica* m^{-2} in Adams Cove. (The projected estimates are conservative because it is assumed that the other size classes of the bivalve population will process half again as much sediment as the measured estimates.)

	Daily (g m^{-2} day^{-1})		Yearly (kg m^{-2} yr^{-1})	
	Pellets	Total	Pellets	Total
Small clams	1.5	13.3	0.5	4.8
Large clams	4.3	37.1	1.6	13.5
Total	5.8	50.4	2.1	18.3
Projected Total	8.7	75.6	3.2	27.5

Discussion

The surface deposit-feeding behavior of *M. balthica* resembles that of the European intertidal tellinacean bivalve *Scrobicularia plana* (Hughes 1969). At low tide both bivalves extend their siphons across the sediment surface to feed, but at high tide feeding is limited to the walls of the siphon hole. This behavior may be interpreted as a defense against predators such as fish and crabs. Both small winter flounder *(Pseudopleuronectes americanus)* and killifish *(Fundulus spp.)* were seen feeding at the advancing edge of the tide in Adams Cove. Tidal periodicity of deposit feeding may be an avoidance response of *M. balthica* to predation by fish.

Bubnova (1972) examined the quantities of sediment "reprocessed" by *M. balthica* from the White Sea. Calculations from her data show that clams 10-14 mm long rejected 7.9 times as much material as they ingested. This estimate compares favorably with estimates of the Adams Cove population despite probable differences in grain size distributions of substrates. The ratio of rejected to ingested material is an indirect means of determining particle selection. Because the ratio is a particle weight comparison, it may not reflect possible sorting by size or specific gravity (Self and Jumars 1978).

Both *Macoma nasuta* and *Scrobicularia plana* have been reported to clear their gut within 4-11 h (Hughes 1969; Hylleberg and Gallucci 1975). Though these bivalves feed on different grain sizes than *M. balthica,* they pass sediment in a range of times similar to *M. balthica.* Gut clearance within 6 h suggests that defecation may follow a tidal periodicity. These tellinid bivalves are active deposit feeders with the ability to process sediment quickly.

Because pellet breakdown rates will vary with the agitation that pellets experience, rate comparisons with other reports (e.g. Risk and Moffat 1977; Levinton and Lopez 1977) may not be meaningful. Risk and Moffat (1977) introduced preserved fecal pellets of *M. balthica* to a circular motion shaker table at 3 cycles s^{-1}. Although their data are recorded in terms of a qualitative index, they report that pelletal breakdown occurs within 12 h in the laboratory and assert that on the tidal flat pellets will only endure for a few tidal cycles. My results in Table 1 are at variance with this assertion. Although preserved pellets did experience relatively rapid rates of decay, unpreserved pellets remained intact for 8 to 14 days.

Breakdown curves for preserved pellets suggest an exponential relationship with time (Figure 2B) (Risk and Moffat 1977). Breakdown of unpreserved pellets, however, does not indicate the same type of relationship. The times-to-breakdown are normally distributed due to a normally distributed parameter within the pellet population. The thickness or durability of the mucous sheath enclosing the pellets (Risk and Moffat 1977) may be that parameter. Preservation of fecal pellets in formalin causes hardening of the mucous sheath (Galigher and Kozloff 1971) so the pellets become more brittle.

Laboratory and field data on rates of breakdown agree at two levels. Both curves of pellet decay reflect the same type of decay function in which times-to-breakdown are normally distributed. Also, the mean time of pellet breakdown in a steady wind compares well with laboratory estimates. The laboratory shaker table provides a suitable approach for measuring relative pellet breakdown rates.

Sand and antibiotic-treated pellets show similar decay patterns. The fine sand, though, may have acted to dampen agitation for pellets not much larger than the sand grains, while agitation of large pellets was unaffected. Because the penicillin-streptomycin mixture did not retard decay, either these antibiotics may not be effective in inhibiting estuarine microbes or microbes are not primary agents in the pellet decay process. The results for sand and antibiotic treatments did not differ dramatically from results for untreated pellets. Assuming the antibiotics inhibited microbial activity, abrasion and microbial decay do not appear to be primary processes affecting pelletal decay. The process that does cause pellet breakdown must be operating in all cases because the shapes of the decay curves are comparable. Agitation from the wind/wave regime markedly influences the rate of pelletal decay (Black and Anderson 1978).

The agitation of the pellets in both laboratory and field experiments may overestimate the motion experienced by pellets on the tidal flat. Both shaker and wave motion were almost continuous during the experiments, whereas pellet movement on the flat occurs only as the tide ebbs and floods over the pellets' location. Conversely, containment of the pellets within flasks may damp agitation from waves and currents, so actual pellet decay rates may not be fully realized.

Anderson (1973) reported that the median grain size of surface sediments from Adams Cove was 23 μm. *M. balthica* compacts these grains into discoid pellets of 284-393 μm diameter. In the water column, however, these pellets behave like fine sand grains of 125-177 μm size. This is the grain size that Sundborg (1956) predicted would be most readily transported by water movement. *M. balthica* feeds on fine-grained sediments which it later releases in the form of pellets the size of which favors transport away from the clam before breakdown.

Most previous estimates of sediment reworking by deposit feeders have been reported in terms of ml wet sediment individual^{-1} day^{-1} (Rhoads 1974). Assuming a water content of 60% and a bulk density of 1.9, Rhoads' (1974) rates range between 73.0 and 301.8 g dry wt individual^{-1} year^{-1}. The mean yearly estimates of *M. balthica* calculated from Table 2 are 56.5 g bivalve^{-1} year^{-1} for small *M. balthica* and 109.0 g bivalve^{-1} year^{-1} for large *M. balthica*. The bivalves and polychaetes cited by Rhoads (1974) are subtidal so they may have longer feeding times.

Risk and Moffat (1977) discussed estimates of defecation by *M. balthica*. Using methods similar to those of Rhoads (1967), they concluded that the clam processes sediment at 0.37 g bivalve^{-1} day^{-1} at 10C and 0.52 g bivalve^{-1} day^{-1} at 15C. Bubnova (1972) reported an estimate of 0.6 g bivalve^{-1} day^{-1}. All three estimates are higher than the values of Table 2. Their estimates depend on a rejection/ingestion ratio of 10 as interpreted from Bubnova (1972). In conclusion, Risk and Moffat (1977) reported the potential for a rate of deflation (erosion) of 28 cm m^{-2} per year for Bay of Fundy tidal flats.

If it is assumed that there is no net transport of pellets to or from Adams Cove (i.e., a closed system), then the number of pellets breaking down in Adams Cove each day will be equal to those produced each day. Then 8.7 g m^{-2} of sediment should be *M. balthica* fecal pellets, about 1.2% by weight of the uppermost 2 mm. Each day the clam population processes 9.9% by weight of the surface layer.

According to the estimates of Table 3, *M. balthica* can turn over the surface layer of Adams Cove about 36 times each year, so it is sorting through that layer at a rate of once every 10 days. Pellets released after digestion require one to two weeks to break down. During this period they will not be re-ingested by the clam (Risk and Moffat 1977; Black pers. obs.). After one week, benthic diatoms recover their previous population levels on pellets undergoing decay (Levinton and Lopez 1977). Bacteria and protozoa also colonize particle surfaces in less than one week (Fenchel and Harrison 1977). Because the mucous sheath enclosing the pellet is a rich substrate favorable to microbial colonization, Hargrave (1977) concluded that deposit feeders enhance their food supply by their reworking activity.

These assertions suggest that the clam populations are grazing the available resources on the tidal flat surface at an efficient rate. After examining carbon and nitrogen pools and assimilation rates for *M. balthica* in the Bay of Fundy, Tunnicliffe and Risk (1977) concluded that the clam's nitrogen requirement may only be met by suspension feeding. The above results indicate that *M. balthica* in Great Bay, New Hampshire is utilizing food resources of the surface sediment effectively at a rate that may be limited by resource renewal, as described by the model of Levinton and Lopez (1977). In Adams Cove, as on other tidal flats, sediment fluxes from water transport and activities of other organisms (i.e. deep deposit feeders) may prevent such a limitation. Here, models of resource utilization that assume no removal or addition of sediment (e.g. Levinton and Lopez 1977) are inappropriate. However, the fact that *M. balthica*'s rate of grazing is similar to reported rates of resource renewal suggests that food limitation is possible.

The influence of other organisms on the food resources of *M. balthica* cannot be overlooked. *M. balthica* accounts for only 8.7 g of 50 g m^{-2} or 18% of the pellets from Adams Cove. The deep deposit-feeding activities of *H. filiformis* produce many of the pellets in the surface sediment (Black and Anderson 1978). This material, brought to the surface by the polychaete, will also offer a rich substrate for microbial colonization and become a food resource for *M. balthica* after the pellets deteriorate. Thus, surface deposit feeders benefit from the sediment turnover of deep deposit feeders.

Acknowledgments

I am indebted to Drs. Franz Anderson and Larry Harris for their inspiration and guidance during this study. Dr. L. David Meeker was most helpful in exploring the mathematical nature of pellet decay. I gratefully acknowledge financial assistance, in part, from the Marine Program, the Zoology Department, the Graduate School and Jackson Estuarine Laboratory of the University of New Hampshire. Drs. J. Kraeuter, T. Loder, and M. Risk made many helpful comments and criticisms on preparatory drafts. Among others, W. Lull and K. Stapelfeldt provided useful suggestions and advice. Particular thanks to W. Beckingham for perseverance in typing this manuscript.

References Cited

Anderson, F. E. 1973. Observations of some sedimentary processes acting on a tidal flat. *Mar. Geol.* 14:101-116.

Anderson, F. E. and L. F. Black. 1980. A method for sampling fine-grained surface sediments in intertidal areas. *J. Sed. Pet.* (in press).

Black, L. F. 1979. Deposit feeding by *Macoma balthica* (L.) (Mollusca: Bivalvia) in a New Hampshire estuarine tidal flat. M.S. Thesis. U. New Hampshire, Durham, NH. 42 pp.

Black, L. F. and F. E. Anderson. 1978. What is the role of pellet producers in estuarine tidal flat sedimentation? *Geol. Soc. Am. Abstract* 3:367.

Brafield, A. E. and G. E. Newell. 1961. The behavior of *Macoma balthica* (L.). *J. Mar. Biol. Ass. U.K.* 41:81-87.

Bubnova, N. P. 1972. The nutrition of the detritus-feeding mollusks *Macoma balthica* (L.) and *Portlandia arctica* (Gray) and their influence on bottom sediments. *Oceanology* 12:1084-1090.

Emery, K. O. 1938. Rapid method of mechanical analysis of sands. *J. Sed. Pet.* 8:105-110.

Fenchel, T. 1972. Aspects of decomposer food chains in marine benthos. *Berh. Deutsch. Zool. Ges. Jah.* 14:14-22.

Fenchel, T. and P. Harrison. 1977. The significance of bacterial grazing and mineral recycling for the decomposition of particulate detritus, pp. 285-299. *In:* J. M. Anderson and A. Mac-Fayden (eds.), *The Role of Terrestrial and Aquatic Organisms in Decomposition Processes.* Blackwell Sci. Publ., Oxford.

Frankenberg, D., S. L. Coles and R. E. Johannes. 1967. The potential trophic significance of *Callianassa major* fecal pellets. *Limnol. Oceanogr.* 12:113-120.

Galigher, A. E. and E. N. Kozloff. 1971. *Essentials of Practical Microtechniques.* Lea and Febiger. Philadelphia, PA. 531 pp.

Gilbert, M. A. 1977. The behavior and functional morphology of deposit feeding in *Macoma balthica* (Linne, 1758) in New England. *J. Moll. Stud.* 43:18-27.

Hargrave, B. T. 1977. The central role of invertebrate feces in sediment decomposition, pp. 301-321. *In:* J. M. Anderson and A. MacFayden (eds.), *The Role of Terrestrial and Aquatic Organisms in Decomposition Processes,* Blackwell Sci. Publ., Oxford.

Haven, D. S. and R. Morales-Alamo. 1968. Occurrence and transport of fecal pellets in suspension in a tidal estuary. *J. Sed. Geol.* 2:141-151.

Hughes, R. N. 1969. A study of feeding in *Scrobicularia plana. J. Mar. Biol. Ass. U.K.* 49:805-823.

Hylleberg, J. and V. F. Gallucci. 1975. Selectivity in feeding by the deposit-feeding bivalve *Macoma nasuta. Mar. Biol.* 32:167-178.

Levinton, J. S. and G. R. Lopez. 1977. A model of renewable resources and limitation of deposit-feeding benthic populations. *Oecologia* 31:177-190.

Newell, R. C. 1965. The role of detritus in the nutrition of two marine deposit feeders, the prosobranch *Hydrobia ulvae,* and the bivalve *Macoma balthica. Proc. Zool. Soc., Lond.* 144:25-45.

Newell, R. C. 1970. *Biology of Intertidal Animals.* American Elsevier, New York. 555 pp.

Rhoads, D. C. 1967. Biogenic reworking of intertidal and subtidal sediments in Barnstable Harbor and Buzzards Bay, Massachusetts. *J. Geol.* 75:461-476.

Rhoads, D. C. 1973. The influence of deposit-feeding benthos on water turbidity and nutrient recycling. *Am. J. Sci.* 273:1-22.

Rhoads, D. C. 1974. Organism-sediment relations on the muddy sea floor. *Oceanogr. Mar. Biol. Ann. Rev.* 12:263-300.

Risk, M. J. and J. S. Moffat. 1977. Sedimentological significance of fecal pellets of *Macoma balthica* in the Minas Basin, Bay of Fundy. *J. Sed. Pet.* 47:1425-1436.

Self, R. F. L. and P. A. Jumars. 1978. New resource axes for deposit feeders? *J. Mar. Res.* 36:627-641.

Sundborg, F. A. 1956. The river Klaralven, a study of fluvial processes. *Geogr. Ann. Arg.* 37:125-315.

Tunnicliffe, V. and M. J. Risk. 1977. Relationships between the bivalve *Macoma balthica* and bacteria in intertidal sediments: Minas Basin, Bay of Fundy. *J. Mar. Res.* 35:499-507.

Wernstedt, C. 1942. Studies on the food of *Macoma balthica* and *Cardium edule*. *Viden. Medd. Dansk Naturhist. Furen., Kobenh.,* 106:241-252.

Yonge, C. M. 1949. On the structure and adaptations of the Tellinacea, deposit-feeding Eulamellibranchia. *Phil. Trans. Roy. Soc., B,* 234:29-76.

ECOSYSTEM DYNAMICS

THE IMPACT OF PREY-PREDATOR WAVES FROM ESTUARIES ON THE PLANKTONIC MARINE ECOSYSTEM

W. Greve and F. Reiners

Biologische Anstalt Helgoland
Meeresstation, Haus A
2192 Helgoland
Federal Republic of Germany

Abstract: In the South East of the North Sea, Heligoland Bight, the inner part of the German Bight contains a hydrographically complex pelagic ecosystem which is influenced by three estuaries, the Jade, the Weser and the Elbe. The existing temperature and salinity gradients there correspond with the population dynamics of small copepods, *Pleurobrachia pileus,* and *Beroe gracilis.* The earliest growth phases for copepods and ctenophores occur in the estuaries, with these organisms immigrating in different fashion from the lower estuaries. The spreading out of these populations according to their trophic relationships can be described as prey-predator waves. The predatory effect of *Pleurobrachia pileus* on small copepods and their nauplii is obvious; the predatory effect of *Beroe gracilis* on *P. pileus* is a dominant character of this ecosystem. The interpretation of these phenomena has to take into account scattering processes of organisms supported by multidirectional currents, the seasonal migration of temperature isoclines from the coast into the marine biota, and the parameters of population interaction in the dominant ecological groups. The impact of these processes consists mainly of a temporary melioration of the trophic basis of copepod-feeding fish (larvae as well as adults). The intensity of this impact conforms with the intensity of the self-regulatory system in ctenophore dynamics, by which the detrimental effect of *Pleurobrachia pileus* is converted by *Beroe gracilis* into a long-lasting protection against *P. pileus'* negative impact on copepod populations.

Introduction

"So our impression is further supported that Pleurobrachia *comes and goes without respect to season, temperature or depth".* Krumbach (1926).

"Because of the reduction in zooplankton, however, swarms of Pleurobrachia *could be responsible for larval fish mortalities, even if the fish themselves are only rarely eaten, and they could affect also adult herring and other plankton feeders".* Fraser (1970).

Rauck (1974) has shown the movement of plaice (*Pleuronectes platessa*) eggs and larvae from the southern bight of the North Sea with the sea currents into German coastal waters. The investigations conform with further descriptions of the transport of early life history stages of fish as well as with 0-group fish migration (e.g. Korringa 1973; Cushing 1976). As the highest mortality in the life cycle of marine fish occurs during the young stages, minimum changes in the rate of their mortality may be as important for the population as the spawner-recruitment relationship.

In the light of these statements, the abundance of *Pleurobrachia pileus* in Heligoland Bight is of major interest. Investigations in the last 10 years have shown a high local transient stability of the dynamics of this population (Greve 1971 and unpublished data) at Heligoland. Further investigations (Greve and Kinne 1973) reveal patterns which suggest that the population dynamics of ctenophores depend on coastal or estuarine processes. For the preparation of this paper these past data were intended to be the data base. Fortunately, a series of research cruises in 1979 made it possible to carry out a more detailed investigation upon which this paper is now based.

Methods and Materials

On five cruises with the research vessels "Uthörn" and "Friedrich Heincke", a quasisynoptic survey of the ecological conditions in Heligoland Bight was made. The dates (1979) of the cruises were: 23-25 April; 14-16 May; 05-07 June; 25-27 June; 23-25 July. The positions of the stations are given in Fig. 1. The stations were spaced as a compromise between max-

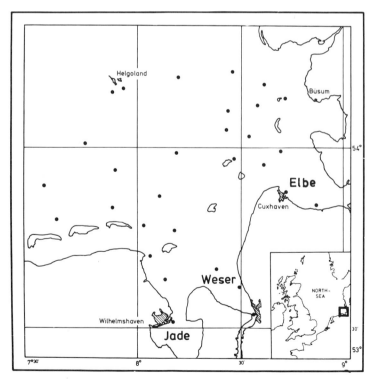

Figure 1. *Area investigated. Insert: Position in the North Sea. Black dots represent stations.*

imum resolution at places of high salinity gradients, time, and availability of navigable water in the shallow coastal zone (mean depth at stations: 20.4 m). On each of the 30 stations (twice on the cruises with the larger vessel one station in the shallows had to be left out) water-samples (5L) were taken for the determination of temperature, salinity, phytoplankton and microzooplankton, net plankton >150µm (mean sample size: 0.29 m³) and macroplankton >500µm (mean sample size: 67 m³).

The methods used aboard were: thermometer and areometer measurements for temperature and salinity; secci disc transparency; lugol fixation of phytoplankton samples (bottom and surface); formalin fixation of the small zooplankton (oblique haul with standard net, 18 cm front diameter, 150µm mesh size, and current meter) as well as of the larger zooplankton (oblique haul with CalCOFI net, 1m front diameter, 500µm mesh size, and current meter). Before fixation the large zooplankton was investigated alive to determine the size distribution and degree of reduction of cilia in the living ctenophores; also microzooplankton determination and counts using a projection microscope were carried out on board.

In the laboratory, the preserved samples were sorted and the individuals of each species counted if possible to a maximum of 60 individuals per sample (Lenz 1968). The data were then fed into a computer (DEC PDP 11/34). The figures given here were plotted by a Calcomp 1039 plotter. The decision to use a log scale histogram representation made it possible to have a parallel representation of up to ten separate informations on the populations over several orders of magnitude in each figure. By giving two absolute numbers on the bar in the lower right of each figure, all numbers can be recalculated. The raw data are available from the authors as computer-printout tables. Further, the decision to use abundance as the main parameter stresses the fact that interactive processes are abundance dependent. However, the number per representative m² differs from the number per m³ up to one order of magnitude. The total population therefore is not equally represented by the chosen presentation. Figures 2 and 3 demonstrate this fact. The interpretation of the data in this paper is regarded as a preliminary analysis of the available information.

Results and Discussion

Hydrography

Heligoland Bight, the inner part of the German Bight, is a hydrographically complex system. The major components include a coastal current coming from the southern North Sea and bending north within the Bight, and the inflow of river water from estuaries with different characteristics. The Jade is a habitat with almost marine conditions (Gillbricht 1956) whereas the Weser and Elbe have a strong freshwater inflow (Lucht and Gillbricht 1978).

The tidal currents in the three estuaries are in the range of 1.5 to 5

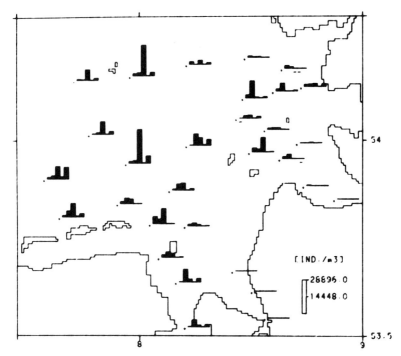

Figure 2. Temora longicornis *(copepodites) in Heligoland Bight during five successive cruises given in abundance (number m⁻³). In this and all similar figures, cruise data are represented by bars reading from left to right at each station, and all columns are sized to the same log scale. Missing data are indicated by a — below the base line. Small white structures represent islands. Numbers on axes represent latitude and longitude.*

knots maximum velocity (Eisma 1973). The residual time of the water passing the German Bight was calculated by Meier Warden (1979); the result, 36 months mean residual time, shows a relatively low exchange rate for the continental coastal water of the German Bight. Further complications occur due to the bottom structure. South of Heligoland, a channel of up to 60 m in depth is cut into the surrounding sea floor with a mean depth of 20 m. Backhaus (1979) showed that water movement in this channel often differs in direction from the surface waters, thereby compensating for the wind-driven surface currents. Upwellings during winds from the east have been occasionally registered.

The temperature and salinity measurements during the five cruises have been used to calculate the mean thermohaline conditions, as given in Figs. 4 and 5. From these it can be seen that, due to the hydrographic conditions, the impact of estuaries on the marine system has been shifted to the northeast. Here temperatures are higher and salinities lower for the months of the investigations.

Biology

From the data obtained during the investigation the following were regarded as the most important components of the zooplankton: small copepods (2 mm), which are the major food for *Pleurobrachia pileus*; their nauplii, which are the food of *P. pileus* after hatching; *Pleurobrachia pileus*; *Beroe gracilis* which feeds exclusively on *P. pileus* (Greve 1970); and fish larvae (Künne 1952).

Meroplanktonic larvae, especially trochophorae of spionid polychaetes, veligers of lammellibranchs and gastropods, nauplii of barnacles, echinoderm larvae, zoeae, hydro- and scyphomedusae have been sorted and counted but not included in this selection of data from our investigation. After a rough calculation they seemed to be of minor im-

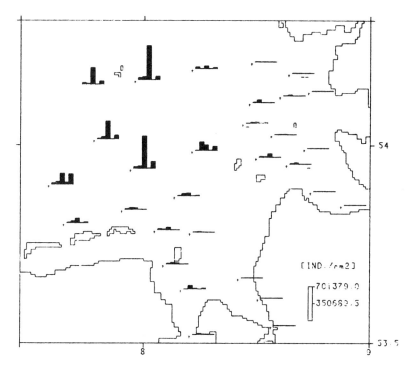

Figure 3. Temora longicornis *(copepodites) during five successive cruises given as local populations (number below a representative* m²; *this was determined by the mean depth of the water column of the 20 km² surrounding the sampling station). All columns are sized to the same log scale. A comparison with Fig. 2 shows the impact of depth on the histograms.*

portance with respect to the processes investigated here. They will be taken into account in a later publication.

Copepods

These include *Acartia longiremis/clausi; Centropages typicus/hamatus; Eurytemora affinis; Oithona helgolandica/similis; Paracalanus parvus; Pseudocalanus elongatus; Temora longicornis* (Fig. 6). On the first two cruises the estuaries contained many more copepods than did truly marine waters. This was mainly due to the populations of *Eurytemora affinis* in the estuaries, where detritus provides a food source independent of the phytoplankton bloom (Heinle et al. 1977). The high level of abundance is maintained in the inner estuary while the marine species *Temora longicornis* and *Acartia longiremis/clausi* have a growing population. On the fourth cruise a drastic decline in population size was registered as a general phenomenon with the exception of the low salinity regions. On the fifth cruise the populations had almost regained their previous abundance.

The nauplii of small copepods (Fig. 7) did not reach the abundances of the adults. This was partially due to the fact that the mesh size of

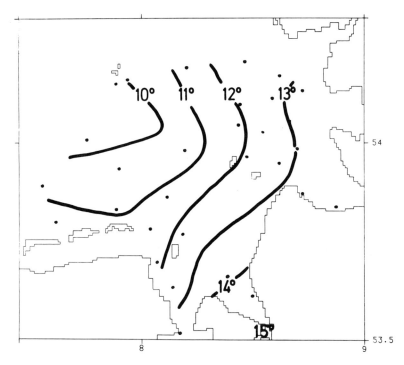

Figure 4. Mean temperature distribution in Heligoland Bight during five successive cruises from April to July 1979.

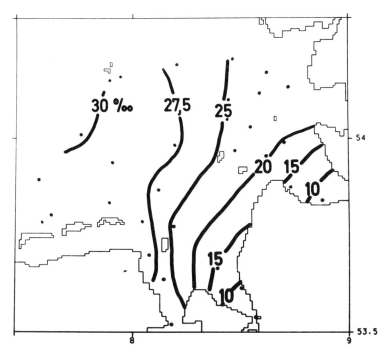

Figure 5. Mean salinity distribution in Heligoland Bight during five successive cruises from April to July 1979.

150μm permitted the escape of the early stages. The dynamics of the nauplii corresponded with that of the adults.

Pleurobrachia pileus (Fig. 8) was only found in small numbers in April during the first cruise. The population then grew by up to four orders of magnitude to reach a peak during the third cruise at the eastern stations and during the fourth cruise at the western stations. During the last cruise, only very few *P. pileus* were found in the northeast.

A more detailed analysis of the population development looking at the population structure of *P. pileus* (Fig. 9) shows juvenile stages only outside the estuaries; considerable growth then occurred in the outer Elbe region (Fig. 10). This growth continued in population numbers as well as in organism size. Juveniles started to appear in the truly marine station (Cruise 3) (Fig. 11). During the next three weeks, the population structure changed (Fig. 12). Mean size and abundance ceased to increase in the estuary while they were continuously increasing in the sea. Three weeks later, a population without juveniles remained in some of the coastal waters. The rest of the population disappeared (Fig. 13).

Beroe gracilis could not be found in a total volume of approximately 2000 m³ of water filtered during the first cruise throughout the investigated region (Fig. 14). Three weeks later, only two findings represented the

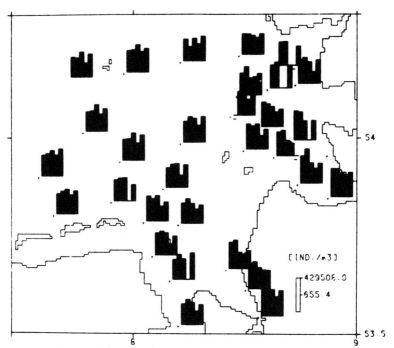

Figure 6.   Copepod (see text) abundance during five successive cruises.

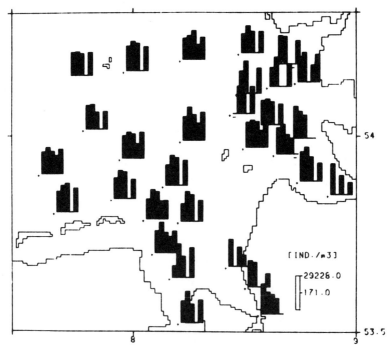

Figure 7.   Copepod nauplii (see text) abundance during five successive cruises.

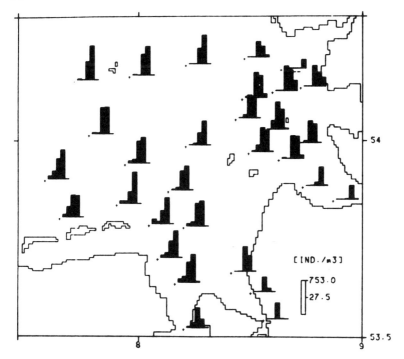

Figure 8. Pleurobrachia pileus *abundance during five successive cruises.*

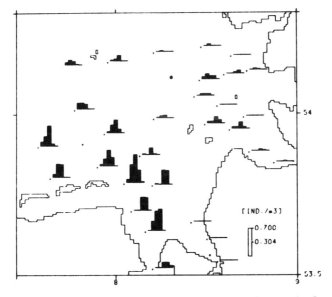

Figure 9. Pleurobrachia pileus *abundance during Cruise 1. Size classes (1-3, 3-5, 5-10, 10-15, 15-20 mm in diameter) are represented by the bars reading from left to right at each station.*

whole population; a single individual was sampled on a central station, and four specimens represented the population in the inner Jade. Three weeks later the Jade population had grown as had the population in the outer Elbe region where earlier the population increase of P. pileus had been registered. These two regions remained in a dominant position when three weeks later the coastal abundance of B. gracilis had grown about two orders of magnitude. Four weeks later, the numbers of P. pileus had diminished and B. gracilis was the dominant ctenophore. It was declining already in the Elbe estuary, where the impact on the P. pileus population had reached its peak three weeks earlier. During the whole time B. gracilis was not found in the upper estuary.

Fish larvae (Fig. 15) were least abundant on the first cruise. In the estuaries, populations of less than 1 individual per m³ were registered. With the increase in the number of copepods, the number of fish larvae grew. At maximum abundance of P. pileus when the copepods decreased in number up to four orders of magnitude, this decline and the dominance of predators could not be correlated with changes in number of fish larvae. A detailed taxonomic analysis of data has not yet been carried out but the over-representation of clupeiform larvae (mainly Clupea harengus, Sprattus sprattus) compared with flatfish larvae (mainly Arnoglossus laterna, Hippoglossoides platessoides, Limanda limanda, Pleuronectes platessa, Platichthys flesus, Psetta maxima, Solea solea) in the estuaries is obvious. The flatfish larvae also disappeared when P. pileus was abundant. Still, whether life history, season, or ecological conditions cause this change can-

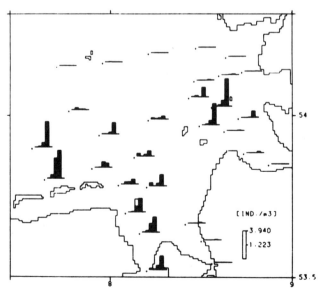

Figure 10. Pleurobrachia pileus *abundance during Cruise 2. Size classes are as Fig. 9.*

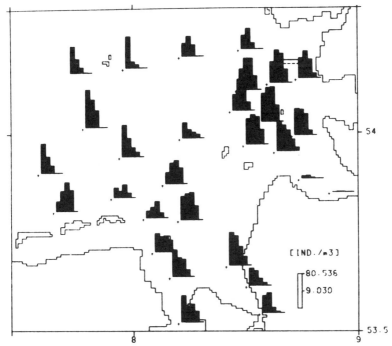

Figure 11. Pleurobrachia pileus *abundance during Cruise 3. Size classes are as Fig. 9.*

not be decided from the data analysis at this date. With regard to fish larvae it should be remembered that the total population size is better represented in plots of the number of individuals below a representative square meter (Fig. 16).

The dynamics of the ecosystem as investigated here (Fig. 17) show a development of two successive prey-predator waves which are characterized by a time lag of about three weeks from the Elbe estuary to the outer marine waters of Heligoland Bight. These prey-predator waves are not uniform with respect to the species of copepods which first develop at the coast, then in the marine habitat, before their populations are heavily affected by *Pleurobrachia pileus*. This ctenophore is present mainly in water of higher salinity in spring, then propagates fastest in the outer estuaries and coastal waters. The population then spreads out to the northwest. *Beroe gracilis* again develops in the outer estuaries when *P. pileus* has reached maximum population growth. At these locations, *B. gracilis* has a population increase of up to three orders of magnitude in three weeks. The population of *P. pileus* is first extinguished in the outer estuary, then the population decline moves from the outer estuary into the marine biota. The population of *P. pileus* then becomes almost extinct. It must be added here that, for the succeeding two months when no more cruises could be

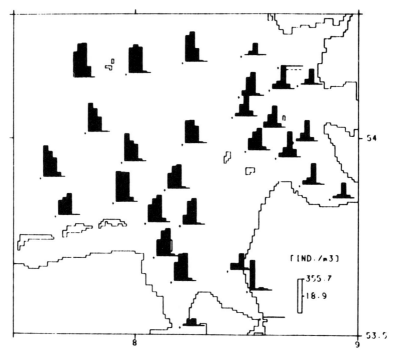

Figure 12. Pleurobrachia pileus *abundance during Cruise 4. Size classes are as Fig. 9.*

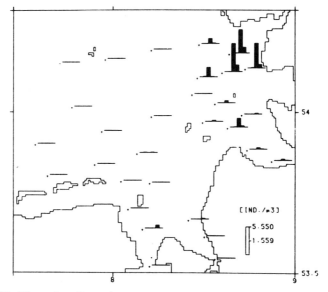

Figure 13. Pleurobrachia pileus *abundance during Cruise 5. Size classes are as Fig. 9.*

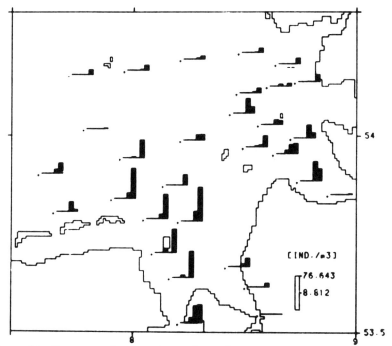

Figure 14. Beroe gracilis *abundance during five successive cruises.*

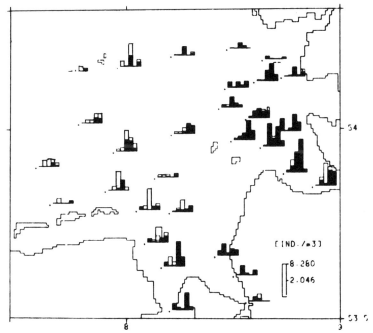

Figure 15. Fish larvae (total histogram) and clupeiform larvae (dark histogram) abundance (per m³) during five successive cruises.

carried out, the continuous plankton investigations at Heliogoland registered the dominance and slow destruction of the *B. gracilis* population due to starvation or other adverse impacts.

The planktonic ecosystem of Heligoland Bight thereby can be characterized as a temporarily predator-controlled system. Phytoplankton is continuously increasing from February to August (Hagmeier 1978). Presumably, decline of copepods thereby results from carnivorous impacts. The decline can be correlated with the increase of *P. pileus* (correlation coefficient −0.97 for Cruises 2 to 4). System complexity does not permit excluding other carnivorous impacts such as nectonic carnivory, but calculations of trophic efficiency of *Pleurobrachia* (Greve 1972; Hirota 1972; Reeve et al. 1978) can be taken as evidence.

The effect of *B. gracilis* on *P. pileus* is supported by experimental ecological investigations (Greve 1970) leaving little evidence for any other explanation. The correlation coefficient for Cruises 3 to 5 is −0.77.

The origin of the time lag between estuarine and marine processes may be the temperature gradient as well as the detritus-based higher abundances of copepods and nauplii in spring. The first assumption conforms with Uhlig's (1972, 1978) data on *Noctiluca miliaris* which spreads from

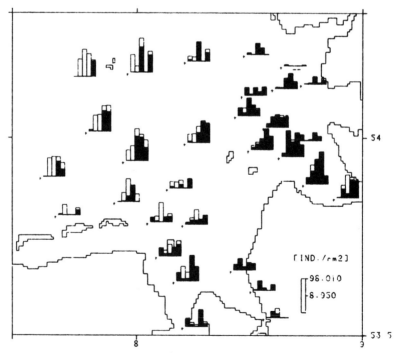

Figure 16. Fish larvae (total histogram) and clupeiform larvae (dark histogram) abundance per m² during five successive cruises.

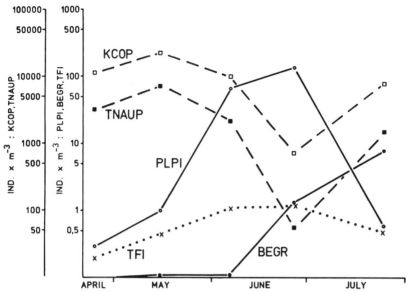

Figure 17. Dynamics of small copepods (KCOP), nauplii (TNAUP), Pleurobrachia pileus (PLPI), Beroe gracilis (BEGR) and fish larvae (TFI). Mean abundance from 30 stations.

the coast into the open sea according to their reproduction rate increasing with temperature. This is confirmed by our data from this investigation (Fig. 18). The second assumption cannot be quantified from detritus measurements, but secchi disc data indicate high values of detritus in the upper estuaries where *Eurytemora affinis* was abundant.

Implications

The role of carnivorous zooplankton in marine systems has been interpreted differently by several authors. Greve and Parsons (1977) see a possible shift from fish to ctenophores in systems with relatively high proportions of flagellates compared with diatoms as described by Gieskes and Kraay (1975) for the southern North Sea. Jones (1979) sees the possibility of predation preventing a larval cohort from outgrowing its food supply. Reeve et al. (1978) evaluate the excretory products of ctenophores as a basis of further phytoplankton growth. These assumptions of adverse effects, combined with the complex hydrography as given earlier, prevent any simple management rules. Yet, the following reflections should be considered: the marine habitats of Heligoland Bight are influenced by the estuaries. These may serve as a reservoir for populations, e.g. as *Beroe gracilis* in the Jade, pouring the "seed" out into the system. This process may be supported by detritus feeders, which provide a trophic background to carnivorous species. Further, the fit of the estuarine temperature-regime

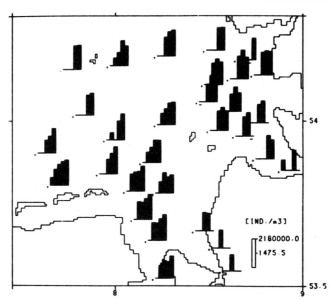

Figure 18. Noctiluca miliaris *abundance during five successive cruises.*

to the succeeding marine temperatures may be relevant. Any managerial impact on the system that changes the size of the estuaries, the availability of detritus, or the temperature/salinity regime as such will have an effect on the ctenophore populations in the marine habitat. This again affects the trophic basis and possible survival of fish. Managerial action therefore should take into account these major interactions within the system. As adequate means for such measures, simulation models of the main processes based on further aggregated information are suggested.

Acknowledgments

We wish to express our thanks for their help in this investigation to Captains R. Klings and H. Falke and the crews of the "Uthörn" and "Friedrich Heincke"; to N. Dubilier, F. Keller, and J. Krollpfeiffer who helped to sort the samples; and to V. Kartheus and L. Heitmann who programmed the plot routines. Financial support was given by the German Minister for Research and Technology MFU-0328/1.

References Cited

Backhaus, J. 1979. On currents in the German Bight, a three-dimensional model. Proc. of Conf. on Mathematical Modelling of Estuarine Physics, Hamburg. Springer, New York (in press).

Cushing, D. H. 1976. Biology of fishes in the pelagic community, pp. 317-340. *In:* Cushing, D. H. and J. J. Walsh, *The Ecology of the Seas.* Blackwell, Oxford.

Eisma, D. 1973. Sediment distribution in the North Sea in relation to marine pollution, pp. 131-152. *In:* Goldberg, H. D. (ed.), *North Sea Science.* MIT Press, Cambridge, MA.

Fraser, J. H. 1970. The ecology of the ctenophore *Pleurobrachia pileus* in Scottish waters. *J. Cons. Int. Explor. Mer.* 33:149-168.

Gieskes, W. W. and G. W. Kraay. 1977. Continuous plankton records: Changes in the plankton of the North Sea and its eutrophic Southern Bight from 1948 to 1975. *Neth. J. Sea Res.* (3/4): 334-364.

Gillbricht, M. 1956. Die Hydrographie des Jadebusens und der Innenjade. *Veröffentl. Inst. Meeresk.* NF IV: 153-170.

Greve, W. 1970. Cultivation experiments on North Sea ctenophores. *Helgoländer wiss. Meeresunters.* 20:304-317.

Greve, W. 1971. Okologische Untersuchungen an *Pleurobrachia pileus* 1. Freilanduntersuchungen. *Helgoländer wiss. Meeresunters.* 22:303-325.

Greve, W. 1972. Okologische Untersuchungen an *Pleurobrachia pileus* 2. Laboratoriumsuntersuchungen. *Helgoländer wiss. Meeresunters.* 23:141-164.

Greve, W. and O. Kinne. 1973. Untersuchungen zum Einfluss von industriellen Abwässern auf Makrozooplankter der Deutschen Bucht, pp. 74-82. *In:* DFG (ed.): 3. DFG Kolloquium "Litoralforschung-Abwässer in Küstennähe", Bonn-Bad Godesberg.

Greve, W. and T. R. Parsons. 1977. Photosynthesis and fish production: Hypothetical effects of climatic change and pollution. *Helgoländer wiss. Meeresunters.* 30:666-672.

Hagmeier, E. 1978. Variations in phytoplankton near Helgoland. *Rapp.P.-v. Réun.Cons.int.Explor.Mer.* 172:361-363.

Heinle, D. R., R. P. Harris, J. F. Ustach and D. A. Flemer. 1977. Detritus as food for estuarine copepods. *Mar. Biol.* 40:341-353.

Hirota, J. 1972. Laboratory culture and metabolism of the planktonic ctenophore *Pleurobrachia bachei* A. Agassiz, pp. 465-484. *In:* Tagenouti et al. (eds.), *Biological Oceanography of the Northern North Pacific Ocean*, Idemitsu Shoten.

Jones, R. 1979. Simulation studies of the larval stage and observations on the first two years of life, with particular reference to the haddock. ICES Early Life History Symposium. Woods Hole, MA. mimeo.: 1-18.

Korringa, P. 1973. The edge of the North Sea as nursery ground and shellfish area, pp. 361-382. *In:* Goldberg, E. D. (ed.), *North Sea Science*. MIT Press, Cambridge, MA.

Krumbach, T. 1926. Ctenophora, pp. 1-50. *In:* Grimpe et al. (eds.), *Tierwelt der Nord- und Ostsee, 8*.

Künne, C. 1952. Untersuchungen über das Grossplankton in der Deutschen Bucht und im Nordsylter Wattenmeer. *Helgoländer wiss. Meeresunters.* 4:15-44.

Lenz, J. 1968. Bestandsaufnahme - Plankton, pp. 48-61. *In:* Schlieper, C. (ed.), *Methoden der meeresbiologischen Forschung*. Fischer, Jena.

Lucht, F. and M. Gillbricht. 1978. Long term observations on nutrient contents near Helgoland in relation to nutrient input of the river Elbe. *Rapp.P.-v.Réun.Cons.Int.Explor.Mer.* 172:358-360.

Meier Warden, E. 1979. Some effects of the Atlantic circulation and river discharges on the residual circulation of the North Sea. *Deutsche Hydrographische Zeitschrift* 32:126-130.

Rauck, G. 1974. The arrival of different groups of young plaice in the German Wadden Sea. *Ber. dt. wiss. Kommn. Meeresforsch.* 23:273-288.

Reeve, M. R., M. A. Walter and T. Ikeda. 1978. Laboratory studies of ingestion and food utilisation in lobate and tentaculate ctenophores. *Limnol. Oceanogr.* 23:740-751.

Uhlig, G. 1972. Experimentell-ökologische und in-situ Untersuchungen an *Noctiluca miliaris*, pp. 20-21. *In:* Biologische Anstalt Helgoland (ed.), *Jahresbericht 1971*, Hamburg.

Uhlig, G. 1978. Experimentall-ökologische Untersuchungen am Meeresleuchttierchen, pp. 31-32. *In:* Biologische Anstalt Helgoland (ed.), *Jahresbericht 1977*, Hamburg.

ONTOGENETIC TROPHIC RELATIONSHIPS AND STRESS IN A COASTAL SEAGRASS SYSTEM IN FLORIDA

Robert J. Livingston

Department of Biological Science
Florida State University
Tallahassee, Florida

Abstract: Pinfish *(Lagodon rhomboides)* undergo progressive, temporally conservative, ontogenetic changes in food habits which traverse major trophic levels. Detailed changes in this progression were used to test the effectiveness of trophic relationships as indicators of environmental changes. Stomach contents of certain carnivorous and omnivorous growth stages of pinfish were altered by pollution, probably as a result of habitat-related disruptions of benthic prey assemblages. Such changes included disjunct ontological sequences of prey use, substitution, and reduced numbers of prey items. Partial habitat recovery was reflected in selective dietary changes and a return to feeding patterns similar to those of fish in control areas. Although this approach was considered useful in explaining system interactions, the results indicate problems in current ideas of generalized trophic levels and the use of the species as an ecological entity. If the ontogenetic spectrum of feeding patterns is regrouped so as to place closely related life stages in functional trophic units, a more realistic appraisal of system reaction to pervasive habitat alteration can be made.

Introduction

The food habits of fishes have been broadly analyzed (Darnell 1958; Carr and Adams 1972, 1973; de Sylva 1975; Stickney et al. 1975; Adams 1976; Hobson and Chess 1976; Chao and Musick 1977). Various aspects of feeding ecology have been documented. These include changes with time (Hickey 1975; Hobson and Chess 1976; Love and Ebeling 1978; Robertson and Howard 1978), variation in space (Feller and Kaczynski 1975; Stoner 1979), and ontogenetic changes in food preferences (Carr and Adams 1972, 1973; Hobson and Chess 1976; Stoner 1979). Group interactions, specifically food resource partitioning, have also been studied (Tyler 1972; McEachran et al. 1976; Gascon and Leggett 1977; Ross 1977; Desselle et al. 1978; Sheridan and Livingston 1979). There has been, nevertheless, relatively little analysis of the relationship of feeding to causal interactions between changes in individual fish populations and controlling habitat phenomena.

The basic approach presented was developed as part of an eight-year study of Apalachee Bay, a coastal system in the northeast Gulf of Mexico. The project compared the unpolluted Econfina drainage system with the Fenholloway system, an area polluted by kraft pulp mill effluents from 1954 through 1974 (Fig. 1). Permanent stations were established to

423

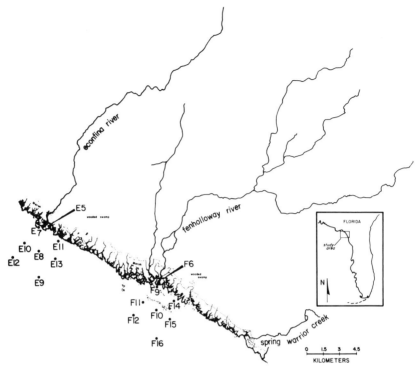

Figure 1. Chart showing Econfina and Fenholloway drainage systems of Apalachee Bay (north Florida, U.S.A.). Also shown are permanent sampling stations.

compare long-term physical, chemical, and biological trends. Analysis from 1971 through 1973 showed extensive changes in grassbed associations caused by pulp mill effluents (Livingston 1975; Heck 1976; Zimmerman and Livingston 1976). A study (1974-1979) following establishment of a pollution abatement program indicated partial recovery of the system. The feeding biology of the 28 most abundant species of fishes in each drainage system was analyzed. This study assessed spatial, temporal, and ontogenetic aspects of the food habits of fishes in Apalachee Bay from 1971-1977.

Data on a sparid fish, the pinfish (*Lagodon rhomboides*), were taken to test the effectiveness of using trophic relationships as an index of impact and recovery. This species is one of the most abundant fishes along the Florida Gulf coast (Reid 1954; Cameron 1969; Hastings 1972; Livingston 1975); in our study area, it represents 55% of all fishes taken from 1971 through 1979. The pinfish is a grassbed organism (Caldwell 1957) with definite ontogenetic patterns of feeding habits (Carr and Adams 1973; Adams 1976). Stoner (1979) detailed a series of developmental feeding stages which included, in chronological order, planktivory (calanoid and cyclopoid copepods, invertebrate eggs), benthic carnivory (amphipods, harpacticoid copepods), omnivory (microepiphytes, shrimp, amphipods, polychaetes, isopods), and herbivory (epiphytes, macrophytes). The considerable variability of such food habits reflected changes in feeding behavior and morphology (Stoner 1979). The pinfish appeared to be well adapted to seasonal patterns of coastal productivity. This broad adaptability indicated a response to ecological change which could be used to test the relationships between habitat-controlled feeding behavior and population distribution in space and time.

Methods and Materials

Paired sampling sites (e.g. E7, F9; E8, F10; Fig. 1) were selected in comparable locations in the two drainage systems so that physicochemical components such as temperature and salinity did not differ substantially between paired stations (Livingston 1975). Initial habitat stratification was based on preliminary field analyses. Such studies were used to locate subareas homogeneous in terms of water quality and depth, benthic macrophyte distribution and other ecological conditions. This sampling approach restricted any differences detected between the two systems to conditions caused by pulp effluent loading. The methods used for physicochemical and biological sampling are detailed by Livingston (1975), Livingston et al. (1976), and Zimmerman and Livingston (1976). Surface and bottom water samples were taken with a 1-l Kemmerer bottle. Temperature (C) was measured with a dissolved oxygen meter while salinity ($^o/_{oo}$) was determined with a temperature-compensated refractometer. Water color (Pt-Co units) was measured by the American Public Health Association platinum-cobalt method. Fishes were collected with 5-m otter trawls (1.9 cm mesh wing, 0.6-cm mesh liner); seven 2-min trawl tows were taken at each of several stations on a monthly basis from June, 1971, through May, 1979. Daytime trawling was during maximal pinfish feeding (Kjelson et al. 1975). Supplementary sampling was by beach seines during the day and trammel nets at night.

Fish were preserved in buffered 10% formalin, identified to species, and measured (SL = standard length, mm). Stomachs of larger fishes were removed in the field and injected with 10% formalin. Pinfish were sorted into 5-mm size classes up to 40 mm SL, 10-mm size classes from 41 to 100 mm SL, and 20-mm size classes for those larger than 100 mm SL. Fish stomachs from the preserved samples were pooled (no less than 15 fish were used in the final analysis of each size class) and labelled by size class, date, and station. Each stomach sample was preserved in 70% isopropanol and a dilute solution of rose bengal stain. Stomach contents were then analyzed according to the gravimetric sieve fractionation procedure (Carr and Adams 1972). Each sample was analyzed as percent dry weight composition. Items were identified to species wherever possible and categorized according to type (Table I). These categories were used for all statistical analyses.

Data for groups of stations within each system were pooled over 2-yr periods (1971-1972; 1976-1977). This analysis allowed comparisons between control and polluted areas before (1971-1972) and after (1976-1977) the pollution control program. All statistical analyses were made with a computer program (Woodsum 1979) used with the Florida State University computing system. Cluster analyses used the ϱ similarity coefficient (Matusita 1955; Van Belle and Ahmad 1974) and the flexible

Table 1. List of food types taken by *Lagodon rhomboides* in Apalachee Bay from 1971 through 1977, and codes which appear in cluster analyses.

am	Amphipods	**hy**	Hydroids
ba	Barnacles	**ie**	Invertebrate eggs
bc	Bryozoans	**il**	Insect larvae
bi	Bivalve mollusks	**is**	Isopods
bn	Barnacle nauplii	**it**	Invertebrate tubules
br	Brachiurans	**my**	Mysids
cc	Copepods (calanoid, cyclopoid)	**ne**	Nematodes
ch	Chaetognaths	**nm**	Nemerteans
cj	Crab juveniles	**nu**	Nudibranchs
cm	Crab megalops	**os**	Ostracods
cr	Crabs	**pl**	Polychaete larvae
cs	Crustacean remains	**po**	Polychaetes
ct	Chitons	**pr**	Plant remains
cu	Cumaceans	**sc**	Scallops
cz	Crab zoea	**sh**	Shrimp
de	Detritus	**sl**	Spicules
di	Diatoms	**so**	Sponge
ec	Echinoderms	**sp**	Shrimp postlarvae
ei	Sipunculids	**st**	Stomatopods
fe	Fish eggs	**sy**	*Syringodium filiforme*
fl	Fish larvae	**ta**	Tanaids
fo	Foraminiferans	**th**	*Thalassia testudinum*
fp	Fecal pellets	**tr**	Trematodes
fr	Fish remains	**tu**	Turbellarians
ga	Gastropod mollusks	**vl**	Veliger larvae
hc	Harpacticoid copepods	**•m**	Miscellaneous (used in food habit diagrams for all items composing less than 3% of the total mass)

grouping cluster analysis (beta = −0.25) (Lance and Williams 1967) as qualified by Sneath and Sokal (1973) and Livingston and Meeter (1977).

Results and Discussion

Water Quality Trends

Increased water color (Fig. 2) is an indicator of pulp mill effluent distribution (Livingston 1975). Such color changes are superimposed over seasonal and long-term trends in local precipitation which determine background color conditions in nearshore coastal waters (Livingston and Loucks 1978). During 1971-1972, the Fenholloway system was characterized by relatively dark water. Subsequent to the pollution abatement program (1974-1979), there was a substantial reduction in the Fenholloway color values, although they were still above those in the control area. Thus, while there was a recent amelioration of Fenholloway water quality, recovery apparently was not complete. This conclusion is consistent with other water quality features measured in various portions of the Fenholloway drainage system (unpubl. data).

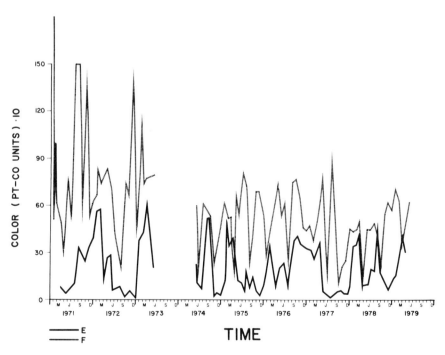

Figure 2. Color (Platinum-Cobalt units) measured monthly at stations E5 and F6 in Apalachee Bay from January, 1971, through May, 1979.

Pinfish Distribution

As an indicator of recovery, the intensive (monthly) sampling effort was restricted to stations within a 2-km radius of the respective river mouths. Long-term variation in pinfish numbers tended to follow established water quality changes. Consistently higher numbers of pinfish were taken in the Econfina system than in the Fenholloway. There was some increase in pinfish numbers with time in the Fenholloway system, especially during 1976-1977. This trend was more apparent when the outer stations (sampled quarterly) were included in the analysis. While this analysis did not rigorously treat the long-term trends of this species, there has been a comparative increase in pinfish numbers in the Fenholloway area relative to changes at control stations. These trends are supported by other data concerning changes in benthic macrophytes and fish assemblages in the areas of study (unpubl. data). The relative partial recovery of pinfish numbers in the Fenholloway system follows changes in habitat features and water quality.

Trophic Relationships

Stoner (1979) found that *L. rhomboides* undergoes ontogenetic changes in its feeding habits. This pattern is evident in the Econfina system during both sampling periods and in the Fenholloway system after establishment of the pollution-abatement program (Fig. 3). There is thus a certain temporal stability in the feeding strategy of this species. The feeding patterns tend to clarify these relationships (Fig. 4). The early growth stages (11-15 mm SL) feed largely on calanoid and cyclopoid copepods. As the fish grow from 16 to 35 mm SL, there are gradual changes in the diet with increased amounts of harpacticoid copepods, amphipods, shrimp postlarvae, invertebrate eggs, and plant matter. This stage gradually develops into a series of omnivorous stages (36-80 mm SL) in which the diet is balanced between plant matter (largely microepiphytes) and macrobenthic organisms such as amphipods, shrimp, harpacticoid copepods, and other small invertebrates. With further growth (81-120 mm SL), plant matter and detritus become proportionately more important. Bivalve mollusks such as *Brachidontes exustus* are also eaten at this stage, along with invertebrates such as amphipods, isopods, and small crabs. Pinfish larger than 120 mm SL are almost entirely herbivorous with seagrasses such as *Syringodium filiforme* and *Thalassia testudinum* particularly prominent in the diet.

The ontogenetic progression actually represents a continuum with some overlap between adjacent growth stages. While these differences reflect specific spatial and temporal variation in feeding patterns of local populations (Stoner 1979), the overall sequential pattern did not change substantially with time in the control (Econfina) area.

Pinfish feeding habits in the Fenholloway system during 1971-1972 were not uniform, especially during the carnivore and early omnivore growth stages. No sequential feeding pattern was evident in size groups

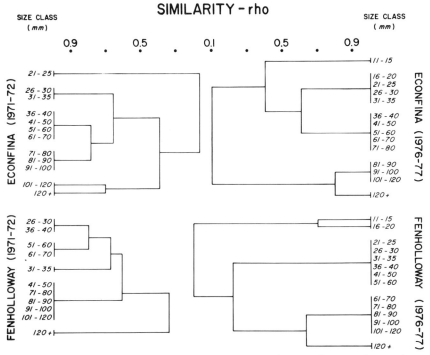

Figure 3. Comparison of ontogenetic changes in feeding habits of Lagodon
rhomboides *taken from the Econfina (E) and Fenholloway (F)
systems (Apalachee Bay, Florida) during 1971-1972 and
1976-1977. Summed food item data were clustered according to
size classes. Groups were associated by similarity coefficients
>0.80.*

measuring 20-70 mm SL. With the exception of fishes of the 31-35 mm
size class, which tended to resemble earlier stages, there was no consistent
pattern like that noted in control areas in 1971-1972 or 1976-1977.
However, the more mature growth stages of the early Fenholloway sequence tended to be more orderly and were somewhat similar in food
habits to the patterns found in control areas. This conclusion is qualified by
the fact that mature pinfish in the Fenholloway area in 1971-1972 tended
to substitute bivalves for seagrasses.

By 1976-1977, the trophic sequence in the Fenholloway pinfish
resembled that in control areas, although bivalves and polychaete worms
were more prevalent in the diet of more mature (70-120+ mm SL) fishes
of the Fenholloway system. Such differences were relatively minor,
however, in comparison with earlier (1971-1972) findings. The return to
herbivory in the largest size classes by 1976-1977 was substantial though
not complete. With minor exceptions, the qualitative composition and tem-

LAGODON RHOMBOIDES

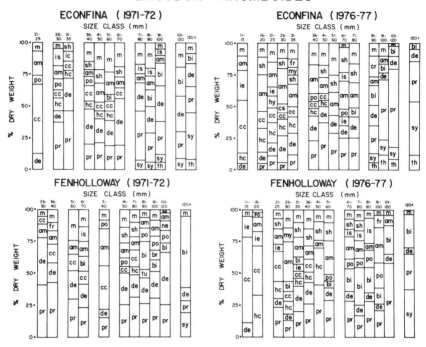

Figure 4. Relative proportions of major dietary components of Lagodon rhomboides taken from the Econfina (E) and Fenholloway (F) systems (Apalachee Bay, Florida) during 1971-1972 and 1976-1977. Components are arranged according to cluster analysis of similarity among size classes. Codes for food items are shown in Table 1.

poral relationships of the pinfish diet in the Fenholloway system in 1976-1977 resembled that in the control areas (1971-1972; 1976-1977).

    The data were further ordered by a cluster analysis of the summed stomach contents by system (Econfina, Fenholloway), time (1971-1972, 1976-1977), and size class (Fig. 5). Grouping of similar feeding classes more clearly defined the differentiation of spatial and temporal variation of food habits. Planktivorous stages (11-25 mm SL) and late omnivore-herbivore stages (> 71 mm SL) showed relatively little system- or time-based food differentiation, although there were some spatial differences in 1977. Fishes of later growth stages (26-70 mm SL) in the Fenholloway system taken during 1971-1972 were clearly differentiated from those taken in the Econfina area during both time periods and in the Fenholloway area in 1976-1977. The grouping of Econfina trophic groups of 1971-1972 and 1976-1977 with Fenholloway groups taken after partial

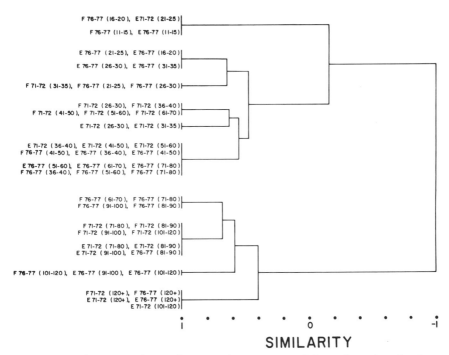

SIMILARITY

Figure 5. Cluster analysis of stomach contents of Lagodon rhomboides *summed by system (Econfina = E, Fenholloway = F), time (1971-1972, 1976-1977), and size class (mm SL). Groups associated by similarity coefficients >0.80 were clustered together.*

recovery (1976-1977) tends to substantiate qualitative differences in food habits noted previously. Although certain overall time-based system-specific feeding trends were evident in this analysis, the most clear-cut changes occurred in fish taken from the polluted (1971-1972) Fenholloway system. Such differences were restricted mainly to certain (carnivore-omnivore) trophic stages.

The Trophic Unit

A cluster analysis of the summed data from both systems (4333 stomachs) indicated a well-ordered progression of food habits by size class, including the following: planktivore, 11-15 mm; mixed planktivore-benthic carnivore, 16-25 mm; omnivore stage 1 (plant remains, detritus, amphipods), 26-60 mm; omnivore stage 2 (plant remains, detritus, bivalve mollusks), 61-120 mm; and herbivore, > 120 mm. While some overlap occurred, this classification by size represents functional, ontogenetic changes in food habits. Hereafter, the ordered groups will be referred to as trophic units.

After identification of the trophic units according to river system and time (E 71-72, E 76-77, F 71-72, F 76-77), the numbers of food items taken per feeding stage were calculated and then grouped by trophic unit with particular attention to those units (i.e. 26-60 mm, 61-120 mm) with high enough replication (n > 4 observations) for a statistical comparison. A series of one-way analyses of variance were run with the data. For trophic unit 26-60 mm, the number of food items taken by pinfish in the Fenholloway system in 1971-1972 ($\bar{X}$ = 19.0, S.E. = 2.39) was significantly lower (p < 0.0001) than that of other combinations. The group mean for F 76-77 ($\bar{X}$ = 28.2, S.E. = 1.85) was intermediate between that of F 71-72 and the two Econfina units (E 71-72, E 76-77), which were statistically indistinguishable ($\bar{X}$ = 34.7, S.E. = 1.45; $\bar{X}$ = 34.0, S.E. = 0.84, respectively). The same analysis for trophic unit 61-120 mm showed no significant difference (p < 0.1691) even though there were somewhat higher mean values in the Econfina data. In both cases, the multiple range test (p < 0.050) corroborated the above associations. These results tended to reinforce previous findings which indicate that pinfish feeding patterns reflected trends of pollution and recovery, but such changes were specific to certain growth stages, in this case the early omnivore unit.

While various studies have established that coastal fishes change their diet throughout development, there has been little systematic use of such information in studies of population response to environmental variables. In this paper I have tried to show that the species is not necessarily the optimal ecological unit. In fact, our studies on the feeding habits of 28 species of fishes (Livingston 1980) indicate that there is a broad spectrum of feeding habits which range from the very narrow for some species to continua which span most of the so-called trophic levels for others (e.g. the pinfish). By first defining a spectrum of ontogenetic feeding habits and then ordering these spectra into trophic units, we can define trophic interactions without returning to untenable assumptions regarding the basis of the trophic structure of the system. By examining ontogenetic trophic stages, we can thus determine the overall adaptive processes of a given species. The habitat-species relationship can thus be based on a functional entity, the trophic unit.

The approach taken here would allow the use of community indices to establish feeding habits of a given population. In this case, pinfish vulnerability to habitat alteration varies from one stage of its development to the next. Since the basic feeding pattern remains temporally stable within the spatial limits of the system, habitat changes due to pollution are reflected in altered trophic relationships. Possibly, the impact of natural or unnatural stress can be evaluated through measurement of incremental changes in trophic response. If so, the tropho-dynamic concept may be applicable to a definition of species adaptability, which, ultimately, is the key to an understanding of the impact of environmental change on natural populations.

Summary and Conclusions

1) An analysis was made of the feeding habits of pinfish *(Lagodon rhomboides)* along the north Florida Gulf coast. This study included detailed observations of spatial and temporal variation of trophic response as a function of incremental growth stages.

2) During the life history of the pinfish in shallow coastal grassbeds, there is an ontogenetic continuum of planktivory, benthic carnivory, omnivory, and herbivory. This condition is typical of other species under study, and tends to contradict certain established models of trophic organization.

3) By ordering identifiable ontogenetic feeding groups into trophic units, I found that such units could be used to evaluate impact from pollution and subsequent recovery. The feeding habits of the pinfish remained temporally stable within limits of spatial variation, but were altered by extreme habitat modification.

4) Analysis of trophic response as a function of habitat alteration indicated disruption only at specific developmental stages (in this case, the carnivore-early omnivore trophic units). There were modifications of the sequence of feeding as well as changes in the composition and number of food items taken. Such trends tended to follow observed water quality modification and recovery over a seven-year period.

5) The use of developmental feeding stages to determine community interactions is suggested as an alternative to more traditional approaches which regard the species as the basic unit of ecological interactions.

Acknowledgments

I wish to thank various students who participated in the bay studies over the years, with particular attention to H. S. Greening and J. D. Ryan. Portions of the 1976-1977 pinfish feeding data were collected by A. W. Stoner. Thanks also go to G. C. Woodsum for assistance in the development of computational methodology. A. B. Thistle edited and typed the manuscript. Portions of the project were funded by the Buckeye Cellulose Corporation (Perry, Florida) and the U.S. Environmental Protection Agency (EPA Program Element No. 1BA025; R. C. Swartz).

References Cited

Adams, S. M. 1976. Feeding ecology of eelgrass fish communities. *Trans. Amer. Fish. Soc.* 105:514-519.

Caldwell, D. K. 1957. The biology of the pinfish, *Lagodon rhomboides* (Linnaeus). *Bull. Fla. St. Mus. Biol. Sci.* 2:77-173.

Cameron, J. N. 1969. Growth, respiratory metabolism and seasonal distribution of juvenile pinfish *(Lagodon rhomboides* Linnaeus) in Redfish Bay, Texas. *Contrib. Mar. Sci.* 14:19-36.

Carr, W. E. S. and C. A. Adams. 1972. Food habits of juvenile marine fishes: evidence of cleaning habit in the leatherjacket, *Oligoplites saurus,* and the spottail pinfish, *Diplodus holbrooki. Fish. Bull.* 70:1111-1120.

Carr, W. E. S. and C. A. Adams. 1973. Food habits of juvenile marine fishes occupying seagrass beds in the estuarine zone near Crystal River, Florida. *Trans. Amer. Fish. Soc.* 102:511-540.

Chao, L. N. and J. A. Musick. 1977. Life history, feeding habits, and functional morphology of juvenile sciaenid fishes in the York River Estuary, Virginia. *Fish. Bull.* 75:657-702.

Darnell, R. M. 1958. Food habits of fishes and larger invertebrates of Lake Pontchartrain, Louisiana. *Contrib. Mar. Sci.* 5:353-416.

Desselle, W. J., M. A. Poirrier, J. S. Rogers and R. C. Casgner. 1978. A discriminant functions analysis of sunfish *(Lepomis)* food habits and feeding niche segregation in the Lake Pontchartrain, Louisiana estuary. *Trans. Amer. Fish. Soc.* 107:713-719.

de Sylva, D. P. 1975. Nektonic food webs in estuaries, pp. 420-447. *In:* L. E. Cronin (ed.), *Estuarine Research, Vol. 1.* Academic Press, New York.

Feller, R. J. and V. W. Kaczynski. 1975. Size selective predation by juvenile chum salmon *(Oncorhynchus keta)* on epibenthic prey in Puget Sound. *J. Fish. Res. Board Can.* 32:1419-1429.

Gascon, D. and W. C. Leggett. 1977. Distribution, abundance, and resource utilization of littoral zone fishes in response to a nutrient/production gradient in Lake Memphremagog. *J. Fish. Res. Board Can.* 34:1105-1117.

Hastings, R. W. 1972. The Origin and Seasonality of the Fish Fauna on a New Jetty in the Northeastern Gulf of Mexico. Ph.D. Dissertation, Florida State U., Tallahassee, FL.

Heck, K. L., Jr. 1976. Community structure and the effects of pollution in seagrass meadows and adjacent habitats. *Mar. Biol.* 35:345-357.

Hickey, C. R. 1975. Fish behavior as revealed through stomach content analysis. *New York Fish. Game J.* 22:148-155.

Hobson, E. S. and J. R. Chess. 1976. Trophic interactions among fishes and zooplankters near shore at Santa Catalina Island, California. *Fish. Bull.* 74:567-598.

Kjelson, M. A., D. S. Peters, G. W. Thayer and G. N. Johnson. 1975. The general feeding ecology of postlarval fishes in the Newport River estuary. *Fish. Bull.* 73:137-144.

Lance, G. N. and W. T. Williams. 1967. A general theory of classificatory sorting strategies. I. Hierarchical systems. *Comput. J.* 9:373-380.

Livingston, R. J. 1975. Impact of kraft pulp mill effluents on estuarine and coastal fishes in Apalachee Bay, Florida. *Mar. Biol.* 32:19-48.

Livingston, R. J. 1980. Trophic relationships and recovery processes in a shallow coastal seagrass system. Final Report, Corvallis Environmental Research Laboratory, U.S. Environmental Protection Agency, Corvallis, OR.

Livingston, R. J., R. S. Lloyd and M. S. Zimmerman. 1976. Determination of sampling strategy for benthic macrophytes in polluted and unpolluted coastal areas. *Bull. Mar. Sci.* 26:569-575.

Livingston, R. J. and O. Loucks. 1979. Productivity, trophic interactions and food web relationships in wetlands and associated systems, pp. 1-19. *In:* Greeson, P. E., J. R. Clark and J. E. Clark (eds.), *Wetland Functions and Values: The State of Our Understanding.* Proceedings, American Water Resources Association. Minneapolis, MN.

Livingston, R. J. and D. A. Meeter. 1977. Analysis of statistical methods used to determine effects of pollutants on aquatic populations and species assemblages. Final Report, Corvallis Environmental Research Laboratory, U.S. Environmental Protection Agency, Corvallis, OR.

Love, M. S. and A. W. Ebeling. 1978. Food and habitat of three switch-feeding fishes in the kelp forests off Santa Barbara, California. *Fish. Bull.* 76:257-271.

Matusita, K. 1955. Decision rules based on the distance, for problems of fit, two samples and estimation. *Ann. Math. Statist.* 26:631-640.

McEachran, J. D., D. F. Boesch and J. A. Musick. 1976. Food division within two sympatric species-pairs of skates (Pisces: Rajidae). *Mar. Biol.* 35:301-317.

Reid, G. K., Jr. 1954. An ecological study of the Gulf of Mexico fishes in the vicinity of Cedar Key, Florida. *Bull. Mar. Sci.* 4:1-94.

Robertson, A. I. and R. K. Howard. 1978. Diel trophic interactions between vertically-migrating zooplankton and their fish predators in an eelgrass community. *Mar. Biol.* 48:207-213.

Ross, S. T. 1977. Patterns of resource partitioning in searobins (Pisces: Triglidae). *Copeia* 1977(3):561-571.

Sheridan, P. F. and R. J. Livingston. 1979. Cyclic trophic relationships of fishes in an unpolluted, river-dominated estuary in north Florida, pp. 143-161. *In:* R. J. Livingston (ed.), *Ecological Processes in Coastal and Marine Systems.* Plenum Press, New York.

Sneath, P. H. A. and R. R. Sokal. 1973. *Numerical Taxonomy.* W. H. Freeman & Co., San Francisco. 573 p.

Stickney, R. R., G. L. Taylor and D. B. White. 1975. Food habits of five species of young southeastern United States estuarine Sciaenidae. *Chesapeake Sci.* 16:104-114.

Stoner, A. W. 1979. The Macrobenthos of Seagrass Meadows in Apalachee Bay, Florida, and the Feeding Ecology of *Lagodon rhomboides* (Pisces: Sparidae). Ph.D. Dissertation. Florida State U., Tallahassee, FL.

Tyler, A. V. 1972. Food resource division among northern, marine, demersal fishes. *J. Fish. Res. Board Can.* 29:997-1003.

Van Belle, G. and I. Ahmad. 1975. Measuring affinity of distributions, pp. 651-668. *In:* F. Proschan and R. J. Serfling (eds.), *Reliability and Biometry: Statistical Analysis of Lifelength.* S.I.A.M., Philadelphia.

Woodsum, G. C. 1979. Special Program for Ecological Science (SPECS) User Manual. Unpublished.

Zimmerman, M. S. and R. J. Livingston. 1976. The effects of kraft mill effluents on benthic macrophyte assemblages in a shallow bay system (Apalachee Bay, North Florida, U.S.A.). *Mar. Biol.* 34:297-312.

FEEDING STRATEGIES AND PATTERNS OF MOVEMENT IN JUVENILE ESTUARINE FISHES

John M. Miller

Department of Zoology
North Carolina State University
Raleigh, North Carolina

and

Michael L. Dunn

Department of Marine Science and Engineering
North Carolina State University
Raleigh, North Carolina

Abstract: Juvenile fishes in estuaries are trophic generalists; there is little evidence of their dependence on specific prey populations. The energetic costs of obtaining food are unknown since food habit data are rarely coupled with prey availability data, but it appears that locating prey may be more important than prey abundance. Cues for locating prey may be either prey abundance or environmental correlates of prey abundance. Juvenile fish may respond to the environmental rigors of the estuary by 1) increased breadth of tolerance limits or 2) inter- or intra-habitat movements. In hopes of encouraging the development of a data base on juvenile estuarine fishes, the authors list critical research needs.

Introduction

Estuarine environments are generally believed to be important nursery areas for many fishes. Joseph (1973) characterized these nursery grounds as areas that 1) are physiologically suitable in terms of chemical and physical features, 2) provide an abundant food supply, and 3) provide some degree of protection from predators. Yet estuaries are also potentially stressful environments in terms of the often unpredictable fluctuations in their biotic and abiotic features.

There are numerous species of fishes whose juveniles have managed to adapt to estuarine conditions, since many occur in greatest abundance in these areas. One of the most abundant groups of fishes utilizing U.S. Atlantic and Gulf Coast estuarine environments is the family Sciaenidae (Joseph 1972). Because of its importance in commercial and sport fisheries (McHugh 1966), numerous studies have been made concerning the biology of this largely estuarine group (Thomas 1971; Stickney et al. 1975; Chao and Musick 1977). As a result of this information plus the fact that our own work in the Pamlico River estuary of North Carolina is largely concentrated on the ecology of juvenile sciaenids, much of the discussion that follows will concern this group.

Two of the most "successful" sciaenids in terms of numbers in South Atlantic and Gulf Coast estuaries are the spot *Leiostomus xanthurus* (Weinstein 1979) and croaker *Micropogonias undulatus* (Sabins and Truesdale 1974). What factors contribute to the observed success of certain species? The discussion that follows attempts to answer this question by looking at the patterns of movement of juvenile fishes and how their feeding relationships and responses to environmental factors influence these patterns.

Feeding Relationships of Juvenile Fishes in Estuaries

Exploitation of an abundant food supply has been hypothesized as a major determinant in the evolution of the migratory patterns exhibited by numerous juvenile fishes (Hall 1972; Northcote 1978). Estuarine environments are, in fact, among the most productive environments in the world (Schelske and Odum 1961).

Productivity, however, may not be as critical as the standing crop of food organisms in determining the utilization of an area by a transient exploiter such as a juvenile fish. Coupled with this high standing crop of food organisms is a depauperate endemic fish fauna capable of utilizing this food resource (Van Engel and Joseph 1967). For example, McHugh (1967) gives evidence that only two or three species of fish can be considered truly estuarine species for portions of the Chesapeake Bay system. Day (1967) shows similar results for the Knysna estuary in South Africa. Our own data for the Pamlico River estuary indicate that the strictly estuarine resident forms are relatively few in number (mostly anchovies, killifishes and gobies). In addition, there seem to be few piscivorous fishes in these nursery areas (Van Engel and Joseph 1967). Thus, because of abundant food and low predation pressure, estuaries may indeed provide ideal environments in which migratory juveniles can survive and grow.

This agrees with Margalef's (1963) view of resource availability in unpredictable environments characterized by high abiotic stress. He described such environments (which he termed less mature) as having a high resource standing crop coupled with low utilization by endemic species. The corollary to this hypothesis is that there is some gain to be accrued to a population outside such an environment by sending large numbers of early life stages into these areas to exploit the food resources. This appears to be the situation in estuaries.

One of the first steps in determining an organism's functional role in an ecosystem or in relating its feeding ecology to overall population dynamics is a study of food habits. Numerous studies on this subject have covered a wide range of species and geographic areas. Much of the work to date has concentrated on juveniles in the larger size classes (e.g. > 50 mm). Relatively little is known about the feeding of fishes in the 15-35 mm size ranges (Kjelson and Johnson 1976).

Pertinent literature on food habits of juvenile sciaenids has been summarized by Chao and Musick (1977). They concluded that "juvenile sciaenids feed opportunistically" in the nursery ground. Data collected on juvenile sciaenids in our own studies have also shown that juvenile spot and croaker feed on a wide range of prey items including harpacticoid and calanoid copepods, mysids, and epibenthos. Although these prey are major constituents of the diet of many juvenile sciaenids, there is no evidence of any real "dependence" on any one food source. The same can be said for other species of juvenile fishes such as anchovies (Sheridan 1978), pinfish (Carr and Adams 1973), and salmon (Dunford 1975) which also occur in great abundance in certain estuaries.

Based on a review of the available data on food habits, it appears that the characterizations of salt marsh fish fauna and freshwater fish communities discussed by Harrington and Harrington (1961) and Larkin (1956) are equally applicable to the feeding relationships of estuarine juvenile fish assemblages. These generalized characteristics of feeding relations include: 1) flexibility of feeding habits in time and space; 2) omnivory; 3) sharing a common pool of food resources among species; 4) exploitation of food chains at different levels by the same species; 5) ontogenetic changes in diet with rapid growth; and 6) short food chains based on detritus-algal feeders.

Of particular significance in this characterization is the fact that the dominant species of juvenile estuarine fishes (e.g. spot, croaker, anchovies) tend to exhibit even more generalized patterns in food habits and habitat selection. Keast (1970) suggested that in freshwater fishes the most versatile and opportunistic feeder is also the most abundant. Examples of this generalized behavior for estuarine juvenile fishes can be found by comparing diets of sea trout (*Cynoscion* species) and spot (Chao and Musick 1977; Miller et al. unpublished data). Juvenile sea trout tend to rely heavily on one or two categories of prey items (e.g. mysids or small fish). The much more abundant juvenile spot, however, tend to eat many types of food from infauna to macrozooplankton. Based on our studies in North Carolina, a comparison of habitat selection between these species reveals that juvenile trout tend to be more restricted to either the deeper waters of channels (*C. regalis*) or to grass beds (*C. nebulosus*), whereas juvenile spot are found throughout the range of available habitats. This indicates diet and habitat breadth are critical factors in the success of juvenile estuarine fishes and fits the expected feeding behavior of organisms living in other unpredictable and physically stressed ecosystems (Margalef 1963; Odum 1969; Orians 1975).

The ability of generalist feeders to switch prey items in accordance with food availability has important implications regarding the well-being of juvenile fishes inhabiting estuaries as the abiotic stresses prevalent in estuarine environments may also affect the distribution and abundance of their prey items (Carriker 1967). Thus it is likely that the spectrum of available prey items in estuarine environments undergoes rapid and unpredictable changes in relation to factors such as turbidity, water turbulence, variations in runoff from uplands and marshes, and movements of water masses. In any case, the ability of juvenile fishes to switch from one prey item to another provides a selective advantage over species which eat restricted diets. Little information is available, however, on the energetic costs to the predator of switching prey in estuarine systems. The rate of change in abundance of certain prey items may temporarily reduce the efficiency of prey capture and food conversion if there is any period of "readaptation" required by the juvenile fish (Kinne 1967). One case where changes in food items does alter feeding efficiency is in relation to prey size. Presumably, larger prey require less time to capture than does an equivalent biomass of small prey (within limits set by the morphology of the feeding apparatus of the fish) (Parsons and LeBrasseur 1970). Experimental confirmation of this hypothesis for a freshwater fish comes from the work of Werner and Hall (1977) using bluegill. Using handling time as a currency, these authors showed that bluegill were more efficient at capturing the larger size class of zooplankton prey. Availability of a wide range of prey sizes can presumably both help to partition food resources (Carr and Adams 1973; Sheridan 1978) as well as provide less costly (in terms of energy) choices in the form of the larger size classes of prey items.

Thus, in terms of the well-being of juvenile fish, it may not be what the particular food item is *per se*, but rather the energetic costs of obtaining that food item that is important. In this respect, it is important to realize that characteristics of the prey organism such as density, size and availability are important determinants of the relative costs of obtaining a particular prey item (Swenson 1977).

A critical aspect of prey availability which has received relatively little attention in estuarine studies is that of prey distribution. It is likely that prey organisms are not distributed evenly throughout the environment but occur in response to environmental conditions in patches of varying concentrations. The extent of patchiness of the prey population can have important implications for the feeding patterns of juvenile fish. Ivlev (1961) demonstrated that the aggregation of food particles increased the feeding success of fish relative to an even distribution of prey with the same abundance. Furthermore, inferences made concerning juvenile fish feeding and prey concentrations may be inappropriate if the two factors are not coupled in time and space (O'Brien and Vinyard 1974). If prey patchiness occurs, the ability of juvenile fish to locate and exploit these patches of concentrated food will exert an important constraint on their well-being, particularly if the average concentrations of suitable prey are potentially limiting to growth (Thayer et al. 1974). Thus, it is apparent that information concerning the movements of juvenile fish may be critical in this regard.

Patterns of Movement

Movements of fishes in relation to gradients in food supply have been considered by a number of authors. Hansen (1970) suggested that juvenile croaker initiated movements in relation to changes in the available food resources. Reis and Dean (1980) postulated that juvenile anchovies underwent diel movements into and out of tidal creeks for the purpose of feeding. Harrington and Harrington (1961) showed that several species of fish invaded a flooded portion of high marsh in order to exploit a temporarily abundant supply of mosquito larvae. Miller et al. (1980), in a study of marsh utilization in North Carolina, found differences in the stomach contents of juvenile spot moving into a tidal marsh through a culvert on flood tides versus those moving out of the marsh on ebb tides. The pooled results of samples taken at hourly intervals over two consecutive tidal cycles are shown in Fig. 1. It is evident that the marsh-creek system was a much more valuable feeding area for these fish than were the adjacent deeper waters. Such studies provide evidence that juvenile fishes do exhibit short term local movements in order to exploit food resources associated with productive shallow water environments.

Assuming that the movements of juvenile fishes do in some way relate to their food supply (directly as a causative mechanism or indirectly as a matter of change in available prey as a result of some movement), it is important to consider the following questions:

1) What are the cueing mechanisms used to initiate and guide these movements?
2) What are the time and space scales on which these movements occur?
3) What are the costs of such movements relative to the gains?

The first question has two possible solutions if we assume movements in relation to food are in some sense directed. First, juvenile fish may change their position in direct response to their perception of the available food supply. Hyatt (1979) listed the sensory capabilities of the fish coupled with the detection of specific stimuli from the prey items (e.g. movement, size, shape, color and contrast) as being the major determinants of effective food location by a foraging fish. If prey populations are distributed very patchily in the environment, detection of food gradients seems unlikely. It would seem instead that fish would move along a search path until they encountered a patch of sufficient prey abundance and then move on when that patch was depleted to a given level. Ware (1972) suggests that fish may initiate movement from one foraging patch to another if the rate of capture of prey items in the first site falls below some threshold level. Pyke (1979) discusses the need for experimental studies on how fish

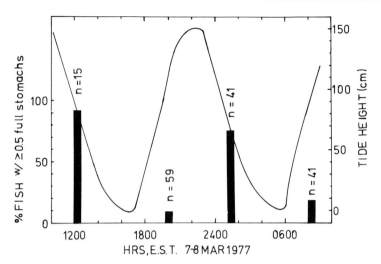

Figure 1. Stomach contents of juvenile spot (Leiostomus xanthurus)
entering (on rising tides) and leaving (on falling tides) a salt
marsh. n = sample size.

respond to patchiness in the spatial distribution of their food. Hunter and
Thomas (1974) provided some evidence in describing a clinokinetic
response (increased turning frequency) in anchovy larvae which kept them
in the vicinity of food patches. Where environmental gradients shift independently of food patches, a fish exhibiting an orthokinetic (or other)
response to the environment may be removed from its food. Presumably
this problem would be most severe in the case of patchily distributed benthic food sources.

　　The other alternative is that fish movements are coupled to environmental cues which might be correlated with abundant food resources.
Norris (1963) postulated that the response of young of the intertidal fish
Girella nigricans to temperature gradients may serve to direct them to
areas in which optimum growth conditions occur and/or in which specific
types of food or shelter are available. Similar coupling of movements to
salinity gradients is believed by some to direct longer term migrations of
juvenile fish to low salinity areas within the estuary which are rich in food
resources (Gunter 1961; Hoese 1965). Most likely, fishes exhibiting
directed movement in relation to food supply respond not just to a single
environmental factor but to a complex array of sensory stimuli (Hyatt
1979).

　　Studies on the time and space scales on which movements occur
have been conducted by Bozeman and Dean (1980) and Shenker and
Dean (1979) in South Carolina; Marshall (1975) and Weinstein (1979) in
North Carolina; Herke (1977) and Hackney (1977) along the Gulf Coast;
and Nixon and Oviatt (1973) in New England. Various sampling methods

including weirs, block nets and seining, tide traps and channel nets were employed in these studies. The total distances involved in many of these movements are not known because of the lack of information on the extent of dispersal outside the sampled environments.

Mark-recapture and trapping techniques similar to those used by Lotrich (1975) in the study of the movements of *Fundulus heteroclitus* in salt marsh creeks would be valuable in determining movements of juveniles suspected of having a rather limited range. Determination of interhabitat movements will require larger scale sampling using a variety of techniques such as tagging (Arnoldi et al. 1974; Knudsen and Herke 1978), intensive sampling at constrictions along habitat boundaries (e.g. cove inlets), or synoptic sampling of habitat grids.

The temporal scale of small-scale movements varies greatly with local conditions and species involved. However, three broad categories can be used to classify most of these movements. First, diel movement patterns have been examined in a number of situations in estuarine environments (Livingston 1976; Reis and Dean 1980). Second, diel movements are frequently coupled with tidal cycles (Shenker and Dean 1979). Third, there are the short-term movements which occur as a result of orthokinetic responses to environmental variations other than daylight and periodic tidal fluctuations. These are varied in nature and may range from very rapid responses to those occurring over several hours. Movements of this type range from those associated with changing levels of dissolved oxygen, salinity or temperature (due to changing weather conditions) to those associated with wind-generated fluctuations in depth and turbulence of the water. Changes such as these may take place very rapidly in shallow water environments (e.g. a salinity drop due to rainstorms) and therefore require rapid response capabilities in juvenile fish (Harrington and Harrington 1961).

All of these considerations point to the need for studies designed to determine the extent (in both time and space) of juvenile fish movements in relation to their food supply. Another factor which requires attention is that of prey distribution in relation to feeding habits. This type of study will require intensive sampling over a wide range of habitat types and will necessitate the use of a variety of gear types. The other major need is for determination of environmental cues used by estuarine fish to initiate and guide their movements. An examination of how abiotic factors influence juvenile fishes may provide some clues in this regard.

Responses of Juvenile Fish to Abiotic Environmental Factors

Estuaries are characterized by greater environmental stresses than nearly any other aquatic habitat. Chief among the potentially stressful environmental factors are salinity, temperature, and oxygen.

Variations in the time and/or space scale of these factors can influence the degree of stress within estuaries. Salinity and temperature gradients are steepest and fluctuations greatest near the edge of the salt wedge in a partially mixed estuary, near the mouths of freshwater discharges, and, if stratified, near the pycnocline in a vertical gradient. Deeper areas in stratified estuaries and the shallow and/or stagnant waters in contact with organic sediments are more likely to be deoxygenated. Since much of the variation in these environmental factors is linked to weather and climate, certain estuaries are potentially more stressful than others. West coast estuaries with greater maritime climatic influence are likely to be more benign than East coast estuaries. And, besides latitudinal climatic differences resulting in greater extremes of these factors, certain basin configurations will be accompanied by steeper spatial gradients than others.

Given the existence of stressful conditions within an estuary, juvenile fish may respond by either tolerating the stress or by seeking less rigorous areas. Both of these options entail some sort of energetic cost to the organism. Most studies on the tolerance of single factor stresses on fish have indicated a non-linear response (e.g. Alderdice 1963). There seems to be an additional burden associated with variations of stress factors. When the rate of change of a factor exceeds the adaptive rate of the organism, additional stress will be imposed. Houston (1959) showed about 30 h were necessary for *Onchorhynchus keta* to adapt to a salinity change from $0^o/_{oo}$ to $22^o/_{oo}$. Rates of thermal acclimation range around 1 C/day in goldfish (Brett 1956). Rates of acclimation to oxygen changes have been little studied, but Sheppard (1955) found that brook trout (*Salvelinus fontinalis*) fingerlings could reduce their critical oxygen concentrations by about 0.4 - 0.8 mg l⁻¹ over 100 - 200 h.

The most important implication of acclimation rates to juvenile estuarine fishes is that the time scale of any change in a potential stress factor makes an important contribution to the costs of tolerating sub-optimal conditions. It would seem that rates of change in the order of 1 ppt salinity, 1°C, or 0.01 mg l⁻¹ O₂ or less per day deserve more attention.

Unfortunately, we have little information on the energetic costs of tolerating even single factor sub-optimal environmental conditions for juvenile estuarine fish. Responses by fish to possible synergistic effects are virtually unknown. Behavioral regulation by moving in response to changes in environmental gradients could be more expensive energetically than tolerance of changes, especially if movements were away from optimal prey densities. It is presently impossible to predict the dynamic distribution of juvenile fish with respect to fluctuations in abiotic conditions within estuarine nursery grounds.

In conclusion, it appears that the most successful estuarine juvenile fish are those with the greatest niche breadth. They are characterized as being trophic generalists exploiting a wide range of habitat types. This apparent flexibility in habitat selection suggests that these species must either be more tolerant of stress or more adept at avoiding it. Determination of patterns of optimal use of estuarine nursery areas by fish awaits further study. Some of the more critical research needs are as follows: 1) synoptic studies of food availability and food habits; 2) data on the energetic costs to the predator of various foraging activities; 3) measurements of fish movements in relation to environmental variations; 4) studies of ontogenetic changes in tolerance or preference; and 5) information on energetic costs of behavioral regulation and tolerance of sub-lethal levels of stress factors. Data such as these would contribute much to our ability to predict the effects on fish populations of perturbations in estuarine environments.

Acknowledgments

This work was sponsored, in part, by the Office of Sea Grant, NOAA, U.S. Department of Commerce under Grant No. NA79AA-D-00048 and the North Carolina Department of Administration and the North Carolina State University Highway Research Program, Grant No. ERSD 76-1, in cooperation with the North Carolina Department of Transportation and the Federal Highway Administration - U.S. Department of Transportation. The U.S. government is authorized to produce and distribute reprints for governmental purposes notwithstanding any copyright that may appear hereon.

References Cited

Alderdice, D. F. 1963. Some effects of simultaneous variation in salinity, temperature and dissolved oxygen on the resistance of young coho salmon to a toxic substance. *J. Fish. Res. Bd. Canada.* 20:525-550.

Arnoldi, D. C., W. H. Herke and E. J. Clairain. 1974. Estimate of growth rate and length of stay in a marsh nursery of juvenile Atlantic croaker, *Micropogon undulatus,* "sand blasted" with fluorescent pigment. *Proc. Gulf Carib. Fish. Inst.* 26:158-172.

Bozeman, E. L., Jr. and J. M. Dean. 1980. The abundance of estuarine larval and juvenile fish in a South Carolina intertidal creek. *Estuaries* (in press).

Brett, J. R. 1956. Some principles in the thermal requirements of fishes. *Quart. Rev. Biol.* 31:75-87.

Carr, W. E. S. and C. A. Adams. 1973. Food habits of juvenile marine fishes occupying seagrass beds in the estuarine zone near Crystal River, Florida. *Trans. Amer. Fish. Soc.* 102:511-540.

Carriker, M. R. 1967. Ecology of estuarine benthic invertebrates: A perspective, pp. 442-487. *In:* G. H. Lauff (ed.), *Estuaries.* Amer. Assoc. Advanc. Science, Publ. No. 83, Washington, D.C.

Chao, L. N. and J. A. Musick. 1977. Life history, feeding habits, and functional morphology of juvenile sciaenid fishes in the York River estuary, Virginia. *Fish. Bull.* 75:657-702.

Day, J. H. 1967. The biology of Knysna estuary, South Africa, pp. 397-407. *In:* G. H. Lauff (ed.), *Estuaries.* Amer. Assoc. Advanc. Science, Publ. No. 83, Washington, D.C.

Dunford, W. E. 1975. Space and food utilization by salmonids in marsh habitats of the Fraser River estuary. MS Thesis. U. British Columbia, Vancouver, B.C. Canada. 67 pp.

Gunter, G. 1961. Some relations of estuarine organisms to salinity. *Limnol. Oceanogr.* 6:182-190.

Hackney, C. T. 1977. Energy flux in a tidal creek draining an irregularly flooded *Juncus* marsh. Ph.D. Dissertation, Mississippi State U., Mississippi State, MS. 83 pp.

Hall, C. A. S. 1972. Migration and metabolism in a temperate stream ecosystem. *Ecology* 53:585-604.

Hansen, D. J. 1970. Food, growth, migration, reproduction and abundance of pinfish, *Lagodon rhomboides,* and Atlantic croaker, *Micropogon undulatus,* near Pensacola, Florida, 1963-1965. *Fish Bull.* 68:135-146.

Harrington, R. W., Jr. and E. S. Harrington. 1961. Food selection among fishes invading a high subtropical salt marsh: From onset of flooding through the progress of a mosquito brood. *Ecology* 42:646-666.

Herke, W. H. 1977. Life history concepts of motile estuarine-dependent species should be re-evaluated. W. H. Herke, Baton Rouge, LA. 97 pp.

Hoese, H. D. 1965. Spawning of marine fishes in the Port Aransas, Texas area as determined by the distribution of young and larvae. Ph.D. Dissertation, U. of Texas, Austin, TX. 144 pp.

Houston, A. H. 1959. Osmoregulatory adaptation of steelhead trout (*Salmo gairdnerii* Richardson) to sea water. *Can. J. Zool.* 37:729-748.

Hunter, J. R. and G. L. Thomas. 1974. Effect of prey distribution and density on the searching and feeding behavior of larval anchovy *Engraulis mordox* Girard, pp. 559-574. *In:* J. H. S. Blaxter (ed.), *The Early Life History of Fish.* Springer-Verlag, Berlin.

Hyatt, K. D. 1979. Feeding strategy, pp. 71-120. *In:* W. S. Hoar, D. J. Randall and J. R. Brett (eds.), *Fish Physiology, Vol. 8.* Academic Press, New York, NY.

Ivlev, V. S. 1961. *Experimental Ecology of the Feeding of Fishes.* Yale U. Press, New Haven, CT. 302 pp.

Joseph, E. G. 1972. The status of sciaenid stocks of the middle Atlantic coast. *Chesapeake Sci.* 13:87-100.

Joseph E. G. 1973. Analyses of a nursery ground, pp. 118-121. *In:* A. L. Pacheco (ed.), *Proceedings of a Workshop on Egg, Larval and Juvenile Stages of Fish in Atlantic Coast Estuaries.* Tech Publ. No. 1, Mid-Atlantic Coastal Fish Ctr., Highlands, NJ.

Keast, A. 1970. Food specializations and bioenergetic inter-relations in the fish faunas of some small Ontario waterways, pp. 377-411. *In:* J. H. Steele (ed.), *Marine Food Chains.* U. California Press, Berkeley, CA.

Kinne, O. 1967. Physiology of estuarine organisms with special reference to salinity and temperature: General aspects, pp. 525-540. *In:* G. H. Lauff (ed.), *Estuaries.* Amer. Assoc. Advanc. Science, Publ. No. 83, Washington, D.C.

Kjelson, M. A. and G. N. Johnson. 1976. Further observations of the feeding ecology of postlarval pinfish, *Lagodon rhomboides,* and spot, *Leiostomus xanthurus. Fish Bull.* 74:423-432.

Knudsen, E. E. and W. H. Herke. 1978. Growth rate of marked juvenile Atlantic croakers, *Micropogon undulatus,* and length of stay in a coastal marsh nursery in Southwest Louisiana. *Trans. Amer. Fish. Soc.* 107:12-20.

Larkin, P. A. 1956. Interspecific competition and population control in freshwater fish. *J. Fish. Res. Bd. Canada* 13:327-342.

Livingston, R. J. 1976. Diurnal and seasonal fluctuations of organisms in a North Florida estuary. *Est. Coastal Mar. Sci.* 4:373-400.

Lotrich, V. A. 1975. Summer home range and movements of *Fundulus heteroclitus* (Pisces: Cyprinodontidae) in a tidal creek. *Ecology* 65:191-198.

McHugh, J. L. 1966. Management of estuarine fisheries, pp. 133-154. *In:* R. F. Smith, A. H. Swartz and W. H. Massmann (eds.), *A Symposium on Estuarine Fisheries.* Amer. Fish. Soc., Spec. Publ. No. 3, Washington, D.C.

McHugh, J. L. 1967. Estuarine nekton, pp. 581-620. *In:* George H. Lauff (ed.), *Estuaries.* Amer. Assoc. Advanc. Science, Publ. No. 83, Washington, D.C.

Margalef, R. 1963. On certain unifying principles in ecology. *Amer. Natur.* 97:357-374.

Marshall, F. L. 1975. Effects of mosquito control ditching on *Juncus* marshes and utilization of mosquito control ditches by estuarine fishes and invertebrates. Ph.D. Dissertation, U. North Carolina, Chapel Hill, NC. 187 pp.

Miller, J. M., B. M. Currin and P. A. Rublee. 1980. Effects of highway culverts on migration and feeding of juvenile estuarine fishes. Draft report to U.S. Department of Transportation, Federal Highway Administration. Located at Zoology Department of North Carolina State University, Raleigh, NC (unpublished manuscript).

Nixon, S. W. and C. A. Oviatt. 1973. Ecology of a New England salt marsh. *Ecology* 43:463-498.

Norris, K. S. 1963. Functions of temperatures in the ecology of the percoid fish, *Girella nigricans. Ecol. Monogr.* 33:23-62.

Northcote, T. G. 1978. Migratory strategies and production in freshwater fishes, pp. 326-359. *In:* S. D. Gerking (ed.), *Freshwater Fish Production.* John Wiley & Sons, New York.

O'Brien, W. J. and G. L. Vinyard. 1974. Comment on the use of Ivlev's electivity index with planktivorous fish. *J. Fish. Res. Bd. Canada* 31:1427-1429.

Odum, E. P. 1969. The strategy of ecosystem development. *Science* 164:262-270.

Orians, G. H. 1975. Diversity, stability and maturity in natural ecosystems, pp. 139-150. *In:* W. H. van Dobben and R. H. Lowe-McConnell (eds.), *Unifying Concepts in Ecology.* Proc. Intl. Congr. Ecol., 1st Wagenigen: Cent. Agric. Publ. Docl, W-Junk, The Hague.

Parsons, T. R. and R. J. LeBrasseur. 1970. The availability of food to different trophic levels in the marine food chain, pp. 325-343. *In:* J. H. Steele (ed.), *Marine Food Chains.* U. California Press, Berkeley, CA.

Pyke, G. H. 1979. Optimal foraging in fish, pp. 199-202. *In:* H. Clepper (ed.), *Predator-Prey Systems in Fisheries Management.* Sport Fishing Institute, Washington, D.C.

Reis, R. R. and J. M. Dean. 1980. Temporal variation in the utilization of an intertidal creek by the bay anchovy (*Anchoa mitchilli*). *Estuaries* (in press).

Sabins, D. S. and F. M. Truesdale. 1974. Diel and seasonal occurrence of immature fishes in a Louisiana tidal pass, pp. 161-170. *In:* William A. Rogers (ed.), *Proc. Twenty-eighth Ann. S. E. Assoc. Game and Fish. Comm.,* White Sulfur Springs, WV.

Schelske, C. L. and E. P. Odum. 1961. Mechanisms maintaining high productivity in Georgia estuaries. *Proc. Gulf and Carib. Fish Inst.* 14:75-80.

Shenker, J. M. and J. M. Dean. 1979. The utilization of an inter-tidal salt marsh creek by larval and juvenile fishes: Abundance, diversity and temporal variation. *Estuaries* 2:154-163.

Sheppard, M. P. 1955. Resistance and tolerance of young brook trout (*Salvelinus fontinalis*) to oxygen lack, with special reference to low oxygen acclimation. *J. Fish. Res. Bd. Canada* 12:387-466.

Sheridan, P. F. 1978. Trophic relationships of dominant fishes in the Apalachicola Bay system (Florida). Ph.D. Dissertation, Florida State U., Tallahassee, FL. 215 pp.

Stickney, R. R., G. L. Taylor and D. B. White. 1975. Food habits of five species of young southeastern United States estuarine Sciaenidae. *Chesapeake Sci.* 16:104-114.

Swenson, W. A. 1977. Food consumption of walleye (*Stizostedion vitreum vitreum*) and sauger (*S. canadense*) in relation to food availability and physical environmental conditions in Lake of the Woods, Minnesota, Shagawa Lake, and western Lake Superior. *J. Fish. Res. Bd. Canada* 34:1643-1654.

Thayer, G. W., D. E. Hoss, M. A. Kjelson, W. F. Hettler, Jr. and M. W. Lacroix. 1974. Biomass of zooplankton in the Newport River estuary and the influence of postlarval fishes. *Chesapeake Sci.* 15:9-16.

Thomas, D. L. 1971. The early life history and ecology of six species of drum (Sciaenidae) in the lower Delaware River, a brackish tidal estuary. Progress Report 3 (Part III), Ichthyological Associates, Middletown, DE.

Van Engel, W. A. and E. B. Joseph. 1967. Characteristics of coastal and estuarine fish nursery grounds as natural communities. Unpublished report to U.S. Fish and Wildlife Service, 43 pp. Available from: Virginia Institute of Marine Science, Gloucester Point, VA.

Ware, D. M. 1972. Predation by rainbow trout *(Salmo gairdneri)*: the influence of hunger, prey density and prey size. *J. Fish. Res. Bd. Canada* 29:1193-1201.

Weinstein, M. P. 1979. Shallow marsh habitats as primary nurseries for fishes and shellfish, Cape Fear River, North Carolina. *Fish. Bull.* 77:339-358.

Werner, E. E. and D. J. Hall. 1977. Competition and habitat shift in two sunfishes (Centrarchidae). *Ecology* 58:869-876.

SEAGRASS HABITATS: THE ROLES OF HABITAT COMPLEXITY, COMPETITION AND PREDATION IN STRUCTURING ASSOCIATED FISH AND MOTILE MACROINVERTEBRATE ASSEMBLAGES

Kenneth L. Heck, Jr.

The Academy of Natural Sciences of Philadelphia
Benedict Estuarine Research Laboratory
Benedict, Maryland

and

Robert J. Orth

Virginia Institute of Marine Science
and School of Marine Science
The College of William and Mary
Gloucester Point, Virginia

Abstract: Seagrass meadows represent a distinct habitat in shallow coastal and estuarine ecosystems. We examine the role of seagrass meadows as an important habitat for fishes and large mobile invertebrates. In particular, we emphasize the importance of the structural complexity of the vegetation and associated algal components. Based on data from a variety of geographical localities we consider how vegetation density, plant morphology and associated sessile colonial animals can influence abundance and diversity of predator and prey species in vegetated areas on both local and regional geographical scales. In so doing we generate hypotheses that lead to predictions concerning: size of populations and the amplitude of their fluctuations in vegetated habitats at different latitudes; success rate of predators using different foraging strategies in vegetation of different densities; and resultant diversity and abundance of invertebrate prey, juvenile fish and adult fish in different densities of vegetation.

Introduction

Seagrass meadows are distinct and important components of marine and estuarine environments. Among their most important attributes are their ability to serve as nursery areas containing high densities and diversities of fishes and invertebrates, as feeding grounds for fish and waterfowl, and as sediment stabilizers (Ginsburg and Lowenstam 1958; Thayer et al. 1975; Orth 1977; Robertson 1977).

Although research on functional ecology of seagrass beds has greatly expanded recently (Adams 1976b; McRoy and Helfferich 1977; Ott and Maurer 1977), past emphasis has been concentrated on the structural components of communities (Thayer et al. 1975; Kikuchi and Peres 1977). Based on existing data, the biota inhabiting seagrass meadows has been subdivided into several structural subunits (Kikuchi and Peres 1977): 1) species living on leaves, including epiphytes, micro- and meiofauna (e.g. ciliates and nematodes), sessile fauna (e.g. bryozoans, colonial tunicates and barnacles), mobile creeping and walking epifauna (e.g. gastropods, isopods and amphipods) and swimming epifauna (e.g. caridean shrimps and small fishes such as syngnathids); 2) species attached to stems and rhizomes (e.g. sessile polychaetes and tubicolous amphipods); 3) mobile species under the leaf canopy (e.g. portunid crabs and fishes). This group can be divided further into permanent seagrass residents, seasonal residents, visitors which forage over a wide area (part of which is the seagrass habitat), and occasional migrants; 4) infaunal species (e.g. polychaetes and bivalves) many of which are found in unvegetated sites as well.

Most studies aimed at elucidating structural aspects of seagrass communities have concentrated on mobile species living under the leaf canopy (3), or on biota living on leaves (1), or on both these groups of species. These groups have been well studied primarily because they contain the economically important fishes and decapod crustaceans and their prey, but also because they may be sampled using relatively simple techniques such as seining and trawling. Because of the comparatively large data base which exists for these groups of species, they are particularly suitable for examining similarities and differences among different types of seagrass habitats. Even though there are differences in sampling efficiencies of different gear types used in previous studies, we believe that general trends shown in these studies are real, and in many cases are intuitive, and assert that they would be similar regardless of the sampling gear used. Therefore, using data on these species, the following discussion will first describe a number of patterns which emerge from studies on structure of seagrass faunal assemblages along the temperate and tropical coasts of the Americas. Second, explanations for the origin of these patterns will be proposed, with particular emphasis on the importance of habitat complexity in seagrass meadows. A final section is devoted to examining hypotheses of the roles of predation and competition as they influence community structure and contribute to the nursery role played by seagrass environments.

The Data Base

Studies of motile epibenthic macroinvertebrates and fishes have been undertaken in subtidal seagrass meadows in lower Chesapeake Bay (Heck and Orth 1980; Orth and Heck 1980); along the coasts of North Carolina (Godfrey 1970; Thayer et al. 1975; Adams 1976a); New York (Briggs and O'Connor 1971); Massachusetts (Allee 1923; Nagle 1968); Florida (Tabb et al. 1962; Livingston 1975; Heck 1976; Hooks et al. 1976; Young et al. 1976; Young and Young 1977; Thorhaug and Roessler 1977; Gore et al. 1980); and Texas (Hoese and Jones 1963); and in the Caribbean along the coasts of Cuba (Murina et al. 1974); Puerto Rico (Zimmerman 1978); the U.S. Virgin Islands (Ogden 1976; Ogden and Zieman 1977); and Panama (Heck 1977, 1979). Recently, preliminary work on eelgrass assemblages of the Pacific coast of the Americas has been reported (Thayer and Phillips 1977) but most work on Pacific eelgrass faunas remains descriptive and as yet unsuitable for detailed analysis.

Because collections have been made in Chesapeake Bay (Heck and Orth 1980; Orth and Heck 1980), northern Gulf of Mexico (Livingston 1975; Heck 1976; Hooks et al. 1976), and in Panama (Heck 1977; Weinstein and Heck 1979) using identical sampling procedures (i.e. otter trawling) and because of our personal participation in these studies, they can be most readily compared. We concentrate here on macroinvertebrate and fish species collected in these studies. The macroinvertebrates include primarily decapod crustaceans and secondarily molluscs, echinoderms and some polychaetes; the fishes are predators on these invertebrates in many cases. It is these organisms which can be most readily collected by trawl, albeit inefficiently (Kjelson and Johnson 1978). Sessile, creeping, or walking epifauna, infauna, or larval or larger fishes are not included in these data because they cannot be collected reliably, if at all, by trawling. However, data on these components of the seagrass biota are included in ensuing discussions based on work of other investigators. Likewise, data from many other seagrass studies are incorporated in the following discussion where appropriate, so that a more complete picture of seagrass communities is portrayed.

Habitat Complexity

Diversity of seagrass-associated faunas varies considerably on both the local and geographical scale. We now consider what we believe to be the single most important factor influencing changes in faunal composition: namely, variation in the physical structure or complexity of the seagrass habitat.

In turtlegrass *(Thalassia testudinum)* meadows, areas of greater plant biomass contain more species and greater abundances of epibenthic invertebrates than do areas of lesser plant biomass (Tabb et al. 1962; Roessler and Zieman 1970; Hooks et al. 1976; Heck and Wetstone 1977). Thus, above ground plant biomass may serve as a reasonable estimate of habitat complexity for benthic macroinvertebrates (Heck and Wetstone 1977; Gore et al. 1980) under most conditions, because as plant biomass increases there apparently is a concomitant increase in habitable living space, and a formation of protected living spaces for species requiring such habitats. However, we suggest that there are conditions when this relationship may not be evident. Because seagrass meadows are located in shallow areas, they may be subject to extreme environmental fluctuations. Since most animals live within a range of conditions (temperature, etc.) that allow them to maintain normal metabolic functions, they are likely to avoid conditions (e.g. high temperatures) which interfere with their activity patterns. For example, Adams (1976b) suggested that the greater night abundances of fish in North Carolina grass beds in summer are due to lower, more favorable water temperatures. Thus, even though seagrass density was high, summer daytime fish abundance was low, presumably because fish avoided the extremely hot nearshore water during the day.

There are also qualitative aspects to the plant-invertebrate relationships, although there does not appear to be a simple relationship between the number of plant species and the species richness or abundance of benthic invertebrates (Heck and Wetstone 1977). Instead, certain species of plants seem to be of disproportionate importance. Most conspicuous in the turtlegrass habitat are species of rhodophytes which make up the "drift algal complex" (Tabb et al. 1962; Hooks et al. 1976; Thorhaug and Roessler 1977; Gore et al. 1980) and the calcareous green alga *Halimeda opuntia* (Heck and Wetstone 1977), both of which we consider below.

Thorhaug and Roessler (1977) and Gore et al. (1980) have recently produced quantitative evidence to support earlier observations that areas with drift algae contain significantly higher animal densities than do areas with seagrasses alone. The two most likely explanations for these higher densities are (1) that large food supplies (red algae and their associated epiphytes) attract animals, and (2) that red algal masses serve as refuges in which small organisms may escape their predators.

Certain amphipods associated with drift algal complexes do utilize the algae for food (Zimmerman 1978) and small fishes such as pipefishes and gobies, which are abundant in the drift algae, actively feed on the associated amphipods (G. R. Kulczycki, Fla. Dept. of Envir. Regulation, Ft. Pierce, FL. pers. comm.). This suggests that, at least for amphipods, drift algae may serve primarily as a food source and not a site of reduced predator effectiveness. However, it seems unlikely that these animals are attracted by food alone, since species of the drift algae are also present on individual grass blades, and other food in the form of detritus is widely distributed throughout seagrass beds. Thus, as Nelson (1979b) suggests, it seems highly unlikely that food is limiting to amphipod populations in seagrass meadows. For large invertebrates such as caridean shrimps, gastropods and echinoderms, as well as for small fishes associated with drift algae, it is likely that masses of drift algae serve as sites of reduced predator effectiveness. This is intuitive since the algae provide cover and thereby make the location and capture of prey more difficult (cf. Huffaker 1958). It has also been demonstrated that marsh plants provide protection for invertebrates from fish predators (Vince et al. 1976; Van Dolah 1978) and that aquatic vegetation provides protection for nonschooling fishes from larger fish predators (Neill and Cullen 1974).

Calcareous algae of the genus *Halimeda*, especially species such as *H. opuntia* that form thick mats, are another element which seems to exert a strong influence on the invertebrates of *T. testudinum* meadows (Heck and Wetstone 1977). When *H. opuntia* is present in dense amounts it forms an understory beneath the seagrass canopy. Within and beneath this understory many species of animals that normally use coral rubble or rocks for shelter can be found (Heck and Wetstone 1977). For example, pistol shrimps, porcelain crabs and several species of majid crabs are found in much greater abundance in turtlegrass with *H. opuntia* mats than in pure stands of turtlegrass (Heck and Wetstone 1977).

Animals living among and in close proximity to seagrasses may also impart increased physical complexity. Most important are hermatypic corals and sponges. On the scale of one to a few meters, isolated coral colonies of *Porites porites*, which commonly occur among seagrasses in the Caribbean reef flats, may provide shelter for invertebrate coral commensals as well as for coral dwelling fishes. In addition, coral rubble provides shelter for many species of decapod crustaceans (Abele 1974) and many invertebrates otherwise found in protected habitats. Also, on a larger scale, coral reefs can be significant factors in the ecology of adjacent seagrass meadows because many species of fishes forage in seagrass meadows at night while they shelter on reefs during daylight hours (Randall 1963, 1967; Starck 1971; Ogden and Ehrlich 1977; Ogden and Zieman 1977). Especially important among these predatory fishes are pomadasyids and lutjanids. There are also some invertebrates such as the urchin *Diadema antillarum* which make similar migrations from reefs into adjacent vegetated areas (Ogden et al. 1973; Ogden 1976), although the number of invertebrates which treat the seagrass-reef complex as a single habitat seems to be much smaller than the number of fishes which do so (Weinstein and Heck 1979; Heck 1979).

Large sponges such as the loggerhead *Spheciospongia vesparia* may also provide shelter for certain invertebrates such as xanthid crabs (Williams 1965) and alpheid shrimps (Chace 1972), as well as being used by a number of fishes usually associated with corals (Collette and Rutzler 1977).

Temperate seagrass meadows are dominated by eelgrass *Z. marina*, although *Ruppia maritima* and *Halodule wrightii* occur in mixed stands in North Carolina waters (Thayer et al. 1975), and *Z. marina* and *R. maritima* occur in some Virginia seagrass meadows (Orth and Heck 1980). In North Carolina, mixed stands of seagrasses, foliose drift algae (Thayer et al. 1975; Nelson 1979a) and sponges should provide substantial habitat heterogeneity, roughly equivalent to that found in the northern Gulf of Mexico. In Chesapeake Bay eelgrass meadows, large masses of bryozoan *Alcyonidium verrilli* may serve to increase habitat complexity as do sponges, but on a smaller scale. Colonies of *A. verrilli* found in eelgrass beds harbor large numbers of xanthid crabs, many more than are usually collected from areas without bryozoan colonies (Heck and Orth 1980).

Another important component of habitat complexity in seagrass meadows is patchiness within a seagrass stand. Many meadows are not uniform in appearance due to biological and physical disturbances, such as wave action, boats, or the feeding activities of cownose rays (Orth 1975; Patriquin 1975; Zieman 1976). Because grass beds are continually subjected to disturbances which vary both in space and time, the fauna in grass beds may exist as a spatial and temporal mosaic of different species diversities. Severe natural disturbances which eliminate seagrasses and associated algae could initiate regressive trends toward a community with lower diversity, while benign periods in disturbed areas would allow recolonization by seagrasses (Patriquin 1975) and eventual increases in the diversity of seagrass-associated species. Thus, severe disturbances can increase overall spatial heterogeneity of the environment, by increasing both within-habitat diversity and pattern diversity (Pielou 1966) and providing opportunities for local differentiation through random colonization (Levin and Paine 1974).

In general, we suggest that there is a pronounced latitudinal gradient in structural complexity and spatial heterogeneity of seagrass environments, progressing from the simple cool temperate systems with essentially only the seagrass themselves to the intermediate warm temperate and subtropical systems with seagrasses, drift algae and sponges, to the most complex tropical systems with seagrasses, red algae, calcareous green algae, sponges and corals. Similar variation in habitat complexity may also occur on a local scale as one moves along gradients in salinity or wave shock. From the preceding discussion, it is clear that there are many different sorts of "turtlegrass habitats," which make it necessary to speak in precise terms about the structure of a given area (Heck 1977; Brook 1978). Not nearly as much information has been compiled on different *Z. marina* habitats, but it is evident that there are distinct kinds of eelgrass habitats just as there are distinct turtlegrass habitats.

Even though the number of herbivorous species inhabitating seagrass meadows is small, the local abundance of these grazing species can be extremely great (Ogden 1976). Additionally, the activity of these grazers can be significant in reducing abundance of plants (Camp et al. 1973) which can effectively reduce habitat complexity of seagrass environments for other species. In this regard, species richness of seagrass-associated invertebrates and fishes can be indirectly controlled by effects of grazing species (cf. Heck 1979).

Predation and Competition

Predator-Prey Relationships

It has been shown many times that predation is a significant biological interaction in aquatic systems and Connell (1975) has urged that predation be considered to be of primary importance in models of community organization except where the intensity of predation is lowered by other factors.

In Fig. 1, we have plotted monthly abundances of epibenthic invertebrate species from Chesapeake Bay, Gulf of Mexico, and Caribbean trawl samples (Heck 1976, 1977; Heck and Orth 1980). In each area, trawl size and mesh, as well as towing procedures, were identical. Population fluctuations were much greater in temperate *Zostera marina* habitat than in the Gulf of Mexico, which in turn were greater than in the Caribbean.

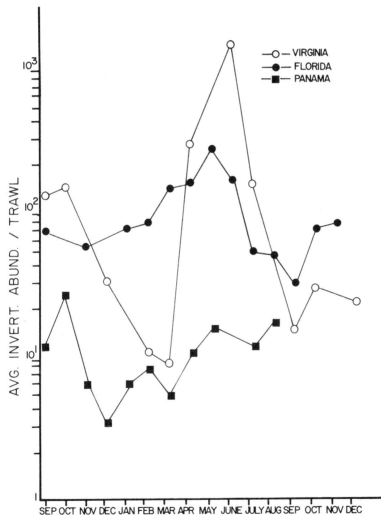

Figure 1. *Average number of individuals/trawl sample taken monthly from the lower Chesapeake Bay in Virginia (38°N; ○), the northern Gulf of Mexico in Florida (30°N ●), and the Caribbean coast of Panama (9°N ■).*

These patterns are similar to those of planktonic populations along boreal, temperate and tropical coasts in nonupwelling areas. These latter patterns are believed to result from a cyclic predator-prey interaction (Cushing 1975).

We believe a similar explanation can account for latitudinal differences in standing crop and the amplitude of invertebrate population fluctuations in Chesapeake Bay, Gulf of Mexico, and Caribbean Sea. In the Chesapeake (38°N), fish populations decline to low levels during December-March (Heck and Orth 1980; Orth and Heck 1980) when water temperatures drop to near freezing. In the Gulf of Mexico, there is a brief winter period of inactivity (approximately January-February) (Livingston 1975) whereas in Panama, populations are present all year (Weinstein and Heck 1979). We suggest that the longer delay period in the cool temperate Chesapeake Bay allows invertebrates to escape their predators for longer periods and thereby reach large abundances early in the season. As predators grow and begin to feed upon macroinvertebrates, population sizes should drop dramatically resulting in cycles of large amplitude (Fig. 1). In this scheme, Gulf of Mexico populations represent an intermediate condition in which standing crops and cycles of abundance fall between those of the cool temperate and tropical regions. Because populations of both invertebrates and fishes are present all year in the tropics, standing crops and amplitudes of invertebrate population fluctuations would be relatively low.

Similar patterns may exist for juvenile fishes. For example, juvenile fishes are consumed by adult fishes and are analogous to the invertebrates in our previous discussion. Therefore, population fluctuations of juvenile (and, consequently, adult) fishes would be greatest in cool temperate, intermediate in warm temperate and lowest in tropical seagrass habitats (as well as in unvegetated habitats) for the reasons discussed above.

If this scheme is correct, we can predict that as yet unstudied seagrass-associated invertebrate and fish populations in the West Pacific Ocean, for example, would show similar patterns as one progresses from cool temperate (e.g. northern Japan) to warm temperate (e.g. southern Japan) to tropical (e.g. Philippines) localities.

There has been at least one published study (Nelson 1979b) and one as yet unpublished study (Heck and Thoman, ms) which have specifically attempted to test the hypotheses that presence of seagrasses reduces predator effectiveness and that denser vegetation provides more protection from predation than does sparser vegetation. The results of both these studies demonstrated that increased density of seagrasses does reduce the effectiveness of fish predators. Nelson's (1979b) study used pinfish *(Lagodon rhomboides)* preying on amphipod species, while Heck and Thoman (unpublished) used killifish *(Fundulus heteroclitus)* preying on grass shrimp *(Palaemonetes pugio)*. In both studies, a threshold density of seagrass was required before any significantly reduced predator effectiveness was noted. For example, at artificial eelgrass densities equivalent to sparse field densities of eelgrass (274 turions/m²), killifish fed on grass shrimp at a rate similar to their feeding rate in experiments done on bare sandy bottoms. Only at densities of 652 turions/m² was the capture rate of killifish significantly reduced (Heck and Thoman unpublished).

Both these experiments used predators which actively pursue their prey. We expect that experiments with fishes using other foraging tactics would show different results. For example, ambush predators such as lizardfish or toadfish which lie in wait inconspicuously would be less likely to have their efficiency reduced by the presence of vegetation. Stalking predators such as pike which slowly approach prey while using vegetation for cover may be more successful in vegetation than on sandy bottoms.

We suggest that plant morphology can influence prey capture rates significantly. For a given plant biomass we hypothesize that those plants with more foliose leaves and therefore greater surface areas per unit weight ought to provide more protection than plants with simpler leaves and lower surface area per weight ratios. Thus, we would predict that a given biomass of foliose drift algae would provide more protection than would turtlegrass or eelgrass. In theory, one ought to be able to rank the amount of protection provided by an equal biomass of individual plant species, based on their surface area per weight ratios. In Fig. 2 we outline the consequences of this hypothesized relationship. We use a histogram in this figure because experimental work suggests that predation intensity is not linearly related to plant density (Nelson 1979a; Heck and Thoman, ms) but may be described more accurately by a step function (Nelson 1979a).

Further, we suggest that diversity and abundance of macroinvertebrates, juvenile fishes and adult fishes will be related to plant surface area per area of bottom substrate in the manner shown in Fig. 3. We reason that vegetation should provide protection from predators for invertebrates and juvenile fish and therefore survivorship and density should increase up to some asymptotic value as vegetation increases. Habitat complexity should also increase with increasing vegetation density thereby leading to increases in species richness among invertebrates and juvenile fish. At some point, however, vegetation may become so dense that conditions may be unfavorable for invertebrates and juvenile fish (e.g. anoxic condition occurring at night) and diversity and abundance would decline to lower levels. Many adult fishes are primarily predators and they should be attracted to grass beds in increasing numbers with increasing plant density until grass density reduces their effective hunting success to a point equal to or lower than that realized in surrounding habitats. While it seems that this increasing number of predatory fish would result in reduced invertebrate abundance, existing data indicate that losses to predators are more than offset by the greater protection provided invertebrate populations by increasing plant surface area. (Of course there are certain fishes which because of their herbivorous or detritivorous feeding habits may show patterns closer to those depicted for invertebrates in Fig. 3, just as

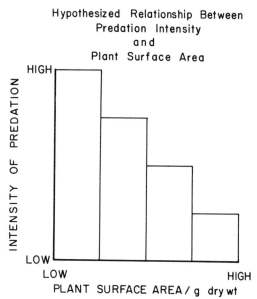

Figure 2. Hypothesized relationship between predation intensity and plant surface area/unit wt. Predators should be more successful at prey capture among plants of relatively simple morphology with low surface area/unit weight ratios than among more foliose plants with higher surface area/unit weight ratios (cf Nelson 1979a, Fig. 7).

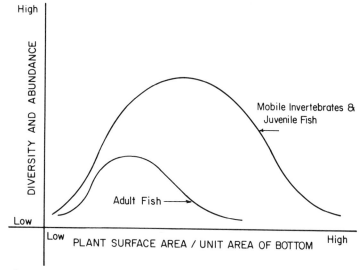

Figure 3. Hypothesized relationships between diversity and abundance of seagrass-associated animals and the amount of plant cover in a given area. See text for explanation.

there are some larger predaceous invertebrates such as large blue crabs which may follow the trends we have described for fishes.) We anticipate that the net benefit (prey intake-searching costs) to foraging predators in different plant densities can be represented by a curve similar to that in Fig. 4. We hypothesize that predators will crop prey profitably at low plant densities because shelter is limited and predator mobility will be hindered only slightly by sparse vegetation. However, high plant densities will offer substantial shelter for prey species, increase their chances of successfully eluding visually hunting predators, and reduce predator mobility and maneuverability. In combination, these factors will limit substantially the net benefit accruing to foraging predators in stands of high density. A similar curve should, for basically the same reasons, describe the general trends in the average size of predators found in different densities of plants.

In lakes, fish predators have been noted to wait near the periphery of vegetated areas and attack when prey become visible near the edge of the vegetation (Johannes and Larkin 1961). Anecdotal information exists for this same phenomenon in the Chesapeake Bay area. Local fishermen

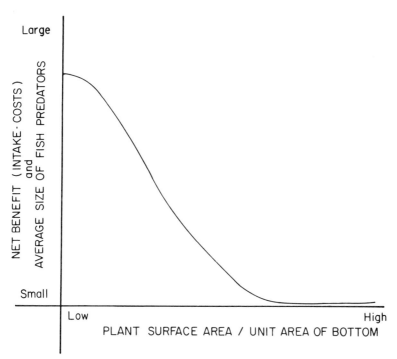

Figure 4. Hypothesized relationships between the net benefit and size of seagrass-associated fishes and the amount of plant cover in a given area.

believe that fishing for large predators such as sea trout, *Cynoscion* spp., and striped bass, *Morone saxatilis*, is most effective along the edges of vegetation and that bare pockets in dense vegetation are more likely to produce large predators. Thus we expect that larger predaceous fish should forage near the periphery of dense stands of vegetation so that they are effectively foraging in intermediate densities of plants and are most likely to capture prey successfully.

Competitive Interactions

Although competition is a central concept in modern ecological theory, little significance has been attributed to competition among seagrass inhabitants. This is because food in the form of detritus, epiphytes, etc. appears to be so abundant in seagrass meadows that it is probably not a limiting resource (Jackson 1972; Mann 1972; Adams 1977; Nelson 1979b).

However, given what we have discussed about the importance of predation in seagrass habitats, food is unlikely to be a limiting resource since prey populations should not reach high enough densities to exhaust food supplies. Instead, cover from predators would seem to be the most important potentially limiting resource, and we suggest that it is interference and not exploitation competition which is important among animals in seagrass meadows. If this is true we should see evidence of aggression and competition for hiding places among potential competitors.

Two studies have found that interference competition can occur among seagrass invertebrates. Nagle (1968) described amphipods maintaining territories by pinching conspecifics when they ventured into inhabited areas on individual eelgrass blades, and Thorp (1976) showed that two species of grass shrimp competed for different substrates, presumably because one (shell) provided more shelter than the other (sand). A recent study has shown that in the laboratory two species of palaemonid shrimps do prefer vegetation over bare sand bottoms and that one species can actively displace the other from vegetation (Coen 1979; Coen, Heck and Abele unpublished ms). Addition of a fish predator to this experimental system showed that the species excluded from the vegetation was eaten in significantly greater numbers than its dominant competitor. In this experimental system, the resource in short supply is cover and the consequence of being competitively inferior is increased susceptibility to predation.

We suggest that in low to intermediate plant densities, interference competition may be quite important in determining which species is protected from predation by plant biomass and therefore which species escapes predators. Interference competition could therefore be significant among most motile invertebrates and nonschooling small fishes. As outlined earlier, we expect predation intensity and, therefore, interference competition to be less important in very dense vegetation.

Acknowledgments

L. G. Abele, L. D. Coen, R. J. Livingston and A. W. Stoner greatly improved the quality of this manuscript by commenting on earlier drafts. Support for our work has come from a grant by the Coastal Coordinating Council of the State of Florida to R. J. Livingston, a Smithsonian Predoctoral Fellowship at the Smithsonian Tropical Research Institute, the Academy of Natural Sciences of Philadelphia, the Virginia Institute of Marine Sciences, and EPA Grants R806151 and R805974. Beverly Knee, Daisy Hills, Elaine Smith and Shirley Sterling aided in the preparation of the manuscript. We are grateful to all.

Contribution No. 926 from the Virginia Institute of Marine Science.

References Cited

Abele, L. G. 1974. Species diversity of decapod crustaceans in marine habitats. *Ecology* 55:156-161.

Adams, S. M. 1976a. The ecology of eelgrass *Zostera marina* (L.) fish communities. I. Structural analysis. *J. Exp. Mar. Biol. Ecol.* 22:269-291.

Adams, S. M. 1976b. The ecology of eelgrass *(Zostera marina)* fish communities. II. Functional analysis. *J. Exp. Mar. Biol. Ecol.* 22:293-311.

Adams, S. M. 1977. Feeding ecology of eelgrass fish communities. *Trans. Amer. Fish. Soc.* 105:514-519.

Allee, W. C. 1923. Studies in marine ecology. I. The distribution of common littoral invertebrates in the Woods Hole region. *Biol. Bull.* 44:167-191.

Briggs, P. T. and J. S. O'Connor. 1971. Comparison of shorezone fishes over naturally vegetated and sand covered bottoms in Great South Bay. *New York Fish and Game J.* 18:15-41.

Brook, I. M. 1978. Comparative macrofaunal abundance in turtlegrass *(Thalassia testudinum)* communities in south Florida characterized by high blade density. *Bull. Mar. Sci.* 28:212-220.

Camp, D. K., S. P. Cobb and J. F. van Breedveld. 1973. Overgrazing of seagrasses by a regular sea urchin, *Lytechinus variegatus. Bioscience* 23:37-38.

Chace, F. A., Jr. 1972. The shrimps of the Smithsonian-Bredin Caribbean expeditions with a summary of the West Indian shallow-water species. (Crustacea: Decapoda: Natantia). *Smithsonian Contr. Zool.* 98:1-179.

Coen, L. D. 1979. An experimental study of habitat selection and interaction between two species of caridean shrimps (Decapoda: Palaemonidae) M.S. Thesis, Florida State U., Tallahassee, FL. 37 p.

Collette, B. B. and K. Ruetzler. 1977. Reef fishes over sponge bottoms off the mouth of the Amazon River. *Proc. Third Int'l. Coral Reef Symp.* 1:305-310.

Connell, J. H. 1975. Some mechanisms producing structure in natural communities, pp. 460-490. *In:* M. L. Cody and J. M. Diamond (eds.), *Ecology and Evolution of Communities.* Harvard U. Press, Cambridge.

Cushing, D. H. 1975. *Marine Ecology and Fisheries.* Cambridge U. Press, Cambridge. 278 p.

Den Hartog, C. 1977. Structure, function and classification in seagrass communities, pp. 89-122. *In:* C. P. McRoy and C. Helfferich (eds.), *Seagrass Ecosystems.* Marcel Dekker, Inc., New York.

Ginsburg, R. M. and H. A. Lowenstam. 1958. The influence of marine bottom communities on the depositional environment of sediments. *J. Geol.* 66:310-318.

Godfrey, M. M. 1970. Seasonal changes in the macrofauna of an eelgrass *(Zostera marina)* community near Beaufort, N.C. M.S. Thesis, Duke U., Durham, NC. 212 p.

Gore, R. H., L. E. Scotto, K. A. Wilson and E. E. Gallaher. 1980. Studies on decapod crustacea from the Indian River region of Florida. XI. Species composition, structure, biomass and species areal relationships of seagrass and drift-algae associated macrocrustaceans. *Est. Coastal Mar. Sci.* (in press).

Heck, K. L., Jr. 1976. Community structure and the effects of pollution in seagrass meadows and adjacent habitats. *Mar. Biol.* 35:345-357.

Heck, K. L., Jr. 1977. Comparative species richness, composition and abundance of invertebrates in Caribbean seagrass *(Thalassia)* meadows. *Mar. Biol.* 41:335-348.

Heck, K. L., Jr. 1979. Some determinants of the composition and abundance of motile macroinvertebrate species in tropical and temperate turtlegrass *(Thalassia testudinum)* meadows. *J. Biogeogr.* 6:183-197.

Heck, K. L., Jr. and G. S. Wetstone. 1977. Habitat complexity and invertebrate species richness and abundance in tropical seagrass meadows. *J. Biogeogr.* 4:135-142.

Hoese, H. D. and R. S. Jones. 1963. Seasonality of larger animals in a Texas turtlegrass community. *Publ. Inst. Mar. Sci.* 9:37-47.

Hooks, T. A., K. L. Heck, Jr. and R. J. Livingston. 1976. An inshore marine invertebrate community: structure and habitat associations in the northeastern Gulf of Mexico. *Bull. Mar. Sci.* 26:99-109.

Huffaker, C. B. 1958. Experimental studies on predation: dispersion factors and predator-prey oscillations. *Hilgardia* 27:343-383.

Jackson, J. B. C. 1972. The ecology of the molluscs of *Thalassia* communities, Jamaica, West Indies. II. Molluscan population variability along an environmental stress gradient. *Mar. Biol.* 14:304-337.

Johannes, R. E. and P. A. Larkin. 1961. Competition for food between redside shiners *(Richardsonus balteatus)* and rainbow trout *(Salmo gairdneri)* in two British Columbia lakes. *J. Fish. Res. Bd. Can.* 18:203-220.

Kikuchi, T. and J. M. Peres. 1977. Consumer ecology of seagrass beds, pp. 147-193. *In:* C. P. McRoy and C. Helfferich (eds.), *Seagrass Ecosystems.* Marcel Dekker, Inc., New York.

Kjelson, M. A. and G. N. Johnson. 1978. Catch efficiencies of a 6.1-meter otter trawl for estuarine fish populations. *Trans. Amer. Fish. Soc.* 107:246-254.

Levin, S. A. and R. T. Paine. 1974. Disturbance, patch formation and community structure. *Proc. Nat. Acad. Sci. U.S.* 71:2744-2747.

Livingston, R. J. 1975. Impact of kraft pulp-mill effluents on estuarine and coastal fishes in Apalachee Bay, Florida. *Mar. Biol.* 32:19-48.

Mann, K. H. 1972. Macrophyte production and detritus food chains in coastal waters. *Mem. First Ital. Idrobiol. 29 Suppl.* :353-383.

McRoy, C. P. and C. Helfferich (eds.). 1977. *Seagrass Ecosystems.* Marcel Dekker, Inc. New York. 314 p.

Murina, V. V., V. D. Chukhckin, O. Gomez and G. Suarez. 1974. Quantitative distribution of bottom macrofauna in the upper sublittoral zone of the northwestern part of Cuba, pp. 242-259. *In:* Investigations of the Central American Seas. Indian National Science Documentation Centre: New Dehli.

Nagle, J. S. 1968. Distribution of the epibiota of macroepibenthic plants. *Contr. Mar. Sci.* 13:105-144.

Neill, S. R. St. J. and J. M. Cullen. 1974. Experiments on whether schooling by their prey affects the hunting behavior of cephalopods and fish predators. *J. Zool.* 172:549-569.

Nelson, W. G. 1979a. Experimental studies of selective predation on amphipods: Consequences for amphipod distribution and abundance. *J. Exp. Mar. Biol. Ecol.* 38:225-245.

Nelson, W. G. 1979b. An analysis of structural patterns in an eelgrass *(Zostera marina)* amphipod community. *J. Exp. Mar. Biol. Ecol.* 39:231-264.

Ogden, J. C. 1976. Some aspects of herbivore-plant relationships on Caribbean reefs and seagrass beds. *Aquatic Bot.* 2:103-116.

Ogden, J. C., R. A. Brown and N. Salesky. 1973. Grazing by the echinoid *Diadema antillarum* Philippi: Formation of halos around West Indian patch reefs. *Science* 182:715-717.

Ogden, J. A. and P. R. Ehrlich. 1977. The behavior of heterotypic resting schools of juvenile grunts *(Pomadasyidae)*. *Mar. Biol.* 42:273-280.

Ogden, J. A. and J. C. Zieman. 1977. Ecological aspects of coral reef-seagrass contacts in the Caribbean. *Proc. Third Int'l. Coral Reef Symp.* 1:377-382.

Orth, R. J. 1975. Destruction of eelgrass, *Zostera marina,* by the cownose ray, *Rhinoptera bonasus,* in the Chesapeake Bay. *Chesapeake Sci.* 16:205-208.

Orth, R. J. 1977. The importance of sediment stability in seagrass communities, pp. 281-300. *In:* B. C. Coull. (ed.), *Ecology of Marine Benthos.* U. South Carolina Press, Columbia, SC.

Orth, R. J. and K. L. Heck, Jr. 1980. Structural components of eelgrass *(Zostera marina)* meadows in Chesapeake Bay——Fishes. *Estuaries* (in press).

Ott, J. and L. Maurer. 1977. Strategies of energy transfer from marine macrophytes to consumer levels: The *Posidonia oceanica* example, pp. 493-502. *In:* B. F. Keegan, P. O'Ceidigh and P. J. S. Boaden (eds.), *Biology of Benthic Organisms,* Pergamon Press, New York.

Patriquin, D. G. 1975. Migration of "blowouts" in seagrass beds at Barbados and Carriacou West Indies and its ecological and geological implications. *Aquatic Bot.* 1:163-189.

Pielou, E. C. 1966. Species-diversity and pattern-diversity in the study of ecological succession. *J. Theor. Biol.* 10:370-383.

Randall, J. E. 1963. An analysis of the fish populations of artificial and natural reefs in the Virgin Islands. *Carib. J. Sci.* 3:31-47.

Randall, J. E. 1967. Food habits of reef fishes of the West Indies. *Stud. Trop. Oceanogr.* 5:665-847.

Robertson, A. J. 1977. Ecology of juvenile King George whiting *Sillaginodes punctatus* (Cuvier and Valenciennes) (Pisces: Perciformes) in Western Port, Victoria. *Aust. J. Mar. Freshw. Res.* 28:35-43.

Roessler, M. A. and J. C. Zieman. 1970. The effects of thermal additions on the biota of Biscayne Bay, Florida. *Proc. Gulf Carib. Fish. Inst.* 22:136-145.

Starck, W. A. 1971. Biology of the gray snapper, *Lutjanus griseus,* in the Florida Keys. *Stud. Trop. Oceanogr.* 10:11-150.

Tabb, D. C., D. Dubrow and R. Manning. 1962. The ecology of northern Florida Bay and adjacent estuaries. *Tech. Ser. Fla. St. Bd. Conserv.* 39:1-79.

Thayer, G. W., S. M. Adams and M. W. LaCroix. 1975. Structural and functional aspects of a recently established *Zostera marina* community, pp. 518-540. *In:* L. E. Cronin (ed.), *Estuarine Research, Vol. 1.* Academic Press, N.Y.

Thayer, G. W. and R. C. Phillips. 1977. Importance of eelgrass beds in Puget Sound. *Mar. Fish. Rev.* 39:18-22.

Thorhaug, A. and M. A. Roessler. 1977. Seagrass community dynamics in a subtropical estuarine lagoon. *Aquaculture* 12:253-277.

Thorp, J. H. 1976. Interference competition as a mechanism of coexistence between two sympatric species of the grass shrimp, *Palaemonetes* (Decapoda: Palaemonidae). *J. Exp. Mar. Biol. Ecol.* 25:19-35.

Van Dolah, R. F. 1978. Factors regulating the distribution and population dynamics of the amphipod *Gammarus palustris* in an intertidal salt marsh community. *Ecol. Monogr.* 48:191-217.

Vince, S., I. Valiela, M. Backus and J. M. Teal. 1976. Predation by the salt marsh killifish, *Fundulus heteroclitus (L.)* in relation to prey size and habitat structure: consequences for prey distribution and abundance. *J. Exp. Mar. Biol. Ecol.* 23:255-266.

Weinstein, M. P. and K. L. Heck, Jr. 1979. Ichthyofauna of seagrass meadows along the Caribbean coast of Panama and in the Gulf of Mexico: Composition, structure and community ecology. *Mar. Biol.* 50:97-107.

Williams, A. B. 1965. Marine decapod crustaceans of the Carolinas. *Fish. Bull.* 65:1-298.

Young, D. K., M. A. Buzas and M. W. Young. 1976. Species densities of macrobenthos associated with seagrass: A field experimental study of predation. *J. Mar. Res.* 34:577-592.

Young, D. K. and M. W. Young. 1977. Community structure of the macrobenthos associated with seagrass of the Indian River estuary, Florida, pp. 359-382. *In:* B. C. Coull (ed.), *Ecology of Marine Benthos.* U. South Carolina Press, Columbia, SC.

Zieman, J. C. 1976. The ecological effects of physical damage from motor boats on turtlegrass beds in southern Florida. *Aquatic Bot.* 2:127-139.

Zimmerman, R. J. 1978. The feeding habits and trophic position of dominant gammaridean amphipods in a Caribbean seagrass community. Ph.D. Dissertation, U. Puerto Rico, Mayaguez, P.R. 93 p.

FISH COMMUNITY STRUCTURE AND FUNCTION IN TERMINOS LAGOON, A TROPICAL ESTUARY IN THE SOUTHERN GULF OF MEXICO

Alejandro Yáñez-Arancibia, Felipe Amezcua Linares

Universidad Nacional Autónoma de México
Centro de Ciencias del Mar y Limnologia
México, D.F.

and

John W. Day, Jr.

Coastal Ecology Laboratory
Center for Wetland Resources
Louisiana State University
Baton Rouge, Louisiana

Abstract: A total of 21,734 fish from 121 species was taken in 173 collections from July 1976 to March 1979. Seventeen species had a broad distribution and comprised 82% of total numbers. Twelve species (10%) were thought to be permanent residents, 54 (45%) probably used the lagoon as a nursery or feeding ground or both, and 55 (45%) were occasional visitors. Twenty-two percent were herbivores, detritivores, or omnivores, 51% were second order consumers, and 27% were higher consumers. Clustering was used to construct a dendrogram of faunal similarity; it corresponded to major differences in the physical makeup of the lagoon and demarcated southern and northern shore groups of fish. There is an apparent pattern of migration within the lagoon reflecting prevailing currents. The season of maximum juvenile influx from the sea (Aug. to Oct.) was during maximum river flow. Diversity indices (H', D, J') reflected changes in numbers of species and individuals during the year, and some were highly correlated with temperature and salinity. Biomass at different sampling stations ranged from 0.95 to 11.3 g wet wt m^{-2} and H' ranged from 0.53 to 2.50. Fishery productivity was as high as 12.0 g m^{-2} year^{-1}.

Introduction

This paper is a summary of the status of knowledge of ecology of fish of Terminos Lagoon. Detailed accounts are presented in the following works and in a number of references in them: Bravo Núñez and Yáñez-Arancibia (1979); Amezcua Linares and Yáñez-Arancibia (1980); Yáñez-Arancibia (1980). The interest in estuaries as frontiers between the sea and fresh water, and the poor information that there is on the fauna of Mexican estuaries, led us to carry out this study. The objective was to investigate the composition and ecology of the fish of Terminos Lagoon, the largest coastal lagoon in the southern Gulf of Mexico ($\sim$2500 km²).

Study Area

Terminos Lagoon (Fig. 1) is located in the southern Gulf of Mexico, off Campeche Bay on the Mexican coast. The area has a warm humid climate with three climatic seasons. From June until the end of September there are almost daily afternoon and evening showers. From October into March is the season of the *nortes* or winter storms. These storms are generally strongest and usually associated with rains during October, November, and December. January through May is the dry season. Water temperatures generally range from 36C in summer to 25C in winter and precipitation ranges between 1,200 and 2,000 mm per year. The lagoon is shallow (mean depth 3.5 m) with a limited channel system at the two wide inlets. Prevailing easterly winds and longshore currents cause a strong net flow into the eastern inlet and out from the western inlet (Mancilla and Vargas 1980). Three rivers enter the lagoon: the Palizada, Chumpan, and Candelaria, and in part, the Atasta system forming important Fluvial-Lagoon Systems. The season of high river flow is from August to

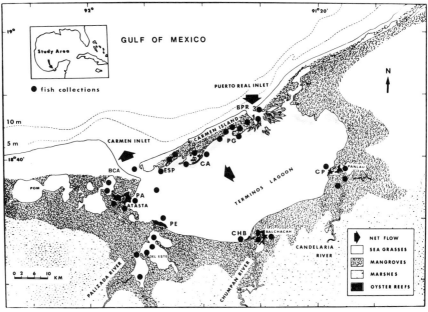

Figure 1. Terminos Lagoon in Campeche, México. Localities of fish collections are: BPR = Puerto Real Inlet; Carmen Island inner areas, PG = Punta Gorda, CA = Bajos del Cayo, ESP = Estero Pargo; BCA = Carmen Inlet; Fluvial-Lagoon Systems, CP = Candelaria-Panlau, CHB = Chumpan-Balchacah, PC = Palizada-del Este, PA = Pom-Atasta.

November. Tidal range is 0.5 to 0.7 m. More detailed descriptions of the lagoon are found elsewhere (Bravo Núñez and Yáñez-Arancibia 1979; Amezcua Linares and Yáñez-Arancibia 1980).

Materials and Methods

Fish were sampled from July 1976 until March 1979 during 22 collecting trips: monthly from July-Dec. 1976; Jan., Mar., May, July, Sept., Nov. 1977; Jan., Feb., Apr., June, Aug., Oct. 1978; and Jan. and Mar. 1979. One hundred and seventy-three trawl collections (9 m trawl; mouth opening while fishing was 5 m; mesh size was 3/4") were made at the stations listed in Fig. 1 and Table 1. In addition, salinity, temperature, oxygen, and transparency were also measured (Table 1). Field observations of substrate, vegetation, and common macroepifauna were made to verify earlier reports from the same areas (see references in Table 1). In the laboratory, fish were identified, counted, weighed, and measured. The state of gonadic maturity was determined according to the method of Nikolsky (1963). Stomach contents were quantified volumetrically for 2,314 individual fish of different sizes and food habits.

Trophic position was determined on the basis of gut content, and morphology of gut, mouth, and gill raker size. Fish were placed in the following groups (see Table 2) using the criteria of Yáñez-Arancibia (1978). The criteria used were developed based on earlier studies by Darnell (1961), Odum (1971), Wagner (1973), and Day et al. (1977). (1) *First order consumers:* included in this category are plankton feeders (phyto- and/or zoo-), and feeders on detritus and other vegetable remains, and omnivores (feeders on detritus, vegetable material and small animals); (2) *second order consumers:* included in this category are predominantly carnivorous fish, even when they consume small amounts of plant material and detritus. These fish consume primarily macro- and microbenthos and small fishes; (3) *third order consumers:* fish that are exclusively carnivorous and feed primarily on macrobenthos and second order consumer fishes.

Clustering was used to construct a dendrogram of faunal similarity using the criteria discussed by Horn and Allen (1976), Warburton (1978), and Daniels (1979). Three kinds of cluster analyses were run. The first was based on the presence or absence and numerical abundance of each species at each station (Table 2). The second was based on the habitat characteristics at each station (Table 1). For the third analysis a combination of both species abundance and habitat characteristics was used. A number of ecological indices were calculated. The information function H' of Shannon and Weaver (1963), which increases as both the number of species (richness) and the equitability of species abundance (evenness) increase, was calculated. It is desirable to consider indices that treat richness and evenness separately because these two components of diversity may react differently to certain types of factors. For the "species richness" component of diversity we selected the function (D) suggested by Margalef

Table 1. General features for study area. Ranges of parameters and characteristics of habitats[1].

Localities	Salinity (ppt)	Temperature (°C)	Oxygen ml l-1	Depth (m)	Transparency (%)	Substrate[2]	Food availability[3] (Zoobenthos)	Vegetation[4]	Turbulence
Puerto Real Inlet (BPR-1)	29-38	21-29	2.6-6.4	4	64	Muddy with fine sand and clayed silt. 50 to 60% CaCO3. High organic content	Macroepifauna varied and abundant	Banks of Thalassia testudinum	Moderate waves
Puerto Real Inlet (BPR-3)	27-38	22-29	3.4-5.4	5	48	Sandy with 60 to 70% CaCO3. Low organic content	Macroepifauna variable and low abundance	Rhodophiceae and Phaeophiceae	Strong wave motion
Carmen Island Inner Area (ESP,CA,PG)	CA 24-26 ESP 21-25	24-28 25-28	1.8-5.5	1.5	80	Muddy with fine sand and clayed silt. 40 to 50% CaCO3. High organic content	Macroepifauna varied and abundant	Banks of Thalassia testudinum and swamps of Rhizophora mangle (also macroalgae)	Low wave motion
Carmen Inlet (BCA)	14-36	26-35	4.0-5.8	8	10-30	Muddy with fine sand and clayed silt. 20 to 30% CaCO3. Low organic content	Macroepifauna almost absent, low abundance	Without submerged vegetation	Very strong waves and currents
Fluvial-Lagoon Systems	CP 5-26 CHB 10-28 PE 2-14 PA 2-25	28-31 28-35 24-31 20-30	1.8-7.2 2.2-2.4 2.9-6.8 2.9-6.0	2	20-40	Silty-clay with less 25% CaCO3. Variable organic content	Crassostrea virginica reefs and community	Rhizophora mangle swamps	River motion flow ranging seasonality

[1]See Figure 1 for distribution and key of localities.

[2]Ayala-Castañares 1963; Phleger and Ayala-Castañares 1971.

[3]Garcia-Cubas 1963; and pers. communication, Centro de Ciencias del Mar., UNAM, Marron 1975.

[4]Hornelas 1975.

Table 2. Type of inhabitant (T.I.), trophic category (T.C.), and frequency of occurrence in different sampling areas of each species. Occ = Occasional visitor, Cv = cyclical or seasonal visitor, Sed = Resident (typically estuarine) species. 1, 2, and 3 refer to first, second, and third order consumers, respectively.

Species	BPR-1	BPR-3	PG	CA	ESP	CP	CHB	PE	PA	T.I.	T.C.
Urolophus iamaicienses	2	1	5	3	7					Sed	3
Dasyatis sabina						10		1	7	Cv	3
Himatura schmardae			3	1						Occ	3
Dasyatis hastatus				1						Occ	3
Elops saurus					3					Occ	3
Albula vulpes			4							Occ	2
Sardinella macrophthalmus		1208	38	4	81	3	18		6	Cv	1
Sardinella humeralis								1		Occ	1
Opisthonema oglinum		38	2			1		15	1	Cv	1
Anchovia sp.		2								Occ	1
Anchoa mitchilli mitchilli			52	9	3751	8	23	8	44	Cv	1
Anchoa hepsetus hepsetus	1	7							4	Occ	1
Anchoa lamprotaenia			49	770	20				18	Cv	1
Centengraulis edentulus		3	108	3	38	29	11	4	115	Cv	1
Anchoviella sp.		2								Occ	1
Synodus foetens		1	3	2	3				8	Cv	3
Cyprinodon variegatus		7	1	285						Occ	1
Ictalurus meridionalis								9		Occ	2
Bagre marinus						7	5	3	14	Cv	2-3
Arius felis	8	207	137	54	75	5	1	9	74	Sed	2-3
Arius melanopus		2	600		91	257	418	225	141	Sed	2-3
Gobiosox strumosus	1									Occ	2
Opsanus beta		1	61	38	18	1	5			Cv	2
Nautopaedium porosissimun							4	1	5	Cv	2
Hemirhamphus brasiliensis	11		1							Occ	2
Hyporhamphus unifasciatus		26	16				1		1	Occ	2
Tylosurus raphidoma	4		1	1	1					Occ	3
Tylosurus acus	2				1					Occ	3
Strongylura marina			6							Occ	3
Poeciliopsis sp.					1					Occ	1
Chriodorus aterinoides			9	1						Cv	2
Syngnathus rosseau	1		5	1						Cv	3
Syngnathus machayi		1		2		2				Cv	3
Syngnathus scoveili				2	2					Cv	3
Hippocampus hudsonius punctulatus	1		2		5					Cv	3
Scorpaena plumieri	2									Occ	3
Scorpaena grandicornis			1							Occ	3
Prionotus carolinus			10	21	2				1	Occ	3
Dactylopterus volitans	6									Occ	3
Centropomus undecimalis			3		30	1	3		8	Cv	3
Centropomus paralellus					2					Cv	3
Epinephellus guttatus	3		2	5	1	1	1			Cv	3
Diplectrum radiale		3								Occ	3
Echeneis neucrates			10	1						Occ	3
Caranx hippos					9	3			7	Cv	2
*Caranx ruber					1					Occ	2
Caranx latus					1					Occ	2
Chloroscombrus chrysurus			24	2	1	8	5	1	12	Cv	2
Selene vomer				41	4				3	Cv	2
Oligoplites saurus			1	8	1	2	1		2	Cv	2
*Trachinotus falcatus			2	9	1					Occ	2
Lutianus griseus	3		13	16	39	2				Cv	3
Lutianus synagris	3	5	1							Cv	3
Lutjanus analis	1				2					Cv	3
Lobotes surinamensis					1					Occ	3
Eucinostomus gula	36	5	1434	903	621	393	5	1	275	Sed	1
Eucinostomus argenteus		54	60	86	43	22			14	Cv	1
Eucinostomus melanopterus			1	3	3					Occ	1

Table 2.   (Continued)

Species	BPR-1	BPR-3	PG	CA	ESP	CP	CHB	PE	PA	T.I.	T.C.
LOCALITIES											
Gerres cinereus				2						Cv	1
Eugerres plumieri			25		12	1	20	13	1	Cv	1
Diapterus rhombeus			6	1	171	343	1	1	11	Cv	1
Diapterus evermani					24					Occ	1
Orthopristis chrysopterus	84	4	154	27	94					Cv	2
Orthopristis poeyi	24		92	17						Cv	2
Anisotremus virginicus	6			3						Cv	2
Anisotremus spleniatus	2		15							Cv	2
Bathystoma rimator	25		1							Cv	2
Haemulon plumieri	68		145	171	104					Cv	2
Haemulon bonariense			35	28	8					Cv	2
Calamus penna		1	1							Occ	2
Archosargus probatocephalus	2		10	1	9	3	1		1	Cv	2
Archosargus unimaculatus	9	300	793	128	426					Cv	2
Calamus calamus	1									Occ	2
Menticirrus martinicences		1	1							Cv	2
Menticirrus saxantilis		24								Cv	2
Micropogon furnieri				4	35	214	2	36	274	Cv	2
Bairdiella chrysura	41	271	179	179	168	401	13	4	51	Cv	2
Bairdiella rhonchus				10	47	117	35	9	12	Cv	2
Cynoscion nebuiosus	3		19	17	7	61		2	6	Cv	3
Cynoscion nothus						9		3	18	Cv	3
Cynoscion arenarius									22	Cv	3
Equetus acuminatus	1			1						Occ	2
Odontoscion dentex	42	2	3							Occ	2
Corvula sancta-lucia	30									Occ	2
Chaetodipterus faber		9	35	17	12	12	2	2	32	Sed	1
Chaetodon ocellatus	1									Occ	1
Pomacanthus arcuatus			1							Occ	1
Cichlasoma urophthalmus			221	363	3	22				Sed	1-2
Cichlasoma fenestratum								9		Occ	1-2
Mugil curema					32				4	Cv	1
Polynemus octonemus						42		5	17	Cv	2
Novaculichthys infirmus	1									Occ	1
Scarus noyesi	1				4					Occ	2
Nicholsina ustus	20		3		19					Cv	2
Hypsoblennius hentz			1							Occ	2
Gobionellus oceanicus							1		1	Occ	1
Gobiosoma bosci									1	Sed	1
Trichiurus lepturus		1			12			2	25	Cv	2
Scomberomorus maculatus									1	Occ	3
Cytharichthys spilopterus			1	4	26	2	5	3	122	Sed	3
Ancyclopsetta quadrocellata			2							Occ	3
Etropus crossotus			1						2	Cv	3
Achirus lineatus		1	15	2	1	4		5	15	Sed	2
Gymnachirus melas			1					5		Occ	2
Trinectes maculatus								5		Occ	2
Symphurus plagiusa			1					1		Occ	2
Balistes capriscus	1									Occ	2
Monacanthus hispidus	3	5	24	1	1					Cv	2
Monacanthus ciliatus	1									Occ	2
Alutera schoefil		5	2		3					Occ	2
Lactoprhys tricornis	5		40	11	6					Sed	2
Lactoprhys bicaudalis	1									Sed	2
Sphoeroides marmoratus	1		8	122	3					Cv	2-3
Sphoeroides testudineus	19	45	148	14	148	20	8	24	172	Sed	2-3
Sphoeroides nephelus			6	5	7					Occ	2
Sphoeroides spengleri	1									Occ	2-3
Sphoeroides sp.				1	5			2	3	Occ	2
Lagocephalus lavigeatus		1	1							Occ	2-3

Table 2.   (Continued)

| Species | LOCALITIES | | | | | | | | | | |
	BPR-1	BPR-3	PG	CA	ESP	CP	CHB	PE	PA	T.I.	T.C.
Diodon hystrix	45									Occ	2-3
Chilomycterus schoepfil	˙4		30	22	3					Occ	2-3
Chilomycterus antennatus			9	31	12					Cv	2-3

Note: Castro (1978) reported the following additional species for Terminos Lagoon:
Carcharhinus limbatus, Carcharhinus porosus, Carcharhinus leucas, Sphyrna tiburo,
Pristis pectinatus, Pristis perotteti, Rhinobatos lentiginosus, Narcine
brasillensis, Dasyatis americana, Megalops atlanticus, Brevortia guntheri,
Strongylura timucu, Menidia beryllina, Syngnathus louisianae, Syngnathus
floridae, Oostethus lineatus, Lutjanus cyanopterus, Lutjanus apodus, Ulaema
lefroyi, Eucinostomus habana, Rypticus saponaceus, Promicrops itaiara,
Epinephellus morio, Diplodus caudimacula, Lagodon rhomboides, Cynoscion
jamaiciensis, Umbrina coroides, Stillifer lanceolatus, Mugil cephalus, Mugil
tricodon, Sphyraena barracuda, Gobiomorus dormitator, Dormitator maculatus,
Awaous tajasica, Gobionellus hastatus, Astroscopus ygraecum, Batrachoides
surinamensis, Stephanolepis hispidus, Atherinomorus stipes.   Toral and Resendes
(1974) also reported the following species: Cichlosoma meeki, Cichlosoma
aureum, Cichlosoma pearsi, Cichlasoma sextafaciatum, Cichlosoma friedrichsthali,
Cichlosoma heterospilum, Petenia splendida.

*Species collected with different methods and not processed quantitatively.

(1968). The "evenness" (J') index of Pielou (1966) was the ratio used for the equitability of species abundance. All diversity calculations were based on use of natural logs. The abundance of fish was calculated by number, density (individuals by area), and biomass (standing crop). The Wilhm (1968) function (H'w) was also used. Secondary fish production was calculated following methods described by Lowe-McConnell (1975) and Gerking (1978).

Results

Taxonomical Knowledge

The fish are a very conspicuous and important component of the fauna of Terminos Lagoon. The taxonomic composition is probably understood as well for the lagoon as any other comparable geographic area. Based on the references and our studies, we present the total record of fish identified from the lagoon (Table 2).

Distribution, Abundance, Diversity and Dominant Species

A total of 21,734 fish from 121 species was taken. Seventeen species made up 82% of the catch. They were: *Anchoa mitchilli* (3,895 individuals), *Eucinostomus gula* (3,673), *Arius melanopus* (1,734), *Archosargus unimaculatus* (1,656), *Sardinella macrophthalmus* (1,358), *Bairdiella chrysura* (1,307), *Anchoa lamprotaenia* (857), *Cichlasoma urophthalmus* (609), *Sphoeroides testudineus* (598), *Arius felis* (570), *Diapterus rhombeus* (534), *Cetengraulis edentulus* (311), *Eucinostomus argenteus* (279), *Cytharichthys spilopterus* (163), *Opsanus beta* (124), *Chaetodipterus faber* (121), *Cynoscion nebulosus* (115). These species were broadly distributed over the lagoon and were found at most of the sampling stations. Only 15 species (12%) were common to all localities. A number of other species were numerically abundant but their distribution was restricted within the lagoon. These included *Micropogon furnieri* (565), *Haemulon plumieri* (488), and *Sphoeroides marmoratus* (134). For example, *M. furnieri* was found mainly in the Fluvial-Lagoon Systems while *H. plumieri* occurred only in Puerto Real inlet and the Carmen Island inner areas. All of the above species are either estuarine residents or regular seasonal visitors.

It must be remembered that this information is based on trawl-caught fish. Trawls catch some fish very well and others very inefficiently. Types of fishes which are often underrepresented are fast swimmers and fishes which live in very shallow habitats. For example, *Cyprinodon variegatus* is a common inhabitant of shallow areas in the lagoon. However it was caught in only three areas and 285 of a total of 293 specimens were taken in a single sample. This is not a reflection of either the true distribution or abundance of this species, but of the fact that it lives in very shallow water and is most often missed by the trawl. The mullet, *Mugil curema*, was also encountered very rarely. Because it often occurs in shallow habitat and is a very fast swimmer, we believe that it is very much underestimated. In fact some authors think that mullet are among the most abundant estuarine species (see for example, Gunter 1945, 1967).

Table 3 shows abundance, diversity, biomass, and size in the different sampling areas. In the Fluvial-Lagoon Systems the presence of a few large specimens explains the inverse relationships between density and biomass in these areas, especially during late summer. Seasonal changes in density and biomass resulted primarily from the influx of juveniles in late summer and autumn (Amezcua Linares and Yáñez-Arancibia 1980).

Types of Community Inhabitants

Based on frequency of occurrence, gonadic maturity, food habits, patterns of migration, and size, we classified the fish species as: (a) Occasional or accidental visitors (Occ), (b) cyclical or seasonal visitors (Cv), and (c) residents or typical estuarine species (Sed) (see Table 2). These groupings are based on earlier classifications described by McHugh (1967), Dovel (1971), and Haedrich and Hall (1976). Residents were found in the lagoon at all times. They reproduce and grow to maturity in the lagoon and appear to rarely, if ever, leave. Seasonal visitors use the lagoon in a regular, repeating pattern and seem to be dependent on it at some stage of their life cycle. Common examples in this group are estuarine-dependent species which spawn offshore, move into the lagoon as young, and return to the

Table 3. Distribution of ranges of diversity, abundance, and size of fish populations in localities of study area.

Localities	Species Number	Diversity[1] (H')	Density[1] Indiv. (m^-2)	Biomass[1] Range (g m^-2)	Biomass[1] (H'w)	Total Weight (g)	Mean Biomass (g m^-2)	Range of Lengths (mm)
Puerto Real Inlet								
(BPR-1)	44	1.43 (Aug)-2.19 (Oct)	0.001 - 0.054	0.97 - 2.23	1.05 (May)-2.31 (Oct)	54,150	2.8	30-305
(BPR-3)	37	1.17(Sept)-1.65 (Jan)	0.004 (Aug)-0.11(Sept)	0.18 (Aug)-11.0(Sept)	1.06 (Oct)-1.77 (Mar)	63,957	4.4	20-380
Carmen Island Inner Area								
(PG)	69	-	-	-	-	145,104	11.2	-
(CA)	49	1.0 (Oct)-1.6 (Mar)	0.07 (Oct)-0.19 (Mar)	1.77 (Oct)-3.75 (Mar)	1.0 (Oct)-2.0 (Mar)	90,208	8.0	25-326
(ESP)	61	1.19 (Oct)-2.27 (Mar)	0.05 (Oct)-0.26 (Mar)	1.0 (Oct)-9.8 (Mar)	1.71 (Oct)-2.27 (Mar)	80,198	6.8	20-473
Fluvial-Lagoon Systems								
(CP)	33	1.80 (May)-2.22 (Aug)	0.06 (Sept)-0.16 (Aug)	0.87 (May)-4.96(Sept)	1.09 (May)-2.30 (Aug)	56,643	8.6[1]	25-490
(CHB)	21	0.53(Sept)-2.23 (Oct)	0.01 (May)-0.13(Sept)	0.21 (May)-7.04(Sept)	0.73(Sept)-2.14 (Oct)	30,014	2.7[1]	23-489
(PE)	29	0.98 (Oct)-2.14 (Aug)	0.01 (Aug)-0.29 (Oct)	0.46 (Mar)-1.52 (Oct)	0.62 (Oct)-2.56 (Mar)	15,561	2.9[1]	23-510
(PA)	43	1.90 (Jan)-2.50 (Oct)	0.01 (Mar)-0.06 (Aug)	0.25 (Jan)-1.93 (Mar)	1.08 (Mar)-2.52 (Aug)	47,694	8.7[1]	25-560

[1]Data from Bravo Núñez and Yáñez-Arancibia (1979); Amezcua Linares and Yáñez-Arancibia (1980).

sea as adults. Occasional visitors had no regular pattern of use of the lagoon.

In placing each species into one of the three categories, we took trawl selectivity into consideration. As we noted earlier, some species are underrepresented in trawls because of habitat or swimming speed. If a species was represented by only a few specimens, we used literature information in determining the appropriate category. For example, only two specimens of *Menticirrus martinicences* were taken. This species is in the family Scianidae and practically all Scianidae have an estuarine dependent life cycle (Springer and Woodburn 1960). Thus we classified this species as a cyclic visitor. In addition, the trawl probably underestimated its abundance because it is a fast swimmer. As mentioned, *Cyprinodon variegatus* was collected at only three stations and practically all the specimens were taken in one tow. However we know that this species is a common estuarine resident (Odum 1971) which lives in water normally too shallow for trawling.

Trophic Structure, Biomass and Fishery Productivity

It appears that organic detritus is a major dietary component of first order consumer fish. Microbenthic animals are utilized by the majority of second order consumers, and by some first order consumers. The microbenthos includes small crustaceans, molluscs, and polychaetes. The top carnivores feed largely on the first and second order consumer fish and macrobenthic forms. These results are very similar to reports of trophic structure of other estuaries (Darnell 1961; Odum 1971; Day et al. 1973).

In Puerto Real Inlet area BPR-1, and in the Carmen Island inner areas (ESP, CA, PG), both located in banks of *Thalassia testudinum*, the benthic fauna is varied and abundant. Invertebrate food resources of the lagoon are described by Garcia-Cubas (1963, 1980), Signoret (1974), Marron (1975), and Caso (1980). These together with the large quantity of detritus and epiphytes present provide a readily available food source. There were 31 species of fish in the BPR-1 area (Bravo Núñez and Yáñez-Arancibia 1979): 4 species (13%) were first order consumers with a mean production of 0.06 g m^{-2} year^{-1}, 17 (55%) were second order consumers with a mean production of 4.71 g m^{-2} year^{-1}, and 10 (32%) were third order consumers with a mean production of 2.65 g m^{-2} year^{-1}. Total biomass ranged from 0.97 to 2.23 g m^{-2}. Total fishery productivity for this area was 7.4 g m^{-2} year^{-1}.

In the Fluvial-Lagoon Systems (CP, CHB, PE, PA), 47 species were collected (Amezcua Linares and Yáñez-Arancibia 1980): 16 species (34%) were first order consumers with a mean production of 2.51 g m^{-2} year^{-1}, 21 (45%) were second order consumers with a mean production of 4.95 g m^{-2} year^{-1}, and 10 (21%) were third order consumers with a mean production of 1.0 g m^{-2} year^{-1}. Total secondary fish production for the system was estimated at 8.5 g m^{-2} year^{-1}.

Only total secondary production was calculated for the other areas sampled. For areas BPR-3, CA, and ESP, estimated secondary production values were 12.0, 5.6, and 8.0 g m^{-2} year^{-1}, respectively (Bravo Núñez and Yáñez-Arancibia 1979).

Affinity of Subsystems

The analysis of affinity showed a high degree of statistical significance among the following subsystems: (1) the four Fluvial-Lagoon Systems, (2) the different localities in the Carmen Island inner areas, and (3) between the easternmost Fluvial-Lagoon System (CP) and Carmen Island inner areas, especially BPR and PG (Fig. 3). These affinities are the consequence of similar ecological characteristics and thus characteristic species and ranges of length and biomass (Table 3, Figs. 2 and 3). The cluster dendrogram of faunal similarity corresponded to major differences in the physical makeup of the lagoon and demarcated southern and northern shore groups of fish. These groups were similar at the eastern end of the lagoon (Fig. 3).

These results suggest a pattern of migration and/or colonization beginning from Puerto Real Inlet (Fig. 3). Bravo Núñez and Yáñez-Arancibia (1979) and Amezcua Linares and Yáñez-Arancibia (1980) present additional evidence that supports the idea that there is a preferential immigration through Puerto Real Inlet by species with estuarine dependent life cycles. The greatest numbers of juvenile forms occurred in Puerto Real Inlet and in the easternmost areas of the lagoon (PG and CP). These results seem to indicate that the dominant migratory route is from east to west. The relationship between fish size and salinity (Fig. 2) tends to support this idea. This migratory pattern follows the prevailing currents within the lagoon suggesting that the young fish utilize the currents to move through the lagoon.

The greatest number of juvenile fishes occurred in August and September. This corresponds to the first half of the season of high river flow. Nutrient levels in the lagoon are significantly higher during high river flow (Botello and Mandelli 1975) as is aquatic primary productivity (Richard Day, Center for Wetland Resources, LSU, personal communication). Our conclusion based on this information is that young migratory fish enter the lagoon during the season of highest productivity and follow prevailing currents through the lagoon. The arrows in Fig. 3 indicate the direction of these migratory patterns.

Discussion

The ichthyological fauna of Terminos Lagoon is composed of more than 121 species which are found occasionally, seasonally, or permanently in the lagoon-estuarine system. Occasional visitors represent 55 species (45%) of the total fauna of the lagoon. Seasonal species are 54 in number (45%) and make regular incursions into the estuarine waters. These penetrations are probably either of a trophic nature or are linked to reproductive cycles. The presence of these species in the estuarine waters is variable in time, being often transient in nature and limited to particular areas and periods of the year. Finally, resident (typically estuarine) species (12 species, 10%) rarely leave estuarine waters.

More than 50% of the fish species were second order consumers. These species showed the greatest affinity for a particular habitat (Table 2; Fig. 3; Vargas, Yáñez-Arancibia, and Amezcua Linares, unpublished data; Bravo Núñez and Yáñez-Arancibia 1979). This is not surprising since most of these species are demersal and many have fairly specific feeding habits. Thus, they would correlate more closely with particular habitats than would higher consumers that range widely over the lagoon. Also, because of their numerical abundance and habitat specificity, second order consumers are more important in determining characteristics of particular fish communities.

The actual values of H' are quite low reflecting a skewed distribution where a few species are very abundant. The variability of salinity, available food, turbidity, dissolved oxygen, and temperature limits the number of species able to live permanently inside the lagoon (Table 3, Fig. 4). H' was positively correlated with salinity ($r = 0.75$, $P < .01$) and temperature ($r = 0.71$, $P < .01$).

No trend was discernible using the relative species abundance or "evenness" index J' (Fig. 4). While species certainly move in and out of the lagoon during late summer and autumn, this movement was not detectable with this index due to changes in the frequency of occurrence of other species. Therefore, the distribution over species remain unchanged. Values for H'w were generally lower during the summer rainy season (late May through October) and highest during the dry season (February-May) for the riverine systems (Table 3). This reflects the bimodal distribution of biomass during the period of high juvenile immigration and the presence of

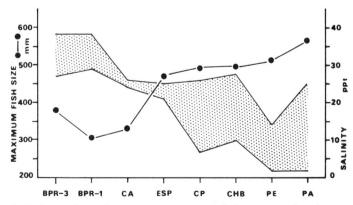

Figure 2. Relationships between maximum fish size and salinity in the study areas for fishes common to all areas (see Table 2).

a more uniform population of larger fish during the dry season (Amezcua Linares and Yáñez-Arancibia 1980). The highest biomass in the riverine system also occurs during the rainy season while the lowest values occur during the dry season. The opposite is true for the Carmen Island inner areas. This suggests that the grass beds and mangrove swamps in these areas serve as a refuge during the dry season. A seasonal change in relative abundance in some areas of the lagoon, i.e., the Fluvial-Lagoon Systems (Fig. 4), resulted primarily from the influx of juveniles (Amezcua Linares and Yáñez-Arancibia 1980).

Further evidence suggesting replacement of species is provided by the "species richness" index D. This index relates the total number of species to the total catch and weights each species equally. It shows no obvious seasonal trend either. Since there are definite seasons, this result probably reflects a sequential use of the lagoon by different species.

These tropical waters carry many more fish species than comparable waters in the temperate zone (Wagner 1973; Moore 1978), but biomass and fish production may not be higher (Haedrich and Hall 1976); however, in tropical lagoons with ephemeral inlets, fishery productivity can be high for a short period of time (Yáñez-Arancibia 1978). Only small areas can be sampled comprehensively, and conclusions based on such data have to be regarded with caution when results are extrapolated for larger areas. Fish move about a good deal and the presence of migrant and schooling fish at the time of sampling will greatly affect the results.

Fisheries research in the tropics is complicated by (1) the large number of species present in a community, and (2) the difficulties of determining fish ages and growth rates. In the absence of true production data, yield expressed as weight/area is often used as an index of production. Yields tend to be greater in shallow than in deep waters, partly for biological reasons (higher temperatures and greater light penetration in shallow waters support higher primary production), and partly because fish are easier to catch in shallow waters (a higher proportion of the production is cropped as yield). In Terminos Lagoon fish grow faster and life cycles are shorter with most fish maturing in less than one year (Vargas, Yáñez-Arancibia, and Amezcua Linares, unpublished data). The growing season extends over the whole year and young fish are available for stocking (recruitment into the system) throughout the year. Ecologically complementary species are available to increase yields; many species can withstand poorly oxygenated waters; and fish in the lagoon have high availability of food (Garcia-Cubas 1963, 1980; Signoret 1974; Marron 1975; and Caso 1980).

In summary, fish represent an important part of the lagoon ecosystem. The ecological role of the fish includes the following aspects. In coastal lagoons fish act to: (1) transform energy from primary sources, (2) actively conduct energy through the food web, (3) exchange energy with neighboring ecosystems through exportation and importation, (4) constitute a form of energy storage within the ecosystem, and (5) function as agents

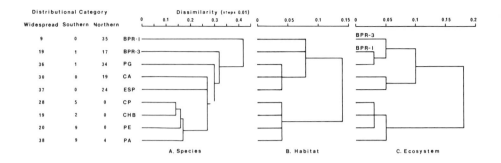

Distributional Category				Dissimilarity (steps 0.01)		
Widespread	Southern	Northern		A. Species	B. Habitat	C. Ecosystem
9	0	35	BPR-I			
19	1	17	BPR-3			
36	1	34	PG			
30	0	19	CA			
37	0	24	ESP			
28	5	0	CP			
19	2	0	CHB			
20	9	0	PE			
38	9	4	PA			

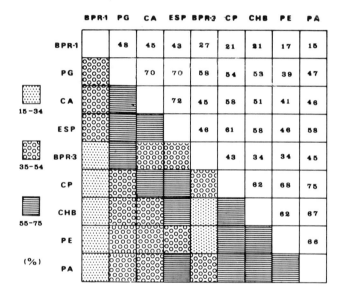

	BPR-1	PG	CA	ESP	BPR-3	CP	CHB	PE	PA
BPR-1		48	45	43	27	21	21	17	15
PG			70	70	58	54	53	39	47
CA				72	45	58	51	41	48
ESP					46	61	58	46	58
BPR-3						43	34	34	45
CP							62	68	75
CHB								62	67
PE									66
PA									

15-34

35-54

55-75

(%)

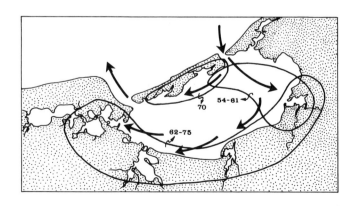

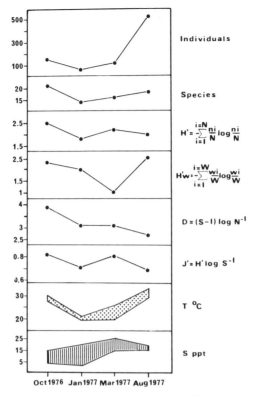

Oct1976 Jan1977 Mar1977 Aug1977

Figure 4. Temporal variation and comparison of diversity, abundance for selected months and the possible relationships with temperature (°C) and salinity (ppt) in Fluvial-Lagoon System PA. (Modified from Amezcua Linares and Yáñez-Arancibia 1980).

Figure 3. (opposite). Dendrogram of the clustering of localities in study area using the Canberra-metric index of dissimilarity and group-average sorting. A: based on presence/absence of fish species (Table 2). B: based on characteristics of habitat (Table 1). C: based on habitat plus presence/absence and numerical abundance of fish species. Fish species were placed in three broad distributional categories (indicated by the three enclosed areas) based on a two-way table (localities vs. species) generated with the cluster analysis. The trellis diagram shows the percent affinity based on presence/absence of fish species The dendrogram and trellis diagram reflect a probable pattern of migration within the lagoon (indicated by arrows) reflecting prevailing currents and the diversity of estuarine habitats.

of energetic regulation (Yáñez-Arancibia and Nugent 1977). The ecological role of the lagoon is to provide food, spawning and nursery areas, protection, and serve as pathways for migration.

Acknowledgments

Many contributed directly or indirectly to the success of this study. The authors thank Agustin Ayala-Castañares, Alfredo Laguarda Figueras, and Jose Luis Rojas Galaviz for institutional and financial support; Rodolfo Cruz Orozco and the staff of the Estacion de Investigaciones Marinas "El Carmen" for field support; Violeta Leyton, Irma Vargas, Patricia Sanchez Gil, Margarito Alvarez for technical assistance, statistical analysis of data, help with computer programs and preparation of figures; and Bruce Thompson for critically reading the manuscript. This work is a result of investigation sponsored by the Regional Program of Scientific and Technological Development of the Organization of American States through the Special Projects Office, Washington, D.C. Partial support was also provided by the National Science Foundation Program and the Louisiana Sea Grant College Program maintained by NOAA, U.S. Department of Commerce. Contribution No. 201 of the Centro de Ciencias del Mar y Limnologia, Universidad Nacional Autónoma de México, and No. LSU-CEL-80-03 from the Coastal Ecology Laboratory, Louisiana State University.

References Cited

Amezcua Linares, F. and A. Yáñez-Arancibia. 1980. Ecología de las sistemas fluvio-lagunares asociados a Laguna de Términos. El habitat y estructura de las comunidades de peces. *An. Centro Cienc. del Mar y Limnol.*, Univ. Nac. Autón México 7 (in press).

Ayala-Castañares, A. 1963. Sistemática y distribución de los foraminiferos recientes de la Laguna de Términos, Campeche, México. *Univ. Nac. Autón. México, Bol. Instit. Geol.* 67:1-30.

Botello, A. V. and E. F. Mandelli. 1975. A study of variables related to water quality of Terminos Lagoon and adjacent coastal areas, Campeche, Mexico. Final Report, Proj. GU 853, Centro Cienc. del Mar y Limnol., Univ. Nac. Autón. México, 92 p.

Bravo Núñez, E. and A. Yáñez-Arancibia. 1979. Ecología de la Boca de Puerto Real, Laguna de Términos. I. Descripción del área y análisis estructural de las comunidades de peces. *An. Centro Cienc. del Mar y Limnol.*, Univ. Nac. Autón. México 6:125-182.

Caso, M. E. 1980. The echinoderms of Terminos Lagoon, Campeche, Mexico. *Centro Cienc. del Mar y Limnol.*, Univ. Nac Autón. México, Spec. Publ. No. 3. 190 p.

Castro, J. L. 1978. Catálogo sistemático de los peces marinos que penetran a las aguas continentales de México con aspectos zoogeográfico y ecológicos. Departamento de Pesca, México. *Ser. Cient.* 19:1-298.

Daniels, K. 1979. Habitat designation based on cluster analysis of ichthyofauna, pp. 317-324. In: J. W. Day, Jr., D. D. Culley, R. E. Turner and A. J. Mumphrey (eds.), *Proc. Third Coastal Marsh and Estuary Management Symposium.* Louisiana State U., Div. Continuing Education, Baton Rouge, LA.

Darnell, R. N. 1961. Trophic spectrum of an estuarine community based on Lake Pontchartrain, Louisiana. Ecology 42:553-568.

Day, J. W., Jr., W. Smith and C. Hopkinson. 1977. Some trophic relationships of marsh and estuarine areas, p. 115-135. In: R. Chabreck (ed.), *Proc. Second Coastal Marsh and Estuary Managment Symposium,* Div. Continuing Education, Louisiana State U., Baton Rouge, LA.

Day, J. W., Jr., W. Smith, P. Wagner and W. Stowe. 1973. Community structure and carbon budget of a salt marsh and shallow bay estuarine system in Louisiana. Louisiana State U. Center for Wetland Resources, Sea Grant Publ. No. LSU-SG-72-04. 79 p.

Dovel, W. 1971. Fish eggs and larvae of the upper Chesapeake Bay. U. Maryland, Natural Resources Inst., Spec. Rep. No. 4. 71 p.

Garcia-Cubas, A. 1963. Sistemática y distribución de los micromoluscos de la Lagona de Términos, Campeche, México. *Univ. Nac. Autón. México, Bol. Instit. Geol.* 67:1-55.

Garcia-Cubas, A. 1980. Molluscs of a tropical lagoon system in the southern Gulf of Mexico, Laguna de Términos, México. *Centro Cienc. del Mar y Limnol.,* Univ. Nac. Autón. México, Spec. Publ. No. 6. 200 p.

Gerking, S. D. (ed.). 1978. *The Biological Basis of Freshwater Fish Production.* Blackwell Scientific Publ., Oxford.

Gunter, G. 1945. Studies of marine fishes of Texas. *Publ. Inst. Mar. Sci. U. Texas* 1:1-190.

Gunter, G. 1967. Some relationships of estuaries to the fisheries of the Gulf of Mexico, p. 621-638. *In:* G. H. Lauff (ed.), *Estuaries.* Amer. Assoc. Adv. Sci. Publ. No. 83, Washington, D.C.

Haedrich, R. and C. Hall. 1976. Fish and estuaries. *Oceanus* 19:55-63.

Horn, M. H. and L. G. Allen. 1976. Number of species and faunal resemblance of marine fishes in California bays and estuaries. *Bull. S. Cal. Acad. Sci.* 75:159-170.

Hornelas, Y. 1975. Comparacion de la Fanerogama marina *Thalassia testudinum* Konig 1805, en tres diferentes areas geograficas del Golfo de México. Tesis profesional. Fac. Ciencias, Univ. Nac. Autón. México. 54 p.

Lowe-McConnell, R. H. 1975. *Fish Communities in Tropical Freshwaters.* Longman, Inc., New York. 337 p.

McHugh, J. L. 1967. Estuarine nekton, p. 581-620. *In:* G. H. Lauff (ed.), *Estuaries.* Amer. Assoc. Adv. Sci. Publ. No. 83, Washington, D.C.

Mancilla, M. and J. Vargas. 1980. Los primeros estudios sabre el flujo neto de aqua a trazes de la Laguna de Términos, Campeche. *An. Centro Cienc. del Mar y Limnol.* Univ. Nac. Autón. México 8 (in press).

Margalef, R. 1968. *Perspectives in Ecological Theory.* U. Chicago Press, Chicago, IL. 111 p.

Marron, M. A. 1975. Estudio cuantitativo y sistemático de los poliquetos (Annelida, Polychaeta) bentonicos de la Laguna de Términos, Campeche, México. Tesis doctoral. Fac. Ciencias, Univ. Nac. Autón. México. 143 p.

Moore, R. 1978. Variations in the diversity of summer estuarine fish populations in Aransas Bay, Texas, 1966-1973. *Est. Coastal Mar. Sci.* 6:495-501.

Nikolsky, G. V. 1963. *The Ecology of Fishes.* Academic Press, New York. 352 p.

Odum, W. E. 1971. Pathways of energy flow in a south Florida estuary. Univ. of Miami Sea Grant Program Tech. Bull. No. 7, 162 p.

Phleger, F. B. and A. Ayala-Castañares. 1971. Processes and history of Terminos Lagoon, Mexico. *Bull. Amer. Assoc. Petrol. Geol.* 55:2130-2140.

Pielou, E. C. 1966. The measurement of diversity in different types of biological collections. *J. Theoret. Biol.* 13:131-144.

Shannon, C. E. and W. Weaver. 1963. *The Mathematical Theory of Communication.* U. Illinois Press, Urbana. 117 p.

Signoret, M. 1974. Abundancia, tamaño, y distribution de camarones (Crustacea, Peneidae) de la Laguna de Términos, Campeche, y su relacion con algunos factores hidrologicos. *An. Inst. Biol. Univ. Nac. Autón. México, Ser. Zoologia* 45:119-140.

Springer, V. G. and T. Woodburn. 1960. An ecological study of the fishes of the Tampa Bay area. *Fla. State Board of Conservation. Marine Laboratory Professional Paper* 1:1-104.

Toral, S. and A. Resendez. 1974. Los ciclidos (Pisces: Perciformes) de la Laguna de Términos y sus afluentes. *Rev. Biol. Trop.* 21:254-274.

Wagner, P. 1973. Seasonal biomass, abundance, and distribution of estuarine dependent fishes in the Caminada Bay system of Louisiana. Ph.D. dissertation, Louisiana State U., Baton Rouge, LA. 193 p.

Warburton, K. 1978. Community structure, abundance and diversity of fish in a Mexican coastal lagoon system. *Est. Coastal Mar. Sci.* 7:497-519.

Wilhm, J. L. 1968. Biomass units versus numbers of individuals in species diversity indices. *Ecology* 49:153-156.

Yáñez-Arancibia, A. 1978. Taxonomía, ecología y estructura de las communidades de peces en lagunas costeras con bocas efímeras del Pacífico de México. *Centro Cienc. del Mar y Limnol.*, Univ. Nac. Autón. México. Spec. Publ. No. 2:1-300.

Yáñez-Arancibia, A. 1980. Ecology in the entrance of Puerto Real, Terminos Lagoon. II. Discussion on trophic structure of fish communities in banks of *Thalassia testudinum*. In: P. Lasserre and H. Postma (eds.), *Present and Future Research in Coastal Lagoons*. Proc. UNESCO/IABO Seminar, Duke University Marine Laboratory, Tech. Paper Mar. Sci. UNESCO-33.

Yáñez-Arancibia, A. and R. S. Nugent. 1977. El papel ecologico de los peces en estuarios y lagunas costeras. *An. Centro Cienc. del Mar y Limnol.*, Univ. Nac. Autón. México 4:107-114.

HYPOTHESES OF ESTUARINE ECOLOGY

THE STATUS OF THREE ECOSYSTEM-LEVEL HYPOTHESES REGARDING SALT MARSH ESTUARIES: TIDAL SUBSIDY, OUTWELLING, AND DETRITUS-BASED FOOD CHAINS

Eugene P. Odum

Institute of Ecology
University of Georgia
Athens, Georgia

Abstract: During the first 10 years of salt marsh research at Sapelo Island (1952-1962), three general hypotheses emerged as follows: (1) tides provide an energy subsidy that enhances productivity, (2) organic matter is exported from productive estuaries to offshore waters (outwelling), and (3) detritus rather than grazing food chains predominate in the salt marsh ecosystem. These hypotheses, which we judge to be "emergent properties" of the salt marsh estuary as a whole, have now been challenged and tested in many places up and down the coast. The tidal subsidy hypothesis has been verified sufficiently to stand as a general principle. Outwelling seems to be strictly a local question depending on relative productivity of inshore and offshore waters and the magnitude of water flow in and out of the estuary; some estuaries export while some import, and the material exported (or imported) may involve nutrients, organic matter, or organisms. Although dominance of detritus-based food chains has been verified for most shallow water estuaries (in contrast to dominance of the grazing food chain in open water marine habitats), recent work has indicated that detritus complexes are like autotroph-heterotroph microcosms with algae, protozoa, fungi and bacteria providing major energy sources for detritus consumers which in turn are the chief food for fish and higher trophic levels in general.

Introduction

During the 1950's when studies on Georgia salt marsh estuaries began at Sapelo Island, a number of general hypotheses emerged which served as focal points for individual research projects. Many hypotheses that were advanced then have subsequently been challenged and tested in many other areas. Some proved to be untenable and were abandoned, but a number seem to have stood the test of time. In this article I would like to review the status of three ecosystem-level hypotheses, namely, (1) the tidal subsidy concept, (2) the outwelling idea, and (3) the concept of a detritus-based food chain as the dominant channel of energy flow in estuaries.

First, a word about system-level properties. Salt (1979) in a recent commentary suggests that properties-of-whole (holistic attributes) are best considered in two quite different categories, namely *collective properties* which are merely summations of the behavior of components, and *emergent properties* which are new properties resulting from the integration of components and, therefore, are not discernible from observation of components alone. In other words, a collective property is a sum of parts while an emergent property is more than a sum of parts. Just as the properties of water are not predictable from the properties of hydrogen and oxygen (the components), so emergent properties of biological systems are discernible only from observation of the whole unit (the ecosystem in the cases under discussion).* Species diversity of a community is an example of a collective property since it is a summation of the relative abundances of species components. The symbiosis of coelenterate animals and algal plants in a coral reef that produces a unique mineral recycling system that enhances productivity is an example of an emergent property as the coupling is manifested only at the level of organization of the reef ecosystem as a whole. I consider the three ecosystem-level properties discussed in this paper to be emergent properties because they "emerge" only at the level of the salt marsh estuary as a whole.

The Tidal Subsidy Hypothesis

Traveling in a small boat along a marsh creek, one feels like a corpuscle moving along an arterial capillary in some vast circulatory system whose distant heart is the pumping action of the tides. Tidal action appears to be a good example of the subsidy-stress syndrome, as outlined by Odum et al. (1979). At the population level, tidal action can certainly be a stress for intertidal organisms which must cope with alternate flooding and exposure that brings abrupt changes in temperature, salinity and moisture conditions. However, it occurred to us in the early days that, from the standpoint of the ecosystem as a whole, tidal action might prove to be a subsidy rather than a stress because the most active marsh communities seemed to occur where tidal action was most pronounced. Thus, it was our contention that adapted, coevolved populations in estuarine communities might actually utilize tidal power, as it were, as an auxiliary energy source coupled with solar power, the principle driving force. I believe that we can now say that this idea has endured. There are a number of good studies which verify the fact that, at least within the moderate range of tidal amplitudes, the greater the amplitude, the higher will likely be the primary production of the estuarine community. For example, Steever et al. (1976) compared the terminal standing crop (which is a rough indicator of primary production in the temperate zone) in a number of Atlantic coastal marshes which varied greatly in degree of tidal flux. As shown in Fig. 1, they found that there was a significant positive correlation between tidal range and the terminal standing crop of marsh grasses. Within the range from 0.5 to 2 m, an increase in tidal amplitude of 1 m approximately doubled standing crop. It should be emphasized that consistent with the subsidy-stress hypothesis, the stress component will likely exceed the subsidy component at the ex-

---

*For this reason species attributes, as discussed by Slobodkin in another paper in this symposium, are not by themselves an adequate description of an ecosystem.

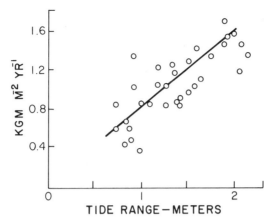

Figure 1. *Terminal standing crop (kg m⁻² yr⁻¹) and tidal range for* Spartina *sp. grasses in Atlantic coast salt marshes. The positive correlation is statistically significant (after Steever et al. 1976).*

tremes of tidal amplitude. Thus, very high tides, such as occur in the Bay of Fundy, are not likely to enhance productivity; likewise for very weak tidal action.

The scatter of points in Fig. 1 is rather wide, as might be expected, as, of course, tide is not the only factor which would affect productivity of a salt marsh community. Light, temperature, nature of substrates, and, especially, input of nutrients (such as fertile river inflow or sewage outfall) would be some of the factors that could be expected to affect productivity. That waterflow can provide energy subsidy for other kinds of wetlands is indicated by Louisiana data reported by Connor and Day (1976). They found that swamp forests with flowing water conditions were much more productive than swamps where the water is more or less stagnant. Highest productivity occurred where there was a pronounced seasonal "tide" as in the case of bottomland hardwoods on floodplains.

There are a number of causal relationships between waterflow and productivity which could be mentioned. Tides and other flows function to increase nutrient fluxes, perform recycling work, and remove waste materials, all to the benefit of adapted organisms. With such maintenance work performed by tidal power, organisms can divert more converted solar energy (photosynthate) to productive channels.

Also, in tidal marshes the pumping action of the tide provides a powerful irrigation force for root systems and for the general microbial communities in sediments. To demonstrate this subsurface tidal action, Riedeburg (1978) placed a fluorescent dye, rhodamine-B, in pipe wells and then monitored dye movement by sampling from monitoring wells located in concentric rings around the injection well. Figure 2 shows the general results of this study. Dye moved freely and quickly vertically from the site of injection (25 cm below the sediment surface) in the banks of the tidal

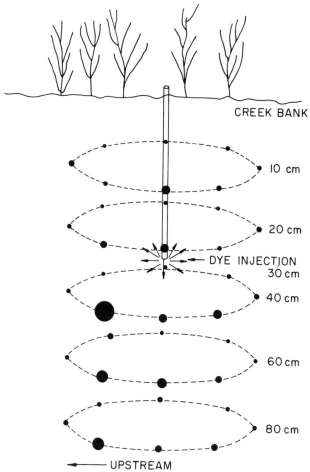

CREEK BANK

10 cm

20 cm

DYE INJECTION
30 cm

40 cm

60 cm

80 cm

UPSTREAM

Figure 2. Diagram summarizing results of an experimental study in which dye was introduced 30 cm below surface of a creek bank Spartina alterniflora marsh at Sapelo Island and movement monitored by sampling from 8 pipe wells located in one-meter circles at 5 depths, as shown. Relative concentration of dye collected during a spring tide period is indicated by the size of the black circles. (Data from Riedeburg 1975).

creeks both upward to surface and downward to deepest monitoring wells (80 cm). Dye also moved readily in horizontal directions at almost all depths which were sampled. Major movement was parallel with the tidal creek, with more dye moving upstream on spring tides and a major flow downstream during neap tide periods. In the high marsh at some distance from tidal creeks, tidal inundation is, of course, less frequent and less vigorous. In these areas dye movement was much more restricted; no dye could be detected more than 0.5 m from the injection well. Even in the high marsh, however, there was a distinct movement vertically.

Thus, the tidal pump, as it were, results in extensive irrigation of sediments in the marsh that brings oxygen and nutrients to root systems and to the microbial populations that colonize the highly organic soils. Without such tidal action, growth of marsh grasses and other primary producers and the microbial recycling of nutrients would be greatly reduced under stagnant, waterlogged, and salty soil conditions.

Outwelling

In 1968, I stated the idea of outwelling as follows: "Most fertile zones in coastal waters capable of supporting expanded fisheries result either from the 'upwelling' of nutrients from deep water or from 'outwelling' of nutrients or organic detritus from shallow water nutrient traps such as reefs, banks, seaweed or seagrass beds, algal mats, and salt marshes." The idea was that fertile estuaries might contribute significantly to productivity of offshore waters. The first evidence of this in the Georgia region came from a study by Thomas (1966) who found that primary production of offshore waters, as measured by the ^{14}C bottle method, was much higher just offshore from the openings between barrier islands than was the case further out to sea. He found further that this productivity was not necessarily due to materials coming down large rivers because high values were obtained along the shore at some distance from large rivers. But in all cases, offshore enriched zones were close to the salt marsh estuarine zone, which in Georgia may be up to 10 miles or more in width. Recently, Turner et al. (1979) have confirmed this enrichment along the Georgia shore and they also documented an outwelling effect along the coast of Louisiana. Offshore productivity and the density of zooplankton, fish eggs, and fish larvae were strongly coupled with the extent and productivity of local estuaries. They concluded that the influence of estuaries on continental shelf ecology was extensive.

Figure 3 indicates that the enriched zone, at least off the south Atlantic coast, might be quite narrow (Turner et al. 1979). In other words, production rate is quite high just offshore and then drops within about 10 km from the mouths of the estuaries. This narrow, enriched zone can be designated as the *coastal front*. In Georgia, it is noticeable that shrimpers drag this narrow zone very intensively; it is evident that as the young shrimp come out of the estuaries they concentrate here, evidently finding good feeding areas.

Dr. Don Kinsey, Director of the Sapelo Laboratory, has become interested in this zone and has begun studies with an objective of making the coastal front a focus for future team research. He is finding very high production rates which almost equal that of the estuary itself. Rather than using bottle techniques, which only measure plankton productivity, he is employing the diurnal curve method for measuring metabolism of the en-

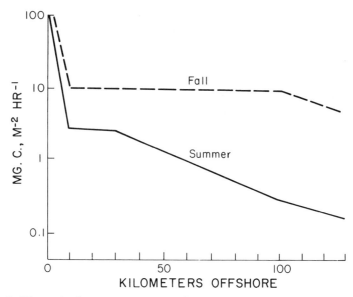

Figure 3. Phytoplankton primary production as a function of distance offshore from salt marsh estuaries (Barrier Islands) of Georgia (Redrawn from Turner et al. 1979).

tire water column, including benthic organisms. He is also finding that respiration is even higher than primary production, indicating that the system is definitely heterotrophic, which is strong evidence that there is a considerable export of organic matter from estuaries to the coastal front. Because of the apparent narrowness of this frontal zone, it would be easy to overlook the marked effect which an estuary may have on offshore waters. Obviously, there is much to be learned from studies of this ecotone which lies between estuaries proper and continental shelf waters.

Not everyone has found evidence for outwelling. On some areas of the coast, notably in some New England areas, estuaries seem to import rather than export nutrients. In reviewing this situation, W. E. Odum et al. (1979) have suggested that geomorphometry of the coastal bays and estuaries, tidal amplitude, and magnitude of freshwater inputs are the three key factors which determine whether there is outwelling, or as we might say, "inwelling." Three contrasting situations are diagrammed in Fig. 4. Where channels between offshore and inshore waters are narrow and blocked by a sill, as shown in diagram A, or where tidal action is weak, one would not expect to find extensive outwelling. This is apparently the case in the well-studied Flax Pond situation (Woodwell et al. 1977, 1979) which is often cited as evidence against outwelling. In contrast, where there is a more open estuary with extensive exchanges between estuarine and continental shelf waters, as would likely be the case with geomorphological configurations diagrammed in Fig. 4 B and C, then one would expect outwelling if estuaries are richer than shelf waters.

Also, there is considerable question, and probably considerable variation, as to what is exported from estuaries into coastal waters. Originally, we had assumed that organic detritus was probably the main material which flows out (perhaps because of highly visible racks of dead marsh grass that move out of the marshes) but it seems likely from other studies that the exported material might take many different forms ranging from mineral nutrients to organic aggregates of various sorts to small fish or other organisms which directly or indirectly enhance either primary or secondary productivity, or both, of coastal waters. Based on $^{13}C/^{12}C$ ratios, Haines (1980) suggests that more coastal seston may originate from algae and terrestrial plant material than from *Spartina sp.* marsh grasses.

Finally, it seems likely that outwelling in many, if not most, situations

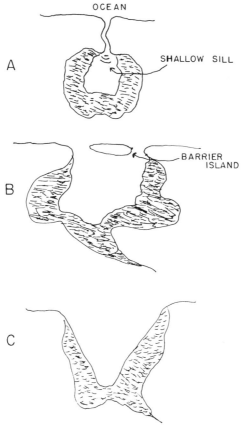

Figure 4. Three geomorphometric patterns of estuaries which are thought to influence the degree of outwelling. See text for explanation (Redrawn from W. E. Odum et al. 1979).

is periodic or seasonal, associated especially with high spring tides and storms. Just as export of nutrients from a forested watershed occurs mostly during storms, so one would expect a marsh "watershed" to export during periods of heavy rain, high tides, etc.

Returning to my statement cited at the beginning of this section, we can conclude that outwelling and upwelling interact in such a way as to enhance productivity of coastal waters. Relative importance of the two will likely vary widely from locality to locality.

The Detritus Food Chain

When we started investigations on Georgia estuaries, thinking about marine productivity was very much dominated by the elegant work of Riley on Long Island Sound. In this coastal shallow-water system, phytoplankton are heavily grazed by zooplankton so that much of the energy passes along what we now call a grazing food chain. It did not take us long to find out that this was not the case for the salt marsh estuaries. *Spartina sp.* and other marsh grasses are very little grazed and most of the heterotrophic part of the marsh food chain involves utilization of the ungrazed, decaying material at some later date. Thus, the concept of the detritus food chain was born (Odum and de la Cruz 1963; Odum 1963). While few will challenge the concept that organic detritus is the link between autotrophic and heterotrophic parts of the ecosystem in most estuaries, there is considerable disagreement about how the producer detritus energy is actually utilized by heterotrophs themselves.

The food source for small animals such as the meiofauna, polychaetes, shrimp, and small fish is particularly uncertain. It is becoming evident that detritus complexes are little microcosms with self-contained cycles that sequester nutrients, and within which the more resistant plant residues are reprocessed over and over again until all nutrients and potential energy are released. Furthermore, small benthic-type algae are an integral part of the detritus particle along with heterotrophic microorganisms (bacteria, fungi, and protozoa) and meiofauna (nematodes, etc).

In Fig. 5, I present a graphic model of the coupling of estuarine food chains that incorporates these concepts. The unique feature of an estuary, perhaps an "emergent property," is the integration of three distinct classes of primary producers, macrophytes, phytoplankton, and benthic (i.e., "mud") algae, all of which are linked together by detritus and tidal action, as previously discussed. The diagram shows how food chains originating from each of the three types of autotrophs may be linked together to form an estuarine food web. Because benthic algal components are so tightly coupled to heterotrophic microorganisms, both structurally and functionally, "detritus" is depicted as a single unit subsystem which is the primary energy source for macrofauna in the detritus food chain, which as already indicated is the dominant energy-flow pathway in most estuaries. Detritivores, mostly small invertebrates, but also a few fish, ingest these

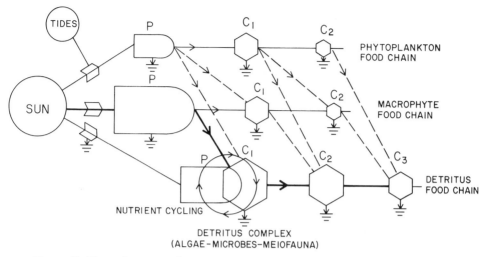

Figure 5. Flow diagram of an estuarine food web showing three coupled food chains, namely, two grazing food chains (phytoplankton, macrophyte) and the detritus food chain. In Georgia salt marsh estuaries, the detritus pathway via macrophytes is the dominant one, as shown.

detrital particles, digest and assimilate the microorganisms (and in some cases some of the macrophyte substrate) and eliminate the more resistant material which is again colonized by bacteria, fungi, algae and protozoa. Thus, detritus originating from marsh grass or seagrass or large algae can be processed, so to speak, over and over again. This would seem to form a very efficient system with considerable nutrient conservation possibilities. We present this as possibly a useful model for the detritus food chain.

Retrospect

It has been our experience that setting up ecosystem-level hypotheses promotes team research at field laboratories. Individual investigators with their differing skills and approaches are greatly stimulated by having some common hypothesis to test, and soon become enthusiastic in developing their own approaches to a basic problem. Thus, the emphasis on detritus in the early days resulted in creation of an almost completely new field of microbial ecology. It soon became evident that traditional microbiological techniques used in the laboratory were completely unsuited for the study of natural systems. Isolating organisms in pure cultures has practically no meaning when it comes to understanding how microorganisms actually function in ecosystems. Thus, a whole new set of techniques had to be developed before the roles of various organisms, both aerobic and anaerobic, could be understood. The microbial component of estuaries is probably now the most important focus for estuarine research.

In summary, the examples of the three system-level hypotheses presented in this paper serve to illustrate one approach to integrating physical and biological sciences in the search for a better understanding of such priceless resources as our estuaries and other wetlands. In a sense then, the search for emergent properties, that is, phenomena which are discoverable only when the whole system is considered, becomes the great challenge of environmental science today. Ultimately, the assessment of pollution and other effects of man must be made at the systems level, as well as at the organism or species level.

Acknowledgments

Figure 1 is used with permission from Estuarine and Coastal Marine Science. Copyright by Academic Press Inc. (London) Ltd.; Figure 3 is redrawn from one copyright 1979 by the American Association for the Advancement of Science; Figure 4 is used with permission from Plenum Publishing Corporation.

References Cited

Connor, W. H. and J. W. Day. 1976. Productivity and composition of a bald cypress-water tupelo site and a bottomland hardwood site in a Louisiana swamp. *Amer. J. Bot.* 63:1334-1364.

Haines, E. B. 1980. On the origins of detritus in Georgia salt marsh estuaries. *Oikos* (in press).

Odum, E. P. 1963. Primary and secondary energy flow in relation to ecosystem structure. *Proc. 16th. Int. Cong. Zool.* 4:336-338.

Odum, E. P. 1968. A research challenge: evaluating the productivity of coastal and estuarine water. Proc. 2nd. Sea Grant Conf., Grad. School Oceanography, Univ. Rhode Island, Kingston, RI.

Odum, E. P. and A. A. de la Cruz. 1963. Detritus as a major component of ecosystems. *AIBS Bulletin* (now *BioScience*) 13:39-40.

Odum, E. P., J. T. Finn and E. H. Franz. 1979. Perturbation theory and the subsidy-stress gradient. *BioScience* 29:349-352.

Odum, W. E., J. S. Fisher and J. C. Pickral. 1979. Factors controlling the flux of particulate organic carbon from estuarine wetlands, pp. 69-80. *In:* R. J. Livingston (ed.), *Ecological Processes in Coastal and Marine Systems.* Plenum Publ. Co., New York.

Riedeburg, C. H. 1975. The intertidal pump in a Georgia salt marsh. MS Thesis, Univ. of Georgia, Athens, GA. 81 pp.

Salt, G. W. 1979. A comment on the use of the term emergent properties. *Amer. Nat.* 113:145-148.

Steever, E. Z., R. S. Warren and W. A. Niering. 1976. Tidal energy subsidy and standing crop production of *Spartina alterniflora*. *Est. Coastal Mar. Sci.* 4:473-478.

Thomas, J. P. 1966. The influence of the Altamaha River on primary production beyond the mouth of the river. MS Thesis, Univ. of Georgia, Athens, GA. 88 pp.

Turner, R. E., S. W. Woo and H. R. Jitts. 1979. Estuarine influences on a continental shelf plankton community. *Science* 206:218-220.

Woodwell, G. M., C. A. S. Hall, D. E. Whitney and R. A. Houghton. 1979. The Flax Pond ecosystem study: Exchanges of inorganic nitrogen between an estuarine marsh and Long Island Sound. *Ecology* 60:695-702.

Woodwell, G. M., D. E. Whitney, C. A. S. Hall and R. A. Houghton. 1977. The Flax Pond ecosystem study: Exchanges of carbon in water between a salt marsh and Long Island Sound. *Limnol. Oceanogr.* 22:833-838.

ON THE EPISTEMOLOGY OF ECOSYSTEM ANALYSIS

L. B. Slobodkin

Department of Ecology and Evolution
State University of New York
Stony Brook, New York

D. B. Botkin

Environmental Studies
University of California
Santa Barbara, California

B. Maguire, Jr.

Department of Biology
University of Texas
Austin, Texas

B. Moore, III

Complex Systems Group
O'Kane House
University of New Hampshire
Durham, New Hampshire

and

H. Morowitz

Department of Molecular Biophysics
and Biochemistry
Yale University
New Haven, Connecticut

Abstract: It is impossible to construct a general theory or model of any particular ecosystem which will be useful for answering all possible questions about that system, although if we know enough about any ecosystem it is possible to construct such models once a specific question has been posed. This knowledge cannot be gained entirely from the system at issue, due to restrictions in time and resources, as well as to the fact that certain kinds of thorough ecological analysis may damage the system analyzed. Therefore, it is advisable to use relevant information from ecosystems other than the one of immediate interest. A partial list of species present in an ecosystem permits access to the information gained by naturalists working on other systems. We therefore justify the usual practice of making species lists because such a list is the best (i.e. cheapest and most useful) preliminary step in answering questions about any ecosystem. While explicit measurements must also be made in the object ecosystem in order to usefully model it, it is likely that the number of such necessary measurements may be reduced and their usefulness enhanced by the background natural history information implicit in a partial species list. To demonstrate that the information of natural history can be communicated in a relatively complete way, we provide a partial representation of an adaptive response surface for *Hydra sp.* in which much of the kind of information about these organisms that might be useful for model construction can be presented in a relatively simple diagram.

The Problem for the Naturalist

Ecology has progressed to the point that certain epistemological problems, i.e. problems in the theory of knowledge, of a non-trivial kind have emerged as real and pressing. We have a great deal of information about ecosystems. Obviously we can use more, but we can no longer use logical structures borrowed from Physics, Mathematics, or Computing Science as a framework for our knowledge without considering the unique properties of ecological systems.

We will consider a somewhat simplified description of the relation between the naturalists, who are the source of detailed factual knowledge about ecosystems, and the mathematically oriented theoreticians, who are to a large degree responsible for the systemization which makes this knowledge generally accessible.

To discover intimate facts about nature is extremely difficult, expensive, and time-consuming. Naturalists know that a fruitful lifetime of research can often be represented by a half-dozen facts about a few species, often with no assurance that they apply to other species. By contrast, the normal format of mathematical system construction in a physical or engineering context is first to provide an essentially complete definition of the system, in which it is made clear which portions of the world are, and are not, relevant to the problem at hand. An airplane or a tub of water may be the system defined. A state description suitable to the problem at hand is then constructed. The speed of an airplane, combined with relevant measures of wind velocity, load, and turbulence may be such a state descriptor as may the temperature and pressure surrounding the tub of water. A theory of airplane flight involves feeding the definition of the airplane, that is, its weight and metal structure and so on, and an appropriate state descriptor into a model which will describe how the state of the airplane will change with time or with alterations in ambient circumstances.

When the theoretician concerned with an ecosystem requests from the naturalist a definition and a state descriptor he is often given a compilation of facts which is, by and large, grossly inadequate for most theoretical purposes. The theorist, perhaps with management waiting for an answer, can do no better than attempt to build a plausible approximation to what the definition and state descriptor might be and proceed with theorization and prediction in the usual way, as if his knowledge were adequate to the task. As we know, the results are often dubious.

For various reasons, the problem cannot be solved meaningfully by head-on crash program methods. Consider that the problem of studying a small salt marsh cannot be alleviated by putting 200 naturalists to work on that marsh, since the repeated passage of their rubber boots would probably be a sufficiently great perturbation to destroy many interesting or valuable properties that the marsh might have possessed. We cannot even escape from our problem when critical definitional properties have been exceeded. Once an airplane crashes, its parts are a problem for someone other than the pilot but a drastically altered erosystem is still a problem for ecologists.

So far we claim to have demonstrated that a problem exists. The remainder of the paper will attempt to demonstrate one set of possible steps to the solution of the problem.

The Constraints on Ecological Descriptors

We must distinguish between learning *about* an ecosystem and learning *from* an ecosystem. To learn from an ecosystem implies that we would use information gained from a particular system to build a model complete enough to permit answering a specific question about that system. Even if we replace rubber-booted naturalists with daintier information gathering systems, such as electronic sensors, the construction of a complete model of an ecosystem, using only information from that system itself, is often difficult. First, data over time are required to model such systems, and regardless of the subtlety of data collection this time must be expended. Second, when attempts have been made to count the number of kinds of organisms in particular ecosystems (Elton 1966), numbers in the thousands are reached. A complete model of such a system would involve minimally a matrix of enormously high dimension, particularly when we consider that the number of kinds of between-species interactions may be quite large.

It is useful for some purposes to consider the multiplicity of species and organisms in an ecosystem as statistical populations analogous with those of statistical mechanics (Kerner 1972). The relatively small number of kinds of particles dealt with in statistical mechanics and the possibility of assuming equivalance between samples taken from the system, i.e. the property of ergodicity, is a major contributor to the relative simplicity and generality of theories in physics and chemistry. By contrast, the parts of ecological systems differ from each other in so many interesting ways that we cannot assume the statistical equivalence of samples separate from each other in either space or time, except under very constrained circumstances, and we certainly cannot assume that a relatively small number of kinds of entities are interacting; therefore for most purposes an approach analogous to that of statistical mechanics is not available.

If complete data could be collected without destroying the system under study, the construction of a complete model would be too difficult to handle by present computational methods, and, if it could be handled, the resulting model might be as difficult to interpret as nature itself. We conclude, in part, that if we are to understand an ecosystem we must bring information to that system from other sources. We omit discussion of the problem of verifying a model that was constructed from information taken from a particular ecosystem by using the same ecosystem in the process of verification.

If we are to abandon the hope of making a complete description, what is the best possible descriptor available to us? From here on we will advance a positive program rather than make negative arguments. We can state some of the properties which a workable, realistic description of an actual ecosystem must have. It must be concise enough to be completely stated. It must contain a definition of the system so that we can tell different ecosystems apart. It must include information about the system, but it cannot constrain the questions that might be asked about the system.

The last point is non-obvious. We do not know in advance what questions will be necessary to be asked about ecosystems, but we would like, if possible, to describe ecosystems in advance of the occurrence of these questions to facilitate our reply. We therefore would prefer a description which is theoretically neutral.

An Ecosystem Descriptor

We now suggest how to construct such a descriptor, and then we will give an example of its use. We must first request a kind of tolerance from the reader. What we will present is neither difficult nor esoteric. It corresponds to the naturalists' sense of nature and uses the intuitions and attitudes of biology in a simple way. This simplicity is no reason to reject it, *a priori*, because it can be shown, despite our apparent reversion to 19th Century standards, that it is really the best system of description for actual problem solving.

We suggest most strongly and seriously that the optimal definition available for any ecosystem is the list of species that has been found in that system. We will now defend this apparently trivial assertion.

A species name, despite being subject to change as taxonomic revisions occur, is enormously rich in information in the sense that it provides access, through the scientific literature, to the work of naturalists studying organisms of that kind and similar kinds over the past three centuries. It also provides information about all of the properties these organisms share with the broader taxonomic groups to which the particular species belongs. This pool of information also includes statements about necessary environmental conditions and biological prerequisites of the organisms. In fact, there is no cheaper or more effective way of gaining flexible, useful information about an ecosystem than making a list of the species that can relatively easily be found there. We must immediately make it clear that we are not asserting that information about the individual component species in an ecosystem can replace observation of that system itself, since to understand ecosystems requires that we understand interactions of many kinds as well as understanding the interacting components. We are as-

serting that knowledge about the component species is extremely useful as
a first step to the understanding of any ecosystem for any purpose.

The total amount of information that may be available about the
organisms in a species list varies but it may be so enormous and unwieldy
as to make its usefulness suspect. This is not an unsurmountable problem
since actual investigation of actual ecosystems are undertaken with a
specific goal. The particular goal of any particular study defines the way
the information associated with the species list is to be used and also dic-
tates the subset of the total available information that need be considered.

Consider what is learned from the first species named on the list. Of
the approximately 2 million species on Earth we have found a particular
one. We know roughly its temperature and food requirements, and we
know whether or not it is terrestrial or marine or aquatic. We can therefore
eliminate, as possible complicators of our consideration of this ecosystem,
more than 1 million species, as having requirements incompatible with the
known property of the organisms we have found. In almost all cases, finding that first organism had essentially no impact on the ecosystem itself.
There are some ecosystems in which the acquisition of an incomplete
species list may involve damage to the system of a fairly serious kind.
These include attempts to make a list of the species living in coral heads or
in soil. They also include situations in which rather rare species may be
seriously damaged by the attempt to determine whether or not they are
present. But in almost all cases, finding the first species of the species list
has essentially no impact on the ecosystem itself.

Now imagine collecting another specimen, avoiding taxonomic
duplication. This second species will reduce the class of possible species yet
to be found to the logical product of the species conceivably coexistent
with both it and the first species, adding more to our list of impossible new
finds. After 20 or 30 kinds of organisms have been found, the 21st or
perhaps 31st new species may add very little new information about the
characteristics of that ecosystem.

At this point, the 2 million possible kinds of inhabitants that might
have been present have been divided into four portions. The largest portion consists of organisms one can be sure are absent. Then there will be
several thousand species that may or may not be expected to be found on
further investigation. There is a third class of organisms that have not yet
been found but we believe must be present, as their presence is a prereq-
uisite for the occurrence of some organism we have actually seen. Finally,
there is a small list of actually observed species.

We also will have, from the data associated with past investigations
of the observed species, a reasonably good idea of the physical and
chemical properties of the ecosystem, probably a much better idea than we
would have had if we had spent a corresponding amount of time and effort on chemical, geological, and meteorological analyses. The short species

list is therefore seen as a cheap, rapid, information-laden approach to defining an ecosystem.

This procedure works for two basic reasons. First, each species is related to the world in a slightly different way, but all share the property of temporal continuity. Each species can be thought of as an environmental sensor, whose persistence in a particular place is a reliable indication of the condition and history of that place. To acquire knowledge of each species' properties would be prohibitively time-consuming, expensive and possibly environmentally perturbing, but we have free and quick access to information gained by past studies, generally in different ecosystems. Thus external information can be brought to the system in which we are interested.

So far we claim a species list to be a uniquely valuable definition of any ecosystem for any purpose. The more complete the list the better, but even an incomplete list is very good. Note by contrast how little we know about the microstructure of an ecosystem if we know only its diversity index, biomass or energy flow. None of these would permit us to distinguish between a forest and a coral reef.

If we conclude that a species list is a proper definition of an ecosystem, there remains the problem of how this definition may prove useful. For example, the relative abundance of species in a community also carries information. Compare a forest with one wild cherry tree and a thousand poplars with a forest that contains a thousand cherry trees and only one poplar! There are other descriptions which also each carry information of a different kind—flux diagrams of all kinds, biomass calculations, and so forth. Each of these is particularly suitable for some specific question about the ecosystem, but, as indicated above, we really don't know what question will be asked in advance and each of these descriptions may prove to be irrelevant to some set of questions. We claim that the information associated with a species list will be of value for answering almost any question that may be asked.

In general, there is a distinction between defining a system and defining the state of that system. For most imaginable questions some state descriptor, in addition to the species list, is necessary, which will consist of a vector of measurements made in the ecosystem plus quantities, such as temperature range, which might be inferable from information about species in the observed species list. The general usefulness of a species list is of course dependent on how much we actually know about the individual species on the list so that natural history, i.e. basic information on the lives of the organisms concerned, is of general and permanent value. It also follows that a species list is less useful in areas where the preliminary natural history studies have not been undertaken. The general value of natural history and of the species list is not contingent on any attempt to model the ecosystem in advance of a specific question. It is our assertion that regardless of what question is asked in the future, this information will prove to be of value.

What we have described is actually the way the best of ecosystem studies in ecological management programs have proceeded. All we have done is dissect out the methodological structure and recommend its general utilization. Further amplification of the rationale for this procedure and of the more formal properties of species lists and state descriptors in general can be found in Botkin et al. (1979) and Maguire et al. (1980).

Once it is conceded that a species list plus a state vector are of enormously high information content, it is possible to consider the list itself to be an interesting object for theoretical study. Freckman et al. (1980) were concerned with how one might test the hypothesis that communities with larger numbers of species will differ in their response to perturbation from communities with smaller numbers of species. This relates to the concept of species packing and to other theories of diversity. In order to avoid preconceptions and selection of information, they considered it advisable to study communities which were not well known to anyone, so that no preconceptions were possible. Therefore they collected 14 samples of soil at different locations near Riverside, California and identified the genera of living Nematodes in each. The samples were divided into two classes; those with more than 12 genera (species-rich) and those with 7 or fewer (species-poor). A relative frequency count of the first approximately 100 live specimens was made for each sample. Each sample was treated as a separate community described by a species list and the differential count vector (a kind of state vector).

The whole samples were then subjected to a nematacide until approximately 50% of the specimens were immobilized. The Nematodes were washed out of the nematacide and a new generic list and differential count was made for the motile survivors.

Freckman et al. (1980) then compared the state of each of the communities before and after perturbation. They found no significant difference between species-rich and species-poor samples in the proportion of species eliminated by the nematacide. But, considering the differential species count as a vector in N space, where N was the initial number of genera, both the degree of rotation and Euclidian distance between the vectors was greater after perturbation if the species list was initially longer. This supports the theoretical assertion that species-rich Nematode communities are composed of more sensitive individuals than are species-poor communities.

In this example, a taxonomic list and simple state vector could be used as a means to test a theoretical assertion on the assumption that it is the best descriptor available. It may prove to be the case that use of this descriptor may bring a new simplicity into some kinds of ecological theory.

Complete Description in Natural History

One of the assumptions in the above discussion was that one can actually start with a species' name and use it as a starting point for a search of the scientific literature to discover what naturalists have learned about that species. Running counter to this assumption, however, is what might be called the *naturalists' mystique,* namely, that there is an esoteric quality to the knowledge of naturalists. If what is known by a naturalist cannot be stated in a reasonably complete and clear form, it is not very valuable by normal scientific standards. There exists a very real problem of making sure that the assertions of naturalists are publicly verifiable.

We believe for two reasons that lengthy species-by-species descriptions are not necessary. The obvious reason is that, by and large, the classical system of classification works reasonably well, so that while certain details of muskrat biology may be the esoteric knowledge of only an experienced muskrat expert, many of the properties associated with rodents, mammals, chordates and eucaryotes in general are well known. We will not belabor this point further.

The second, more speculative point we will illustrate with an example. Slobodkin (1980) attempted to summarize studies of various aspects of the biology of hydra in as complete and explicit a way as possible. He started with several facts, valid for some hydra, although they have not been demonstrated for all species (support for these assertions is presented in Hecker and Slobodkin (1977), Otto and Campbell (1977) and Slobodkin (1980) and in references cited therein):

1. There is an asymptotically increasing relation between food supply and both growth rate and budding rate at any particular temperature, kind of food, and water chemistry;

2. Between species, there is an inverse relation between growth rate and budding rate. This is intimately related to the apical control mechanism. That is, if there is strong dominance by the hypostome, budding is restricted and the food that would have supported the budding process can only be used for growth;

3. Increasing temperature lowers body size and increases budding rate for several species. It may, therefore, rotate the response surface towards small body size and rapid budding. Alterations in water chemistry may also distort the surface in a reasonably simple manner.

From this information a partially hypothetical adaptive response surface for hydra can be constructed. This is presented in Figs. 1 and 2. The three axes are *food level* and the *body size* and *budding rate* attained by individual animals all at constant temperature and kept at constant food until an approximate steady state in their budding rate has been reached. The surface flattens against the food level-body size plane at low food levels, since hydra can be maintained at food levels which permit survival but not budding. At food levels below A, hydra die. As food level increases, the smaller species of hydra begin to bud before the larger ones. At sufficiently high food levels, the growth and reproductive rates approach an asymptote. The meridional lines (As$_i$, etc.) on Fig. 1 represent the assumed trace of particular genotypes on this surface. That these lines do

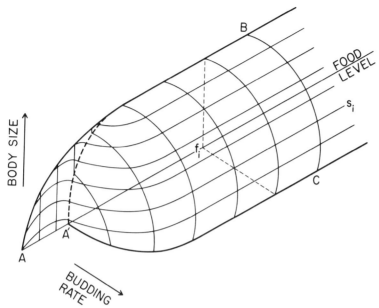

Figure 1. The adaptive response surface for hydra. On the basis of preliminary information it seems likely that the interaction between food level, budding rate and body size in hydra can be represented by this surface. Meridional lines (s_i) on this three dimensional surface represent the response of different genotypes of hydra. For further description see text.

not cross is an explicit hypothesis. The surface shown in Fig. 2 is identical with that shown in Fig. 1 except that the meridional lines have been omitted for clarity. Closed regions have been mapped onto Fig. 2 indicating particular properties of hydra. For example, green hydra all bud rapidly but are relatively small. Floating occurs in relatively large species only, and only when they are hungry. Sexuality occurs primarily in intermediate sized hungry animals. Large well fed animals are predacious on small green animals. Slobodkin (1980) has suggested that essentially all of the ecologically interesting properties of hydra may prove to be mappable onto closed discrete regions of Fig. 2. If this is so, then at least in this case the possibility of a simple presentation of the complete natural history of a group of animals will have been demonstrated.

The surfaces drawn in Figs. 1 and 2 are adaptive response surfaces which arise purely from observation of hydra. They are not a consequence of any more general theory. That is, no optimization theory or theory of population genetics or niche theory or adaptive landscape theory predicts how to draw the surface, although there is reason to suspect that a surface of this type is possible in general. While it is possible to rationalize the par-

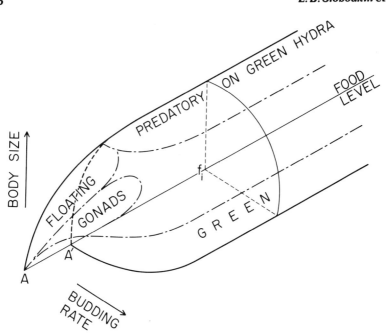

Figure 2. Mapping of ecologically significant properties onto the adaptive response surface. For further description see text.

ticular responses, the rationalizations are independent of the description. That is, the transmission of information about hydra is free of optimality assumptions or models, but can be used to infer optimizations or models. We have just provided the information to permit the reader to "think like a hydra" to some extent.

Conclusions

We have suggested that it is both feasible and interesting to attempt to describe nature in the most information laden way possible on both the levels of the ecosystem and of natural history. The properties of the natural world may then dictate to us not only the facts of ecology, but also the form of ecological theory. We have given a few brief examples of particular descriptions.

We believe that the procedure of attempting to produce rich descriptions is not only useful but that it generates important theoretical questions, different from those generated by attempting to fit ecology into the epistomelogical nexus generated by the physics and engineering of inanimate systems. This is not vitalism but is a refusal to relegate our intellectual problems to strangers.

Acknowledgments

Interaction between the co-authors was made possible by the support of NASA Space Biology Program. The studies of *Hydra sp.* from the laboratory of L. B. Slobodkin were supported by the National Science Foundation Ecology program.

Contribution No. 297, Department of Ecology and Evolution, State University of New York at Stony Brook.

References Cited

Botkin, D., B. Maguire, B. Moore, H. Morowitz and L. B. Slobodkin. 1979. Closed regenerative life support systems for space travel: Their development poses fundamental questions for ecological science. (COSPAR) Life Sciences and Space Research. XVII:3-12.

Elton, C. 1966. *The Pattern of Animal Communities.* Methuen, London, 432 pp.

Freckman, D. W., L. B. Slobodkin and C. Taylor. 1980. Pesticide use and the stability of species-rich and species-poor communities of nematodes. *Proc. VII Int. Soils Zoology Colloquium* (in press).

Hecker, B. and L. B. Slobodkin. 1977. Responses of *Hydra oligactis* to temperature and feeding rate, pp. 175-183. In: G. O. Mackie (ed.) *Coelenterate Ecology and Behavior.* Plenum Press, New York.

Kerner, E. 1972. *Gibbs Ensemble, Biological Ensemble.* Gordon Breach, New York.

Maguire, B., D. Botkin, B. Moore, H. Morowitz and L. B. Slobodkin. 1980. A new paradigm for the examination of (closed) ecosystems. In: J. Giesy (ed.), *Symposium on Microcosms in Ecological Research.* Technical Information Center, U.S. Dept. of Energy (in press).

Otto, J. J. and R. C. Campbell. 1977. Tissue economics of *Hydra*: Regulation of cell cycle, animal size and development by controlled feeding rates. *J. Cell. Sci.* 28:117-132.

Slobodkin, L. B. 1980. Problems in ecological description. I. The adaptive response surface of hydra. *Mem. Inst. Ital. Idrobiol. (Suppl.)* 37:77-95.

GENETIC DIVERGENCE IN ESTUARIES

J. S. Levinton

Department of Ecology and Evolution
State University of New York
Stony Brook, New York

Abstract: Studies of several invertebrate and vertebrate species argue for the common occurrence of genetic divergence of estuarine populations from conspecific populations in open coastal marine waters. Estuarine differentiation raises many unresolved questions: (1) Does this common divergence indicate the isolation of estuarine populations from open marine populations?; (2) Is natural selection of sufficient intensity to cause divergence, despite constant influxes and exports of propagules to and from the estuary?; (3) What are the most significant agents of natural selection? Differences among estuaries of differing salinity and temperature regimes may also promote extensive genetic differentiation. Pollutants are a potentially potent among-estuary source of natural selection. Polluted estuaries may harbor populations of animals and plants with evolved tolerance to pollution. This could be easily verified through a broad study of species' tolerances and the genetic basis of the tolerance. The consequences of such adaptations to the establishment of environmental monitoring programs and to the study of the transfer of pollutants up estuarine food chains deserves some reconsideration in the light of these conclusions.

Introduction

The world's estuaries are a series of unique natural laboratories, nearly ideally suited for the study of natural selection. Unfortunately this assertion cannot be followed with the citation of a rich literature documenting the extent and basis of among- and within-estuary genetic differentiation. In this paper, I will (1) introduce the scope of our present understanding of genetic divergence in estuaries; (2) discuss the interesting ecological, genetic, physiological and evolutionary questions raised by these data; and (3) attempt to outline important areas for future research that are now, or have been recently, initiated.

A variety of conceivable barriers to gene flow occurs along estuarine and latitudinal gradients. These may be geographic barriers such as a sill that restricts water flow (and the movements of larvae) or simply an unfavorable environment that impedes dispersal (e.g. an area of coarse sand separating two mud areas favorable to a deposit-feeding species). But "steep" selective gradients might also modulate gene flow. The salinity gradient, for example, might act as a strong inhibitor of gene flow between estuarine and open marine conspecific populations. This would hold particularly for situations where estuarine populations are strongly adapted to the estuary, whereas conspecific open marine populations are similarly

509

adapted to open marine habitats. These respective local adaptations would strongly reduce the probability of interlocality gene flow. The implications of this possibility are discussed below.

It would be useful to focus on those areas of the estuary where natural selection is likely to be intense. It seems that there are two such locations. First, the mouth of the estuary is the first area where organisms must adjust physiologically to lowered salinity. Therefore, the vicinity of the opening of an estuary may provide an interesting focus for studies of natural selection. Unfortunately, our knowledge of estuarine-open ocean exchange processes is often inadequate to provide a strong environmental context within which to study natural selection. In any event, most evidence of genetic divergence in estuaries is centered around the estuary-open ocean boundary region (e.g. Boyer 1974; Christiansen and Frydenberg 1974; Lassen and Turano 1978; Theisen 1978).

Second, the horohalinicum or critical salinity is a probable region of intense selection (Khlebovich 1969; Kinne 1971). This region encompasses an approximate salinity range of 5-8^0/$_{00}$ and marks a pronounced minimum of benthic invertebrate species richness (Remane and Schlieper 1958). A relatively rich fauna of bivalve molluscs resides in fresh water. However, freshwater species decrease in numbers at a maximum salinity of 5-8^0/$_{00}$. Estuarine-marine bivalves are also rare at this salinity, but increase steadily in diversity with increasing salinity. Khlebovich (1968, 1969) maintained that this critical salinity range was the threshold above which ion regulation was not necessary in most marine organisms.

The horohalinicum marks a zone where Ca/Na and K/Na increases dramatically with decreasing salinity (Khlebovich 1968). The critical salinity also marks an empirically elucidated break in physiological adaptation, at least in bivalve molluscs (Gainey and Greenberg 1977). Marine euryhaline and stenohaline bivalves can regulate cell volume in higher salinities by changes in cellular free amino acids (e.g. Potts 1958; Lange 1963), but are generally incapable of the ion regulation necessary at and below the critical salinity. Freshwater stenohaline species have lost the capacity for extensive volume regulation but can regulate ionic concentration and maintain a hyperosmotic state in fresh water.

The absolute value of the critical salinity range may be relatively independent of the geologic history of sea water and peculiarities of phylogenetic origin. However, the differential success of taxa in invading fresh water seems related to some major differences in physiological function that postdate the divergence of the groups from common ancestors. Whereas the Bivalve orders Paleoheterodonta and, to a lesser extent, Heterodonta have successfully invaded fresh water, the Pteriomorpha have been notably unsuccessful. Ancestors of the living Paleoheterodonta appeared in the Devonian, and freshwater families in the Heterodonta first appeared in the Jurassic. Pteriomorphs in fresh water have apparently derived more recently from marine ancestors (Greenberg and Deaton

1980). Pteriomorph hearts have different physiological and pharmacological properties from other bivalves; this is especially apparent in ionic requirements (Greenberg and Deaton 1980). An extracellular calcium-dependent cardiac excitability and reduced capacity for calcium (Ca^{++}) storage and regulation perhaps make Pteriomorphs maladapted for freshwater invasion (Greenberg and Deaton 1980).

In summary, two salinity thresholds are likely to be of interest in studies of genetic differentiation in estuaries: (1) the salinities which require volume regulation and which separate marine stenohaline from marine euryhaline populations (or species); (2) the critical salinity (horohalinicum).

Salinity may be the most obvious variable but temperature and other factors may play equally important roles in the genetic differentiation observed in estuaries. Although salinity decreases from the North Sea to the Baltic, the colder regime of the Baltic is an equally salient feature. Temperature must also play an important role as it interacts with the same salinity gradient over a broad range of latitudes. The effects of low salinity on organisms often differ greatly with temperature, and salinity-temperature interactions may differ with developmental stage (Kinne 1971).

Evidence of Genetic Divergence in Estuaries

Many characteristics of fully marine organisms differ from those of estuarine conspecifics. However, it is often questionable as to whether there is a genetic basis for the difference. These differences include:

(1) *Allozymes:* differences in allelic and genotypic frequencies between estuaries and open marine waters (e.g. Christiansen and Frydenberg 1974; Lassen and Turano 1978; Theisen 1978).

(2) *Morphology:* differences in counts or measurements of meristic or other morphological characters (e.g. Schmidt 1918).

(3) *Physiology:* differences in tolerance, oxygen consumption or other properties (e.g. Lange 1968).

(4) *Ecological niche breadth:* estuarine populations often show expanded ecological occurrence, relative to open marine conspecifics (e.g. Segerstråle 1965; Fenchel 1975).

(5) *Reproduction:* the timing and peakedness of gametogenesis and spawning can differ between open-coast and estuarine populations (Koehn 1980).

Unfortunately, there is little concrete evidence to demonstrate that most observed differences have a genetic basis. An example is the range of physiological differences observed between estuarine and open marine conspecific populations. Most have been established by bringing individuals from different environments into the laboratory and measuring physiological properties with or without an intervening acclimation routine. But it is not known whether the differences observed between populations taken from different habitats represent ecophenotypic differences acquired in the field, true genetically based differences, or a combination of both. For example, mussels (*Mytilus edulis* L.) collected from the field respire maximally at salinities approximating their field habitat (Remane and Schlieper 1958; Lange 1968). Does this mean that local populations have evolved to optimize metabolism in habitats of differing salinities or do the data indicate that mussels have a broad capacity to acclimatize to local conditions, and that the acclimatization is not easily reversed in the laboratory?

The genetic acclimatization nexus cannot therefore be easily dissected without controlled matings and examination of progeny from populations collected in differing regions but spawned under identical conditions. If progeny obtained under identical laboratory conditions nevertheless differ in some physiological property, we can be confident that the difference is genetical, with the important reservation that maternal effects (e.g. egg cytoplasmic transmission) might maintain an influence. It is equally important to realize that the *absence* of differences does not necessarily imply a lack of between-population genetic difference in physiological response. It may well be that some feature of the home environment is required to actuate expression of the genes involved; this feature may be absent in the laboratory microcosm.

The stringent requirement of comparing progeny from differing field populations that have been derived from laboratory populations spawned under identical conditions has rarely been met. Ideally, it would be desirable to raise individuals spawned in the laboratory from field populations and then examine the next generation. Otherwise the conceivably different environmental and trophic conditions experienced by the parent can influence physiological performance of the offspring. Bayne et al. (1975) found that mussel larvae spawned from starved parents grew more slowly than those spawned from parents with more available food.

Most comparisons of progeny from differing habitats have involved progeny of field-collected gravid females that were spawned in the laboratory. Vernberg and Costlow (1966) examined physiological differences among and within latitudinally wide-spread species of the fiddler crab *Uca*. Differences were found within species. Schneider (1967) compared field-collected adults of the mud crab *Rhithropanopeus harrisii* from North Carolina and Florida and showed that strong differences in metabolic response to acute temperature change were maintained even after individuals were held in the laboratory under constant conditions. However, metabolic differences in progeny were nonexistent. Battaglia (1957) reared populations of the copepod *Tisbe reticulata* from brackish and open marine waters under uniform laboratory conditions for many generations. Despite the uniform rearing, brackish water-derived populations were more tolerant of lowered salinity than were open marine-derived populations. This provides strong support for the hypothesis of physiologically-related genetic divergence in estuaries. Recently, Ament (1978) has shown that progeny of laboratory-spawned *Crepidula* spp.

populations show growth rates consistent with adaptation to latitude and differing levels of gene flow.

It would be of interest to know if genetic divergence between populations is accompanied by genetic polymorphism within populations for traits relevant to selection along the estuarine gradient. Unfortunately, traits of interest (e.g. growth rate, tolerance, metabolic rate) rarely have simple genetic explanations conducive to Mendelian analysis. Bradley (1978) examined tolerance to elevated temperatures in an harpacticoid copepod and found significant within-population variation. Levinton (1980) raised separated progeny of several families of the polychaete *Ophryotrocha costlowi* under identical conditions and found that growth rate differed significantly among families. There is thus some evidence for a potential variation upon which natural selection can operate.

Although evidence for genetically based between-population physiological differences is scanty, we have ample evidence for genetic differentiation between open marine and estuarine populations. Several clines in allozyme allele frequency have been found at the mouths of estuaries. These clines raise questions as to the amount of gene exchange and natural selection in these regions. An excellent example is the entrance to the Baltic Sea. Sharp geographic differentiation has been found at allozyme loci for the eel pout *Zoarces viviparus* and the mussel *Mytilus edulis* between the North Sea and the Baltic Sea (*Z. viviparus:* Christiansen and Frydenberg 1974; *M. edulis:* Theisen 1978). For both species, distinct morphological differences between Baltic Sea and North Sea populations have been recorded as well. Thus, sharp differentiation has been found over a short geographic space in a species with viviparous reproduction (*Z. viviparus*) and one with dispersal *via* relatively long-lived pelagic larvae (*M. edulis*). Differentiation at a hemoglobin locus in cod (Sick 1965) has also been recognized from the North Sea to the Baltic Sea. But in this case, the site of maximum geographic differentiation is not along the Baltic-North Sea interface (Kattegat and Øresund) as in the cases of *M. edulis* and *Z. viviparus*. Rather, a Baltic Sea cod deme has its western geographic margin in the vicinity of Bornholm, well within the Baltic.

The smaller scale distribution of allozyme markers and morphological characters in *Zoarces viviparus* suggests strong localization of genetic differentiation. For both an heritable metric character and an allozyme locus, it is possible to delineate localized, genetically distinct, subpopulations. Similarly, clinal variation in four metric characters and at the *ESTIII* locus occurs along several small fjords (Schmidt 1917a, 1917b, 1918; Christiansen and Frydenberg 1974; Christiansen and Simonsen 1978). In *Mytilus edulis* adults, strong geographic differentiation at the *LAP* locus occurs over a scale of hundreds of meters in creeks on Cape Cod; however, in juveniles, there is little differentiation (Boyer 1974). This implies that some form of selective process enhances dynamically the differentiation over relatively short periods of time.

The distributional data suggest several contributing processes to explain the observed differentiation. Unfortunately, we can never safely exclude any two of these three factors in most of the case studies cited above.

1. *Stochastic Processes.* It is possible that geographic divergence in allele frequencies is due to random forces operating in subpopulations that are relatively isolated from each other. This might explain divergence in a viviparous fish, but would be unlikely in *Mytilus edulis* where dispersal potential is great. Even modest selection or gene flow would probably override the effects of stochastic processes on differentiation.

2. *Complete Isolation and Subsequent Mixing Between Genetically Distinct Populations.* In this case, genetic differentiation has occurred in the (possibly remote) past and geographic isolation maintains the difference. The initial difference may have been caused by natural selection or stochastic processes. This case might be realized in stocks of migrating fishes that maintain geographically circumscribed and separate migration patterns (e.g. Sick 1965; Harden Jones 1968; Jamieson and Turner 1978). There is even some probable isolation within brackish bodies of water (e.g. The Mattituck Sill in Long Island Sound— Hardy 1972).

3. *Dynamic Selection.* The geographic differentiation may be due to active selection against genotypes dispersing into a suboptimal environment. Thus, despite extensive exchange of larvae, selection might maintain clinal distributions dynamically (as in Boyer 1974). Immigrants might suffer selective mortality, or grow and reproduce at a slower rate than indigenous genotypes. Two distinctly different types of natural selection could maintain a cline:

   (i) *Target loci.* Selection operates directly on the locus or loci whose expressed phenotypic variation shows a clinal distribution. These loci might also be linked closely to other loci that actually are the targets of selection. In either case it makes no difference as to the origin of the genotype; in other words we could predict the fitness by knowing only the allozyme genotype for the locus in question.

   (ii) *Ecotypes.* Selection occurs to enforce isolation between an "estuarine ecotype" and an "open marine ecotype." Thus, descendants of larvae dispersing from open marine waters into estuaries would have a fitness less than that of the local estuarine population. This would, in effect, greatly reduce gene flow between populations.

All current evidence points to a combination of hypotheses 2 and 3 to explain estuarine differentiation. Christiansen and Frydenberg (1974) argue forcefully that drift cannot explain geographic differentiation in *Z. viviparus*, which shows clinal change at some loci but homogeneity in allele frequency at others, along the same geographic space. Drift could not account for both types of variation in the same organisms along the same geographic space. Step-clines, smooth clines and lack of differentiation at different loci in *Mytilus edulis* on the east coast of North America support this argument (Koehn et al. 1976). Although circumstantial, recent evidence supports the notion that selection is behind the overall genetic differences between estuarine and open marine populations (e.g. Levinton and Lassen 1978a; Levinton and Suchanek 1978; Koehn 1978, 1980).

Although selection is the important factor, it is difficult to distinguish between the alternative (i) target loci and (ii) ecotypes hypotheses. I believe that the ecotypes hypothesis is most likely. Levinton and Lassen (1978a) present evidence of a lack of among-genotype differences at the *LAP* locus, but show that there are strong differences in tolerance and growth rate response to low salinity between Long Island Sound, New York and adjacent open marine *Mytilus edulis*. These ecotypes may be analogous to the physiological races postulated by many marine invertebrate physiologists (e.g. Loosanoff and Nomejko 1951). This hypothesis is supported by the finding of sharp clinal differentiation at three different allozyme loci in *Mytilus edulis* in the Øresund (Theisen 1978). It seems very unlikely that strong selection is operating at several loci independently—the genetic load so generated would be enormous. Therefore, the allozyme data and other morphological data point to physiological differentiation between Baltic Sea and North Sea populations over a wide part of the genome. It seems no coincidence that in essentially random samples of allozyme loci we can come up with so many examples of strong clines. This argues for differentiation over much of the genome between estuarine and open marine populations.

The significance of the ecotypes hypothesis as proposed by Levinton and Lassen (1978a, b) has been characterized by Christiansen and Simonsen (1978) as *accelerating differentiation*. We assume that estuaries are quite different from the adjoining open marine habitat; selection for adaptive change is initially induced at many loci where strong selection operates. "But this differentiation increases the fjord population, and consequently the fitness of the immigrants are decreased which decreases the effective amount of migration. This increase allows for further differentiation at the already differentiated loci, but in addition it allows for minor adaptive adjustments at other loci. This again strengthens the isolation which in turn allows for more genetic differentiation." (Christiansen and Simonsen 1978, p. 190).

Mixture and dynamic selection between ecotypes can explain adequately the generalities and some specifics of the strong clines seen at the *LAP* locus in Long Island Sound (Lassen and Turano 1978) and at three loci in the Øresund (Theisen 1978). At the entrance to Long Island Sound, a sharp cline at the *LAP* locus over 20 km is observed to steepen between juveniles and adults along the north shore of Long Island (Lassen and Turano 1978; Koehn 1980). Further, deviations from Hardy Weinberg expectations are maximal at the center of the cline. This could indicate population mixing (Wahlund effect) or intense selection among *LAP* genotypes (see Lassen and Turano 1978; Koehn 1978 for discussion). However, the ecotype theory explains the data as well. Data presented by Koehn (1980) show a displacement of the reproductive seasons of estuarine and open marine *Mytilus edulis*. When settlement is maximal along the cline, the great majority of settlers must be from the open marine population (frequency of common allele is ca. 0.55). If we assume: (1) a 9:1 ratio of marine:estuarine settling larvae; (2) an ecotype form of genotype-independent mortality; with (3) an exponentially increasing death rate of marine dispersers with distance into the Long Island Sound estuary, we get the results plotted in Fig. 1. The fit to available data (e.g. Lassen and Turano 1978) is encouraging but note that it only shows "what might be," not what actually has occurred.

Selective Agents

As mentioned above, temperature, salinity, and their interactions provide the most potent agents of selection. Koehn (1978 and work underway) is now investigating the physiological function of "leucine aminopeptidase" in the context of salinity adaptation. Clinal variation along the east coast suggests that temperature must be involved as well.

Regrettably, pollutants may be potent selective agents. Estuaries are often sites of concentrations of heavy metals and other toxicants (e.g. Greig and McGrath 1977; Greig et al. 1977). These pollutants might kill off local flora and fauna, but might also select for resistance to the pollutant. In the case of heavy metals there is evidence for such adaptation in terrestrial plants (Bradshaw 1952) and strong evidence for adaptation in marine organisms as well (Russell and Morris 1970; Bryan and Hummerstone 1973; Stokes 1975; Brown 1979). Inshore phytoplankton seem more resistant to pollutants than are related offshore phytoplankton (Fisher 1977). This creates the possibility of transfer and amplification of dangerous substances through the food web (e.g. Young 1975). Genetic studies are therefore urgently needed for applied as well as for academic reasons. If genetic adaptation to pollutants is common, then many environmental quality monitoring systems using bioassays such as developmental rate or mortality will be flawed. The bioassay will be a function of the extent to which the population may have evolved to detoxify, sequester or avoid uptake of the pollutant. The success of a local individual in dealing with relatively high doses of a pollutant is both a measure of the probability of avoidance of extinction and of the probability the pollutants will be passed up the food web.

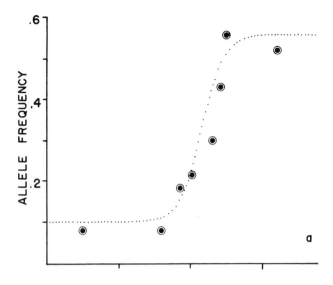

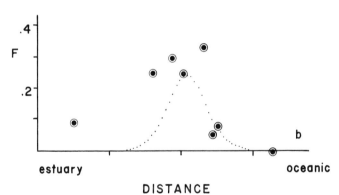

*Figure 1. Predicted cline (a) and deviation of heterozygotes from Hardy
Weinberg equilibrium (b) under the ecotype model. F is Wright's
inbreeding coefficient; positive values imply deficiencies of
heterozygotes.*

Conclusions

Much evidence supports the assertion that estuarine populations are
genetically differentiated from open marine populations. The mechanisms
must include drift, isolation and selection. It is likely that the differentiation
is caused by natural selection occurring over a wide spectrum of the

genome and therefore it probably influences physiological performance, ecology and behavior. The broadscale differentiation probably enforces isolation between marine and estuarine populations.

Note

This paper is contribution number 289, Program in Ecology and Evolution, State University of New York at Stony Brook.

References Cited

Ament, A. S. 1978. Geographic variation in relation to life history in three species of the marine gastropod genus *Crepidula*: growth rates of newly hatched larvae and juveniles, pp. 61-76. *In:* S. Stancyk (ed.), *Reproductive Ecology of Marine Invertebrates.* Univ. South Carolina Press, Columbia, S.C.

Battaglia, B. 1957. Ecological differentiation and incipient intraspecific isolation in marine copepods. *Ann. Biol.* 33:259-268.

Bayne, B. L., P. A. Gabbott and J. Widdows. 1975. Some effects of stress in the adult on the eggs and larvae of *Mytilus edulis* L. *J. Mar. Biol. Ass. U.K.* 55:675-689.

Boyer, J. F. 1974. Clinal and size-dependent variation at the *LAP* locus in *Mytilus edulis. Biol. Bull.* 147:535-549.

Bradley, B. P. 1978. Genetic and physiological adaptation of the copepod *Eurytemora affinis* to seasonal temperatures. *Genetics* 90:193-205.

Bradshaw, A. 1952. Populations of *Agrostis tenuis* resistant to lead and zinc poisoning. *Nature* 169:1089.

Brown, B. E. 1979. Lead detoxification by a copper-tolerant isopod. *Nature* 276:388-390.

Bryan, G. W. and L. G. Hummerstone. 1973. Adaptation of the polychaete *Nereis diversicolor* to manganese in estuarine sediments. *J. Mar. Biol. Assoc. U.K.* 53:859-872.

Christiansen, F. B. and O. Frydenberg. 1974. Geographical patterns of four polymorphisms in *Zoarces viviparus* as evidence of selection. *Genetics* 77:765-770.

Christiansen, F. B. and V. Simonsen. 1978. Geographic variation in protein polymorphisms in the eelpout, *Zoarces viviparus* (L.), pp. 171-194. *In:* B. Battaglia and J. Beardmore (eds.), *Marine Organisms, Genetics, Ecology and Evolution.* Plenum Press, New York.

Fenchel, T. 1975. Factors determining the distribution patterns of mud snails (Hydrobiidae). *Oecologia* 20:1-17.

Fisher, N. S. 1977. On the differential sensitivity of estuarine and open-ocean diatoms to exotic chemical stress. *Amer. Nat.* 111:871-895.

Gainey, L. F., Jr. and M. J. Greenberg. 1977. Physiological basis of the species abundance-salinity relationship in molluscs: A speculation. *Mar. Biol.* 40:41-49.

Greenberg, M. J. and L. E. Deaton. 1980. Possible physiological bases for the phylogenetic and zoogeographic distribution of bivalve molluscs in oligohaline and freshwater. *In:* A. Zhirmunsky (ed.), *Physiology and Biochemistry of Marine Animals.* USSR Academy of Sciences, Far East Science Center, Vladivostok, (in press).

Greig, R. A. and R. A. McGrath. 1977. Trace metals in sediments of Raritan Bay. *Mar. Poll. Bull.* 8:188-192.

Greig, R. A., R. N. Reid and O. R. Wenzloff. 1977. Trace metal concentrations in sediments from Long Island Sound. *Mar. Poll. Bull.* 8:183-188.

Harden Jones, F. R. 1968. *Fish Migration.* Edw. Arnold, London.

Hardy, C. D. 1972. Movement and quality of Long Island Sound waters, 1971. Marine Sciences Res. Ctr., State Univ. New York, Stony Brook. Techn. Report Ser., no. 17, pp. 1-66.

Jamieson, A. and R. J. Turner. 1978. The extended series of Tf alleles in Atlantic Cod, *Gadus morhua*, pp. 699-729. *In:* B. Battaglia and J. Beardmore (eds.), *Marine Organisms, Genetics, Ecology and Evolution.* Plenum Press, New York.

Khlebovich, V. V. 1968. Some peculiar features of the hydrochemical regime and the fauna of mesohaline waters. *Mar. Biol.* 2:47-49.

Khlebovich, V. V. 1969. Aspects of animal evolution related to critical salinity and internal state. *Mar. Biol.* 2:338-345.

Kinne, O. 1971. Salinity: Animals-invertebrates, pp. 821-995. *In:* O. Kinne (ed.), *Marine Ecology, Vol. 1. Environmental Factors, Part 2.* Wiley-Interscience, London.

Koehn, R. K. 1978. Biochemical aspects of genetic variation at the *LAP* locus in *Mytilus edulis*, pp. 211-227. *In:* B. Battaglia and J. Beardmore (eds.), *Marine Organisms, Genetics, Ecology and Evolution.* Plenum Press, New York.

Koehn, R. K. 1980. Marine organisms: the genetics of physiology and the physiology of genetics. *In:* V. Zhirmunsky (ed.), *Physiology and Biochemistry of Marine Animals.* USSR Academy of Sciences, Far East Science Center, Vladivostok, (in press).

Koehn, R. K., R. J. Milkman and J. B. Mitton. 1976. Population genetics of marine pelecypods. IV. Selection, migration and genetic differentiation in the blue mussel, *Mytilus edulis. Evolution* 30:2-32.

Lange, R. 1963. The osmotic function of amino acids and taurine in the mussel, *Mytilus edulis. Comp. Biochem. Physiol.* 10:173-179.

Lange, R. 1968. The relations between the oxygen consumption of isolated gill tissue of the common mussel *Mytilus edulis* L. and salinity. *J. Exp. Mar. Biol. Ecol.* 2:37-45.

Lassen, H. H. and F. J. Turano. 1978. Clinal variation and heterozygote deficit at the Lap-locus in *Mytilus edulis. Mar. Biol.* 49:245-254.

Levinton, J. S. 1980. The hypothesis of evolutionary compensation in physiology of marine invertebrates. *In:* V. Zhirmunsky (ed.), *Physiology and Biochemistry of Marine Animals.* USSR Academy of Sciences, Far East Science Center, Vladivostok, (in press).

Levinton, J. S. and H. H. Lassen. 1978a. Experimental mortality studies and adaptation at the *Lap* locus in *Mytilus edulis*, pp. 229-254. *In:* B. Battaglia and J. Beardmore (eds.), *Marine Organisms, Genetics, Ecology and Evolution.* Plenum Press, New York.

Levinton, J. S. and H. H. Lassen. 1978b. Selection, ecology and evolutionary adjustment within bivalve mollusc populations. *Phil. Trans. Roy. Soc. London, B,* 284:403-415.

Levinton, J. S. and T. H. Suchanek. 1978. Geographic variation, niche breadth, and genetic differentiation at different geographic scales in the mussels *Mytilus californianus* and *M. edulis. Mar. Biol.* 49:363-375.

Loosanoff, V. L. and C. A. Nomejko. 1951. Existence of physiologically different races of oysters, *Crassostrea virginica. Biol. Bull.* 101:151-156.

Potts, W. T. W. 1958. The inorganic and amino acid composition of some lammelibranch muscles. *J. Exp. Biol.* 35:749-764.

Remane, A. and C. Schlieper. 1958. *The Biology of Brackish Water.* Translated 1971, Wiley-Interscience, New York. 372 pp.

Russell, G. and O. P. Morris. 1970. Copper tolerance in the marine fouling alga *Ectocarpus siliculosus. Nature* 228:288-289.

Schmidt, J. 1917a. *Zoarces viviparus* L. and local races of the same. *C. R. Trav. Lab., Carlsberg* 13:277-397.

Schmidt, J. 1917b. Constancy investigations continued. *C. R. Trav. Lab., Carlsberg* 14:1-19.

Schmidt, J. 1918. Racial studies in fishes. I. Statistical investigations with *Zoarces viviparus* L. *J. Genet.* 7:105-118.

Schneider, D. E. 1967. An evaluation of temperature adaptations in latitudinally separated populations of the xanthid crab, *Rhithropanopeus harrisii* (Gould), by laboratory rearing experiments. Ph.D. Thesis, Duke Univ., Durham, N.C. 132 pp.

Segerstrale, S. G. 1965. Biotic factors affecting the vertical distribution and abundance of the bivalve, *Macoma baltica* (L.), in the Baltic Sea. Proc. Fifth Europ. Mar. Biol. Symp., Goteborg., *Bot. Gothoburgensia* 3:195-204.

Sick, K. 1965. Haemoglobin polymorphism of cod in the Baltic and Danish Belt Sea. *Hereditas* 54:19-48.

Stokes, P. M. 1975. Uptake and accumulation of copper and nickel by metal-tolerant strains of *Scenedesmus. Verh. Int. Ver. Limnol.* 14:2128-2137.

Theisen, B. F. 1978. Allozyme clines and evidence of strong selection in three loci in *Mytilus edulis* L. (Bivalvia) from Danish waters. *Ophelia* 17:135-142.

Vernberg, F. J. and J. D. Costlow. 1966. Studies on the physiological variation between tropical and temperate-zone fiddler crabs of the genus *Uca.* IV. Oxygen consumption of larvae and young crabs reared in the laboratory. *Physiol. Zool.* 39:36-52.

Young, M. L. 1975. The transfer of ^{65}Zn and ^{59}Fe along a *Fucus serratus* (L.)→*Littorina obtusata* (L.) food chain. *J. Mar. Biol. Assoc. U.K.* 55:583-610.

FILTER FEEDER COUPLING BETWEEN THE ESTUARINE WATER COLUMN AND BENTHIC SUBSYSTEMS

Richard Dame

Coastal Carolina College and
Belle W. Baruch Institute for Marine Biology and
Coastal Research
University of South Carolina
Conway, South Carolina

Richard Zingmark, Harold Stevenson

Biology Department and
Belle W. Baruch Institute for Marine Biology
and Coastal Research
University of South Carolina
Columbia, South Carolina

and

Douglas Nelson

Coastal Carolina College and
Belle W. Baruch Institute for Marine Biology
and Coastal Research
University of South Carolina
Conway, South Carolina

Abstract: Benthic filter feeders are capable of translocating and transforming large quantities of matter from the estuarine water column. Evidence at the individual, population and ecosystem levels supports the contention that filter feeders are probably a significant coupling between the water column and the benthos. We hypothesize that benthic filter feeders are a major controlling element in marsh-estuarine nutrient cycling.

Introduction

Knowledge of the dynamics of the couplings between benthic and water column subsystems is crucial to our understanding of the structure and function of estuarine systems. In estuaries, filter feeders (oysters, clams and mussels) are probably major heterotrophic components in this coupling. These organisms, which may exist in large numbers, effectively utilize tidal energy as a transportation system for food and wastes.

There is a two-way flow of energy/matter between the water column and benthic subsystems. In one case, filter feeders actively remove suspended particulate matter from the water column and deposit it as feces and pseudofeces, and sediments settle out of suspension to the bottom as a result of gravity. This deposited and settled material from the water column may be incorporated directly as benthic consumer biomass (filter feeders); it may be utilized by deposit feeding microbiota, meiofauna, and/or macrofauna; and sediments may accumulate, thereby increasing bottom elevations. Material also moves from the bottom into the water column. Such upward fluxes can be generated biologically by excretion, bioturbation, spawning, and transient predators. In addition, upward flux may be physically generated by resuspension initiated by tidal currents and wind action.

Recently, Kitchell et al. (1979) reviewed the role of consumer organisms in the control of nutrient cycling. These authors noted that rates of nutrient cycling in ecosystems may be altered by translocation and transformation of matter. These altered rates result from consumption processes and/or the behavior of consumer organisms. Filter feeders translocate matter by removing suspended matter from the water column and depositing this material on the bottom, or incorporating this matter as biomass. Also, filter feeders transform suspended particulate material by changing particulate size distributions in the water column and thus alter the rates of nutrient cycling proportionally to changes in surface/volume relationships.

Assuming the preceding concepts as reasonable, we propose the following hypothesis:

Filter feeders are a major coupling between the water column and benthic subsystems and are a major controlling component in the cycling of nutrients and flow of energy in some estuaries.

Supporting Evidence

Many investigations have established the importance of filter feeding organisms as significant agents in removal and decomposition of suspended particulate matter in estuarine systems. In a series of studies, Haven and Morales-Alamo (1966; 1970; 1972) investigated filtration of suspended particles by oysters and biodeposition of such particles by oysters and a number of other common filter feeders in Chesapeake Bay. Their work showed that filter feeders, and oysters in particular, are capable of removing natural suspended particulate matter between 1 and 12 μm with maximum efficiency at 3 μm. Filter feeders routinely removed small particles (1−10 μm) during feeding and voided them as fecal and pseudofecal (large) pellets (500−3000 μm); these large particles settled faster than did their small component particles (Haven and Morales-Alamo 1972). The authors concluded that the material removed and deposited varied seasonally, reaching a maximum in the early fall and stopping completely below 2.8 C. In addition, they felt that the magnitude of removal might be sufficient to influence deposition, transport, and composition of suspended sediments in estuaries. Their work also supplemented a number of speculations by Damas (1935), Verwey (1952), and Lund (1957) that filter feeders may initiate small particle sedimentation and that filter feeders can be responsible for large quantities of biodeposits.

The work of Tenore and Dunstan (1973) indicated that feeding and decomposition rates of edible mussels, *Mytilus edulis*, the American oyster, *Crassostrea virginica*, and the hard clam, *Mercenaria mercenaria*, are influenced by food concentration. Also, the percentage of available food removed quickly increases to a maximum at food concentrations typical of the natural environment and remains fairly constant for the mussel and oyster, but declines for the clam as food concentrations are further increased. Biodeposition rates increase logarithmically with increased food concentrations (Tenore and Dunstan 1973).

Recently, Bernard (1974) developed a biodeposition and gross energy budget for Pacific oysters, *Crassostrea gigas*, in a British Columbia estuary. His work attempted to expand scientific knowledge on the population level of filter feeders. In his system, annual biodeposition by oysters was $8.9 \, g \, g^{-1}$ oyster y^{-1} or about 1545 kcal m^{-2}. These large quantities of deposited particles, besides forming a nutritional source for other forms, also modified the physical and chemical characteristics of the substrate sediments thereby allowing the establishment of diverse groups of organisms. The attached microorganisms on the deposited particles might be resuspended, becoming available as additional food resources.

The studies reported thus far deal mainly with filter feeders common to subtidal estuarine benthos. Kraeuter (1976) demonstrated the importance of the salt-marsh mussel, *Geukensia demissa*, and the Carolina marsh clam, *Polymesoda caroliniana*, in removing sediments from the water column in a Georgia salt marsh. *G. demissa* was by far the most abundant filter feeder and had a deposition rate of 549 g m^{-2} y^{-1}. Kraeuter's values lie between the estimates of Kuenzler (1961) and Haven and Morales-Alamo (1966). Kraeuter (1976) noted that biodeposition activities may stabilize the sediment surface by the formation of large aggregates, provide a means of retaining nutrients and trace elements in a marsh, and assist in recycling organic materials within detrital food chains.

There have been few studies on the influence of filter feeders on marsh-estuarine nutrient cycling. Kuenzler (1961) developed a phosphorus budget for the marsh mussel *Modiolus demissus* (*=Geukensia demissa*) population in Sapelo Island, Georgia, marshes. His data indicated that mussels have a major effect on particulate phosphorus in the water by removing as much as 1/3 from suspension and depositing it where it is used by other marsh organisms. Keunzler suggested that marsh mussels are more important as agents of sedimentation and nutrient cycling than as agents of energy flow. Oysters probably have a similar function because their turn-over time for phosphorus is of the same magnitude as that of marsh mussels (Pomeroy and Haskin 1954). Physiological evidence collected by Hammen (1968) and Bayne et al. (1976) suggests that oysters,

clams and mussels are capable of excreting considerable quantities of ammonia and organic nitrogen, particularly when metabolic rates are high.
Aside from these studies, there have been few concerted attempts to
delineate the role of estuarine filter feeders in nutrient biogeochemistry.

In summary, filtration rates and efficiencies are known for many
estuarine filter feeders, but some of the laboratory and available field data
are conflicting (Epifanio et al. 1975). There have been no studies actually
describing the interaction of the water column and benthos directly at the
ecosystem level, although many workers have speculated at the
ecosystem level and have suggested that filter feeders are capable of
removing all or most of the material in the water column (Rhodes 1974).

Application to North Inlet Estuary

In North Inlet estuary near Georgetown, South Carolina (79° 12′ N;
33° 20′ W) there are extensive salt marshes dominated by *Spartina alterniflora* and densely populated by intertidal oyster reefs formed by
Crassostrea virginica (Dame 1976). Present studies on the flux of matter in
the North Inlet ecosystem have suggested that there is a coupling between
the intertidal filter feeders and the suspended particulate matter in the
water column. Concentrations of phytoplankton as measured by
chlorophyll-*a*, of phytoplankton plasma volume, and of toal biomass as
estimated by ATP have consistently shown high values at high tide and
rapidly declining values as the tide ebbs (Chrzanowski et al. 1979). On the
ebbing tide, the rates of decline of ATP and chlorophyll-*a* are not the
same. There are indications that the differences are due to other microbiota
(i.e. fungi) which are resuspended and exported (Chrzanowski and Stevenson 1980). Concurrent analysis of suspended sediment indicates an
organic fraction which behaves similarly to ATP and chlorophyll-*a* except
on spring tides. These observations suggest that the North Inlet ecosystem
has a net influx of phytoplankton and possibly bacteria from the coastal
ocean via the water column.

We considered three possibilities. *First,* the incoming material is settling out on the substrate. Unpublished recent observations by Zingmark on
the settlement of phytoplankton in the North Inlet system indicate this
possibility is negligible. *Second,* extensive grazing by zooplankton is occurring. This idea was discounted based on the contention of Williams et al.
(1968) that zooplankton populations in shallow southeastern estuaries are
not large enough to affect phytoplankton populations significantly. In time,
direct evidence for the influence of zooplankton on North Inlet will become
available and this point will be clarified. *Finally,* we considered the idea that
benthic filter feeders are significantly altering the suspended particulate environment in North Inlet. Previous studies have described dense oyster
beds (450 g dry body m^{-2} with 10,000 kcal m^{-2} y^{-1} energy flow) covering
approximately 2½% of the total 30 km^2 of North Inlet (Dame 1976).
These large assemblages of filter feeders must be supported by a large

source of particulate food in the North Inlet water column. At 20 C, oysters are capable of pumping $67-286$ ml g^{-1} min^{-1} depending upon their size (Walne 1972). For the following rough calculations, we will use an intermediate value of 120 ml g^{-1} min^{-1} or 0.0072 m^3 g^{-1} hr^{-1}. By multiplying the pumping rate, the density of oysters, the area of coverage by the oysters, and the time the oysters are submerged (6 h), a value for the total amount of water pumped by oysters in North Inlet can be derived:

$$(7.2 \times 10^{-3} \text{ m}^3 \text{ g}^{-1} \text{ hr}^{-1}) \ (450 \text{ g m}^{-2}) \ (0.75 \times 10^6 \text{ m}^2) \ (6 \text{ h})$$
$$= 1.5 \times 10^7 \text{m}^3$$

The water pumped per tide, 1.5×10^7 m^3, can be compared to unpublished observations by B. Kjerfve and J. Proehl (Baruch Institute, USC-Columbia) that showed the North Inlet ecosystem had a tidal prism of 2.2 $\times 10^7$ m^3 during July 1979. It is evident from these rough calculations that the major filter feeder in North Inlet is capable of influencing the suspended particulate matter environment of the North Inlet ecosystem. Similar calculations of pumping capacity and estuarine volume have shown the same possibility for Flax Pond and for Delaware Bay (C. Hall, Ecol. and Syst., Cornell U., Ithaca, N.Y. and R. Ulanowicz, Chesapeake Biological Laboratory, Solomons, Md., personal communication). These speculations seem to implicate estuarine filter feeders as major components in the coupling between the benthic and water column subsystems.

Filter feeders are a well known economic resource in estuaries. Current knowledge indicates that these organisms may also be significant functional resources through their possible control of energy flux and nutrient cycling between the benthos and the water column. The lack of direct evidence at the ecosystems level points to the need for future studies of estuarine water columns with concurrent measurement of the biogeochemical activity of benthic filtration by filter feeders.

Acknowledgments

The authors are grateful to Don Rice for his most helpful comments in the preparation of this manuscript. This work was supported by Grant No. DEB76-83010 from the National Science Foundation, Ecosystem Section. This is contribution No. 321 from the Belle W. Baruch Institute for Marine Biology and Coastal Research.

References Cited

Bayne, B. L., J. Widdows and R. J. Thompson. 1976. Physiology II, pp. 207-260. *In:* B. L. Bayne (ed.), *Marine Mussels: Their Ecology and Physiology.* Cambridge University Press, New York.

Bernard, F. R. 1974. Annual biodeposition and gross energy budget of mature Pacific oysters, *Crassostrea gigas. J. Fish. Res. Board Can.* 31:185-190.

Chrzanowski, T. and H. Stevenson. 1980. Filamentous fungal propagules as potential indicators of sediment detritus resuspension. *Bot. Mar.* (in press).

Chrzanowski, T., H. Stevenson and B. Kjerfve. 1979. Adenosine 5' triphosphate flux through the North Inlet Marsh System. *App. Env. Microbiol.* 37:841-848.

Damas, D. 1935. Le role des organisms dans la formation des vases marines. *Ann. Soc. Geol. Belg.* 58:143-152.

Dame, R. 1976. Energy flow in an intertidal oyster population. *Est. Coastal Mar. Sci.* 4:243-253.

Epifanio, C., R. Srna and G. Pruder. 1975. Mariculture of shellfish in controlled environments: a prognosis. *Aquaculture* 5:227-241.

Hammen, C. S. 1968. Aminotransferase activities and amino acid excretion of bivalve molluscs and brachiopods. *Comp. Biochem. Physiol.* 26:697-705.

Haven, D. and R. Morales-Alamo. 1966. Aspects of biodeposition in oysters and other invertebrate filter feeders. *Limnol. Oceanogr.* 11:487-498.

Haven, D. and R. Morales-Alamo. 1970. Filtration of particles from suspension by the American oyster *Crassostrea virginica. Biol. Bull.* 139:121-130.

Haven, D. and R. Morales-Alamo. 1972. Biodeposition as a factor in sedimentation of fine suspended solids in estuaries. *Geol. Soc. Mem.* 133:121-130.

Kitchell, J. F., R. V. O'Neill, D. Webb, G. W. Gallepp, S. M. Bartell, J. F. Koonce and B. S. Ausmus. 1979. Consumer regulation of nutrient cycling. *Bioscience* 29:28-33.

Kraeuter, J. N. 1976. Biodeposition in salt-marsh invertebrates. *Mar. Biol.* 35:215-223.

Kuenzler, E. 1961. Phosphorus budget of a mussel population. *Limnol. Oceanogr.* 6:400-415.

Lund, E. J. 1957. A quantitative study of clearance of a turbid medium and feeding by the oyster. *Pub. Inst. Marine Sci. Texas* 4:296-312.

Pomeroy, L. and H. Haskin. 1954. The uptake and utilization of phosphate ions from sea water by the American oyster. *Biol. Bull.* 107:123-129.

Rhoads, D. 1974. Organism-sediment relations on the muddy sea floor. *Oceanogr. Mar. Biol. Ann. Rev.* 12:263-300.

Tenore, K. and W. Dunstan. 1973. Comparison of feeding and biodeposition of three bivalves at different food levels. *Mar. Biol.* 21:190-195.

Verwey, J. 1952. On the ecology of distribution of cockle and mussel in the Dutch Waddensea, their role in sedimentation and the source of their food supply. *Arch. Neerl. Zool.* 10:172-239.

Walne, P. R. 1972. The influence of current speed, body size, and water temperature on the filtration rate of five species of bivalves. *J. Mar. Biol. Assoc. U. K.* 52:345-374.

Williams, R. B., M. B. Murdoch and L. K. Thomas. 1968. Standing crop and importance of zooplankton in a system of shallow estuaries. *Chesapeake Sci.* 9:42-51.

INDEX